Periodic Table of the Elements with the Gmelin System Numbers

1 H 2																1 H 2	2 He 1
3 Li 20	4 Be 26											5 B 13	6 C 14	7 N 4	8 O 3	9 F 5	10 Ne 1
11 Na 21	12 Mg 27											13 Al 35	14 Si 15	15 P 16	16 S 9	17 Cl 6	18 Ar 1
19 * K 22	20 Ca 28	21 Sc 39	22 Ti 41	23 V 48	24 Cr 52	25 Mn 56	26 Fe 59	27 Co 58	28 Ni 57	29 Cu 60	30 Zn 32	31 Ga 36	32 Ge 45	33 As 17	34 Se 10	35 Br 7	36 Kr 1
37 Rb 24	38 Sr 29	39 Y 39	40 Zr 42	41 Nb 49	42 Mo 53	43 Tc 69	44 Ru 63	45 Rh 64	46 Pd 65	47 Ag 61	48 Cd 33	49 In 37	50 Sn 46	51 Sb 18	52 Te 11	53 I 8	54 Xe 1
55 Cs 25	56 Ba 30	57** La 39	72 Hf 43	73 Ta 50	74 W 54	75 Re 70	76 Os 66	77 Ir 67	78 Pt 68	79 Au 62	80 Hg 34	81 Tl 38	82 Pb 47	83 Bi 19	84 Po 12	85 At 8a	86 Rn 1
87 Fr 25a	88 Ra 31	89*** Ac 40	104 71	105 71													* NH_4 23

**Lanthanides 39	58 Ce	59 Pr	60 Nd	61 Pm	62 Sm	63 Eu	64 Gd	65 Tb	66 Dy	67 Ho	68 Er	69 Tm	70 Yb	71 Lu
***Actinides	90 Th 44	91 Pa 51	92 U 55	93 Np 71	94 Pu 71	95 Am 71	96 Cm 71	97 Bk 71	98 Cf 71	99 Es 71	100 Fm 71	101 Md 71	102 No 71	103 Lr 71

A Key to the Gmelin System is given on the Inside Back Cover

Gmelin Handbook of Inorganic Chemistry

8th Edition

Gmelin Handbook of Inorganic Chemistry

8th Edition

Gmelin Handbuch der Anorganischen Chemie

Achte, völlig neu bearbeitete Auflage

Prepared and issued by	Gmelin-Institut für Anorganische Chemie der Max-Planck-Gesellschaft zur Förderung der Wissenschaften Director: Ekkehard Fluck
Founded by	Leopold Gmelin
8th Edition	8th Edition begun under the auspices of the Deutsche Chemische Gesellschaft by R. J. Meyer
Continued by	E. H. E. Pietsch and A. Kotowski, and by Margot Becke-Goehring

Springer-Verlag Berlin Heidelberg GmbH 1987

STAFF GMELIN HANDBOOK

Dr. G. Bär, D. Barthel, Dr. N. Baumann, Dr. W. Behrendt, Dr. L. Berg, Dipl.-Chem. E. Best, K. D. Bonn, P. Born-Heck, Dipl.-Ing. V. A. Chavizon, E. Cloos, Dipl.-Phys. G. Czack, I. Deim, Dipl.-Chem. H. Demmer, R. Dowideit, Dipl.-Chem. M. Drößmar, U. Duggen, M. Engels, Dr. H.-J. Fachmann, Dr. J. Faust, Dr. Chr. Fianda, Dr. W.-D. Fleischmann, V. Frick, G. Funk, Dipl.-Ing. N. Gagel, Dr. U. W. Gerwarth, C. Gorr, Dipl.-Phys. D. Gras, Dipl.-Bibl. W. Grieser, Dr. R. Haubold, Dipl.-Min. H. Hein, G. Heinrich-Sterzel, A. Heise-Schuster, H.-P. Hente, H. W. Herold, U. Hettwer, Dr. W. Hoffmann, Dipl.-Chem. K. Holzapfel, Dipl.-Chem. E.-M. Horn, Dr. W. Huisl, Dr. S. Jäger, Dr. R. Jotter, Dr. J. von Jouanne, Dr. B. Kalbskopf, Dipl.-Chem. W. Karl, H.-G. Karrenberg, Dipl.-Phys. H. Keller-Rudek, A. Klusch, Dipl.-Phys. E. Koch, Dipl.-Chem. K. Koeber, Dipl.-Chem. H. Köttelwesch, R. Kolb, E. Kranz, Dipl.-Chem. I. Kreuzbichler, Dr. A. Kubny, Dr. W. Kurtz, M. Langer, Dr. B. Leduc, Dr. A. Leonard, Dipl.-Chem. H. List, H. Mathis, E. Meinhard, K. Meyer, Dr. M. Mirbach, Dipl.-Chem. B. Mohsin, Dr. U. Neu-Becker, V. Neumann, K. Nöring, Dipl.-Chem. R. Nohl, Dipl.-Min. U. Nohl, Dr. W. Petz, I. Rangnow, Dipl.-Phys. H.-J. Richter-Ditten, Dipl.-Chem. H. Rieger, B. Riegert, E. Rieth, A. Rosenberger, E. Rudolph, G. Rudolph, Dipl.-Chem. S. Ruprecht, Dr. B. Sarbas, V. Schlicht, Dipl.-Chem. D. Schneider, Dipl.-Min. P. Schubert, Dr. B. Schwager, A. Schwärzel, Dipl.-Ing. H. M. Somer, Dr. G. Strametz, U. Taubald, M. Teichmann, Dr. D. Tille, Dr. W. Töpper, Dipl.-Ing. H. Vanecek, Dipl.-Chem. P. Velić, Dipl.-Ing. U. Vetter, H.-M. Wagner, Dipl.-Phys. J. Wagner, R. Wagner, Dr. E. Warkentin, Dr. C. Weber, Dr. M. Winter, Dr. B. Wöbke, K. Wolff

STAFF GMELIN ONLINE DATA SYSTEM

Dr. R. Froböse, Dr. P. Kuhn, Dr. G. Olbrich, Dr. B. Roth

Dipl.-Min. M.-B. Cinarz, Dr. J. Fippinger, Dr. G. Friedrich, Dr. B. Fröhlich, Dr. M. Klöffler, R. Lohmeyer, Dr. M. Pauluth, Dr. K. Schücke

Volumes published on “Manganese” (Syst. No. 56)

Manganese A 1 (in German)
History – 1980

Manganese B (in German)
The Element – 1973

Manganese C 1 (in German)
Compounds (Hydrides. Oxides. Oxide Hydrates. Hydroxides) – 1973

Manganese C 2 (in German)
Compounds (Oxomanganese Ions. Permanganic Acid. Compounds and Phases with Metals of the Main and Subgroups I and II) – 1975

Manganese C 3 (in German)
Compounds of Manganese with Oxygen and Metals of the Main and Subgroups III to VI. Compounds of Manganese with Nitrogen – 1975

Manganese C 4 (in German)
Compounds of Manganese with Fluorine – 1977

Manganese C 5 (in German)
Compounds of Manganese with Chlorine, Bromine, and Iodine – 1978

Manganese C 6 (in German)
Compounds of Manganese with Sulfur, Selenium, Tellurium, Polonium – 1976

Manganese C 7
Compounds of Manganese with Boron and Carbon – 1981

Manganese C 8
Compounds of Manganese with Silicon – 1982

Manganese C 9
Compounds with Phosphorus, Arsenic, Antimony – 1983

Manganese C 10
Electronic Spectra of Manganese Halides. Cumulative Substance Index of C 1 to C 10 – 1983

Manganese D 1 (in German)
Coordination Compounds 1 – 1979

Manganese D 2 (in German)
Coordination Compounds 2 – 1980

Manganese D 3
Coordination Compounds 3 – 1982

Manganese D 4
Coordination Compounds 4 – 1985

Manganese D 5
Coordination Compounds 5 – 1987 (**present volume**)

Gmelin Handbook of Inorganic Chemistry

8th Edition

Mn
Manganese

D 5

Coordination Compounds 5

With 25 illustrations

AUTHORS — Karl Koeber, Helga Köttelwesch, Dietrich Schneider, Gmelin-Institut, Frankfurt am Main

FORMULA INDEX — Helga Köttelwesch, Gmelin-Institut, Frankfurt am Main

EDITORS — Helga Demmer, Helga Köttelwesch, Edith Schleitzer-Rust, Gmelin-Institut, Frankfurt am Main

CHIEF EDITOR — Edith Schleitzer-Rust, Gmelin-Institut, Frankfurt am Main

System Number 56

Springer-Verlag Berlin Heidelberg GmbH 1987

LITERATURE CLOSING DATE: 1985
IN SOME CASES MORE RECENT DATA HAVE BEEN CONSIDERED

Library of Congress Catalog Card Number: Agr 25-1383

ISBN 978-3-662-08177-8 ISBN 978-3-662-08175-4 (eBook)
DOI 10.1007/978-3-662-08175-4

Originally published by Springer-Verlag Berlin Heidelberg New York Tokyo in 1987
Softcover reprint of the hardcover 8th edition 1987

Preface

The present volume, "Manganese" D 5, continues the description of the manganese complexes. The arrangement of the complexes in these D volumes is based on the ligand type. The introduction, on p. 1, shows the classes of complexes, which have already been described in Chapters 1 to 21 in the Volumes D 1 (1979), D 2 (1980), D 3 (1982), and D 4 (1985).

In Chapters 22 to 29 of this volume are treated complexes with amine-N-polycarboxylic acids, hydrazinecarboxylic acids, amides, hydrazides, derivatives of hydroxylamine (e.g., hydroxamic acids), oximes and nitroso compounds, azo compounds, and triazenes. A survey at the beginning of each of these sections gives information on the most characteristic features of the various complex types.

Because of the complexometric relevance of the complexes with amine-N-polycarboxylic acids, there are many studies concerning the existence and the stability of the complexes in solution.

Numerous X-ray investigations, reported for the complexes with urea or with amides and hydrazides of carboxylic acids, show the different structures of the compounds as a result of the varying bonding sites of the ligands. Complexes with hydrazides (e.g., with isonicotinohydrazide) are of special interest, due to their biological activity.

Complexes with hydroxamic acids, oximes or azo compounds have been studied mostly in aqueous organic or pure organic solvents. The characteristic intense colors of many solutions are used for the analytical determination of manganese.

A formula index at the end of this volume lists the ligands by their empirical molecular formulas.

Complexes with Schiff bases and those with ligands containing S, Se, Si, P, or As will be described in "Manganese" D 6.

Frankfurt/Main
March 1987

Edith Schleitzer-Rust

Table of Contents

Coordination Compounds of Manganese
(Continued)

Introduction

Arrangement. In Series D, coordination compounds of manganese, with the exception of the organometallic compounds, are described. The volumes "Mangan" D 1 to D 4 contain the following chapters:

"Mangan" D 1, 1979	1)	Review
	2)	Complexes with H_2O
	3)	Complexes with Alcohols
	4)	Complexes and Salts with Phenols and Other Aromatic Hydroxy Compounds
	5)	Complexes with Aldehydes
	6)	Complexes with Ketones
	7)	Complexes with Quinones
	8)	Complexes with Ethers and O-Heterocycles
"Mangan" D 2, 1980	9)	Complexes and Salts of Carboxylic Acids and Their Derivatives
	10)	Cyanomanganate Complexes
	11)	Cyanato, Thiocyanato, and Selenocyanato Complexes
"Manganese" D 3, 1982	12)	Complexes with Ammonia
	13)	Complexes with Amines
	14)	Complexes with Hydrazine and Its Derivatives
	15)	Complexes with Hydroxylamine
	16)	Complexes with N-Heterocycles
"Manganese" D 4, 1985	16)	Complexes with N-Heterocycles (Continued)
	17)	Complexes with Aminoalcohols, -phenols, and -naphthols
	18)	Complexes with Aminoethers and Aminooxo Compounds
	19)	Complexes with Amino Acids
	20)	Complexes with Peptides
	21)	Complexes with Proteins

The present volume deals with manganese complexes with amine-N-polycarboxylic acids, hydrazinecarboxylic acids, amides and hydrazides of carboxylic acids, urea and related compounds, derivatives of hydroxylamine (e.g. hydroxamic acids), oximes and nitroso compounds, azo compounds and triazenes. Mixed ligand complexes containing different ligands are arranged generally according to the principle of last position, e.g., mixed ligand complexes with amino acids and amine-N-polycarboxylic acids may be found in Chapter 22 on complexes with amine-N-polycarboxylic acids. The index at the end of this volume, which lists the empirical formulas of the ligands, is intended to expedite locating specific compounds.

Rules and Definitions. Generally the names of the ligands correspond to IUPAC nomenclature; trivial names are also used.

The stepwise stability (formation) constants K_n for the formation of complexes in solution from a central atom M and ligands L and the cumulative constants β_n are defined as follows:

$K_n = [ML_n]/[ML_{n-1}] \cdot [L]$ in L/mol for equilibria $ML_{n-1} + L \rightleftharpoons ML_n$ (n = 1, 2, 3, ...)

$\beta_n = [ML_n]/[M] \cdot [L]^n$ in L^n/mol^n for equilibria $M + nL \rightleftharpoons ML_n$ (n = 1, 2, 3, ...)

The formation of complexes with protonated ligands is described by:

$K^M_{MH_pL} = [MH_pL]/[M] \cdot [H_pL]$ for equilibria $M + H_pL \rightleftharpoons MH_pL$ (p = 1, 2, 3, ...)

Enthalpy (ΔH), Gibbs free energy (ΔG), or entropy changes (ΔS) are given the same subscript as the corresponding K: e.g., ΔH_1 for constant K_1. For reactions represented by cumulative constants β, the notation $\Delta H_{\beta n}$ is used. Ionic strengths are given in mol/L.

For partial molal quantities, e.g., $\overline{V}$ or $\overline{C}_p$, the terminology and conventions of Lewis, G. N., Randall, M., et al. (Thermodynamics, 2nd Ed., New York 1961) are used.

Abbreviations and Dimensions. Temperatures are normally given in °C; K stands for Kelvin. Abbreviations used with temperatures are m.p. for melting point and dec. for decomposition. With thermodynamic data, (s) is used to label solids, (g) is used to designate the gaseous state, and (l) is used for liquids.

The vibrational spectra are labeled as IR (infrared) or R (Raman). The symbol ν is used for stretching vibrations and δ for deformation vibrations; wave numbers are given in cm^{-1}. The intensities are placed in parentheses (w = weak, m = medium, s = strong, vs = very strong, etc.); sh means shoulder; br means broad. The UV-visible absorption maxima of the electronic spectra are given in nm (λ_{max}) or cm^{-1} (ν_{max}), the extinction coefficient ε in $L \cdot mol^{-1} \cdot cm^{-1}$.

Abbreviations for methods frequently used in this volume are:

DTA	differential thermoanalysis	ESR	electron spin resonance
TG	thermogravimetry	NMR	nuclear magnetic resonance
DTG	differential thermogravimetry	PRR	proton relaxation rate
DSC	differential scanning calorimetry		

Abbreviations for ligands are listed on p. 321, the first page of the formula index of ligands.

22 Complexes with Amine-N-polycarboxylic Acids

Survey

This chapter compiles formation data, properties, and chemical reactions of manganese complexes with the title compounds. For analytical procedures (or other applications) reported in the recent literature, references are given subsequent to the chemical reactions. With respect to the previous literature on analytical applications, the analytical handbooks and monographs may be consulted, for instance [1 to 4].

Because of the complexometric relevance of the complexes, studies on those in aqueous solution prevail over studies on solid complexes. A large number of equilibrium studies were performed since nitrilotriacetic acid and ethylenediaminetetraacetic acid (H_4edta) have been introduced as "complexones" into analytical chemistry by Schwarzenbach and coworkers in 1945 [1]. The results of these studies now make it possible to appreciate the effect of various modifying groups in the parent ligands on the stability of manganese complexes. For instance, it emerges that elongation of both the alkylenediamine part and the alkylcarboxylic acid part by methylene groups significantly lowers the complex stability, but that substitution of ethylene protons by alkyl or aryl groups has little influence on complex stability. The higher stability of the complexes with *trans*-1,2-cyclohexanediaminetetraacetic acid (H_4cdta) compared with those of ethylenediaminetetraacetic acid is accounted for by the larger positive entropy change on formation of the former. For these highly stable complexes, the usual pH method does not give satisfying stability data. Therefore, exchange equilibria with other metal complexes of the ligands were used, and the concentrations of the metal ions formed were measured by potentiometric or polarographic methods (pM methods [1, 5]).

IR and water proton relaxation studies indicate for the most studied complex, the $Mn^{II}(edta)^{2-}$ ion, a molecular structure with seven-coordinate manganese(II). The hexadentate organic ligand (N, N′, four O) and one water molecule are coordinated, as was also found for the complex unit in solid compounds. The $Mn^{II}(edta)^{2-}$ ion is thermally quite stable over a large pH range, but it is readily oxidized by a variety of oxidizing agents, even by air, to give the intensely red $Mn^{III}(edta)^-$ ion. This color reaction can be utilized for the colorimetric determination of manganese(II). The $Mn^{III}(edta)^-$ ion (log $K_1 \sim 25$ for $I = 0.1$ M) has a much greater complex stability than the $Mn^{II}(edta)^{2-}$ ion (log $K_1 \sim 14$ for $I = 0.1$ M). For the $Mn^{III}(edta)^-$ ion, a seven-coordinate structure is also assumed in contrast to the complex unit in the solid complex. For Mn^{III} complexes with analogues of H_4edta, ligand exchange studies indicate the existence of octahedral structures in solution as well. The $Mn^{III}(edta)^-$ ion is thermally unstable because of an intramolecular redox reaction between the Mn^{III} ion and the coordinated ligand. This is also the case for Mn^{III} complexes with the analogues; however, the $Mn^{III}(cdta)^-$ complex is less subject to the internal redox reaction than the $Mn^{III}(edta)^-$ ion. The kinetics of the decomposition reactions and of the redox reactions with a variety of reducing agents were studied for both ions.

Acid Mn^{II} complex salts, $Mn^{II}(H_2edta) \cdot 4H_2O$ and $Mn_3(Hedta)_2 \cdot 10H_2O$, were prepared from stoichiometric amounts of $MnCO_3$ and H_4edta in hot aqueous solution. Neutral complex salts $M_2Mn(edta) \cdot nH_2O$ with M = Li, Na, K, Rb, Cs, and NH_4 were obtained by adjusting aqueous suspensions of $Mn(H_2edta) \cdot 4H_2O$ to pH 7 with M_2CO_3. There are in addition dinuclear compounds of composition $Mn_2(edta) \cdot 9H_2O$ and $MgMn(edta) \cdot 9H_2O$.

X-ray structural studies and IR spectra show that in all these compounds the anionic $[Mn(edta)(H_2O)]^{2-}$ complex exists, with a seven-coordinate Mn atom in a distorted one-capped trigonal prism structure (NbF_7^{2-} type). These complex units are connected by protons forming H bonds between carboxylate oxygens of adjacent complex units in the acid complex salts. The neutral complexes consist of alternating "chelated" $[Mn(edta)(H_2O)]^{2-}$ and "hydrated" M^I or M^{II} complex units. To the M^I or M^{II} atoms are coordinated, in addition to the water molecules, carboxylate oxygen atoms of adjacent $[Mn(edta)(H_2O)]^{2-}$ units. The strings or layers thus formed are connected by H bonds from water molecules leading to a three-dimensional network.

In another type of compounds, $Mn^{II}M^{II}(edta) \cdot 6H_2O$ with M^{II} = Co, Ni, Cu, and Zn, the M^{II} atom is in the "chelated" complex unit and the Mn^{II} atom in the "hydrated" one.

The manganese(III) complex salt $K[Mn^{III}(edta)] \cdot 2H_2O$ was prepared by reacting freshly prepared MnO_2 in aqueous KOH solution with H_4edta. It contains strongly distorted octahedral $[Mn(edta)]^-$ complex units which are connected with $[K(H_2O)_2O_4]$ octahedra by bridging carboxylate groups. Similar complex salts were prepared with trans-1,2-cyclohexanediaminetetraacetic acid and other analogues of H_4edta.

References:

[1] Schwarzenbach, G., Flaschka, H. (Die Komplexometrische Titration, Enke, Stuttgart 1965).
[2] Přibil, R. (Komplexone in der chemischen Analyse, VEB Deutscher Verlag der Wissenschaften, Berlin 1961).
[3] Bermejo-Martinez, F. (in: Flaschka, H. A., Barnard Jr., A. J., Chelates in Analytical Chemistry, Marcel Dekker Inc., New York 1976, pp. 1/161, 103/4).
[4] Umland, F. (Theorie und praktische Anwendung von Komplexbildnern, Akad. Verlagsges., Frankfurt a. M. 1971).
[5] Schwarzenbach, G., Anderegg, G. (Helv. Chim. Acta **40** [1957] 1773/92).

22.1 With Iminodiacetic Acid $HN(CH_2COOH)_2$ (= $C_4H_7NO_4$)

22.1.1 Manganese(II) Complexes in Solution

The formation constant of the $Mn(C_4H_5NO_4)_2^{2-}$ complex ion in aqueous solution, $\log \beta_2 = 7.1$, was estimated using the K_1 and K_2 values of the corresponding Fe and Cd complexes at 20°C, I = 0.1 M (KNO_3) [1]. A value of $\log \beta_2 = 8.6$ was obtained by calculation of the free energy of formation with a method based on an electrostatic model, taking into account covalent interactions [2]. The reaction with the methyliminodiacetate ion, $Mn(C_4H_5NO_4) + C_5H_7NO_4^{2-} \rightleftharpoons Mn(C_5H_7NO_4) + C_4H_5NO_4^{2-}$, was investigated, and the equilibrium constant and thermodynamic data were calculated: log K = 1.1, $\Delta H = -0.5$ kcal/mol, $\Delta G = -1.4$ kcal/mol, $\Delta S = 3.1$ cal·mol^{-1}·K^{-1} [3]. ESR relaxation parameters of Mn^{2+} in the presence of iminodiacetate ions are given in [4].

References:

[1] Anderegg, G. (Helv. Chim. Acta **47** [1964] 1801/14, 1801).
[2] Münze, R. (Z. Physik. Chem. [Leipzig] **252** [1973] 145/53, 150).
[3] Anderegg, G. (Helv. Chim. Acta **48** [1965] 1718/21).
[4] Reed, G. H., Leigh, J. S., Pearson, J. E. (J. Chem. Phys. **55** [1971] 3311/6).

22.1.2 Isolated Manganese(II) Compounds

$Mn(C_4H_5NO_4)\cdot nH_2O$ (n = 6, 3.5, 1.5, 0). Solubility determinations at 25°C under nitrogen show that the hexahydrate occurs as a solid in the $Mn(C_4H_5NO_4)-M_2(C_4H_5NO_4)-H_2O$ systems with M = Na, Li, K, or Ba/2 in the following concentration ranges: 7.3 to 20.2 wt% $Mn(C_4H_5NO_4)$ and 6.2 to 20.5 wt% $Na_2(C_4H_5NO_4)$; 1.5 to 3.5 wt% $Mn(C_4H_5NO_4)$ and 3.1 to 6.4 wt% $Li_2(C_4H_5NO_4)$; 4.4 to 9.9 wt% $Mn(C_4H_5NO_4)$ and 6.1 to 7.5 wt% $K_2(C_4H_5NO_4)$; 10.5 to 12.7 wt% $Mn(C_4H_5NO_4)$ and 10.1 to 13.0 wt% $Ba(C_4H_5NO_4)$ [2]. $Mn(C_4H_5NO_4)\cdot 6H_2O$ was prepared by reacting 2 M aqueous $MnCl_2$ with 10% excess potassium iminodiacetate under nitrogen. After evaporating the mixture to half the original volume and cooling, rose-colored crystals were formed, which were washed with ethanol and ether and air-dried. The yield was 70% [1, 2].

The magnetic moment of the hexahydrate is 5.56 μ_B. The bands in the IR spectrum of $Mn(C_4H_5NO_4)\cdot 6H_2O$ were assigned as follows (in cm^{-1}): 3330, ν(NH); 1590, $\nu_{as}(COO^-)$; 1410, $\nu_s(COO^-)$. The ν(NH) band of $Mn(C_4H_5NO_4)\cdot 6H_2O$ has the same position as that of $Na_2(C_4H_5NO_4)$, while that of $Mn(C_4H_5NO_4)\cdot 1.5H_2O$ is shifted to 3290 cm^{-1} [1]. The magnetic and spectral properties of the hexahydrate [1] together with its low solubility in water (1.5% at 25°C) [2] suggest a polymeric octahedral structure with partly coordinated water molecules and bidentate bridging carboxylate groups [1].

On prolonged standing, the hexahydrate is dehydrated to $Mn(C_4H_5NO_4)\cdot 3.5H_2O$, which forms $Mn(C_4H_5NO_4)\cdot 1.5H_2O$ on heating at 100 to 140°C. The residual water is split off at 160 to 180°C. The anhydrous compound, $Mn(C_4H_5NO_4)$, decomposes at 190 to 210°C to form manganese oxides [1].

$Mn(C_4H_6NO_4)_2$ and **$[Mn(C_4H_6NO_4)_2(H_2O)_2]$**. To prepare the anhydrous compound, iminodiacetic acid, manganese(II) chloride, and potassium hydroxide were dissolved stepwise in water (mole ratio 2:1:2) and the solution heated to boiling. On cooling, white crystals separated, which were washed with water, ethanol, and ether and air-dried. The yield was 71% [1].

An X-ray diffraction study is reported for the dihydrate (no preparation method given). This compound forms well-faceted, tetragonal, colorless crystals with the lattice parameters a = 8.088(5), c = 9.592(5) Å; Z = 2. The space group is $P\bar{4}2_1c-D^4_{2d}$ (No. 114). The coordination polyhedron is a tetragonal bipyramid with four O atoms of four different $C_4H_6NO_4^-$ residues at its base and two O atoms of water molecules at the apices. The N atoms are not coordinated by the Mn atom, since the iminodiacetate groups have a zwitterion structure (see **Fig. 1**, p. 6). The $C_4H_6NO_4^-$ ions are bridging; the two COO^- groups are each coordinated to different Mn atoms by one O atom, O(1) and O(1′). Thus, layers perpendicular to the z axis are formed with H bonds between the N atom and the carboxylate O atoms. Between these layers, the H_2O molecules form hydrogen bonds with the O(2) atoms of the adjacent layers. The Mn–O(1) distance is 2.193(6) Å, the $Mn-O_{H_2O}$ distance is 2.156(6) Å. The angle $O(1)-Mn-O_{H_2O}$ is 82.4(3)° [4]. The measured density is 1.77 g/cm^3 [3].

The magnetic moment of the anhydrous compound is μ_{eff} = 5.74 μ_B. The IR spectrum of this compound shows absorptions at 3080, ν(NH); 1590, $\nu_{as}(COO^-)$; 1410, $\nu_s(COO^-)$. The ν(NH) band indicates the existence of the noncoordinated NH_2^+ group. A polymeric octahedral structure has been proposed with the manganese atom surrounded by six O atoms of unidentate and bidentate bridging carboxylate groups, the latter linking neighboring molecular units. The compound decomposes at 150 to 220°C [1].

$M_2Mn(C_4H_5NO_4)_2\cdot nH_2O$. Compounds with n = 4 for M = Li, n = 7.5 for M = Na, n = 2 for M = K, and n = 5 for M = Ba/2 were prepared by dissolving 0.03 mol $Mn(C_4H_5NO_4)\cdot 6H_2O$ (see above) in a 1 to 2 M aqueous solution of the appropriate iminodiacetate $M_2(C_4H_5NO_4)$ containing a 10% stoichiometric excess of iminodiacetic acid. The mixture was evaporated to about one third of the initial volume and left to crystallize under nitrogen. The white crystals were washed with ethanol and ether and dried over NaOH under nitrogen. The yield was 55 to 59% with respect to iminodiacetic acid [1]. The optimal molar proportions of $Mn(C_4H_5NO_4)\cdot 6H_2O:M_2(C_4H_5NO_4):H_2O$ for the preparation of the compounds at 25°C are: 1:1.3:13.2 for $Li_2Mn(C_4H_5NO_4)_2\cdot 4H_2O$; 1:1.2:6.5 for $Na_2Mn(C_4H_5NO_4)_2\cdot 7.5H_2O$; 1:1.7:19.2 for $K_2Mn(C_4H_5NO_4)_2\cdot 2H_2O$; and 1:1.4:12.4 for $BaMn(C_4H_5NO_4)_2\cdot 5H_2O$ [2].

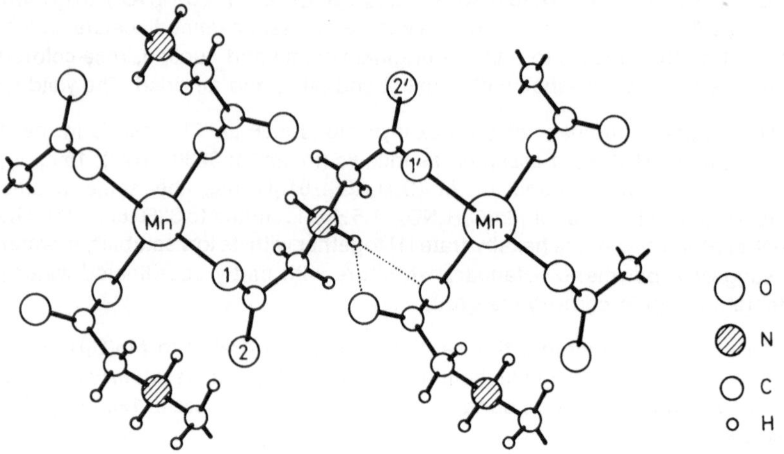

Fig. 1. Essential part of the structure of $[Mn(C_4H_6NO_4)_2(H_2O)_2]$ projected along [001]. The H_2O molecules are omitted for clarity [3].

The concentration ranges of A = $Mn(C_4H_5NO_4)$ and B = $M_2(C_4H_5NO_4)$ for occurrence of the complexes as solids in the $Mn(C_4H_5NO_4)-M_2(C_4H_5NO_4)-H_2O$ systems are 1.6 to 3.5 wt% A, 7.3 to 6 wt% B for $Li_2Mn(C_4H_5NO_4)_2\cdot 4H_2O$, 9.3 to 18.7 wt% A, 24.7 to 12.7 wt% B for $Na_2Mn(C_4H_5NO_4)_2\cdot 7.5H_2O$, 16.8 wt% A, 23.6 wt% B for $K_2Mn(C_4H_5NO_4)_2\cdot 2H_2O$, and 12.9 to 20.2 wt% A, 35.8 to 28.2 wt% B for $BaMn(C_4H_5NO_4)_2\cdot 5H_2O$.

The IR spectra of the complexes (KBr disks or hexachlorobutadiene mulls) reveal a shift of the ν(NH) bands to lower wave numbers compared with those of sodium iminodiacetate (3330 cm^{-1}). This indicates coordination of the nitrogen atom. The positions of the $\nu(COO^-)$ bands suggest monodentate coordination of the COO^- groups. A monomeric octahedral structure has been proposed for all the compounds. The most important IR bands (cm^{-1}) are shown below, together with the observed magnetic moments (μ_{eff} in μ_B) [1]:

complex	μ_{eff}	ν(NH)*)	$\nu_{as}(COO^-)$	$\nu_s(COO^-)$
$Li_2Mn(C_4H_5NO_4)_2\cdot 4H_2O$	5.56	3270 (3280)	1615	1405
$Na_2Mn(C_4H_5NO_4)_2\cdot 7.5H_2O$	5.95	3280 to 3235 3200 (3300, 3325)	1620	1410
$K_2Mn(C_4H_5NO_4)_2\cdot 2H_2O$	5.75	3260 (3250)	1600	1395
$BaMn(C_4H_5NO_4)_2\cdot 5H_2O$	5.66	3270 to 3240 (3290)	1600	1410

*) The ν(NH) bands of the anhydrous complexes are given in parentheses.

The influence of the external cation M on the formation of the complexes $M_2[Mn(C_4H_5NO_4)_2]\cdot nH_2O$ was studied [5].

Thermal decomposition of the hydrates occurs in the following temperature ranges:

complex	t in °C	loss of H_2O	complex	t in °C	loss of H_2O
$Li_2Mn(C_4H_5NO_4)_2\cdot 4H_2O$	30 to 120	4	$K_2Mn(C_4H_5NO_4)_2\cdot 2H_2O$	30 to 120	2
$Na_2Mn(C_4H_5NO_4)_2\cdot 7.5H_2O$	30 to 80	6	$BaMn(C_4H_5NO_4)_2\cdot 5H_2O$	20 to 110	4
$Na_2Mn(C_4H_5NO_4)_2\cdot 1.5H_2O$	100 to 120	1.5	$BaMn(C_4H_5NO_4)_2\cdot H_2O$	150 to 160	1

The anhydrous complexes $M_2Mn(C_4H_5NO_4)_2$ decompose at 170 to 200°C (M = Li), 170 to 240°C (M = Na), 160 to 200°C (M = K), and 170 to 210°C (M = Ba/2) [1].

References:

[1] Lukeš, I., Šmídová, I., Ebert, M. (Collection Czech. Chem. Commun. **47** [1982] 1169/75, 1170, 1173).

[2] Lukeš, I., Šmídová, I., Ebert, M. (Z. Chem. [Leipzig] **21** [1981] 190/1).

[3] Anan'eva, N. N., Artamonova, S. D., Polynova, T. N., Porai-Koshits, M A., Mitrofanova, N. D. (Koord. Khim. **1** [1975] 435; Soviet J. Coord. Chem. **1** [1975] 346).

[4] Anan'eva, N. N., Polynova, T. N., Porai-Koshits, M. A. (Koord. Khim. **11** [1985] 702/5).

[5] Lukeš, I., Ebert, M. (Proc. 10th Conf. Coord. Chem., Prague 1985, pp. 251/6 from C.A. **104** [1986] No. 236143).

22.2 With Derivatives of Iminodiacetic Acid, $RN(CH_2COOH)_2$

22.2.1 Binary Manganese(II) Complexes in Solution

The stability constants of species $\mathbf{MnL_n^{(2-2n)+}}$ and $\mathbf{MnHL^+}$ with ligands H_2L, of species $\mathbf{MnL_n^{(2-3n)+}}$ and **MnHL** with ligands H_3L, and of species $\mathbf{MnL^{2-}}$ or $\mathbf{MnHL^-}$ with ligands H_4L in aqueous solution, determined with a glass electrode unless otherwise stated, are collected in Table 1, pp. 9/11.

Thermodynamic data for 1:1 complexes with several derivatives of iminodiacetic acid, $RN(CH_2COOH)_2$, cited in Table 1, pp. 9/11 (ΔH and ΔG in kcal/mol, ΔS in $cal\cdot mol^{-1}\cdot K^{-1}$):

No.	R	t in °C	medium	method	ΔH_1	ΔG_1	ΔS_1	Ref.
1	CH_3	20	0.1 (KNO_3)	calor.	0.56	−7.24	26.6	[21]
		25	→0	calc.	0	−8.0	26	[22]
4	$HOCH_2CH_2$	25	0.1	?	0.4	—	27	[5]
21	$H_2NC(O)NH$	30	0.1 (KCl)	calor.	3.2	−3.6	22	[23]
22	$H_2NC(S)NH$	30	0.1 (KCl)	calor.	7.2	−2.8	33	[23]

No.	R	t in °C	medium	method	ΔH_1	ΔG_1	ΔS_1	Ref.
23	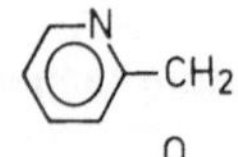	20	0.1 ($NaNO_3$)	calor.	-0.4 ± 0.1	—	31 ± 0.3	[16]
26	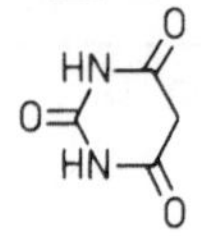	25	0.1 (KNO_3)	calor.	-2.51*)	-13.7	38	[24]

*) A temperature-dependent component ΔH_e and a temperature-independent component ΔH_c were calculated: $\Delta H_e=3.6$ kcal/mol and $\Delta H_c=-6.1$ kcal/mol [24].

The values $\Delta H_2=-0.33$ kcal/mol, $\Delta G_2=-5.59$ kcal/mol, and $\Delta S_2=17.95$ $cal\cdot mol^{-1}\cdot K^{-1}$ were determined calorimetrically at 20°C and $I=0.1(KNO_3)$ for the 1:2 complex with ligand 1 [21]. The values $\Delta H_2=-4.3$ kcal/mol and $\Delta S_2=2$ $cal\cdot mol^{-1}\cdot K^{-1}$ at 25°C are reported for the complex with ligand 4 [5].

Data for the formation of the species $Mn(C_5H_7NO_4)$ by reaction of the iminodiacetato complex $Mn(C_4H_5NO_4)$ with **N-methyliminodiacetic acid** (=$C_5H_9NO_4$) are given on p. 4.

Polarographic studies of the complexes with **N-methyliminodiacetic acid** (=$C_5H_9NO_4$) and **N-(2-hydroxyethyl)iminodiacetic** acid (=$C_6H_{11}NO_5$) are reported in [25, 26]. The Mn^{II}/Mn^0 reduction half-wave potential $E_{1/2}$ for $Mn(C_5H_7NO_4)$ was measured as a function of the free ligand concentration in [26]. For $Mn(C_6H_9NO_5)_2^{2-}$, $E_{1/2}=-1.75$ V vs. SCE at 25°C and I=0.1 M ([$(CH_3)_4N$]Cl) were calculated for the case that the Mn^{II}/Mn^0 reduction is reversible and for a 1 N free ligand concentration. The rate constant of the discharge reaction at the potential zero vs. SCE, $Mn(C_6H_9NO_5)_2^{2-}+2e^-+Hg=Mn(Hg)+2C_6H_9NO_5^{2-}$, is $\log k=-31.2\pm0.4$ (k in s^{-1}). A two-step mechanism with intermediate formation of the 1:1 complex has been proposed [25] for both ligands. Measurements were made as a function of the pH of the solution. The various species occurring were discussed on the basis of the results [26]. A potentiometric study of the hydrolytic behavior of the 1:1 chelate $Mn(C_6H_9NO_5)$ shows that hydrolysis does not take place up to pH 10 [27].

The visible spectrum of the 1:1 complex with **semixylenol orange** (No. 27 in Table 1) shows a maximum at 579 nm with the extinction coefficient $\varepsilon=5.94\times10^4$ $L\cdot mol^{-1}\cdot cm^{-1}$ [20].

Mn^{2+} ions form deep colored chelate complexes with **[[(3,4-dihydroxy-2-anthraquinonyl)-methyl]imino]diacetic acid** (=$C_{19}H_{15}NO_8$) and **[[(3,4,6-trihydroxy-2-anthraquinonyl)methyl]-imino]diacetic acid** (=$C_{19}H_{15}NO_9$). A red 1:1 complex is observed with the dihydroxy compound in aqueous solution of pH 4.3 to 4.6 which shows an absorption maximum at ~500 nm. It is assumed that the N atom, the deprotonated hydroxy group at the 3-position, one O atom of each deprotonated carboxyl group, and two water molecules are involved in complex formation [28]. In alkaline solution, both ligands form 1:2 chelates [29] which are purplish blue [28, 29] ($\lambda_{max}=550$ nm at pH 11.1 to 11.6) in the case of the dihydroxy compound [29], and blueish violet ($\lambda_{max}=530$ to 540 nm at pH 11.4 to 11.8) for the trihydroxy species [30]. The complexes can be used for the spectrophotometric determination of manganese [28 to 30].

The complexation of manganese with ion exchangers containing iminodiacetate [31] (or iminodipropionate) as the chelating groups [32] has been studied [31, 32]. Cation exchange resins containing N-phenyliminodiacetate groups were charged with Mn^{2+} ions and the magnetic susceptibilities of the high polymeric chelate compounds were measured. The magnetic moments (4.7 to 5.9 μ_B) indicate high-spin complex units [33].

Table 1

Stability Constants for 1:1 and 1:2 Complexes of Manganese(II) with Derivatives of Iminodiacetic Acid, $RN(CH_2COOH)_2$.
(Ligands where R contains additional carboxyl groups are treated on p. 22.)

No.	R in $RN(CH_2COOH)_2$	formula	H_nL	t in °C	medium	log K_1	log K_2	log K^{Mn}_{MnHL}	Ref.
1	CH_3	$C_5H_9NO_4$	H_2L	20	0.1 (KCl)	5.40 ± 0.1	4.16 ± 0.1	—	[1]
				20	0	5.87[a)]	—	—	[2]
				25	0.1 ($NaClO_4$)	—	4.0 ± 0.2[b)]	—	[3]
2	C_6H_{13}	$C_{10}H_{19}NO_4$	H_2L	25	0.1 (KNO_3)	7.05	5.50	1.72	[34]
3	$(CH_3)_3CCH_2CH_2$	$C_{10}H_{19}NO_4$	H_2L	20	0.1 (KCl)	5.55 ± 0.1	4.45 ± 0.1	—	[1]
4	$HOCH_2CH_2$	$C_6H_{11}NO_5$	H_2L	20	0.1 (KCl)	5.55 ± 0.1	3.76 ± 0.1	—	[1]
				20	0.1	6.4[c)]	3.3[c)]	—	[4]
				25	0.1	5.56 ± 0.2	3.7 ± 0.4	—	[5]
				30	0.1 (KCl)	5.65	3.93	—	[6]
5	$CH_3OCH_2CH_2$	$C_7H_{13}NO_5$	H_2L	20	0.1 (KCl)	5.53 ± 0.1	4.09 ± 0.1	—	[1]
6	$HSCH_2CH_2$	$C_6H_{11}NO_4S$	H_3L	20	0.1 (KCl)	9.32 ± 0.1	—	—	[1]
7	$CH_3SCH_2CH_2$	$C_7H_{13}NO_4S$	H_2L	20	0.1 (KCl)	5.10 ± 0.1	3.60 ± 0.1	—	[1]
8	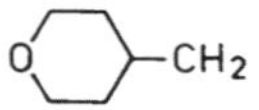	$C_{10}H_{17}NO_5$	H_2L	20	0.1 (KNO_3)	5.89 ± 0.03	4.35 ± 0.02	—	[7]
9	C_6H_5	$C_{10}H_{11}NO_4$	H_2L	20	0.1 (KCl)	1.58 ± 0.1	—	—	[1]
10	$C_6H_5CH_2$	$C_{11}H_{13}NO_4$	H_2L	25	0.1 (KCl)	6.59	—	—	[8]
				—	—	6.6	—	—	[9, 10]

a) Calculated. — b) By Amperometry. — c) By electrophoresis.

Table 1 (continued)

No.	R in $RN(CH_2COOH)_2$	formula	H_nL	t in °C	medium	log K_1	log K_2	log K^{Mn}_{MnHL}	Ref.
11	2-HO–C_6H_5	$C_{10}H_{11}NO_5$	H_3L	20	0.1	7.82	—	2.85	[10, 11]
12	3-HO–C_6H_5	$C_{10}H_{11}NO_5$	H_3L	20	0.1	4.88	—	—	[11]
13	4-HO–C_6H_5	$C_{10}H_{11}NO_5$	H_3L	20	0.1	4.95	—	—	[11]
14	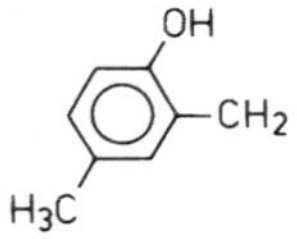	$C_{12}H_{15}NO_5$	H_3L	25	~0	9.00	—	—	[10, 12]
15	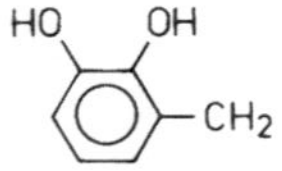	$C_{11}H_{13}NO_6$	H_4L	—	—	—	—	10.3	[10, 13]
16	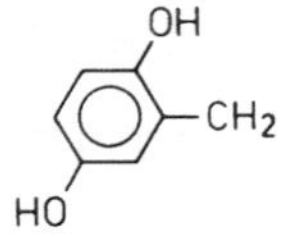	$C_{11}H_{13}NO_6$	H_4L	25	~0	—	—	9.61	[10, 12]
17	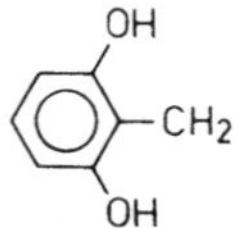	$C_{11}H_{13}NO_6$	H_4L	—	—	—	—	7.4	[10, 13]
18	NC–CH_2	$C_6H_8N_2O_4$	H_2L	20	0.1 (KCl)	3.50 ± 0.1	2.00 ± 0.1	—	[1]
19	$H_2NCH_2CH_2$	$C_6H_{12}N_2O_4$	H_2L	20	0.1 (KCl)	7.71 ± 0.1	3.70 ± 0.1	—	[1]
20	$(CH_3)_3N^+CH_2CH_2$	$C_9H_{19}N_2O_4^+$	H_2L	20	0.1 (KCl)	2.87 ± 0.1	—	—	[1]
21	$H_2NC(O)NH$	$C_5H_9N_3O_5$	H_2L	30	0.1 (KCl)	2.6	—	1.6	[14]

22	$H_2NC(S)NH$	$C_5H_9N_3O_4S$	H_2L	30	0.1 (KCl)	2.0	—	1.5	[14]
23	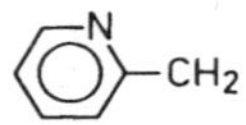	$C_{10}H_{12}N_2O_4$	H_2L	20	0.1 (KNO_3)	6.97 ± 0.01	3.36 ± 0.1	—	[15]
				20	0.1 ($NaNO_3$)	7.10 ± 0.05	3.5 ± 0.1	—	[16]
24	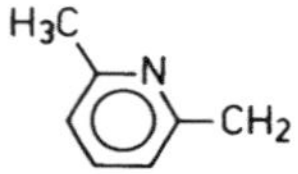	$C_{11}H_{14}N_2O_4$	H_2L	20	0.1 ($NaNO_3$)	6.60 ± 0.05	3.5 ± 0.1	—	[16]
25	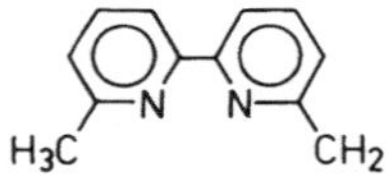	$C_{16}H_{17}N_3O_4$	H_2L	25	0.1 (KCl)	9.4	—	—	[35]
26[a]	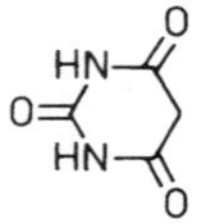	$C_8H_9N_3O_7$	H_3L	20	→0	—	4.0	—	[17]
				25	0.1 ($(CH_3)_4NNO_3$)	10.28 ± 0.01	3.76 ± 0.03	—	[18]
				25	0.1 (KNO_3)	9.87	—	3.48	[10, 19]
27[b]	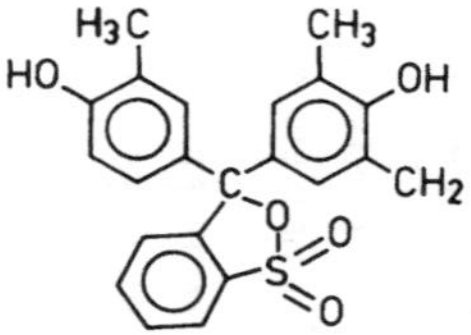	$C_{26}H_{25}NO_9S$	H_4L	25	0.1 (KNO_3)	9.4	—	—	[20]

[a] Uramil-N,N-diacetic acid. – [b] Semixylenol orange.

Manganese(II) complexes of silochrome (modified silica gel) carrying iminodiacetate groups are catalytically active in the O_2 oxidation of sulfide ions and cysteine in aqueous media. The kinetics of the oxidation reactions were studied in [36].

References:

[1] Schwarzenbach, G., Anderegg, G., Schneider, W., Senn, H. (Helv. Chim. Acta **38** [1955] 1147/70, 1150).
[2] Martell, A. E., Smith, R. M. (Critical Stability Constants, Vol. 1, Amino Acids, Plenum, New York 1974, p. 125).
[3] Verdier, E., Piro, J. (Ann. Chim. [Paris] [14] **4** [1969] 213/25, 224).
[4] Jokl, V., Majer, J. (Acta Fac. Pharm. Bohemoslov. **11** [1965] 55/110, 86).
[5] Lewis, R., Nancollas, G. H. (unpublished results from Martell, A. E., Smith, R. M., Critical Stability Constants, Vol. 1, Amino Acids, Plenum, New York 1974, p. 165).
[6] Chaberek, S., Courtney, R. C., Martell, A. E. (J. Am. Chem. Soc. **74** [1952] 5057/60).
[7] Irving, H., Fraústo da Silva, J. J. R. (J. Chem. Soc. **1963** 1144/8).
[8] Hering, R., Krüger, W., Kühn, G. (Z. Chem. [Leipzig] **2** [1962] 374/5).
[9] Dyatlova, N. M., Temkina, V. Ya. (Koord. Khim. **1** [1975] 66/82; Soviet J. Coord. Chem. **1** [1975] 52/64, 53).
[10] Perrin, D. D. (Stability Constants of Metal-Ion Complexes, Part B; Organic Ligands; IUPAC Chemical Data Series No. 22 [1979] 588, 749, 857, 860, 861, 904).

[11] Temkina, V. Ya., Dyatlova, N. M., Rusina, M. N., et al. (Tr. Vses. Nauchn. Issled Inst. Khim. Reaktivov Osobo Chist. Khim. Veshchestv No. 30 [1967] 118/30, 125; C.A. **68** [1968] No. 72873).
[12] Tsirul'nikova, N. V., Temkina, V. Ya., Dyatlova, N. M., et al. (Zh. Analit. Khim. **25** [1970] 839/46; J. Anal. Chem. [USSR] **25** [1970] 724/30, 727).
[13] Nakon, R. (Anal. Biochem. **95** [1979] 527/32).
[14] Goddard, D. R., Nwankwo, S. I. (J. Chem. Soc. A **1967** 1371/5).
[15] Irving, H., Fraústo da Silva, J. J. R. (J. Chem. Soc. **1963** 945/52, 949).
[16] Anderegg, G. (J. Coord. Chem. **11** [1982] 171/5).
[17] Schwarzenbach, G., Biedermann, W. (Helv. Chim. Acta **31** [1948] 456/9).
[18] Jellish, R. M., Thompson, L. C. (J. Coord. Chem. **4** [1975] 199/203).
[19] Fraústo da Silva, J. J. R., Cândida, M., Abreu Vaz, T. (Rev. Port. Quim. **14** [1972] 102/9, 106).
[20] Murakami, S. (J. Inorg. Nucl. Chem. **43** [1981] 335/43, 336, 338).

[21] Anderegg, G. (Helv. Chim. Acta **48** [1965] 1718/21).
[22] Ockerbloom, N., Mihajlov, V. (unpublished results from Martell, A. E., Rec. Trav. Chim. **75** [1956] 781/6).
[23] Goddard, D. R., Nwankwo, S. I., Staveley, L. A. K. (J. Chem. Soc. A **1967** 1376/8).
[24] Cândida, M., Abreu Vaz, T., Fraústo da Silva, J. J. R. (J. Inorg. Nucl. Chem. **43** [1981] 1573/8).
[25] Bennes, R., Piro, J., Verdier, E. (J. Chim. Phys. **64** [1967] 1385/7).
[26] Verdier, E., Piro, J. (J. Chim. Phys. **65** [1968] 1052/9, 1056/7).
[27] Courtney, R. C., Gustafson, R. L., Chaberek, S., Martell, A. E. (J. Am. Chem. Soc. **80** [1958] 2121/8, 2122).
[28] Leonard, M. A., West, T. S. (J. Chem. Soc. **1960** 4477/86, 4479, 4485).
[29] Capitán, F., Román, M., Guiraúm, A. (Quim. Ind. [Madrid] **17** [1971] 15/8).
[30] Capitán, F., Román, M., Guiraúm, A. (Anales Quim. **67** [1971] 147/52).

[31] Hering, R., Haupt, D. (Z. Chem. [Leipzig] **6** [1966] 192/3).
[32] Kühn, G., Hoyer, E. (Makromol. Chem. **108** [1967] 84/94, 89).
[33] Moiseeva, N. P., Sinyavskii, V. G., Romankevich, M. Ya. (Zh. Obshch. Khim. **41** [1971] 943/7; J. Gen. Chem. [USSR] **41** [1971] 951/4).
[34] Chen, C., Guan, L., Li, Y., Ma, S. (Zhongnan Kuangye Xueyuan Xuebao **1984** 51/9; C.A. **102** [1985] No. 173549).
[35] Ohm, C., Voegtle, F. (Chem. Ber. **118** [1985] 22/7).
[36] Gombosuren, O., Berentsveig, V. V., Rudenko, A. P. (Vestn. Mosk. Univ. Ser. II Khim. **26** [1985] 588/92; Moscow Univ. Chem. Bull. **40** No. 6 [1985] 69/73).

22.2.2 Ternary Manganese(II) and Manganese(III) Complexes in Solution

The manganese(II) complexes $Mn(C_5H_7NO_4)$ and $Mn(C_8H_6N_3O_7)^-$ with N-methyliminodiacetic acid ($C_5H_9NO_4$) and uramil-N,N-diacetic acid ($C_8H_9N_3O_7$, see No. 26 in Table 1, pp. 9/11) form mixed ligand complexes with 2,2'-bipyridine. $Mn(C_8H_6N_3O_7)^-$ also forms a mixed ligand complex with 8-hydroxyquinoline (HA). Stability constants K, defined by reactions 1 to 3, and selected kinetic data from temperature-jump measurements (if not otherwise stated) are presented below. Rate constants for the forward and back reactions are k_f (in 10^6 $M^{-1} \cdot s^{-1}$) and k_b (in 10^3 s^{-1}); activation enthalpy $\Delta H^{\neq}$ in kcal/mol, activation entropy $\Delta S^{\neq}$ in $cal \cdot mol^{-1} \cdot K^{-1}$:

No.	reaction	t in °C	medium	log K	Ref.
1	$Mn(C_5H_7NO_4) + bpy \rightleftharpoons Mn(C_5H_7NO_4)bpy$	25	0.3 ($NaClO_4$)	2.84 ± 0.06	[1]
2	$Mn(C_8H_6N_3O_7)^- + bpy \rightleftharpoons Mn(C_8H_6N_3O_7)bpy^-$	25	0.3 ($NaClO_4$)	2.05 ± 0.08	[1]
3	$Mn(C_8H_6N_3O_7)^- + A^- \rightleftharpoons Mn(C_8H_6N_3O_7)A^{2-}$	16	0.1 (KNO_3)	4.67 ± 0.03*)	[2]
		16	0.1 (KNO_3)	5.23 ± 0.04	[2]
		25	0.1 (KNO_3)	5.71	[3]
4	$Mn(C_8H_6N_3O_7)^- + HA \rightleftharpoons Mn(C_8H_6N_3O_7)(HA)^-$	25	0.1 (KNO_3)	—	[3]

*) Spectrophotometrically.

No.	t in °C	k_f	k_b	$\Delta H_f^{\neq}$	$\Delta S_f^{\neq}$	$\Delta H_b^{\neq}$	$\Delta S_b^{\neq}$	Ref.
1	25	0.68 ± 0.05	0.98 ± 0.06	9.9 ± 1.5	3 ± 5	14.3 ± 1.1	3 ± 4	[1]
2	25	1.8 ± 0.2	16 ± 1	5.1 ± 0.9	−11 ± 3	11.1 ± 0.8	−2 ± 3	[1]
3	25	22 ± 3	0.13 ± 0.06	6.4 ± 2.5	−1 ± 8	12.4 ± 2.0	−7 ± 7	[3]
4	25	1.4 ± 0.2	—	4.5 ± 0.5	−13 ± 2	—	—	[3]

Values for reactions 1 and 2 are pH-independent in the 6.5 to 9.5 pH range [1].

The mixed-ligand manganese(III) complex $\mathbf{Mn^{III}HL(H_2P_2O_7)^-}$ is formed by reaction of Mn^{III} pyrophosphate with a large excess of N-(2-hydroxyethyl)iminodiacetic acid (= H_3L) in aqueous solution at pH 1.5 to 4. The stability constant log K = 6.80 ± 0.08 at room temperature and I = 1.0 M ($NaClO_4$) [4, 5] has been determined spectrophotometrically for $K = [MnHL(H_2P_2O_7)^-]/[MnH_2P_2O_7^+][HL^{2-}]$ [4, 5]. The complex shows a maximum at 500 nm in the solution phase absorption spectrum. At pH >7 the absorption band is shifted towards 465 nm, suggesting the formation of an Mn^{III} complex with L^{3-} (with deprotonated hydroxy group). Kinetic studies show that the Mn^{III} oxidation state is not stabilized by the ligand [4].

References:

[1] Hague, D. N., Martin, S. R. (J. Chem. Soc. Dalton Trans. **1974** 254/8).
[2] Hague, D. N., Zetter, M. S. (Trans. Faraday Soc. **66** [1970] 1176/84, 1181).
[3] Hague, D. N., Martin, S. R., Zetter, M. S. (J. Chem. Soc. Faraday Trans. I **68** [1972] 37/46, 41, 43).
[4] Bogdanovich, N. G., Pechurova, N. I., Martynenko, L. I., Spitsyn, V. I. (Zh. Neorgan. Khim. **18** [1973] 2795/801; Russ. J. Inorg. Chem. **18** [1973] 1485/9).
[5] Perrin, D. D. (Stability Constants of Metal-Ion Complexes, Part B; Organic Ligands, IUPAC Chemical Data Series No. 22 [1979] 446).

22.2.3 Isolated Manganese(II) Compounds

$[Mn(C_6H_9NO_5)(H_2O)]\cdot H_2O$. The Mn^{II} complex with N-(2-hydroxyethyl)iminodiacetic acid ($=C_6H_{11}NO_5$) forms pale pink orthorhombic crystals. An X-ray diffraction study yielded the lattice parameters a=10.23(2), b=10.23(2), c=9.81(2) Å; Z=4; space group: $P2_12_12_1-D_2^4$ (No. 19) [1, 2]. The Mn atom is octahedrally coordinated by the N atom, the O atom of the hydroxyethyl group, and two carboxylate O atoms of one ligand, an H_2O molecule and a carboxylate O atom of a similar neighboring moiety (**Fig. 2**). The carboxylate groups containing these O atoms, O′(2) or O(2), are bidentate and bridging thus forming endless $[Mn(C_6H_9NO_5)(H_2O)]_n$ chains proceeding along 2_1(y) screw axes. The chains are joined by van der Waals forces and hydrogen bonds involving water molecules and carboxylate O atoms. The bond lengths around the Mn atom are presented in Fig. 2. The hydroxyethyl ring lies in a plane perpendicular to the common averaged plane of the glycine rings and is deformed to a greater extent. The bond angles are O(1)–Mn–N=70.5(5)°, O(3)–Mn–N=79.4(5)°, and O(5)–Mn–N=79.1(6)°. Other bond lengths, bond angles, as well as d values, intermolecular distances, and atomic positional and temperature parameters, are given in [1].

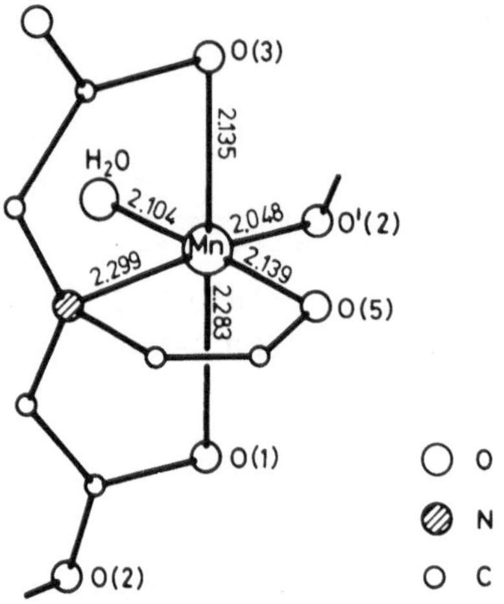

Fig. 2. Element of an $[Mn(C_6H_9NO_5)(H_2O)]_n$ chain in the complex with N-(2-hydroxyethyl)-iminodiacetic acid [1].

$Mn(C_{10}H_9NO_5)$. The complex with N-(2-hydroxyphenyl)iminodiacetic acid ($=C_{10}H_{11}NO_5$) was prepared by reacting equimolar amounts of 0.1 M solutions of $MnCl_2$ and the acid. The pale beige crystals were washed until free of Cl^- ions and dried at 115°C [3].

The IR spectrum of the crystalline compound (in liquid paraffin or hexachlorobutadiene) shows absorptions (in cm^{-1}) at: 1616, 1570, $\nu_{as}(COO^-)$; 1443, ν(C=C); 1410, $\nu_s(COO^-)$; 1270, 1248, ν(C–O). The sodium salt $Na_2(C_{10}H_9NO_5)$ absorbs at 1588, 1500, 1418, and 1253 cm^{-1}. The splitting of the ν(C–O) band indicates participation of the phenol hydroxyl group in complex formation. $Mn(C_{10}H_9NO_5)$ is almost insoluble in water, alcohol, and other organic solvents, but dissolves in alkali solutions with the formation of the $\mathbf{Mn(C_{10}H_8NO_5)^-}$ ion [3].

$\mathbf{Mn(C_{12}H_{13}NO_4)\cdot H_2O}$. The chelate complex with N-(2,6-xylyl)iminodiacetic acid, 2,6-$(CH_3)_2$–$C_6H_3N(CH_2COOH)_2$ (= $C_{12}H_{15}NO_4$) precipitates when aqueous $Mn(NO_3)_2$ is slowly added to an equimolar amount of the ligand in water. The acid must first be made soluble by the addition of a few drops of 2 M NaOH. The precipitate is washed with water and dried in air [4].

The effective magnetic moment is 5.86 μ_B. IR bands (in cm^{-1}) of the compound (KBr pellets) were assigned as follows: 3430, $\nu(OH)_{H_2O}$; 2920, ν(CH); 1590, 1555, $\nu_{as}(COO^-)$; 1390, $\nu_s(COO^-)$. The position of the $\nu_{as}(COO^-)$ band at 1590 cm^{-1} indicates that the Mn–O bond is highly ionic. $Mn(C_{12}H_{13}NO_4)\cdot H_2O$ is dehydrated at 183°C to give the anhydrous compound, $\mathbf{Mn(C_{12}H_{13}NO_4)}$. TG and DTA curves indicate that this compound decomposes in two steps in the 220 to 425°C temperature range to give MnO_2. An enthalpy of dehydration of 29.9 kJ/mol was obtained from the DSC curve [4].

References:

[1] Anan'eva, N. N., Polynova, T. N., Porai-Koshits, M. A. (Koord. Khim. **7** [1981] 1556/61; Soviet J. Coord. Chem. **7** [1981] 771/6).
[2] Anan'eva, N. N., Polynova, T. N., Porai-Koshits, M. A., Mitrofanova, N. D. (Koord. Khim. **1** [1975] 850; Soviet J. Coord. Chem. **1** [1975] 721).
[3] Temkina, V. Ya., Dyatlova, N. M., Zhadanov, B. V., Rusina, M. N. (Zh. Neorgan. Khim. **13** [1968] 1570/3; Russ. J. Inorg. Chem. **13** [1968] 823/5).
[4] Roman-Ceba, M., Avila Roson, J. C., Suarez-Varela, J. (Thermochim. Acta **80** [1984] 115/22).

22.3 With Nitrilotriacetic Acid $N(CH_2COOH)_3$ (= $C_6H_9NO_6$ = H_3L)

22.3.1 Manganese(II) Compounds

22.3.1.1 Binary Complexes in Solution

Mn^{2+} ions form 1:1 and 1:2 complexes in aqueous solution with the trianion of nitrilotriacetic acid [1 to 7]. The $Mn(C_6H_6NO_6)^-$ ion which is formulated $Mn(C_6H_6NO_6)(H_2O)_2^-$ by [7] (p. 16), is stable in the pH range 3.5 to 10. A distribution diagram for this ion in the given pH range is shown in [6]. The stability constants of the $Mn(C_6H_6NO_6)^-$ and $Mn(C_6H_6NO_6)_2^{4-}$ ions in aqueous solution at various temperatures t and ionic strengths I are given below:

t in °C	I in mol/L	log K_1	log K_2	method*)	Ref.
0	→0 (KCl)	8.527	—	gl	[1]
10	→0 (KCl)	8.534	—	gl	[1]
20	→0 (KCl)	8.573	—	gl	[1]
20	0.1 (KNO_3)	—	3.55	gl	[2]
20	0.1 ($KClO_4$)	7.36 ± 0.05	—	dis	[3]
20	0.1 (KCl)	7.44 ± 0.03	—	gl	[4]

t in °C	I in mol/L	log K_1	log K_2	method*)	Ref.
25	0.1 ($NaClO_4$)	8.11	—	pol	[5]
25	0.15 (NaCl)	7.15	3.05	gl	[6]
25	0.15	—	2.58	NMR	[7]
25	0.03 [19]	—	3.02(2)	NMR	[20]
30	→0 (KCl)	8.644	—	gl	[1]
35	0.1 ($HClO_4$)	7.55	—	el	[22]

*) gl = glass electrode, dis = distribution between two phases, pol = polarography, NMR = determined by ^{17}O NMR spectroscopy [7] or by titration with the ligand and use of the ^{1}H nuclear magnetic relaxation rate as indicator [20], el = paper electrophoresis.

The values log $K_1 = 7.44$ [4] and log $K_2 = 3.55$ [2] were evaluated as tentative selected values in the "Critical Survey of Stability Constants of NTA Complexes" [19]. Additional values for stability constants are given in [7 to 10].

A pH-dependent effective stability constant, $K_{eff} = K/\alpha_H$, allows for the protonation of the $C_6H_6NO_6^{3-}$ ions in the equilibrium solution. Values of log α_H, which can be calculated from the pK values of the ligand, are given for various pH in [21].

Thermodynamic parameters of formation, ΔH and ΔG in kcal/mol, ΔS in cal·mol^{-1}·K^{-1}, for the complex ions $Mn(C_6H_6NO_6)^-$ and $Mn(C_6H_6NO_6)_2^{4-}$ in aqueous solution, calculated from stability constants [1], determined by calorimetric measurements [2, 11], or evaluated from ^{17}O NMR line broadening [7]:

t in °C	I in mol/L	ΔH_1	$-\Delta G_1$	ΔS_1	$-\Delta H_2$	$-\Delta G_2$	$-\Delta S_2$	Ref.
0	→0	~0	10.67	39.1	—	—	—	[1]
10	→0	0.8	11.05	49.1	—	—	—	[1]
20	→0	2.0	11.49	46.0	—	—	—	[1]
20	0.1 (KNO_3)	1.14	9.98	37.9	—	—	—	[11]
20	0.1 (KNO_3)	1.44	9.97	38.9	5.58	4.76	1.7	[2]
25	0.15	—	—	—	3.6	—	—	[7]
30	→0	3.5	11.98	51.1	—	—	—	[1]

The rate constants k for the formation reactions of the $Mn(C_6H_6NO_6)^-$ ion in the pH region 3.5 to 6, $Mn^{2+} + (C_6H_6NO_6)^{3-} \rightleftharpoons Mn(C_6H_6NO_6)^-$ (I) or $Mn^{2+} + (C_6H_7NO_6)^{2-} \rightleftharpoons Mn(C_6H_6NO_6)^- + H^+$ (II), have been determined polarographically: log $k_I = 8.7$ and log $k_{II} = 5.3$ at 25°C and I = 0.1 M [12].

The variation of the T_1/T_2 ratio of water protons (T_1 = spin-lattice, T_2 = spin-spin relaxation time) on complexation of Mn^{2+} ions with the ligand was studied [10]. The 9.1 and 35 GHz ESR spectra were recorded for aqueous $Mn(C_6H_6NO_6)^-$ solutions [13].

^{1}H [10] and ^{17}O [7] NMR studies of aqueous solutions containing $Mn(C_6H_6NO_6)^-$ ions suggest that, in addition to the ligand, there are two water molecules complexed to Mn^{II}. The rate constant and the activation parameters for water exchange between complexed and bulk water have been determined from ^{17}O NMR measurements: $k = 1.5 \times 10^9\ s^{-1}$, $\Delta H^{\ddagger} = 6.6$ kcal/mol, $\Delta S^{\ddagger} = 5.57$ cal·mol^{-1}·K^{-1} per mol H_2O at 25°C and I = 0.1 to 0.15 M. The possibility of two kinetically nonequivalent positions of complexed water could not be considered in the above results [7]. In alkaline solutions of pH 10 to 13, $Mn(C_6H_6NO_6)(H_2O)_2^-$ is transformed into the dihydroxo complex [14].

The rate constant for the dissociation of $Mn(C_6H_6NO_6)^-$ into Mn^{2+} and $(C_6H_6NO_6)^{3-}$ in aqueous solution at 25°C and I=0.1 M has been determined polarographically as $k=(9.12\times10^6\cdot[H^+]+5.3)\ M^{-1}\cdot s^{-1}$. The deposition of Mn from $Mn(C_6H_6NO_6)^-$ at the dropping Hg electrode proceeds reversibly; however, the amalgam formed is inactivated [5].

The kinetics of the oxidation reactions of the $Mn(C_6H_6NO_6)^-$ complex with a number of radicals formed by pulse radiolysis in aqueous solutions were studied by spectrophotometry. The rate constants (k in $M^{-1}\cdot s^{-1}$) listed below for the reactions of $Mn(C_6H_6NO_6)^-$ with radicals $R^\bullet$ were obtained at 22±2°C for solutions of two different pH values:

$R^\bullet$	$OH^\bullet$	$OH^\bullet$	$O_2^{\bullet-}$	$O_2^{\bullet-}$	$Br_2^{\bullet-}$	$Br_2^{\bullet-}$	$(SCN)_2^{\bullet-}$	$(SCN)_2^{\bullet-}$
pH	4.5	9.0	4.5	5.5	3.6	4.5 to 5.5	4.5	5.0
10^{-7} k*)	150	250	40	12	0.7	2.0	1.3	4.0

*) Accuracy of k ±15%.

The rate constant for the reaction with the $O_2CH_2C(CH_3)_2OH^\bullet$ radical formed from tert-butanol and the $OH^\bullet$ radical is $k\approx1.5\times10^8\ M^{-1}\cdot s^{-1}$ at pH 4.5 to 9. Kinetic data for the decomposition reactions of the unstable intermediates formed by the above reactions are also given. The data indicate that the $Mn(C_6H_6NO_6)^-$ ion is oxidized to the Mn^{III} complex by all the radicals except for the $OH^\bullet$ radical via an inner-sphere mechanism. With the $OH^\bullet$ radical, a hydrogen abstraction occurs and the $Mn^{II}(OOCCH)N(CH_2COO)_2^{\bullet-}$ ion forms as the first intermediate, which further reacts with decomposition of the ligand radical [15].

The rate constant for the reaction of the $Mn(C_6H_6NO_6)_2^{4-}$ ion with a hydrated electron, e_{aq}^-, formed by the radiolysis of water is $k\leqq5.0\times10^6\ M^{-1}\cdot s^{-1}$ at room temperature [16].

Equilibrium constants log K=1.15, for $Mn^{2+}+Ca(C_6H_6NO_6)^-\rightleftharpoons Mn(C_6H_6NO_6)^-+Ca^{2+}$, and log K=0.56, for $Mn(C_6H_6NO_6)^-+Ca(C_6H_6NO_6)_2^{4-}\rightleftharpoons Mn(C_6H_6NO_6)_2^{4-}+Ca(C_6H_6NO_6)^-$, were determined by simultaneous titration of Mn^{2+} and Ca^{2+} ions with the ligand trisodium salt [20].

Chelates of nitrilotriacetic acid with manganese (and other transition metals) are of interest because of their applicability in analytical chemistry. They can also be used as simple models for complexes in biological systems [15]. Mn^{II} complexes with nitrilotriacetic acid in waste water are quantitatively degradable by specific bacteria [17]. The formation of the $Mn(C_6H_6NO_6)^-$ and $Mn(C_6H_6NO_6)_2^{4-}$ ions had to be considered in the determination of dissociation constants of a manganoprotein (glyoxalase I) using the ligand as a component in a metal-buffer system [18].

References:

[1] Hughes, V. L., Martell, A. E. (J. Am. Chem. Soc. **78** [1956] 1319/24).
[2] Anderegg, G. (Experientia Suppl. No. 9 [1964] 75/83, 76, 78).
[3] Starý, J. (Anal. Chim. Acta **28** [1963] 132/49, 139).
[4] Schwarzenbach, G., Freitag, E. (Helv. Chim. Acta **34** [1951] 1492/502, 1502).
[5] Biernat, J., Koryta, J. (Collection Czech. Chem. Commun. **25** [1960] 38/46, 44).
[6] Jackson, G. E., Kelly, M. J. (Polyhedron **2** [1983] 1313/6).
[7] Zetter, M. S., Grant, M. W., Wood, E. J., Dodgen, H. W., Hunt, J. P. (Inorg. Chem. **11** [1972] 2701/6, 2703, 2705).
[8] Jokl, V. (J. Chromatog. **14** [1964] 71/8, 76).
[9] Schwarzenbach, G., Biedermann, W. (Helv. Chim. Acta **31** [1948] 331/40, 334).
[10] King, J., Davidson, N. (J. Chem. Phys. **29** [1958] 787/91).

[11] Hull, J. A., Davies, R. H., Staveley, L. A. K. (J. Chem. Soc. **1964** 5422/5).
[12] Koryta, J. (Z. Elektrochem. **64** [1960] 196/7).
[13] Reed, G. H., Leigh, J. S., Pearson, J. E. (J. Chem. Phys. **55** [1971] 3311/6).
[14] Hopgood, D., Angelici, R. J. (J. Am. Chem. Soc. **90** [1968] 2508/13).
[15] Lati, J., Meyerstein, D. (J. Chem. Soc. Dalton Trans. **1978** 1105/18, 1113/4).
[16] Meyerstein, D., Mulac, W. A. (Trans. Faraday Soc. **65** [1969] 1818/26, 1823).
[17] Gudernatsch, H. (Invest. Inform. Textil **14** [1971] 407/19, 416; C.A. **76** [1972] No. 17673).
[18] Sellin, S., Mannervik, B. (J. Biol. Chem. **259** [1984] 11426/9).
[19] Anderegg, G. (Pure Appl. Chem. **54** [1982] 2693/758, 2730, 2758).
[20] Schlüter, A., Weiss, A. (Anal. Chim. Acta **99** [1978] 157/66, 164/5).

[21] Schwarzenbach, G., Flaschka, H. (Die Komplexometrische Titration, Enke, Stuttgart 1965, p. 12).
[22] Gupta, D., Singh, S., Yadava, K. L. (Acta Chim. Hung. **120** [1985] 47/55).

22.3.1.2 Ternary Complexes in Solution

The $Mn(C_6H_6NO_6)^-$ ion forms mixed ligand complexes in aqueous solution with the listed neutral ligands L or acids H_nA. Stability constants $K = [Mn(C_6H_6NO_6)L^-]/[Mn(C_6H_6NO_6)^-] \cdot [L]$ or $[Mn(C_6H_6NO_6)A^{(n+1)-}]/[Mn(C_6H_6NO_6)^-] \cdot [A^{n-}]$ were determined potentiometrically, in general with a glass electrode (pot), spectrophotometrically (sp), by paper electrophoresis (el), or kinetically by measurement of the temperature-jump relaxation (kin):

No.	additional ligand L or H_nA	formula	L or H_nA	t in °C	I in M	log K	method	Ref.
1	2,2'-bipyridine	$C_{10}H_8N_2$	L	25	0.3 ($NaClO_4$)	2.63 ± 0.15	sp	[1]
				25	0.3 ($NaClO_4$)	2.44 ± 0.06	kin	[1]
2	8-hydroxy-quinoline	C_9H_7NO	HA	16	0.1 (KNO_3)	4.63 ± 0.06	sp	[2]
				16	0.1 (KNO_3)	5.33 ± 0.04	kin	[2]
				25	0.1 (KNO_3)	5.58	kin	[3]
3	adenosine triphosphate	$C_{10}H_{16}N_5O_{13}P_3$	H_4A	25	0.15 NaCl	9.12 ± 0.07	pot	[11]
			H_3A	25	0.15 NaCl	15.57 ± 0.14	pot	[11]
4	glycine	$C_2H_5NO_2$	HA	25	0.05 KNO_3	2.24 ± 0.05	pot	[4, 5]
				25	0.1 (KNO_3)	1.80 ± 0.10	pot	[5, 6]
5	serine	$C_3H_7NO_3$	HA	25	0.1 ($NaClO_4$)	1.28 ± 0.05	pot	[5, 7]
6	α-amino-butenoic acid	$C_4H_7NO_2$	HA	35	0.1	3.00	el	[12]
7	ethyl valinate	$C_7H_{15}NO_2$	L	25	0.05 KNO_3	2.39 ± 0.02	pot	[4]
8	aspartic acid	$C_4H_7NO_4$	H_2A	25	0.1 ($NaClO_4$)	2.08 ± 0.04	pot	[5, 8]
9	glutamic acid	$C_5H_9NO_4$	H_2A	25	0.1 ($NaClO_4$)	2.22 ± 0.04	pot	[5, 8]
10	proline	$C_5H_9NO_2$	HA	35	0.1 ($NaClO_4$)	3.90	el	[13]

No.	additional ligand L or H_nA	formula	L or H_nA	t in °C	I in M	log K	method	Ref.
11	histidine	$C_6H_9N_3O_2$	HA	25	0.5($NaClO_4$)	2.49 ± 0.05	pot	[9]
12	arginine	$C_6H_{14}N_4O_2$	HA	25	0.1($NaClO_4$)	1.94 ± 0.04	pot	[5, 7]
13	glycylglycine	$C_4H_8N_2$	HA	25	0.1($NaClO_4$)	2.08 ± 0.08	pot	[5, 10]

A distribution diagram for the mixed ligand complexes $Mn(C_6H_6NO_6)A^{5-}$ and $Mn(C_6H_6NO_6)HA^{4-}$ with adenosine triphosphate in the pH range 2 to 10 is presented in [11].

Kinetic data for the reaction 1) $Mn(C_6H_6NO_6)^- + bpy \underset{k_b}{\overset{k_f}{\rightleftharpoons}} Mn(C_6H_6NO_6)bpy^-$ with bipyridine and the reactions 2) $Mn(C_6H_6NO_6)^- + A^- \underset{k_b}{\overset{k_f}{\rightleftharpoons}} Mn(C_6H_6NO_6)A^{2-}$ and 3) $Mn(C_6H_6NO_6)^- + HA \underset{k_b}{\overset{k_f}{\rightleftharpoons}} Mn(C_6H_6NO_6)HA^-$ with 8-hydroxyquinoline (HA) in aqueous solutions were obtained by the temperature-jump relaxation method at 25°C; k_f in 10^6 $M^{-1} \cdot s^{-1}$, k_b in 10^3 s^{-1}, $\Delta H^{\neq}$ in kcal/mol, and $\Delta S^{\neq}$ in cal·$mol^{-1} \cdot K^{-1}$:

No.	medium	k_f*)	k_b	$\Delta H_f^{\neq}$	$\Delta S_f^{\neq}$	$\Delta H_b^{\neq}$	$\Delta S_b^{\neq}$	Ref.
1	0.3($NaClO_4$)	4.17	5.1 ± 0.3	6.1 ± 1.4	−8 ± 5	10.2 ± 1.5	−8 ± 5	[1]
2	0.1(KNO_3)	69.18	0.18 ± 0.03	6.6 ± 2.4	0 ± 8	10.6 ± 2.2	−12 ± 7	[3]
3	0.1(KNO_3)	5.13	—	5.5 ± 0.7	−9 ± 2	—	—	[3]

*) Corrected values obtained on the assumption that the Mn^{2+} ion is hexahydrated and that in $Mn(C_6H_6NO_6)^-$ the ligand is tetradentate.

Measurements were performed in the pH ranges 6.5 to 9 [3] or 6.5 to 10 [1]. Rate constants for the reactions at various temperatures are given in [1, 3] and also in [2] (at 16°C).

References:

[1] Hague, D. N., Martin, S. R. (J. Chem. Soc. Dalton Trans. **1974** 254/8).
[2] Hague, D. N., Zetter, M. S. (Trans. Faraday Soc. **66** [1970] 1176/84, 1181).
[3] Hague, D. N., Martin, S. R., Zetter, M. S. (J. Chem. Soc. Faraday Trans. I **68** [1972] 37/46, 43).
[4] Hopgood, D., Angelici, R. J. (J. Am. Chem. Soc. **90** [1968] 2508/13).
[5] Perrin, D. D. (Stability Constants of Metal-Ion Complexes, Part B; Organic Ligands; IUPAC Chemical Data Series No. 22 [1979] 426).
[6] Israeli, J., Cayouette, J. R. (Can. J. Chem. **49** [1971] 199/201).
[7] Israeli, J., Cecchetti, M. (Can. J. Chem. **46** [1968] 3821/3).
[8] Israeli, J., Cecchetti, M. (Talanta **15** [1968] 1031/4).
[9] Israeli, J., Cecchetti, M. (J. Inorg. Nucl. Chem. **30** [1968] 2709/16, 2712).
[10] Israeli, J., Cecchetti, M. (Can. J. Chem. **46** [1968] 3825/8).

[11] Jackson, G. E., Kelly, M. J. (Polyhedron **2** [1983] 1313/6).
[12] Singh, S., Gupta, D., Yadava, K. L. (Himalayan Chem. Pharm. Bull. **2** [1985] 34/8).
[13] Gupta, D., Singh, S., Yadava, K. L. (Acta Chim. Hung. **120** [1985] 47/55).

22.3.1.3 Isolated Compounds

$Mn_2[Mn(C_6H_6NO_6)_2]\cdot nH_2O$. A compound with n = 2.5 was obtained by dissolving stoichiometric amounts of manganese(II) carbonate and the ligand in boiling water. After a short time, a white precipitate separated, which was washed with hot water and dried in vacuum over P_2O_5 [1]. A compound with n = 2 was obtained in a 70.9% yield using a similar procedure involving washing with absolute ethanol and ether and air-drying at 23°C for 24 h [2]. (However, in a table in [3, p. 893], $Mn_2[Mn(C_6H_6NO_6)_2]\cdot 2.5H_2O$ is cited.) Using MnO_2 instead of $MnCO_3$ gave a compound with n = 1 in a yield of only 35.3% [2].

The important IR bands (cm^{-1}) of the chelate complex in Nujol were assigned as follows: 1678, 1640, 1614, 1584, $\nu_{as}(COO^-)$; 1445, 1414, $\nu_s(COO^-)$; 1120, $\nu(CN)$. The free ligand shows only the $\nu_s(COO^-)$ band at 1434 cm^{-1} in this region. The appearance of three new bands in the 1680 to 1610 cm^{-1} region, which are associated with the coordinated carboxylate group, is indicative of chelate formation. The band at 1584 cm^{-1} is the strongest in the spectrum and rather broad. It is associated with a coordinated carboxylate function of highly ionic character. N coordination is indicated by a shift of the $\nu(CN)$ band to lower wave numbers with respect to $Na_3(C_6H_6NO_6)\cdot H_2O$ (1152 to 1130) [2, 3].

The complex is only sparingly soluble in water. The aqueous solution shows the usual reactions of Mn^{2+} ions. After precipitation of the Mn^{2+} ions with NaOH or $(NH_4)_2CO_3$ the complexed manganese could be detected in the filtrate [1].

References:

[1] Brintzinger, H., Thiele, H., Munkelt, S. (Z. Anorg. Allgem. Chem. **254** [1947] 271/84, 276).
[2] Rajabalee, F. J. M. (J. Inorg. Nucl. Chem. **36** [1974] 557/64, 557, 559, 563).
[3] Rajabalee, F. J. M. (Spectrochim. Acta A **30** [1974] 891/906, 898).

22.3.2 Manganese(III) Compounds

Complexes in Solution. Mn^{III} pyrophosphate reacts with nitrilotriacetic acid in aqueous solution forming the uncharged $Mn(C_6H_6NO_6)$ species. The stability constant has been spectrophotometrically determined at 20°C and I = 1.0 M ($NaClO_4$), pH 3.5: log $K_1 = 20.25 \pm 0.05$ [1, 2]. Aqueous solutions of $Mn(C_6H_6NO_6)$ with pH $\leqq$ 7 show an absorption maximum at 480 nm. Displacement of the band to 475 nm at pH > 7 indicates conversion to a hydroxo species. The complex is slowly reduced by the nitrilotriacetate anion in aqueous solution. The redox reaction is first order with respect to the Mn^{III} concentration. The rate of reduction is lowest in the pH range 3 to 6 [1].

$Mn(C_6H_6NO_6)$ and **$Mn(C_6H_6NO_6)\cdot H_2O$.** The monohydrate was prepared by reacting Mn^{III} chloride or acetate with an equimolar amount of the ligand in hot water. The solution was evaporated to half of the initial volume. The white crystals, which deposited on cooling, were washed with hot water, alcohol, and ether and were dried in vacuum; the yield was 61.1%. By another procedure, freshly prepared Mn^{III} hydroxide was heated with the ligand in boiling water for 2 to 3 h and the compound obtained in a yield of 53.2%.

$Mn(C_6H_6NO_6)\cdot H_2O$ is insoluble in all common solvents. It is dehydrated in vacuum at 180 to 250°C to form the anhydrous compound, $Mn(C_6H_6NO_6)$. A fan-like chelate structure with coordinated N atom and carboxylate O atoms was proposed for $Mn(C_6H_6NO_6)$ [3].

The magnetic moments of the two compounds at 296 K are: $\mu_{eff} = 4.37\ \mu_B$ for $Mn(C_6H_6NO_6)$ and 5.09 μ_B for $Mn(C_6H_6NO_6) \cdot H_2O$ [4].

References:

[1] Bogdanovich, N. G., Pechurova, N. I., Martynenko, L. I., Piunova, V. V. (Zh. Neorgan. Khim. **16** [1971] 2507/11; Russ. J. Inorg. Chem. **16** [1971] 1337/9).

[2] Pechurova, N. I., Spitsyn, V. I., Bogdanovich, N. G., Martynenko, L. I. (Dokl. Akad. Nauk SSSR **198** [1971] 347/9; Dokl. Chem. Acad. Sci. USSR **196/201** [1971] 405/7).

[3] Voronkov, M. G., Mikhailova, S. V. (Khim. Geterotsikl. Soedin. **1969** 49/51; Chem. Heterocycl. Compounds [USSR] **5** [1969] 39/41).

[4] Chetverikova, V. A., Kogan, V. A., Mikhailova, S. V., Osipov, O. A., Voronkov, M. G. (Khim. Geterotsikl. Soedin. **1969** 379/80; Chem. Heterocycl. Compounds [USSR] **5** [1969] 287/8).

22.4 With Derivatives of Nitrilotriacetic Acid

Stability constants for $Mn(C_6H_8N_2O_5)$ and $Mn(C_6H_8N_2O_5)_2^{2-}$ with the monoacetamide, $NH_2COCH_2N(CH_2COOH)_2$ ($= C_6H_{10}N_2O_5$), were determined potentiometrically (glass electrode): $\log K_1 = 4.93 \pm 0.1$, $\log K_2 = 2.30 \pm 0.1$ at 20°C for I = 0.1 M (KCl) [1], $\log K_1 = 4.72 \pm 0.01$ [2], $\log K_2 = 2.21 \pm 0.02$ at 25°C for I = 0.1 M (KNO_3) [3]. On the basis of IR data (not given for Mn complexes) and stability data, an octahedral structure was proposed for both complexes. Coordination of a tetradentate ligand involving the N atom, two carboxylate O atoms, the acetamide O atom and additionally two water molecules was proposed for $Mn(C_6H_8N_2O_5)$. Coordination of two tridentate ligands was assumed for $Mn(C_6H_8N_2O_5)_2^{2-}$, with the acetamide O atom being not involved [3].

The stability constant $\log K_1 = 7.20$ was determined potentiometrically (glass electrode) at 25°C, I = 0.1 M (KCl), for the complex $Mn(C_{13}H_{12}NO_6)^-$ with the (±)-2-benzyl derivative, $C_6H_5CH_2CH(COOH)N(CH_2COOH)_2$ ($= C_{13}H_{15}NO_6 = H_3L$) [4].

The IR spectrum of the complex $Mn(C_8H_{11}NO_6)$ with nitrilotriacetic acid monoethyl ester, $C_2H_5OOCCH_2N(CH_2COOH)_2$ ($= C_8H_{13}NO_6$), in D_2O exhibits stretching frequencies at 1608 (carboxylate group) and 1710 cm^{-1} (ester carbonyl group). The spectrum gives no evidence of ester carbonyl oxygen coordination. Therefore it is assumed that the N atom, two carboxylate O atoms, and three water molecules are coordinated to the Mn atom to form an octahedral complex in aqueous solution. The hydrolysis of the ester group in the complex in the pH range 4.4 to 8.0 obeys a second-order rate law, first order in both complex and hydroxyl ion concentration. The second-order rate constant is $k = 418 \pm 5\ M^{-1} \cdot s^{-1}$ at 25°C. Attack of a hydroxyl ion at the carbonyl carbon of the ester results in the formation of the complex with nitrilotriacetic acid [5].

References:

[1] Schwarzenbach, G., Anderegg, G., Schneider, W., Senn, H. (Helv. Chim. Acta **38** [1955] 1147/70, 1150).

[2] Nakon, R. (Anal. Biochem. **95** [1979] 527/32).

[3] Lance, E. A., Rhodes III, C. W., Nakon, R. (Anal. Biochem. **133** [1983] 492/501).

[4] Hering, R., Krüger, W., Kühn, G. (Z. Chem. [Leipzig] **2** [1962] 374/5).

[5] Leach, B. E., Angelici, R. J. (J. Am. Chem. Soc. **90** [1968] 2504/8).

22.5 With Nitrilotripropionic Acid $N(CH_2CH_2COOH)_3$ $(= C_9H_{15}NO_6)$

Thermodynamic data of formation of $Mn(C_9H_{12}NO_6)^-$ from Mn^{2+} and $C_9H_{12}NO_6^{3-}$ (corrected values) were determined by direct calorimetric measurements: $\Delta H = 0.112$ kcal/mol, $\Delta G = -3.766(6)$ kcal/mol, $\Delta S = 13.01$ cal·mol^{-1}·K^{-1} at 25°C, pH 8.56 for $I = 0.1$ M (KNO_3), Gonzales Garcia, S., Sanchez Santos, F. J., Niclos Guttierrez, J., Fernandez Martinez, M. T. (Ars Pharm. **24** [1983] 257/65).

22.6 With Other Monoamine-N-polycarboxylic Acids

22.6.1 Manganese(II) Complexes in Solution

Stability constants of complexes with the listed ligands in aqueous solution were determined by pH-potentiometric titration, generally with a glass electrode:

No.	ligand	formula	t in °C	I in mol/L	log K_1	log K_{MnHL}^{Mn}	Ref.

1	HOOCCHNHCH$_2$COOH
	   ∣
	   CH$_2$CH$_2$COOH
		$C_7H_{11}NO_6 = H_3L$	25	0.5 (NaCl)	2.42	—	[1]

2	N(CH$_2$COOH)$_2$ ∣ CH$_2$CH$_2$COOH	$C_7H_{11}NO_6 = H_3L$	25	0.1 (KNO$_3$)	7.33	1.51	[2]
3	N(CH$_2$COOH)$_2$ ∣ CH$_2$CH$_2$SCH$_2$COOH	$C_8H_{13}NO_6S = H_3L$	25	0.1 (NaClO$_4$)	6.71 ± 0.02	1.52 ± 0.08	[3]
4	HOOCCHN(CH$_2$COOH)$_2$ ∣ CH$_2$CH$_2$COOH	$C_9H_{13}NO_8 = H_4L$	25	0.5 (NaCl)	5.41	1.78	[4]
5	HOOCCHCH$_2$CHN(CH$_2$COOH)$_2$ ∣ ∣ NH$_2$ COOH	$C_9H_{14}N_2O_8 = H_4L$	22	0.1 (KCl)	4.54	*)	[5]
6	CH$_3$–S$^\oplus$–(CH$_2$)$_2$–CH–COO$^\ominus$ ∣ ∣ HOOCCH$_2$ N(CH$_2$COOH)$_2$	$C_{11}H_{17}NO_8S = H_4L$	25	0.5 (NaCl)	5.88	5.89	[6]
7	⬡—N(CH$_2$COOH)$_2$ (COOH ortho)	$C_{11}H_{11}NO_6 = H_3L$	25 20	0.1 (KNO$_3$) 0.1 (NaNO$_3$)	5.85 5.37	<1 —	[2] [7]
8	HO—⬡—N(CH$_2$COOH)$_2$ (COOH, COOH)	$C_{12}H_{11}NO_9 = H_4L$	25	0.1 (KNO$_3$)	6.49	1.41	[2]
9	⬡N—CH$_2$COOH (COOH, COOH)	$C_9H_{13}NO_6 = H_3L$	25	0.1 (KNO$_3$)	7.40 ± 0.01	—	[8]

*) A stability constant for a protonated complex not clearly defined is also given.

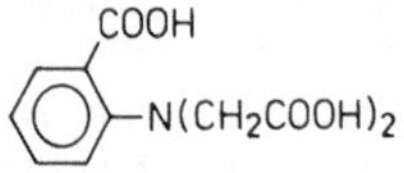

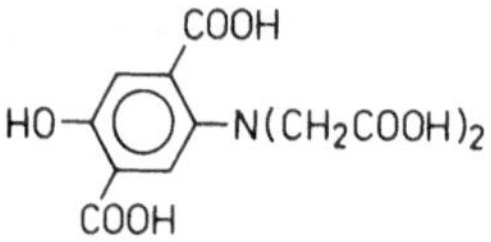

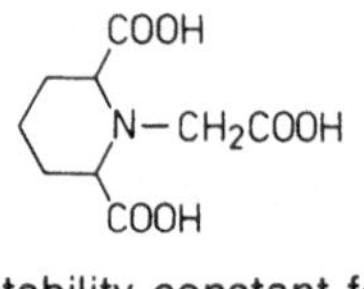

Thermodynamic data of formation for the complexes with ligand 2 and 7 were calorimetrically determined in aqueous solution at 25°C and I = 0.1 M (KNO_3) (ΔH and ΔG in kcal/mol, ΔS in cal·mol^{-1}·K^{-1}): ΔH_1 = 1.1, ΔG_1 = −10, ΔS_1 = 37.2 for the $Mn(C_7H_8NO_6)^-$ ion and ΔH_1 = 4.7, ΔG_1 = −8.0, and ΔS_1 = 42.4 for the $Mn(C_{11}H_8NO_6)^-$ ion [9].

Aqueous solutions of the complex with ligand 3 show the following electronic absorption maxima (cm^{-1}) which were assigned as follows (excited states of the transitions from the ground state $^6A_{1g}(S)$ given): 26900, $^4T_{2g}(D)$; 24100, $^4A_{1g}$, $^4E_g(G)$; 22600, $^4T_{2g}(G)$; 18000, $^4T_{1g}(G)$. Tetradentate coordination of the ligand through the N atom and the O atoms of the three carboxylate groups, but not through the S atom, is assumed [3]. The complex with ligand 7 in aqueous solution of pH 9 shows a maximum at 292 nm [10].

References:

[1] Rodriguez Rios, B., Fuentes Diaz, J. (Anales Quim. B **72** [1976] 428/36, 436).
[2] Uhlig, E., Krannich, R. (J. Inorg. Nucl. Chem. **29** [1967] 1164/8).
[3] Podlahová, J. (Collection Czech. Chem. Commun. **40** [1975] 3306/14, 3309).
[4] Rodriguez Rios, B., Fuentes Diaz, J., Rodriguez Bravo, M. R. (Anales Quim. B **77** [1981] 168/74, 173).
[5] Popova, V. A., Alekseeva, L. V., Ivakin, A. A., Podgornaya, I. V. (Tr. Inst. Khim. Ural'sk. Nauchn. Tsentr Akad. Nauk SSSR No. 30 [1974] 100/4, 103; C.A. **84** [1976] No. 50566).
[6] Rodriguez Rios, B., Fuentes Diaz, J., Sierra Rodriguez, R. (Anales Quim. B **80** [1984] 200/5).
[7] Drăgulescu, C., Simonescu, T., Menessy, I., Anton, R. (Acad. Rep. Populare Romine Baza Cercetari Stiint. Timisoara Studii Cercetari Stiinte Chim. **8** [1961] 9/15, 10; C.A. **57** [1962] 399).
[8] Kundra, S. K., Thompson, L. C. (J. Inorg. Nucl. Chem. **30** [1968] 1847/53, 1850).
[9] Martin, A., Uhlig, E. (Z. Anorg. Allgem. Chem. **375** [1970] 166/70).
[10] Drăgulescu, C., Menessy, I., Pîrlea, M., Ferencz, A. (Bull. Soc. Chim. France **1967** 4740/4).

22.6.2 Manganese(III) Complex in Solution

The uncharged complex $Mn(C_8H_{10}NO_6S)$ with ligand 3 was obtained by reaction of manganese(III) acetate with the ligand in 50% acetic acid. (In water, the Mn^{III} is instantaneously reduced to Mn^{II}.) The absorption spectrum of the solution exhibits maxima at 20300 cm^{-1} (ε = 420 M^{-1}·cm^{-1}) and 21200 cm^{-1} (sh), which were assigned to the $^5E_g \rightarrow {}^5T_{2g}$ transition, Podlahová, J. (Collection Czech. Chem. Commun. **40** [1975] 3306/14, 3309).

22.7 With Ethylenediamine-N,N'-diacetic Acid and Related Compounds
$HOOCCH_2N(R)CH_2CH_2N(R)CH_2COOH$

The equilibrium constants listed on p. 24 were determined at 25°C for I = 0.1 M (KNO_3) [1 to 4] potentiometrically with a glass electrode (gl) or potentiometrically not specified (pot); the temperature and medium were not specified by [5]:

No.	R	formula	H_nL	K	method	log K	Ref.
1	H	$C_6H_{12}N_2O_4$	H_2L	$[MnL]/[Mn^{2+}][L^{2-}]$	gl	6.85 ± 0.05	[1]
					gl	7.05 ± 0.02	[2, 6]
				$[Mn(OH)L^-] \cdot [H^+]/[MnL][H_2O]$	gl	-11.5 ± 0.1	[1]
2	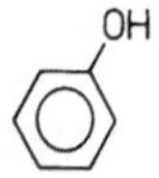	$C_{18}H_{20}N_2O_6$	H_4L	$[MnL^{2-}]/[Mn^{2+}][L^{4-}]$	pot	7.89	[5, 6]
				$[MnHL^-]/[Mn^{2+}][HL^{3-}]$	pot	5.49	[5, 6]
				$[MnH_2L]/[Mn^{2+}][H_2L^{2-}]$	pot	3.91	[5, 6]
3	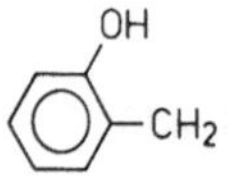	$C_{20}H_{24}N_2O_6$	H_4L	$[MnL^{2-}]/[Mn^{2+}][L^{4-}]$	pot	14.78	[3]
				$[MnHL^-]/[Mn^{2+}][HL^{3-}]$	pot	9.98	[3]
				$[MnH_2L]/[Mn^{2+}][H_2L^{2-}]$	pot	5.56	[3]
4	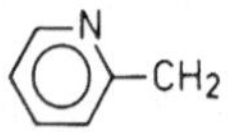	$C_{18}H_{22}N_4O_4$	H_2L	$[MnL]/[Mn^{2+}][L^{2-}]$	gl	12.7	[4]
				$[MnL][H^+]^2/[Mn^{2+}][H_2L]$ [a]	gl	−1.8	[4]
				$[MnL][H^+]/[Mn^{2+}][HL^-]$ [b]	gl	3.9	[4]

[a] For pH range 3 to 5. – [b] For pH range 6.5 to 8.

Thermodynamic data of formation for the complex $Mn(C_6H_{10}N_2O_4)$ with ethylenediamine-N,N'-diacetic acid (ligand 1), were determined calorimetrically at 25°C and I = 0.1 M (KNO_3) or were calculated; ΔH and ΔG in kJ/mol, ΔS in $J \cdot mol^{-1} \cdot K^{-1}$: $\Delta H_1 = -2.9 \pm 0.4$ [1], -0.85 ± 0.1 [2, 6]; $\Delta G_1 = -9.62$ [2, 6]; $\Delta S_1 = 122 \pm 2$ [1], 29.4 [2, 6].

$Mn(C_6H_{10}N_2O_4)$ reacts with ethylenediamine forming $Mn(C_6H_{10}N_2O_4)en$. The formation constant log K = 2.1 ± 0.2 was determined at 25°C and I = 0.10 M (KNO_3) with a glass electrode, and the free energy of formation ΔG = −2.9 kcal/mol was calculated [2, 6]. The equilibrium constant of the reaction with 2,2'-bipyridine, $Mn(C_6H_{10}N_2O_4) + bpy \underset{k_d}{\overset{k_f}{\rightleftharpoons}} Mn(C_6H_{10}N_2O_4)bpy$ in the 7.5 to 9.5 pH range, log K = 2.60 ± 0.07 (measured value) and log K = 3.08 (corrected value) at 25°C and I = 0.3 M ($NaClO_4$) was obtained by kinetic measurements (temperature-jump method). Kinetic parameters under the same conditions are: $k_f = 2.6 \times 10^5\ L \cdot mol^{-1} \cdot s^{-1}$, $\Delta H^{\neq} = 8.0$ kcal/mol, and $\Delta S^{\neq} = -5\ cal \cdot mol^{-1} \cdot K^{-1}$ for the formation reaction and $k_d = 650\ s^{-1}$, $\Delta H^{\neq} = 13.1$ kcal/mol, and $\Delta S^{\neq} = -2\ cal \cdot mol^{-1} \cdot K^{-1}$ for the decomposition reaction [7].

A stability constant of the complex with ethylenediamine-N-acetic acid, $Mn(C_4H_9N_2O_2)^+$, $\log K_1 = 3.63$ at 25°C for I = 0.5 M KCl, was determined with a glass electrode. The $Mn(C_4H_9N_2O_2)^+$ ion is the only complex in the Mn^{2+}– ligand system (Mn : ligand ca. 1:4). It reaches a maximum concentration of 94.7% of the total metal concentration at pH 11. A square planar geometry with coordination of the tridentate ligand, (N, N', O) and of a water O atom was proposed [8].

References:

[1] Gualtieri, R. J., McBryde, W. A. E., Powell, H. K. J. (Can. J. Chem. **57** [1979] 113/8).
[2] Degischer, G., Nancollas, G. H. (Inorg. Chem. **9** [1970] 1259/62).
[3] L'Eplattenier, F., Murase, I., Martell, A. E. (J. Am. Chem. Soc. **89** [1967] 837/43, 840).

[4] Lacoste, R. G., Christoffers, G. V., Martell, A. E. (J. Am. Chem. Soc. **87** [1965] 2385/8).
[5] Temkina, V. Ya., Rusina, M. N., Krinitskaya, L. V., Lastovskii, R. P. (Zh. Obshch. Khim. **38** [1968] 2207/12; J. Gen. Chem. [USSR] **38** [1968] 2136/40).
[6] Perrin, D. D. (Stability Constants of Metal-Ion Complexes, Part B; Organic Ligands, IUPAC Chemical Data Series No. 22 [1979] 1108).
[7] Hague, D. N., Martin, S. R. (J. Chem. Soc. Dalton Trans. **1974** 254/8).
[8] Leporati, E. (J. Chem. Soc. Dalton Trans. **1985** 1605/8).

22.8 With Other N,N'-Dicarboxylic Acids Derived from Ethylenediamine

1) $HOOCCH(CH_3)NHCH_2CH_2NHCH(CH_3)COOH$ (= $C_8H_{16}N_2O_4$)
2) $HOOCCH_2CH_2NHCH_2CH_2NHCH_2CH_2COOH$ (= $C_8H_{16}N_2O_4$)
3) $HOOCCH(2\text{-}C_6H_4OH)NHCH_2CH_2NHCH(2\text{-}C_6H_4OH)COOH$ (= $C_{18}H_{20}N_2O_6$)

Stability constants of the isomeric complexes with ligand 1 and 2, **$Mn(C_8H_{14}N_2O_4)$**, were determined potentiometrically: log $K_1 = 3.4$ at 30°C and I = 0.1 M (KCl) for the complex with ligand 1 [1] and log $K_1 = 6.10 \pm 0.01$ at 20°C and I = 0.1 M (KNO_3) for the complex with ligand 2 [2].

Normal pH-titration curves of Mn^{2+} ions in the presence of ligand 3 were obtained only at low pH, while at high pH, oxidation to Mn^{III} occurred [3] at the expense of ligand reduction [4]. Chelate formation of Mn^{II} at pH ≧6 was reported from a study of the uptake of chelated transition metals by plants [5].

The **(±)-$Na[Mn(C_{18}H_{16}N_2O_6)]\cdot 1.5H_2O$** complex was synthesized by reacting a solution of the racemic form of the ligand and NaOH in degassed water with a stoichiometric amount of Mn^{III} acetate under N_2. Alternatively, $MnCl_2$ can be used as the manganese source, but the reacting solution must be exposed to the air and stirred overnight to effect the oxidation of the Mn^{II} to Mn^{III}. The solution was filtered and evaporated to dryness, and the product was recrystallized from 70% ethanol-water. The yield of purified product was low due to the highly oxidizing nature of Mn^{III}. Extensive recrystallization led to decomposition. The compound forms thin hexagonal plates (not suitable for X-ray analysis). The magnetic moment, $\mu_{eff} = 5.1\ \mu_B$, confirms the presence of trivalent manganese (spin-only value 4.9 μ_B) [4].

The manganese(III) complex ***meso*-$Mn(C_{18}H_{16}N_2O_6)^-$**, with the *meso*-form of ligand 3 was prepared by mixing $MnCl_2$ and the ligand in 1.5 M aqueous NH_3, stirring the solution in the air for 3 d, and adding a small amount of H_2O_2 to oxidize residual Mn^{II} to Mn^{III}. The product was filtered, evaporated to dryness, and taken up in methanol. Further purification was achieved by chromatography on silica gel with methanol as the eluant and on Bio-Gel P-4 (Bio-Rad) with water as the eluant [4]. Electronic spectra of the *meso*-$Mn(C_{18}H_{16}N_2O_6)^-$ ion and the **(±)-$Mn(C_{18}H_{16}N_2O_6)^-$** ion in solution are dominated by a charge-transfer band located at 427 nm (log ε = 3.07) for the *racemic* complex and at 432 nm (3.04) for the *meso*-complex. In addition, both spectra display a shoulder at 480 nm (2.76) assigned to the d-d, $^5E_g \rightarrow {}^5T_{2g}$ transition, and another band at 350 nm (3.48). The (±)-$Mn(C_{18}H_{16}N_2O_6)^-$ ion is readily reducible on a dropping mercury electrode at a potential of −240 mV vs. SCE at pH 11 to 12, where a nearly reversible one-electron reduction occurs [4].

References:

[1] Courtney, R. C., Chaberek, S., Martell, A. E. (J. Am. Chem. Soc. **75** [1953] 4814/8).
[2] Majer, J., Kotouček, M., Dvořáková, E. (Chem. Zvesti **20** [1966] 242/51, 248; C.A. **65** [1966] 3304).

[3] Frost, A. E., Freedman, H. H., Westerback, S. J., Martell, A. E. (J. Am. Chem. Soc. **80** [1958] 530/6).
[4] Patch, M. G., Simolo, K. P., Carrano, C. J. (Inorg. Chem. **21** [1982] 2972/7).
[5] Kroll, H., Knell, M., Powers, J., Simonian, J. (J. Am. Chem. Soc. **79** [1957] 2024/5).

22.9 With Derivatives of Ethylenediamine-N,N′,N′-triacetic Acid

1) $HOOCCH_2N(CH_3)CH_2CH_2N(CH_2COOH)_2$ (= $C_9H_{16}N_2O_6$)
2) $HOOCCH_2N(C_2H_4OH)CH_2CH_2N(CH_2COOH)_2$ (= $C_{10}H_{18}N_2O_7$)
3) $C_2H_5OOCNHCH_2CH_2N(CH_2COOH)_2$ (= $C_9H_{16}N_2O_6$)

22.9.1 Manganese(II) Complexes

Formation in Aqueous Solution. Stability constants were determined by potentiometric titration with a glass electrode: log K_1 = 10.9 at 25°C and I = 0.2 M for $Mn(C_9H_{13}N_2O_6)^-$ with ligand 1 [1]; log K_1 = 10.7 at 29.6°C and I = 0.1 M (KCl) for $Mn(C_{10}H_{15}N_2O_7)^-$ with ligand 2 [2]; log K_1 = 4.60 ± 0.1, log K_2 = 2.96 ± 0.1 at 20°C, I = 0.1 M (KCl), for the $Mn(C_9H_{14}N_2O_6)$ and $Mn(C_9H_{14}N_2O_6)_2^{2-}$ species with ligand 3 [4]. The following thermodynamic data for the $Mn(C_{10}H_{15}N_2O_7)^-$ complex with ligand 2 have been determined calorimetrically or were calculated: $\Delta H_1 = -5.2$ kcal/mol at 25°C and I = 0.1 M (KNO_3) (calorimetrically); $\Delta G_1 = -14.7$ kcal/mol, $\Delta S_1 = 32$ cal·mol^{-1}·K^{-1} (calculated) [3]. The volume change on formation of $Mn(C_{10}H_{15}N_2O_7)^-$ at infinite dilution, $\Delta V = 25.2$ cm^3/mol at 25°C, was estimated from the ionic partial molal volume of the $Mn(C_{10}H_{15}N_2O_7)^-$ ion, $\overline{V}^\infty = 183.7$ cm^3/mol at 25°C, which had been obtained from density data of aqueous solutions of $K[Mn(C_{10}H_{15}N_2O_7)(H_2O)] \cdot 3H_2O$. A seven-coordinate complex geometry with two coordinated water molecules was assumed for the species with ligand 2 on the basis of the $\Delta\overline{V}^\infty$ value obtained [5] and the ΔS_1 value [3] compared to those of other divalent metal complexes. The lower ΔS and $\Delta\overline{V}^\infty$ values for the (hydroxyethyl)ethylenediaminetriacetato complex than those for the ethylenediaminetetraacetato species suggest that the 2-hydroxyethyl group is not bonded to the metal ion [5].

$K[Mn^{II}(C_{10}H_{15}N_2O_7)(H_2O)] \cdot 3H_2O$ was prepared by reacting basic manganese(II) carbonate with free ligand 2 in water at 90°C for 30 min. Then, $KHCO_3$ was added and the reaction continued for a further 10 min at 90°C. A precipitate began to form after addition of absolute ethanol and warming the reaction mixture almost to the boiling point. The solution was then cooled in an ice bath. The product was washed with ethanol and ether, recrystallized from an ethanol-water mixture and dried over silica gel.

The standard partial molal volume at 25°C of $K[Mn(C_{10}H_{15}N_2O_7)(H_2O)]$ in aqueous solution, $\overline{V}^\infty = 187.3$ cm^3/mol, was calculated from density data (not reported) of aqueous $K[Mn(C_{10}H_{15}N_2O_7)(H_2O)] \cdot 3H_2O$ solutions. A plot of apparent molal volume vs. concentration is presented in [5].

References:

[1] Blackmer, G. L. (Diss. Washington State Univ. 1969, pp. 1/88, 42/71; Diss. Abstr. Intern. B **30** [1969/70] 4538).
[2] Chaberek, S., Martell, A. E. (J. Am. Chem. Soc. **77** [1955] 1477/80).
[3] Wright, D. L., Holloway, J. H., Reilley, C. N. (Anal. Chem. **37** [1965] 884/92).
[4] Schwarzenbach, G., Anderegg, G., Schneider, W., Senn, H. (Helv. Chim. Acta **38** [1955] 1147/70, 1150).
[5] Yoshitani, K. (Bull. Chem. Soc. Japan **58** [1985] 1646/50).

22.9.2 Manganese(III) Complexes

Formation and Properties in Aqueous Solution. Spectrophotometric studies indicate that three different complex species exist with ligand 1 and 2 depending on the pH of the solution:

ligand	pH range	species	λ_{max}	log ε	color	Ref.
1	1.3 to 2.5	$Mn(C_9H_{14}N_2O_6)^+$	480	a)	a)	[1]
	2.5 to 4.5	$Mn(C_9H_{13}N_2O_6)$	461	2.57	bright red	[1]
	4.5 to 7	$[Mn(C_9H_{13}N_2O_6)OH]^-$	455 b) 470(sh)	2.50	orange-brown	[1]
2	1.5 to 3.5	c)	475	2.49	—	[4]
	4 to 7	$Mn(C_{10}H_{15}N_2O_7)$ [2]	467	2.53	red [2, 6]	[4]
	7.5 to 9.5	c)	449	2.53	yellow-orange [6]	[4]

a) Not reported because of the instability of $Mn(C_9H_{14}N_2O_6)^+$. – b) At pH 5. – c) Formula not reported, but presumably analogous to the species with ligand 1.

The stability constant of the species with ligand 1, $Mn(C_9H_{13}N_2O_6)$, was determined potentiometrically (glass electrode) at 25°C and I = 0.2 M: log K_1 = 23.0 [1]. A value of log K_1 = 22.7 was obtained for the complex with ligand 2, $Mn(C_{10}H_{15}N_2O_7)$, by spectrophotometric measurements at 25°C and I = 0.2 M ($NaClO_4$) [4]. The equilibrium constant for the reaction $Mn^{3+} + C_{10}H_{17}N_2O_7^- \rightleftharpoons Mn(C_{10}H_{15}N_2O_7) + 2H^+$ at 20°C, log K = 13.55, was determined spectrophotometrically by [2]. The spectrophotometrically determined dissociation constant for the complex with ligand 3, $[Mn(C_9H_{13}N_2O_6)(H_2O)]$, forming $[Mn(C_9H_{13}N_2O_6)(OH)]^-$ is pK = 4.19 at 25°C [1].

The standard redox potential for the reaction $Mn^{III}L + e^- \rightleftharpoons Mn^{II}L^-$ in a sodium acetate-acetic acid buffer solution at 25°C is E° = 0.775 V vs. NHE at I = 0.2 ($LiClO_4$) for $Mn(C_9H_{13}N_2O_6)$ [1] and E° = 0.782 V vs. NHE at I = 0.2 ($NaClO_4$) for $Mn(C_{10}H_{15}N_2O_7)$ [4].

The decomposition of $Mn(C_9H_{13}N_2O_6)$ and $Mn(C_{10}H_{15}N_2O_7)$ in a sodium acetate-acetic acid medium was studied spectrophotometrically. The decomposition products are the Mn^{II} complexes and oxidation products of the coordinated ligands [1, 4]. The reactions are first order in complex concentration. The kinetic parameters are: $k = 3.6 \times 10^{-4}$ s^{-1} at 25°C, $\Delta H^{\neq} = 20.20 \pm 0.35$ kcal/mol, $\Delta S^{\neq} = -6.38 \pm 1.18$ cal·mol^{-1}·K^{-1} for $Mn(C_9H_{13}N_2O_6)$ [1]; and $k = 4.2 \times 10^{-5}$ s^{-1} at 25°C, $\Delta H^{\neq} = 22.0 \pm 0.4$ kcal/mol, $\Delta S^{\neq} = 0.3 \pm 0.9$ cal·mol^{-1}·K^{-1} for $Mn(C_{10}H_{15}N_2O_7)$ [4]. The reduction of $Mn(C_9H_{13}N_2O_6)$ in the presence of an excess of free ligand 1 is considerably faster than the reduction of the complex in the absence of free ligand. Kinetic parameters have also been reported for this case. The proposed mechanism for the ligand independent decomposition involves an initial electron transfer from the coordinated carboxylate group to the metal ion as the rate-determining step. The free radical intermediate then immediately reacts with another complex to produce ethylenediamine-N,N'-diacetic acid, carbon dioxide, and formaldehyde as the oxidation products [1]. The reduction of $Mn(C_{10}H_{15}N_2O_7)$ by $VO(C_{10}H_{15}N_2O_7)^-$ or $VO(edta)^{2-}$ forming Mn^{II} and V^V complexes was studied using the stopped-flow technique at 25°C and I = 0.1 M (CH_3COONa). The rate constant is dependent on the H^+ ion concentration: $k \cdot [H^+] = 0.420$ $M^{-1} \cdot s^{-1}$ at pH 3 and 6.66×10^{-2} $M^{-1} \cdot s^{-1}$ at pH 3.76 for the reaction with $VO(C_{10}H_{15}N_2O_7)^-$, and $k \cdot [H^+] = 0.235$ at pH 3 for the reaction with $VO(edta)^{2-}$ [5].

$[Mn(C_9H_{13}N_2O_6)(H_2O)] \cdot 2H_2O$ was synthesized by a modification of the method [4] used to prepare $Mn(C_{10}H_{15}N_2O_7) \cdot 2.5H_2O$ (p. 28). A slurry containing equimolar amounts of freshly prepared MnO_2 and $Mn(ClO_4)_2$ was reacted with a cold solution of ligand 1 at −10°C. After 1 h, the excess MnO_2 was filtered off, and a deep red oil was obtained upon addition of cold ethanol and cooling. Trituration of the oil with cold ethanol gave the crystalline compound in a ~65% yield. The infrared spectrum is quite similar to that of the corresponding ethylenediaminetetraacetato complex in that a characteristic strong absorption of the coordinated carboxylate group is observed in the 1700 to 1600 cm^{-1} region. There is no absorption of an uncoordinated carboxylate group, which would be expected around 1750 cm^{-1} [1].

$Mn(C_{10}H_{15}N_2O_7)\cdot 2.5H_2O$ was prepared by reacting the neat ligand 2 with a mixture of KOH and MnO_2 (obtained by heating powdered $KMnO_4$ in aqueous ethanol and distilling off the excess ethanol and acetaldehyde after reaction completion) at 0°C. In another preparation, a slurry containing the ligand and $Mn(NO_3)_2$ (2:1 mole ratio) was used instead of the neat ligand. After ~1 h, the residual MnO_2 was filtered off, and cold ethanol was added to the filtrate. After cooling the mixture in a dry ice-acetone bath and warming again, dark red crystals remained which were washed with absolute ethanol and dried in vacuum [4].

$Na[Mn(C_{10}H_{15}N_2O_7)(CH_3COO)]\cdot H_2O\cdot 0.5(CH_3CO)_2O$. To a solution of manganese(III) acetate in acetic anhydride at ~0°C, $Na_3(C_{10}H_{15}N_2O_7)$ was added. The mixture was stirred at this temperature for ~50 min, then filtered. Addition of ether to the filtrate gave rise to an oily product. The ether was removed, and the residue was then dissolved in methanol. After cooling of the solution to 0°C, a reddish orange powder was obtained, which was washed with ethanol and ether, and dried over silica gel. The complex was recrystallized from 95% methanol containing a small amount of acetic anhydride [3].

The infrared spectrum of the complex in KBr shows absorptions at 1735, 1700, 1645(sh), and 1625 cm^{-1}; the first two may be due to acetic anhydride. The visible absorption spectra of the solutions in methanol and dimethyl sulfoxide show maxima at 470 ($\log \varepsilon = 2.59$) and 490 nm (2.56), respectively. In aqueous solution, the compound is reduced to an Mn^{II} complex, but it dissolves in methanol or dimethyl sulfoxide to form fairly stable solutions. The complex decomposes in methanol solutions with pH>10 to form MnO_2. The red-colored complex anion is adsorbed from a neutral methanol solution of the complex by anion exchange resins (Cl^- form). The molar conductance of solutions in 87 wt% aqueous methanol or in dimethyl sulfoxide at 25°C is $\Lambda_\infty = 59.1$ and 24.1 $cm^2\cdot\Omega^{-1}\cdot mol^{-1}$, respectively, indicating a 1:1 electrolyte [3].

References:

[1] Blackmer, G. L. (Diss. Washington State Univ. 1969, pp. 1/88, 42/71; Diss. Abstr. Intern. B **30** [1969/70] 4538).
[2] Mikhailova, T. V., Cherepanova, L. G., Zhirnova, N. M., Astakhov, K. V. (Zh. Fiz. Khim. **46** [1972] 515/7; Russ. J. Phys. Chem. **46** [1972] 299/300).
[3] Shirakashi, T., Tanaka, N. (Chem. Letters **1975** 685/6).
[4] Hamm, R. E., Suwyn, M. A. (Inorg. Chem. **6** [1967] 139/42).
[5] Nelson, J., Shepherd, R. E. (Inorg. Chem. **17** [1978] 1030/4).
[6] Alvarez Bartolomé, M. L., Pérez Pérez, I., Arribas Jimeno, S. (Quim. Anal. **30** [1976] 395/8).

22.10 With Ethylenediamine-N,N,N′,N′-tetraacetic Acid
$(HOOCCH_2)_2NCH_2CH_2N(CH_2COOH)_2$ ($= C_{10}H_{16}N_2O_8 = H_4$edta)

22.10.1 Manganese(II) Complexes in Aqueous Solution

22.10.1.1 Formation

Measurements of the variation of the water proton relaxation times of solutions containing 1×10^{-3} mol Mn/L and a metal:ligand ratio of 1:2 with pH show that complex formation begins above pH 2 and is complete at pH 4 [1]. Complexation of 0, 3, 50, 80, 93, and 100% of the total manganese(II) ions at pH 2.7, 3, 3.35, 3.5, 4, and 4.8 was observed during a sorption study of Mn^{2+} ions on a strongly acidic cation exchanger from a solution containing 7.2×10^{-3} M Mn^{II} and 1.8×10^{-2} M H_4edta [2]. Unchanged spectra associated with the **$Mn(edta)^{2-}$ ion**, which was found to be the seven-coordinate **$Mn(edta)(H_2O)^{2-}$** complex (see p. 32) were obtained from an IR study in the pH range 4 to 6 [3] and from a Raman study in the pH range 3 to 5 [4]. The $Mn(edta)^{2-}$ ion in solutions containing Mn^{2+} ions and the ligand was also detected over wide pH ranges by extensive titration studies, e.g. [5, 6] and polarographic measurements, e.g. [7] aimed at the analytical application of these complexes. For acidic solutions [7] at pH ~ 2 [3], evidence was obtained for the protonated complex ion, **$Mn(Hedta)^-$** [3, 7], while for basic solutions (pH > 11), the existence of a hydroxo complex, **$Mn(edta)(OH)^{3-}$**, was indicated [6]. The formation of a precipitate at pH 9 (NH_3 solution) was observed by [8].

The table lists logarithmic values for the stability constant K_1 of the $Mn(edta)^{2-}$ ion determined by polarography (pol), electrophoresis (el), potentiometric measurement with a mercury electrode (Hg), with a glass electrode (gl), distribution between two phases (dis), or by theoretical calculation of the free energy of formation (cal):

t in °C	20	20	20	20	20	20
I in mol/L	0.1(KNO_3)	0.1(KNO_3)	0.1(KNO_3)	0.1(KCl)	0.1($KClO_4$)	0.1
log K_1	14.20	14.2±0.2[b)]	14.5	13.88±0.05[c)]	12.88±0.05	13.9
method[a)]	gl, pol(Cd)	gl, pol(Cd)	el	gl(Cu, tren)	dis	cal
Ref.	[9]	[7, 37]	[10]	[11, 37]	[12]	[18]
t in °C	20	25	25	25	30	—
I in mol/L	0.1($NaNO_3$)	0.1($NaClO_4$)	0.2(KNO_3)	0.1(KNO_3)	0.1(KNO_3)	—
log K_1	14.01±0.02	13.8	13.64	13.4	13.92	14.76
method[a)]	Hg	Hg	pol(Cu)	gl	Hg	pol(Cd)
Ref.	[13]	[14]	[15]	[16]	[17]	[19]

[a)] Auxiliary metals and/or ligands used are given in parentheses; tren = $N(CH_2CH_2NH_2)_3$. – [b)] In original paper 14.04; corrected with a better K_{CdL} value by [37]. – [c)] In original paper 13.47; corrected with a better K_{CuL} value by [37].

The value log K_1 = 13.95 ± 0.07 [11, 13], I = 0.1 and 20°C, was evaluated as a tentative selected value in the "Critical Survey of Stability Constants of EDTA Complexes" [37] (categories of selected values: recommended, tentative, doubtful and rejected). The equilibrium constant log K = – 2.58 for the reaction $Mn^{2+} + H_2edta^{2-} \rightleftharpoons Mn(edta)^{2-} + 2H^+$ in aqueous solution was calculated from literature data [34].

An effective stability constant, K_{eff}, introduced by [33], allows for the protonation of the $edta^{4-}$ anions in the equilibrium solution: $K_{eff} = [MnL]/[Mn] \cdot [L]'_H$ where $[L]'_H$ is the total concentration of the noncomplexed ligand in all its forms at the given pH (charges omitted). $[L]'_H$ can be calculated from the total ligand concentration, [L], the pK values of the ligand, and the pH value [33]. A graph of log K_{eff} as a function of pH [2] calculated by use of literature values [33] is given in **Fig. 3**, p. 30. The presence of metal hydroxides and of protonated and hydroxy complexes in addition to the protonated ligand species is allowed for by another effective (or "apparent") stability constant $K'_{eff} = [MnL]'/[Mn]' \cdot [L]'_H$. The concentrations of [MnL]' and [Mn]' can be calculated from [MnL] or [Mn], the pH value, and the stability constants of the complexes occurring in addition to the MnL^{2-} or Mn^{2+} ions. At pH 2, 4, 6, 8, 10, 12, or 14, log K'_{eff} is 1.60, 5.65, 9.30, 11.70, 13.55, 12.20, or 8.56, respectively [20]. A third effective stability constant, which allows for the presence of protonated $edta^{4+}$ anions and of Mn hydroxides, is similar to K'_{eff} in the lower pH range, but has a maximum of ~11 at pH 7 and abruptly decreases at higher pH values (from plot in [35]). Specially defined effective stability constants for the complexes of ^{54}Mn with ethylenediaminetetraacetate in seawater or 0.55 M NaCl at pH 8 were determined by paper electrophoresis [22]. The effective stability constants are of importance for the analytical application of these complexes [2, 20, 22, 33].

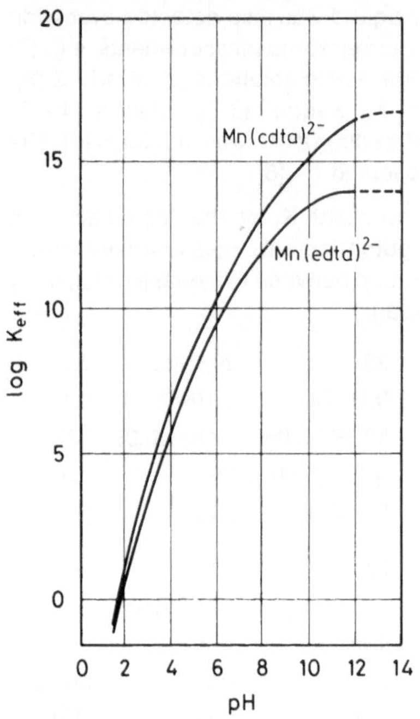

Fig. 3. The pH-dependence of the effective stability constant, log K_{eff}, of $Mn(edta)^{2-}$ in comparison to the $Mn(cdta)^{2-}$ complex (p. 66) [2]. (Definition of K_{eff} see text, p. 29.)

Thermodynamic data of the reaction $Mn^{2+} + edta^{4-} \rightleftharpoons Mn(edta)^{2-}$ in aqueous solution obtained by calorimetric measurement of ΔH and calculation of ΔG and ΔS (calor) or by calculation from literature values (lit) are given in the following table:

t in °C	20	20	20	25	25	25
I in mol/L	0.1(KNO_3)	0.1(KNO_3)	0.1(KNO_3)	0.1(KNO_3)	~0.05 $Mn(NO_3)_2$	—
$-\Delta H_1$ in kcal/mol	4.56	4.56	5.45	4.6[b]	5.2	4.9
in kJ/mol	19.08[a]	19.08	22.80	19.25	21.76	20.50
$-\Delta G_1$ in kcal/mol	—	18.51	18.6	19.2	17.2	—
ΔS_1 in $cal \cdot mol^{-1} \cdot K^{-1}$	46.6	47.6	44.5	49	41	45
method	calor	calor	calor	lit	calor	lit
Ref.	[23]	[24]	[25]	[26]	[27]	[29]

[a] This value corresponds to −17.8 kJ at 25°C [34]. – [b] Adopted from [23].

Electrostatic (el) and nonelectrostatic (crat) components of ΔH_1 and ΔG_1 were calculated: $\Delta H_{el} = 4.6$ and $\Delta H_{crat} = -9.5$ kcal/mol, $\Delta G_{el} = -11.2$ and $\Delta G_{crat} = -7.1$ kcal/mol at 25°C. The electrostatic component of ΔS_1 is $\Delta S_{el} = 53\ cal \cdot mol^{-1} \cdot K^{-1}$ at 25°C. The ΔH_1 value is lower than expected from the high stability of the complex because the exothermic ΔH_{crat} value is compensated by the endothermic ΔH_{el} value [29]. The entropy change for the $Mn(edta)^{2-}$ formation reaction in acid medium, taking into account side reactions of the $edta^{4-}$ anion with protons, was estimated from stability constants at various pH values [28].

The volume change for the reaction $Mn^{2+} + edta^{4-} \rightleftharpoons Mn(edta)^{2-}$ in aqueous solution at infinite dilution, $\Delta\bar{V}^{\infty} = 32.8\ cm^3/mol$ at 25°C, was estimated from the partial molal volume of the $Mn(edta)^{2-}$ ion (p. 33) and literature data [36]. For the reaction $Mn^{2+} + H_2edta^{2-} \rightleftharpoons Mn(edta)^{2-} + 2H^+$ in aqueous solution at 25°C, I = 0.1m, the volume change $\Delta\bar{V} = 22.4\ cm^3/mol$ and the heat capacity change $\Delta\bar{C}_p = 1.1\ J \cdot K^{-1} \cdot mol^{-1}$ ($\bar{V}$ = partial molal volume, $\bar{C}_p$ = partial molal heat capacity) were obtained from the partial molal quantities of the $Mn(edta)^{2-}$ ion (p. 33), H_2edta^{2-} ion, and literature data. For the same reaction at the above conditions, the enthalpy change $\Delta H = 22.4 \pm 3.0$ kJ/mol and the entropy change $\Delta S = 25.7\ J \cdot K^{-1} \cdot mol^{-1}$ were calculated from literature values [34].

Equilibrium constants for the reactions $Mn^{2+} + M(edta)^{2-} \rightleftharpoons Mn(edta)^{2-} + M^{2+}$ in aqueous solution were determined: log K = 3.20 at 25°C, calculated for M = Ca [34]; log K = −2.45 at 20°C, I = 0.1 M (KNO_3), pH 4.35, polarographically for M = Cd [7]; log K = −7.81 at 21.7°C, I = 0.1 M ($NaNO_3$), pH 4.5, potentiometrically with the Hg electrode for M = Hg [13]. Equilibrium constants for the reaction $Mn^{2+} + M(edta)^{2-} + H_3tren^{3+} \rightleftharpoons Mn(edta)^{2-} + M(tren)^{2+} + 3H^+$ (M = Co, Cu) are given in [11]. Spectrophotometric studies gave log K = −0.8 at 20 ± 2°C for the reaction $Mn^{2+} + Cr(edta)^- \rightleftharpoons Mn(edta)^{2-} + Cr^{3+}$ [31]. The formation of $Mn(edta)^{2-}$ by reduction of the Mn^{III} complexes $Mn(edta)^-$ or $Mn(edta)(N_3)^{2-}$ is treated on pp. 47/9 and 53, respectively. Because of its high stability, the $Mn(edta)^{2-}$ ion is readily formed on reaction of Mn^{II} complexes MnY of lower stability with ethylenediaminetetraacetate ions. Data concerning such ligand exchange reactions are reported along with the chemical reactions of the complexes MnY.

The stability constant of the protonated complex ion, $Mn(Hedta)^-$, formed by the reaction of Mn^{2+} with $Hedta^{3-}$ ions was determined potentiometrically with a glass electrode: log K = 6.9 at 20°C and I = 0.1 M (KNO_3) [23]. A lower value, log K = 5.47 at 25°C, I = 0.1 M (KNO_3), was calculated [32] from experimental data [26]. The following thermodynamic parameters for the reaction $Mn^{2+} + Hedta^{3-} \rightleftharpoons Mn(Hedta)^-$ were calculated from spectroscopic data (see below) and literature values: $\Delta H = -0.1$ kcal/mol, $\Delta G = -9.4$ kcal/mol, and $\Delta S = 31\ cal \cdot mol^{-1} \cdot K^{-1}$ at 25°C, I = 0.10 M (KNO_3) [26]. A value of log K = 3.1 at 20°C and I = 0.1 M (KNO_3) for the protonation reaction $Mn(edta)^{2-} + H^+ \rightleftharpoons Mn(Hedta)^-$ was determined by polarography [7], and log K = 3.07 at 25°C and I = 0.1 M (KNO_3) was calculated [32] from experimental data [26]. (These log K values are not consistent with dissociation constants of the $Mn(H_2edta)$ species reported on p. 39.) The following thermodynamic parameters for the protonation reaction were obtained from potentiometric and calorimetric measurements: $\Delta H = -1.2 \pm 0.3$ kcal/mol, $\Delta G = -4.2 \pm 0.1$ kcal/mol, and $\Delta S = 10 \pm 2\ cal \cdot mol^{-1} \cdot K^{-1}$ at 25°C and 0.1 M (KNO_3) [26].

A bimetal chelate is assumed to form on addition of a base to a solution containing Mn^{2+} and $Hg(edta)^{2-}$ ions. The stability constant $K = [MnHg(edta)(OH)_2^{2-}]/[Mn^{2+}][Hg(edta)^{2-}][OH^-]^2$ was determined potentiometrically (Hg electrode): log K = 16.1 at 25°C and I = 0.1 M ($NaClO_4$) [14].

References:

[1] Oakes, J., Smith, E. G. (J. Chem. Soc. Faraday Trans. II **77** [1981] 299/308, 307).
[2] Povondra, P., Přibil, R., Šulcek, Z. (Talanta **5** [1960] 86/91).
[3] Zetter, M. S., Grant, M. W., Wood, E. J., et. al. (Inorg. Chem. **11** [1972] 2701/6).
[4] Krishnan, K., Plane, R. A. (J. Am. Chem. Soc. **90** [1968] 3195/200).
[5] Schwarzenbach, G., Biedermann, W. (Helv. Chim. Acta **31** [1948] 459/65).
[6] Schwarzenbach, G. (Helv. Chim. Acta **32** [1949] 839/53).
[7] Schwarzenbach, G., Gut, R., Anderegg, G. (Helv. Chim. Acta **37** [1954] 937/57, 951).
[8] Reilley, C. N., Scribner, W. G., Temple, C. (Anal. Chem. **28** [1956] 450/4).
[9] Novák, V., Lučanský, J., Svičeková, M., Majer, J. (Chem. Zvesti **32** [1978] 19/26, 24).
[10] Jokl, V., Majer, J. (Chem. Zvesti **19** [1965] 249/58, 254; Acta Fac. Pharm. Bohemoslov. **11** [1965] 55/110, 91).

[11] Schwarzenbach, G., Freitag, E. (Helv. Chim. Acta **34** [1951] 1503/8).
[12] Starý, J. (Anal. Chim. Acta **28** [1963] 132/49, 139).
[13] Schwarzenbach, G., Anderegg, G. (Helv. Chim. Acta **40** [1957] 1773/92, 1787).
[14] Schmid, R. W., Reilley, C. N. (J. Am. Chem. Soc. **78** [1956] 2910/1, 5513/8).
[15] Ogino, K., Tanaka, N. (Bull. Chem. Soc. Japan **38** [1965] 771/7, 776).
[16] Lacoste, R. G., Christoffers, G. V., Martell, A. E. (J. Am. Chem. Soc. **87** [1965] 2385/8).
[17] Yen, C.-H., Liu, S.-C. (Huaxue Xuebao **30** [1964] 546/56, 553; C.A. **62** [1965] 11199).
[18] Muenze, R. (Z. Physik. Chem. [Leipzig] **252** [1973] 145/53, 150).
[19] Sochevanov, V. G., Volkova, G. A. (Zh. Neorgan. Khim. **14** [1969] 118/23; Russ. J. Inorg. Chem. **14** [1969] 61/4).
[20] Amsheeva, A. A. (Zh. Analit. Khim. **33** [1978] 1054/61; J. Anal. Chem. [USSR] **33** [1978] 814/20, 817).

[21] Körös, E. (Proc. Intern. Symp. Microchem., Birmingham Univ. 1958 [1959], pp. 474/8; C.A. **1960** 19269).
[22] Musani-Marazović, L., Pučar, Z. (Marine Chem. **5** [1977] 229/42, 239).
[23] Anderegg, G. (Helv. Chim. Acta **47** [1964] 1801/14, 1805, 1808).
[24] Anderegg, G. (Helv. Chim. Acta **46** [1963] 1833/42, 1837, 1839).
[25] Staveley, L. A. K., Randall, T. (Discussions Faraday Soc. No. 26 [1958] 157/63, 158).
[26] Brunetti, A. P., Nancollas, G. H., Smith, P. N. (J. Am. Chem. Soc. **91** [1969] 4680/3).
[27] Charles, R. G. (J. Am. Chem. Soc. **76** [1954] 5854/8).
[28] Ogura, K., Takatu, K., Yosino, T. (Talanta **23** [1976] 872/3).
[29] Murakami, S., Yoshino, T. (J. Inorg. Nucl. Chem. **43** [1981] 2065/70).
[30] Kern, D. M. H. (J. Am. Chem. Soc. **81** [1959] 1563/8).

[31] Rud, V. T., Zhirnova, N. M., Astakhov, K. V. (Zh. Fiz. Khim. **43** [1969] 607/10; Russ. J. Phys. Chem. **43** [1969] 332/4).
[32] Perrin, D. D. (Stability Constants of Metal-Ion Complexes, Part B; Organic Ligands, IUPAC Chemical Data Series No. 22 [1979] 767).
[33] Schwarzenbach, G., Flaschka, H. (Die Komplexometrische Titration, Enke, Stuttgart 1965, pp. 2/3).
[34] Hovey, J. K., Tremaine, P. R. (J. Phys. Chem. **89** [1985] 5541/9).
[35] Rad'ko, V. A., Yakimets, E. M. (Zh. Neorgan. Khim. **7** [1962] 683/6; Russ. J. Inorg. Chem. **7** [1962] 348/9).
[36] Yoshitani, K. (Bull. Chem. Soc. Japan **58** [1985] 1646/50).
[37] Anderegg, G. (IUPAC Chem. Data Ser. No. 14 [1977] 23).

22.10.1.2 Structure and Physical Properties

Raman [2] and IR studies [1] on aqueous $Mn(edta)^{2-}$ solutions at pH 3 to 5 [2] and pH 4 to 6 [1] indicate that the N atoms [2] and four O atoms, each of one carboxylate group [1, 2] are coordinated to the Mn atom, i.e., the $edta^{4-}$ anion is hexadentate in the complex [2]. Water proton relaxation studies indicate that, in addition to the organic ligand one water molecule is coordinated directly to the Mn atom [3]. This means that the manganese atom is seven-coordinate. This was suggested earlier by ^{17}O NMR and IR studies [1], by the low enthalpy of formation of the $Mn(edta)^{2-}$ ion in solution compared with the corresponding complexes of other metals [4], and by the definite spectral changes [5] accompanying the protonation of these complexes [6]. (The presence of one water molecule in an octahedral complex

$Mn(edta)^{2-}$, i.e., with a quinquedentate $edta^{4-}$ ligand, had been inferred from the acid properties of the complex (pK value ~3) by previous authors [7, 8].) At pH 2, IR spectra indicate that a protonated (unbound) carboxyl group is present [1], i.e., one carboxyl group is not coordinated in the $Mn(Hedta)^{-}$ complex ion.

Measurements of spin-lattice (T_1) and spin-spin relaxation times (T_2) of water protons in $Mn(edta)^{2-}$ solutions were performed as a function of NMR frequency, metal-to-ligand ratio, and pH. Relaxation showed a significant contribution from the water molecule at the seventh coordination site. Secondary contributions arise from water molecules hydrogen-bonded to the carboxylate oxygen atoms of the $edta^{4-}$ ligand [3]. The variation of the T_1/T_2 ratio of water protons upon complexation of Mn^{2+} ions with the ligand [16] and the ratio of deuteron and proton spin-lattice relaxation times [17] were studied by previous ^1H NMR measurements on aqueous $Mn(edta)^{2-}$ solutions [16, 17]. 9.1 GHz ESR spectra were recorded for aqueous $Mn(edta)^{2-}$ solutions at 50 and 3°C and for a sample frozen in a polydextran matrix [18].

The $Mn(edta)^{2-}$ ion in D_2O solution gives an IR spectrum similar to other divalent transition metal complexes with $edta^{4-}$; a graph of the spectrum for the Zn complex in solutions of various pH is presented in [10]. The $\nu_{as}(COO^-)$ band in the IR spectrum of $Mn(edta)^{2-}$ in D_2O is located at 1580 cm^{-1} [11]. The main features of the IR spectrum of $Mn(edta)^{2-}$ in D_2O are discussed in [1], but no frequencies are given.

The Raman spectra of aqueous complex solutions at pH 3 and 5 are highly fluorescent and show strong maxima at 1465, 1445, 1408, 1335, 920, and 445 cm^{-1}. The single band at 920 cm^{-1} (H_4edta 912 and 930 cm^{-1}), which was assigned to a C–C stretching frequency, indicates the equivalence of the four carboxylate groups. The strong band at 445 cm^{-1} was assigned to the Mn-N vibration [2]. Raman spectra of Mn complexes with $edta^{4-}$ in solution [12] (no frequencies given) agree with those of [2]. The electronic absorption spectra in the visible and UV region of $Mn(edta)^{2-}$ solutions at pH 5 show maxima at 210 [5], 290 (log $\varepsilon = 3.45$), and 430 nm (2.91) [13]. The absorbance of the band at 210 nm strongly decreases with decreasing pH, indicating structural changes upon protonation [5]. The capillary tube isotachophoresis of $Mn(edta)^{2-}$ was studied. Bivalent and trivalent metal complexes with ethylenediaminetetraacetate were separated using this method [14].

The partial molal heat capacity and partial molal volume of the $Mn(edta)^{2-}$ ion at 25°C were obtained from the apparent molal quantities of aqueous $Na_2[Mn(edta)] \cdot 4H_2O$ solutions (p. 41): $\bar{C}_p^\infty = 75.5\ J \cdot mol^{-1} \cdot K^{-1}$, $\bar{V}^\infty = 161.97\ cm^3/mol$ at infinite dilution, and $\bar{C}_p = 121.2\ J \cdot mol^{-1} \cdot K^{-1}$, $\bar{V} = 164.43\ cm^3/mol$ for I = 0.1 M [9]. Another author obtained $\bar{V}^\infty = 175.9\ cm^3/mol$ at 25°C [15].

References:

[1] Zetter, M. S., Grant, M. W., Wood, E. J., et al. (Inorg. Chem. **11** [1972] 2701/6).
[2] Krishnan, K., Plane, R. A. (J. Am. Chem. Soc. **90** [1968] 3195/200).
[3] Oakes, J., Smith, E. G. (J. Chem. Soc. Faraday Trans. II **77** [1981] 299/308, 307).
[4] Sacconi, L., Paoletti, P., Ciampolini, M. (J. Chem. Soc. **1964** 5046/51).
[5] Ogura, K., Takatu, K., Yosino, T. (Talanta **23** [1976] 872/3).
[6] Wilkins, R. G., Yelin, R. (J. Am. Chem. Soc. **89** [1967] 5496/7).
[7] Schwarzenbach, G. (Helv. Chim. Acta **32** [1949] 839/53, 839).
[8] Higginson, W. C. E. (J. Chem. Soc. **1962** 2761/3).
[9] Hovey, J. K., Tremaine, P. R. (J. Phys. Chem. **89** [1985] 5541/9).
[10] Sawyer, D. T., Talkett, J. E. (J. Am. Chem. Soc. **85** [1963] 2390/4).

[11] Pechurova, N. I., Spitsyn, A. V. I., Bogdanovich, N. G., Martynenko, L. I. (Dokl. Akad. Nauk SSSR **198** [1971] 347/9; Proc. Acad. Sci. USSR Chem. Sect. **196/201** [1971] 405/7).

[12] Nuttal, R. H., Stalker, D. M. (Inorg. Nucl. Chem. Letters **12** [1976] 639/41).
[13] Lati, J., Meyerstein, D. (J. Chem. Soc. Dalton Trans. **1978** 1105/18, 1113).
[14] Yoshida, H., Nukatsuka, I., Hikime, S. (Bunseki Kagaku **28** [1979] 382/5; C.A. **91** [1979] No. 116649).
[15] Yoshitani, K. (Bull. Chem. Soc. Japan **58** [1985] 1646/50).
[16] King, J., Davidson, N. (J. Chem. Phys. **29** [1958] 787/91).
[17] Rivkind, A. I. (Zh. Eksperim. Teor. Fiz. **34** [1958] 1007/9; Soviet Phys.-JETP **7** [1958] 695/6).
[18] Reed, G. H., Leigh, J. S., Pearson, J. E. (J. Chem. Phys. **55** [1971] 3311/6).

22.10.1.3 Chemical Reactions

The redox potential of the $Mn(edta)^{2-}/Mn(edta)^{-}$ couple is given on p. 49. Reactions of the $Mn(edta)^{2-}$ complex with other ligands to give mixed ligand complexes or with complexes of other ligands to give dinuclear compounds are treated on p. 54.

The rate constant $k=(43.8\pm3)\times10^{7}\,s^{-1}$ and the enthalpy and entropy of activation $\Delta H^{\neq}=7.68\pm0.1$ kcal/mol and $\Delta S^{\neq}=6.76\pm0.29\ cal\cdot mol^{-1}\cdot K^{-1}$ for the exchange of one water molecule in $Mn(edta)(H_2O)^{2-}$ were determined by ^{17}O NMR measurements at 25°C (locked frequency 11.5 MHz, field ~20 kG) [1].

The $Mn(edta)^{2-}$ complex is stable in aqueous solution from pH 3 to 10. The $Mn(Hedta)^{-}$ complex exists at pH ~ 2 [1, 14]. Decomposition occurs at lower pH values [14]. The behavior of the $Mn(edta)^{2-}$ ion with acids and bases is covered in the formation section on pp. 28/9. An autoclave experiment with a 0.1M complex solution showed that, in contrast to the Mn^{2+} ion, the $Mn(edta)^{2-}$ ion is also stable under hydrothermal conditions at 200°C [15].

The $Mn(edta)^{2-}$ ion photodegrades in dilute aqueous solution to give CO_2 and HCHO. Kinetic studies using ^{14}C marked complex solutions show that the photodegradation is strongly pH-dependent: 33% of the ^{14}C was converted into $^{14}CO_2$ at pH 3.5 and 1.7% at pH 6.9 after 96 h [2].

The reduction of $Mn(edta)^{2-}$ by a hydrated electron, e^{-}_{aq}, was studied by pulse radiolysis [3, 4]. The rate constant of the reaction is $k=6.0\times10^{6}\,M^{-1}\cdot s^{-1}$ at pH 11.5 for I = 0.22 M and 20°C [3], and $k=1.5\times10^{6}\,M^{-1}\cdot s^{-1}$ at pH 11.3 for $I\approx0$ and 25°C. The activation enthalpy (4.0 ± 0.6 kcal/mol) indicates a diffusion-controlled reaction [4].

Mn^{II} ethylenediaminetetraacetato complexes are readily oxidized to give the corresponding Mn^{III} complexes by a number of oxidizing agents, e.g., cerium(IV) sulfate, potassium dichromate, lead dioxide, or sodium bismuthate [16]. Reaction with air [17] or pure O_2 gas [6] also forms the Mn^{III} complexes, with the intermediate formation of Mn^{IV} complexes [6, 17]. Ozone destroys the complex. The second-order rate constant of the reaction with ozone is $k=(1.39\pm0.20)\times10^{5}\,cm^{3}\cdot mol^{-1}\cdot s^{-1}$ at room temperature [8]. The reaction with the hyperoxide ion (O_2^{-}) in dimethyl sulfoxide and in aqueous solution was studied by stopped-flow kinetic measurements, by rapid-scan and ESR spectroscopy, and by cyclic voltammetry. In dimethyl sulfoxide, a green intermediate is observed (λ_{max} at 430 and 580 nm), which changes to a yellow-brown final product. Addition of acid to the product solution generates the $Mn^{III}(edta)^{-}$ ion. The green intermediate is believed to be $Mn^{II}(edta)O_2^{3-}$, while the yellow-brown product presumably is an oxo Mn^{IV} complex. In basic aqueous solution, a blue intermediate (λ_{max} at 620 nm) is formed upon addition of the hyperoxide ion. The intermediate decays to give a final solution having no absorption in the visible spectrum [6]. Reaction of Na_2O_2 with $Mn(edta)^{2-}$ also produces a green solution, which turns yellow-brown on standing. The course of reaction is therefore assumed to be similar to that with the hyperoxide ion [6]. At pH 10, H_2O_2 does not oxidize $Mn(edta)^{2-}$ to form $Mn(edta)^{-}$ [6, 16]. However, the reaction (catalyzed by nitrogen oxides) is reported to proceed in acid solution [7]. The $Mn(edta)^{2-}$ ion is oxidized by bromine according to the parallel reactions (1) and (2). The overall reaction is (3). A spectrophotometric study at 25°C, I=1.1 M ($NaClO_4$), pH<4, in the presence of an excess of Mn^{2+} and Br^- ions, yielded the listed second-order rate constants k (in $M^{-1}\cdot s^{-1}$), the free energy changes ΔG, and kinetic parameters (energies in kJ/mol, activation entropy in $J\cdot mol^{-1}\cdot K^{-1}$) [9 to 11]:

oxidation reaction	k	ΔG	$\Delta H^{\ddagger}$	$\Delta S^{\ddagger}$
(1) $Mn(edta)^{2-}+Br_2 \rightleftharpoons Mn(edta)^{-}+Br_2^-$	$(7.8\pm0.2)\times10^{-3}$	30.1	4.2 ± 1.7	-2.4 ± 1.5
(2) $Mn(edta)^{2-}+Br_3^- \rightleftharpoons Mn(edta)^{-}+Br_2^-+Br^-$	$(2.5\pm0.2)\times10^{-4}$	37.2	4.5 ± 0.9	-1.8 ± 0.5
(3) $2Mn(edta)^{2-}+Br_3^- \rightleftharpoons 2Mn(edta)^{-}+3Br^-$	—	−42.7	—	—

The reaction is catalyzed by free Mn^{2+} ions [9]. In the proposed mechanism, a one-electron reduction of the bromine takes place after inner-sphere coordination of Br_2. The reaction rate is limited by the electron-transfer process [10]. Oxidation of the $Mn(edta)^{2-}$ ion by the radicals listed in the following table was studied by pulse radiolysis of complex solutions saturated with N_2O or N_2O+O_2. The $X_2^{\bullet -}$ and $HOC(CH_3)_2CH_2O_2^{\bullet}$ radicals in these solutions were formed by secondary reactions of the initially formed $OH^{\bullet}$ radicals with added X^- anions or t-butyl alcohol. The rate constants of the reaction of $Mn(edta)^{2-}$ with $R^{\bullet}$ to give the unstable addition products were obtained at 22±2°C (accuracy of k±15%):

$R^{\bullet}$	$OH^{\bullet}$		$O_2^{\bullet -}$		$Br_2^{\bullet -}$	$(SCN)_2^{\bullet -}$	$HOC(CH_3)_2CH_2O_2^{\bullet}$	
pH	4.5	9	4.5	5.5	5.5	5	5.5	6.5
k in 10^6 $M^{-1}\cdot s^{-1}$	1500	3000	30	7.5	5×10^4 to 9×10^{5} *)	5×10^4 to 7×10^{6} *)	6	17

*) These values were estimated by experiments with ^{60}Co γ-irradiation.

Rate constants for the decomposition of the unstable intermediates range from 20 to 200 s^{-1}. The results indicate similar reaction mechanisms as given on p. 17 for the complex with nitrilotriacetic acid [5].

$Mn(edta)^{2-}$ is oxidized by the trans-1,2-cyclohexanediamine-N,N,N',N'-tetraacetato complex, $Mn(cdta)^-$, to give the Mn^{III} complex, $Mn(edta)^-$. Details of this reaction are given on p. 72. The kinetics of the reaction $Mn(edta)^{2-}+Al^{3+} \rightleftharpoons Al(edta)^-+Mn^{2+}$ were studied by ^1H NMR measurements of relaxation rates. The reaction is first order in Al^{3+} ion concentration for Mn^{2+} ion concentrations $\geqq 5\times10^{-4}$ M. The first-order rate constant is $(3.32\pm0.2)\times10^{-2} s^{-1}$ at 23°C, pH 3.25 and I=0.25 M. The rate constant decreases when the pH of the medium is raised to 4.2 and increases with subsequent increases in pH (pH>4.2). The reasons for this pH dependence are discussed. The activation enthalpy and entropy at 23°C are: $\Delta H^{\ddagger}=6.9$ kcal/mol and $\Delta S^{\ddagger}=-41.7$ $cal\cdot mol^{-1}\cdot K^{-1}$ [12]. An analogous exchange reaction takes place when Cr^{3+} ions are added to an equimolar amount of an $Mn(edta)^{2-}$ solution at pH 5.2. Spectrophotometric measurements show that at 29°C about 25% of the Cr^{3+} ions are transformed into $Cr(edta)^-$ in 1 h [13].

References:

[1] Zetter, M. S., Grant, M. W., Wood, E. J., et al. (Inorg. Chem. **11** [1972] 2701/6).
[2] Lockhart, H. B., Blakeley, R. V. (Environ. Letters **9** No. 1 [1975] 19/31, 26, 30).
[3] Buitenhuis, R., Bakker, C. M. N., Stock, F. R., Louwrier, P. W. F. (Radiochim. Acta **24** [1977] 189/92).
[4] Anbar, M., Hart, E. J. (J. Phys. Chem. **71** [1967] 3700/2).
[5] Lati, J., Meyerstein, D. (J. Chem. Soc. Dalton Trans. **1978** 1105/18, 1113/4).

[6] Stein, J., Fackler, J. P., McClune, G. J., et al. (Inorg. Chem. **18** [1979] 3511/9, 3514/5, 3518).
[7] Klochkovskii, S. P., Neimysheva, L. P. (Zh. Analit. Khim. **29** [1974] 929/32; J. Anal. Chem. [USSR] **29** [1974] 793/6).
[8] Shambaugh, R. L. (J. Water Pollut. Control Fed. **50** [1978] 113/21, 119).
[9] Vierling, F. (Bull. Soc. Chim. France **1977** 404/9).
[10] Woodruff, W. H., Margerum, D. W. (Inorg. Chem. **13** [1974] 2578/85, 2581).

[11] Vierling, F. (Bull. Soc. Chim. France **1980** I 144/8).
[12] Popel, A. A., Saprykova, Z. A., Galeeva, S. I. (Kinetika Kataliz **15** [1974] 1079/80; Kinet. Catal. [USSR] **15** [1974] 963/4).
[13] Rao, T. S. P., Ramam, V. A., Sastri, M. N. (Natl. Acad. Sci. Letters [India] **2** [1979] 93/4; C.A. **91** [1979] No. 101301).
[14] Schwarzenbach, G., Gut, R., Anderegg, G. (Helv. Chim. Acta **37** [1954] 937/57, 937).
[15] Katsurai, T., Sone, K. (Bull. Chem. Soc. Japan **41** [1968] 519/20).
[16] Přibil, R., Hornichová, E. (Collection Czech. Chem. Commun. **15** [1950] 456/62).
[17] Herculano de Carvalho, A., Gonçalves Calado, J., Legrand de Moura, M. (Rev. Port. Quim. **5** [1963] 15/9).

22.10.1.4 Applications

There are numerous publications on the analytical determination of manganese with ethylenediaminetetraacetate. Reviews on the various methods are given, e.g., in [1 to 4]. For colorimetric determinations by the oxidation of $Mn(edta)^{2-}$ to form the $Mn(edta)^{-}$ ion, see p. 45. The separation of the $Mn(edta)^{2-}$ complex and ethylenediaminetetraacetato complexes of other metals was investigated using paper chromatography [6, 7, 27], thin layer chromatography [8, 9, 28], and paper electrophoresis [6]. R_f values [6 to 9] and migration spaces [6] are given for various solvent systems [6 to 9]. Possible relations between the R_f values and other properties of the complexes are discussed in [7]. Manganese(II) carbonate was extracted to a large extent from carbonate-silicate ores using hot 5% Na_2H_2edta solutions at pH 7 or 9 [26]. The solvent extraction of the $Mn(edta)^{2-}$ complex (and other metal complexes with the $edta^{4-}$ ion) with trioctylmethylammonium chloride in 1,2-dichloroethane was studied in [10]. The masking effect of H_4edta on the solvent extraction of metal acetylacetonates with chloroform or butyl acetate was studied in [11]. A study on the relative masking abilities of Na_3Hedta and other complexing agents toward the hydrolytic precipitation of manganese(II) ions is reported in [12]. The sorption of Mn^{2+} ions on ion exchangers with the functional groups of the ligand was investigated in [13, 14]. The $Mn(edta)^{2-}$ ion strongly inhibits the oxidation of ethylbenzene. A kinetic study of this effect is reported in [15].

The removal of the radionuclide ^{54}Mn from mammalian bodies is intensified by chelation after application of the ligand or its complexes [31]. The decorporation of manganese by application of $Na_2Ca(edta)$ or $Ca_2(edta)$ complexes was also studied in [16]. A protective effect of ethylenediaminetetraacetate on poisoning with manganese has been proved [31]. The preparation and water relaxation properties of proteins labeled with Mn chelates of ethylenediaminetetraacetate were studied in [29]. The $Mn(edta)^{2-}$ complex was less effective than the Mn^{2+} ion as a relaxation modifier in NMR studies of biological systems. However, it is significantly less toxic than the uncomplexed ion [17]. The $Mn(edta)^{2-}$ complex is a suitable paramagnetic compound for studies of plant tissue by ^1H NMR spectroscopy. The uptake of paramagnetic ions by plant roots and their transport to the leaves was demonstrated by use of the complex [18]. The toxicity on culture plants of the $Mn(edta)^{2-}$ complex alone and together with $Zn(edta)^{2-}$ was studied in [19 to 21]. Studies of the behavior of the complex against various soils show that it is not an effective Mn source for plants [22 to 25]. The behavior of the radionuclide ^{54}Mn in seawater or 0.55 M NaCl solution containing 10^{-6} to 10^{-2} M ethylenediaminetetraacetate was studied by high-voltage paper electrophoresis [30]. The destruction of the manganese chelates by ozone in waste water to recover the free metal ions which can be removed by precipitation and subsequent separation is treated in [5].

References:

[1] Přibil, R. (Komplexometrie, Vol. I, Leipzig 1963, p. 75).
[2] Schwarzenbach, G., Flaschka, H. (Die Komplexometrische Titration, Enke, Stuttgart 1965, pp. 185/7).
[3] Kopanica, M., Doležal, J., Zýka, J. (Chelates Anal. Chem. **1** [1967] 226/31).
[4] Umland, F., Janssen, A., Thierig, D., Wünsch, G. (Theorie und praktische Anwendung von Komplexbildnern, Frankfurt a. M. 1971, pp. 321/2).
[5] Shambough, R. L., Melnyk, P. B. (J. Water Pollut. Control Fed. **50** [1978] 113/21).
[6] Qureshi, M., Rawat, J. P. (Separ. Sci. **6** [1971] 451/4).
[7] Sýkora, J., Eybl, V. (Collection Czech. Chem. Commun. **32** [1967] 352/7, 354).
[8] Masoomi, Z., Haworth, D. T. (J. Chromatog. **48** [1970] 581/4).
[9] Vanderdeelen, J. (J. Chromatog. **39** [1969] 521/2).
[10] Irving, H. M. N. H., Al-Jarrah, R. H. (Anal. Chim. Acta **74** [1975] 321/31, 323).

[11] Tabushi, M. (Bull. Inst. Chem. Res. Kyoto Univ. **37** [1959] 252/9, 258).
[12] Wang, K. (Huaxue Xuebao **29** [1963] 78/85; C.A. **59** [1963] 12150).
[13] Brouchek, F. I., Gotsiridze, Sh. P. (Soobshch. Akad. Nauk Gruz. SSR **65** [1972] 325/8; C.A. **77** [1972] No. 52694).
[14] Polyak, N. A., Zimina, I. F., Starobinets, G. L., Malashevich, Zh. V. (Vestsi Akad. Navuk Belarusk. SSR Ser. Khim. Navuk **1971** No. 2, pp. 42/5; C.A. **75** [1971] No. 53633).
[15] Tochina, E. M., Postnikov, L. M., Shlyapintokh, V. Ya. (Izv. Akad. Nauk SSSR Ser. Khim. **1968** 71/6; Bull. Acad. Sci. USSR Div. Chem. Sci. **1968** 65/9).
[16] Koutenský, J., Jonáková, M., Eybl, V., Sýkora, J., Mertl, F. (Prac. Lek. **19** [1967] 52/6; C.A. **67** [1967] No. 10005).
[17] Brown, M. A., Johnson, G. A. (Med. Phys. **11** [1984] 67/72).
[18] Bacic, G., Ratkovic, S. (Biophys. J. **45** [1984] 767/76).
[19] Wallace, A., Wallace, G. A. (J. Plant Nutr. **6** [1983] 465/71).
[20] Wallace, A., Wallace, G. A., Samman, Y. (J. Plant Nutr. **6** [1983] 473/89).

[21] Wallace, A., Wallace, G. A., Cha, J. W. (J. Plant Nutr. **6** [1983] 503/5).
[22] Norvell, W. A., Lindsay, W. L. (Soil Sci. Soc. Am. Proc. **33** [1969] 86/91, 90/1).
[23] Wallace, A., Lunt, O. R. (Soil Sci. Soc. Am. Proc. **20** [1956] 479/82).
[24] Mortvedt, J. J. (Soil Sci. Soc. Am. J. **44** [1980] 621/6).
[25] Aboulroos, S. A. (Z. Pflanzenernähr. Bodenk. **144** [1981] 164/73, 169).
[26] Ganev, P., Nikolova, B. (Tr. Nauchnoizsled. Inst. Cherna Met. 1968/69 **4** [1970] 251/6; C.A. **79** [1973] No. 142584).
[27] Deguchi, T. (Nippon Kagaku Zasshi **92** [1971] 458/9 from C.A. **75** [1971] No. 837806).
[28] Tsunoda, Y., Takeuchi, T., Yoshino, Y. (Sci. Papers Coll. Gen. Educ. Univ. Tokio **14** [1964] 55/62 from C.A. **61** [1964] 153256).
[29] Lauffer, R. B., Brady, T. J. (Magn. Reson. Imaging **3** [1985] 11/6 from C.A. **103** [1985] No. 101218).
[30] Musani-Marazović, L., Pučar, Z. (Marine Chem. **5** [1977] 229/42, 231/2).

[31] Kuhn, A. (Strahlentherapie **137** [1968] 101/9).

22.10.2 Isolated Manganese(II) Compounds

Mn(H_2edta)·4H_2O and Mn[Mn(Hedta)(H_2O)]$_2$·8H_2O. Mn(H_2edta)·4H_2O was prepared by reacting stoichiometric amounts of manganese metal or $MnCO_3$ with a hot aqueous saturated solution of H_4edta. The mixture was heated until the solids were completely dissolved and was then cooled [1]. Mn[Mn(Hedta)(H_2O)]$_2$·8H_2O was obtained by reacting $MnCO_3$ with a warm dilute aqueous solution of H_4edta and evaporating the resulting mixture [2]. Both compounds were recrystallized from water [1, 2].

An X-ray study was made of Mn[Mn(Hedta)(H_2O)]$_2$·8H_2O. The compound is monoclinic; space group $P2_1/c$-C_{2h}^5 (No. 14) with the lattice parameters a = 9.212 ± 0.005, b = 16.16 ± 0.01, c = 11.88 ± 0.01 Å, β = 90°36′ ± 3′; Z = 2; R = 0.16. It contains two crystallographically nonequivalent Mn atoms. Mn(1) (in the general 4e position) is seven-coordinate and forms [Mn(edta)(H_2O)]$^{2-}$ units (**Fig. 4**) which have roughly the same geometry as NbF_7^{2-} ions. Mn(2) (in 2d position) is six-coordinate (**Fig. 5**). All O, N, and C atoms, as well as H_2O molecules, are also on position 4e. The oxygen atoms O(8) are directly coordinated to the Mn(2) atom forming O(8)-Mn(2)-O(8) bridges. The Mn(2)-O(8) bond length is 2.199 Å. The octahedral coordination sphere of the Mn(2) atom is completed by four oxygen atoms of water molecules. Bond lengths within the inner coordination sphere of the [Mn(edta)(H_2O)]$^{2-}$ ion are given in Fig. 4. Bond angles including Mn(1) are: O(1)-Mn-O(2) 85.1°; O(3)-Mn-O(2) 102.7°; O(4)-Mn-O_{H_2O} 106.7°; O(3)-Mn-O_{H_2O} 78.6°; O(1)-Mn-O(3) 154.8°, O(2)-Mn-O(4) 194.1°; O(3)-Mn-O(4) 80.3°; O(1)-Mn-O(4) 97.6°; O(2)-Mn-O_{H_2O} 86.2°; O(1)-Mn-O_{H_2O} 78.0°; N(1)-Mn-N(2) 75.3°. Individual bond lengths (also for H bonds) and bond angles, as well as atomic parameters, are given in the publication. The [Mn(edta)(H_2O)]$^{2-}$ ions are arranged in quasi-infinite chains parallel to the b axis. The ions are connected by protons (the acidic hydrogens) which form H bonds between carboxylate O atoms, O(5) and O(7), of adjacent ions. This bonding is very tight; the O–H–O bond length is 2.469 Å. Further complexing and bridging utilizing all the water molecules and the carboxylate O atoms lead to a three-dimensional network [2] (preliminary report [3]).

Mn(H_2edta)·4H_2O has a density of 2.34 g/cm^3 and a magnetic moment of 6.04 μ_B at 20°C [1]. The measured and calculated densities of Mn[Mn(Hedta)(H_2O)]$_2$·8H_2O are 1.72 and 1.73 g/cm^3, respectively [2].

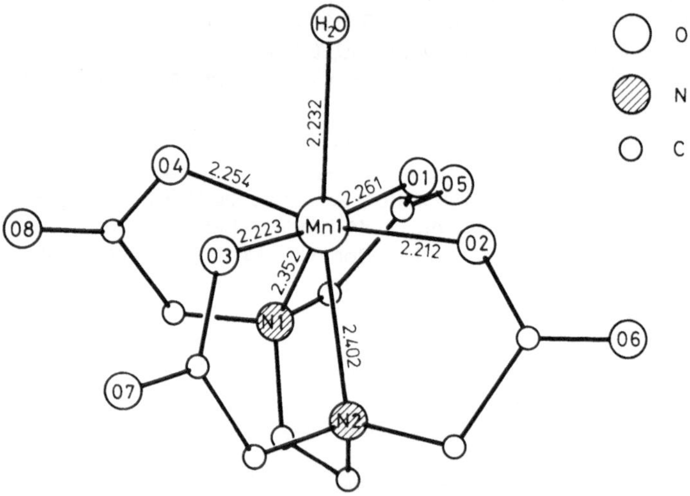

Fig. 4. Idealized model of the [Mn(edta)(H_2O)]$^{2-}$ ion in the Mn[Mn(Hedta)(H_2O)]$_2$·8H_2O complex [2].

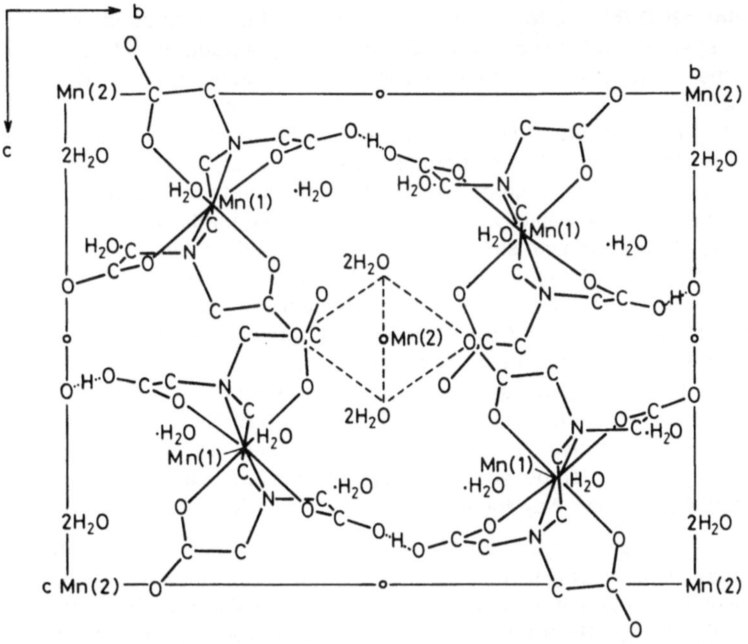

Fig. 5. Structure of $Mn[Mn(Hedta)(H_2O)]_2 \cdot 8H_2O$ [2].

The infrared spectra of $Mn(H_2edta) \cdot 4H_2O$ and $Mn[Mn(Hedta)(H_2O)]_2 \cdot 8H_2O$, the anhydrous complexes, and of the corresponding deuterated compounds in paraffin oil or hexachlorobutadiene are presented in [4]. For the hydrates, two ν(OH) bands occur in the 3600 to 3100 cm^{-1} region. The one located at about 3500 cm^{-1} indicates that one of the water molecules has a different chemical nature. For both compounds, two ν(CH) bands appear in the 3200 to 2800 cm^{-1} region. The ν(CH) bands are located at higher wave numbers than those of $K_4(edta)$ and $K_3(Hedta)$ (2800 to 2510 cm^{-1}) [4]. Other bands for $Mn(H_2edta) \cdot 4H_2O$ [1] or for both compounds [4] were assigned as follows: 1700 cm^{-1} [1], 1710 to 1680 cm^{-1} [4], ν(COOH) (H bonded carboxyl group) [4]; 1600 [1], 1620 to 1580 cm^{-1}, $\nu_{as}(COO^-)$ (coordinated carboxylate); ~1400 cm^{-1}, $\nu_s(COO^-)$; 1135, 1115 cm^{-1}, ν(CN). The splitting of the ν(CN) bands, observed for the hydrates, indicates distortion due to H bonds. Only one band in the same region is observed for the anhydrous compounds. The high frequency of the ν(CN) bands indicates a weak Mn-N interaction corresponding to an ionic Mn-N bond. The similarity of the IR spectra of both compounds indicates a molecular structure for $Mn(H_2edta) \cdot 4H_2O$ similar to that of $Mn[Mn(Hedta)(H_2O)]_2 \cdot 8H_2O$. This means that one proton of the H_2edta^{2-} ion in the first compound forms a hydrogen bond and the other one possibly an H_3O^+ ion [4].

$Mn(H_2edta) \cdot 4H_2O$ is stable at room temperature [1]. A thermogravimetric study shows that for both compounds, the loss of water takes place in one step, at about 130 to 150°C (heating rate 3°C/min) to form the anhydrous compounds. Degradation begins at ~200°C for the remaining complex, $Mn_3(Hedta)_2$, and at ~260°C for $Mn(H_2edta)$ [4]. The solubility of $Mn(H_2edta) \cdot 4H_2O$ is 0.8 wt% at 20°C and 6.1 wt% at 100°C. $Mn[Mn(Hedta)(H_2O)]_2 \cdot 8H_2O$ is only slightly soluble in water [1]. The dissociation constants of $Mn(H_2edta) \cdot 4H_2O$ in aqueous solution are: $\log K_1 = -2.68$ and $\log K_2 = -8.42$, potentiometrically determined at 20°C [1]. Aqueous $Mn(H_2edta) \cdot 4H_2O$ reacts with excess aqueous $N_2H_4 \cdot H_2O$ to form $[Mn_2(H_2edta)_2(N_2H_4)_3] \cdot 8H_2O$ (p. 53) and with $MHCO_3$ or M_2CO_3 (M = Li, Na, K, Rb) to form the salts $M_2Mn(edta) \cdot nH_2O$.

$M_2Mn(edta)\cdot nH_2O$ (M = Li, Na, K, Rb, Cs, and NH_4). All the compounds were prepared by adjusting a suspension of $Mn(H_2edta)\cdot 4H_2O$ to pH 6.5 to 7 by addition of M_2CO_3 or $MHCO_3$ and evaporating the resulting solution to a third of its initial volume. The Li and NH_4 compounds were obtained by free crystallization. The dried crystals had the compositions $Li_2Mn(edta)\cdot 4H_2O$ and $(NH_4)_2Mn(edta)\cdot 4H_2O$ [1]. $Li_2Mn(edta)\cdot 5H_2O$ was obtained by an analogous procedure [5, 6]. The Na, K, and Rb compounds were precipitated by diluting the evaporated solution with five times its volume of a 1:2 vol% methanol-ether mixture. Crystals of the Na compound separated immediately. The K and Rb compounds separated as oils, which crystallized after triturating with small quantities of the methanol-ether mixture. The crystals of $Na_2Mn(edta)\cdot 8H_2O$, $K_2Mn(edta)\cdot 2H_2O$, and $Rb_2Mn(edta)\cdot 3H_2O$ were washed with the solvent mixture and ether and dried in an air stream [1]. $Na_2Mn(edta)\cdot 6H_2O$, $K_2Mn(edta)\cdot 2.5H_2O$, $Rb_2Mn(edta)\cdot 4H_2O$, and $Cs_2Mn(edta)\cdot 4H_2O$ were obtained using a similar procedure [4]. Modifications of this procedure are reported for the Na compound using the following Mn compounds, neutralizing agents, and precipitating solvents: $MnCO_3$, H_4edta, NaOH, absolute methanol [7]; $MnCO_3$, $Na_2H_2(edta)$, ethanol [8]; $MnCl_2\cdot 4H_2O$, $Na_2H_2(edta)$, $NaHCO_3$, absolute ethanol [9]. The resulting crystals were washed with ethanol [8], recrystallized from a water-ethanol mixture [9], and dried between filter paper [8], for 72 h in a vacuum desiccator [9] or in vacuum at 100°C [7] to give $Na_2Mn(edta)\cdot nH_2O$ with n = 6 [8], n = 2 [9], or n = 0.5 [7].

Crystals of $Li_2Mn(edta)\cdot 5H_2O$ belong to the orthorhombic system [5], and prismatic crystals of $Na_2Mn(edta)\cdot 6H_2O$ belong to the monoclinic system [8]. An X-ray study of the lithium compound yielded the lattice parameters a = 11.76(2), b = 9.04(2), c = 16.72(3) Å; Z = 4. Space group Pbcn-D_{2h}^{14} (No. 60); R = 0.126. The structure is composed of corrugated layers perpendicular to the [010] direction consisting of $[Mn(edta)(H_2O)]^{2-}$ units and distorted LiO_4 tetrahedrons, both of which are connected by common carboxylate C atoms. The $[Mn(edta)(H_2O)]^{2-}$ unit has the distorted mono-capped trigonal prism (NbF_7^{2-} type) structure that was described on p. 38 for this unit in $Mn[Mn(Hedta)(H_2O)]_2\cdot 8H_2O$. Bond lengths (in Å) are: Mn-O(1) = 2.28(1), Mn-O(2) = 2.14(1), Mn-O_{H_2O}(1) = 2.25(1), Mn-N = 2.44(1). Bond angles are: O(1)-Mn-N = 68.3(5)°; O(2)-Mn-N = 72.8(5)°, O(1)-Mn-O(2) = 97(1)°, N-Mn-N′ = 71(1)° [5]. Bond lengths and bond angles of the coordination polyhedron of Li and lengths of the H bonds existing between the carboxylate O atoms and the water O atoms, as well as atomic positional parameters and temperature factors are given in [5]; preliminary report [6].

The experimental density D_{exp} in g/cm³ at 20°C of the various hydrates is 1.84 for $Li_2Mn(edta)\cdot 4H_2O$ [1], 1.67 for $Li_2Mn(edta)\cdot 5H_2O$ [5], 1.98 for $Na_2Mn(edta)\cdot 8H_2O$ [1], 2.41 for $K_2Mn(edta)\cdot 2H_2O$ [1], 2.02 for $(NH_4)_2Mn(edta)\cdot 4H_2O$ [1], and 2.11 for $Rb_2Mn(edta)\cdot 3H_2O$ [1]. The calculated density of $Li_2Mn(edta)\cdot 5H_2O$ is 1.69 g/cm³ at room temperature [5].

The magnetic moment of $Na_2Mn(edta)\cdot 6H_2O$ is 6.25 μ_B, corresponding to a high-spin d^5 configuration of the Mn atom. The temperature dependence of the magnetic susceptibility obeys the Curie-Weiss law with the Weiss constant $\Theta = -19$ K [8]. $Na_2Mn(edta)\cdot 6H_2O$ forms white prismatic crystals with a pink tinge, which are biaxial with oblique extinction, $c\cdot n_\gamma = 45°$. The refractive indices are $n_\alpha = 1.528$ and $n_\gamma = 1.549$ [8].

The IR spectra are similar to the IR spectra of the acid salts. Absorption assignments are reported on p. 39. IR spectra for $M_2Mn(edta)\cdot nH_2O$ are given in [4, 9, 11, 12]. The major IR absorptions for $Na_2Mn(edta)\cdot 2H_2O$ (KBr) were assigned as follows: 2920 cm⁻¹ to ν(CH); 1600 cm⁻¹ to $\nu_{as}(COO^-)$; 1420, 1345, 1330(sh), 1320 cm⁻¹ (sh) to $\nu_s(COO^-)$; 1125, 1105 cm⁻¹ to ν(CN) [9, 12]. Some corresponding bands for $K_2Mn(edta)\cdot 2.5H_2O$ (KBr) are at 2895; 1600; 1405, 1340; 1135, and 1110 cm⁻¹. The $\nu_{as}(COO^-)$ band for $K_2Mn(edta)$ in D_2O was observed at 1580 cm⁻¹ [13].

The visible spectrum of $Na_2Mn(edta)\cdot 2H_2O$ in DMSO has an absorption maximum at 500 nm [14].

The compounds are stable at room temperature [1]. Thermogravimetric [4] and differential thermoanalytic investigations [15] at a heating rate of 3 [4] or 8°C/min [15] showed that the loss of water takes place in one step (at 110 to 130°C) for $Li_2Mn(edta)\cdot5H_2O$ and $Na_2Mn(edta)\cdot6H_2O$, and in two steps (at ~50 and ~130°C) for $K_2Mn(edta)\cdot2.5H_2O$, $Rb_2Mn(edta)\cdot4H_2O$, and $CsMn(edta)\cdot4H_2O$. Partially dehydrated products $M_2Mn(edta)\cdot2H_2O$ were isolated for M = Na, K, Rb, Cs. However, it was not possible to obtain completely dehydrated products for all the compounds. $Rb_2Mn(edta)\cdot2H_2O$ and $Cs(edta)\cdot2H_2O$ decomposed with dehydration. Pyrolysis of the (virtually dehydrated) Li, Na, and K compounds began at 220 to 250°C [4]. Isothermal heating of $Na_2Mn(edta)\cdot6H_2O$ at 105 to 107°C for 15 h produced $Na_2Mn(edta)\cdot H_2O$, which started to decompose at 187 to 190°C [8].

The solubility in water at 20°C (in wt%) is 18.9 for $Li_2Mn(edta)\cdot4H_2O$; 58.0 for $Na_2Mn(edta)\cdot8H_2O$; 59.9 for $K_2Mn(edta)\cdot2H_2O$; 43.4 for $(NH_4)_2Mn(edta)\cdot4H_2O$; and 29.3 for $Rb_2Mn(edta)\cdot3H_2O$ [1]. The solubility of $Na_2Mn(edta)\cdot6H_2O$ (in g anhydrous salt per 100 g of water) increases from 74.6 at 13.5°C to 82.5 at 30°C and 107.0 at 60°C [8]; a graph is presented in [10]. The insoluble phase is $Na_2Mn(edta)\cdot6H_2O$ [8]. The compounds are only sparingly soluble in organic solvents [16].

The density and viscosity of a solution of $Na_2Mn(edta)$ (0.1007 M) and $NaNO_3$ (0.2014 M) were measured as $D^4_{30} = 1.0300$ g/cm³ and $\eta_{30} = 0.8976$ [28]. The partial molal heat capacity and the partial molal volume of aqueous $Na_2Mn(edta)$ at 25°C are $\overline{C}^\infty_p = 161.1 \pm 0.2$ J·mol⁻¹·K⁻¹ and $\overline{V}^\infty = 159.55 \pm 0.09$ cm³/mol at infinite dilution, and $\overline{C}_p = 213.6 \pm 0.3$ J·mol⁻¹·K⁻¹ and $\overline{V} = 162.27 \pm 0.14$ cm³/mol for I = 0.1 mol/kg. Apparent molar volumes and apparent molar heat capacities as well as density increments for solutions of various concentrations are also reported [29].

$Mn_2(edta)\cdot9H_2O$ and $MgMn(edta)\cdot9H_2O$. The complex $Mn_2(edta)\cdot9H_2O$ was obtained by adding manganese(II) carbonate to a boiling aqueous solution of H_4edta until pH 6 to 7 was attained and evaporating the solution to a small volume. On cooling, large, faintly pink crystals separated, which were dried in the air [18]. The magnesium compound was prepared by mixing stoichiometric amounts of aqueous suspensions of $Mg_2(edta)\cdot9H_2O$ and $Mn_2(edta)\cdot9H_2O$. The mixture was heated until all of the solid had dissolved and then evaporated to a small volume. Crystallization occurred at room temperature in the air [17].

X-ray structural studies indicate that both compounds are orthorhombic, space group $Pna2_1$-C^9_{2v} (No. 33) with the lattice parameters (in Å): a = 15.18 ± 0.02, b = 16.10 ± 0.02, c = 18.11 ± 0.02; Z = 8 for $Mn_2(edta)\cdot9H_2O$, and Pbcn-D^{14}_{2h} (No. 60) with a = 11.8 ± 0.05, b = 9.50 ± 0.01, c = 19.39 ± 0.02; Z = 4 for $MgMn(edta)\cdot9H_2O$. $Mn_2(edta)\cdot9H_2O$ has the same space group as $MZn(edta)\cdot6H_2O$ (M = Mg, Zn), $MCo(edta)\cdot6H_2O$ (M = Mg, Co), MNi(edta) (M = Mg, Ni), and $MgCu(edta)\cdot6H_2O$ [19]. It is composed of $[Mn(edta)(H_2O)]^{2-}$ ions with seven-coordinate Mn atoms and octahedral complex units of the second Mn atom, which is presumably coordinated by two O atoms from the carboxylate groups of the adjacent $[Mn(edta)(H_2O)]^{2-}$ units and four water O atoms. The resulting chains are connected by a network of hydrogen bonds [4, 20, 23]. $MgMn(edta)\cdot9H_2O$ is isostructural with $Mg_2(edta)\cdot9H_2O$ [19]. It is made up of a three-dimensional framework consisting of $[Mn(edta)(H_2O)]^{2-}$ anions and octahedral $[Mg(H_2O)_6]^{2+}$ cations connected by hydrogen bonds involving the water of crystallization [20 to 22]. The $[Mn(edta)(H_2O)]^{2-}$ unit is a distorted mono-capped trigonal prism (NbF_7^{2-} type) in both compounds [4, 22]. It is described in more detail for $Mn[Mn(Hedta)(H_2O)]_2\cdot8H_2O$ on p. 38 and is the same as in $Li_2Mn(edta)\cdot5H_2O$ (p. 40). The experimental density of $Mn_2(edta)\cdot9H_2O$ is 1.63 g/cm³ [19], 1.98 g/cm³ [1], and of $MgMn(edta)\cdot9H_2O$, 1.52 g/cm³ [19].

The magnetic moment of $Mn_2(edta)\cdot9H_2O$ is 5.87 μ_B at 20°C [1]. Its IR spectrum is similar to those of the acid salts (p. 39) and that of the Na compound (p. 40). The IR spectrum of $Mn_2(edta)\cdot9H_2O$ is presented in [4]. Thermal investigations (TG, DTA, and DTG, heating rate

5°C/min) show that Mn_2(edta)·9H_2O is dehydrated at ~100°C to give **Mn_2(edta)**, which decomposes at ~300°C to give Mn_2O_3. A tetrahydrate and a dihydrate were formed as intermediates under a self-generated atmosphere of ~1 atm H_2O pressure (quasi-isobaric conditions), which was attained by delayed departure of the gaseous reaction products by using a labyrinth crucible. Other intermediates were observed under H_2O partial pressures <1 atm. MgMn(edta)·9H_2O dehydrates at ~115°C to give **MgMn(edta)**. A trihydrate and a monohydrate were observed as intermediates. The anhydrous compound decomposes at ~400°C to give $MgMn_2O_4$ (temperatures of DTA peaks given) [21]. The dehydration process for both compounds under ~1 atm H_2O pressure at ~100°C (the first dehydration step has an activation enthalpy of ~20 kcal) was described in [20]. Investigations of the dehydration kinetics by the analysis of DTA curves for MgMn(edta)·9H_2O in a stream of nitrogen (I) or in a self-generated atmosphere of water vapor (II) show that for case I, in a conversion range between $\alpha = 0.5$ and 83%, the rate-controlling process is the chemical reaction at the interface between the aquo complex and the anhydrous compound ("shrinking sphere" model). At $\alpha > 83\%$ diffusion of water molecules through the layer of the anhydrous product becomes the rate-determining process. In case II, the "shrinking sphere" model describes the process only for $\alpha = 8$ to 12%, while at a higher conversion, nucleation or growth of nuclei becomes rate controlling. The "shrinking sphere" model was used to evaluate the activation enthalpy: $\Delta H^{+} = 20.6$ kcal/mol for case I, and $\Delta H^{+} = 26.3$ kcal/mol for case II, $\alpha = 0.5$ to 37% [24].

The solubility of Mn_2(edta)·9H_2O in water at 25°C is 0.112 mol/L, and that of MgMn(edta)·9H_2O is 0.090 mol/L [17]. Other authors found 5.8% at 20°C for Mn_2(edta)·9H_2O [1]. The relation between the solubility of the solid complexes and the stability of the complex $Mn(edta)^{2-}$ is discussed in [17]. The solubility of Mn_2(edta)·9H_2O at 25°C and I = 1 mol/L in $NaClO_4$-$HClO_4$ mixtures is ~0.120 mol/L at pH 4 and ~0.123 mol/L at pH 6 [26]. Treatment of the complexes with a 1:2 mixture of 2,2-dimethoxypropane and methanol yielded Mn_2(edta)·3CH_3OH·2H_2O or MgMn(edta)·CH_3OH·2H_2O. The IR spectra of the mixed solvates are presented [30].

CaMn(edta)·3H_2O. (Preparation not given.) Dehydration of the compound in a helium stream (120 mL/min) begins at 119°C (heating rate 4 to 6°C/min); the enthalpy of activation is 34 kcal/mol. Isokinetic temperatures were established for the series CaM(edta)·nH_2O, M = Mn, Co, Ni, Cu, Zn, Cd, Ca. It is assumed that the Ca atom is mainly coordinated by the O atoms of water molecules, whereas the central transition metal atom is mainly coordinated by the donor atoms of the organic ligand [25].

MnM^{II}(edta)·nH_2O. Compounds of composition MnM((edta)·6H_2O, with M^{II} = Zn, Ni, Co, Cu, were prepared by mixing the appropriate metal salts (generally nitrates, 2 mol) and the ligand (1 mol) in water. The complexes were precipitated from the solutions as crystalline solids by careful addition of ethanol. The solids were washed with a water-ethanol mixture and dried in a flow of dry air during 30 min. The compounds, which were characterized by X-ray powder diffraction data, by IR and visible spectra, and by thermal analysis, are best described by the formula $[M^{II}(edta)Mn(H_2O)_4]$·2H_2O. These compounds consist of zig-zag strings of alternating hydrated and chelated complex ions. The manganese(II) ions are in the cationic "hydrated" $[Mn(H_2O)_4O_IO_{II}]^{2+}$ complex position. They are coordinated to four water molecules and two *cis*-oxygen atoms belonging to two bridging carboxylate groups of side-chelated centers [31, 32].

The crystal structure of another bimetallic complex containing manganese and cadmium is reported in [33]. The compound, formulated as $[Cd(edta)Mn(H_2O)_4]_n$·2nH_2O, is triclinic, space group $P\bar{1}$-C_i^1 (No. 2) with a = 10.852(3), b = 10.535(3), c = 9.233(2) Å, $\alpha = 96.44(4)°$, $\beta = 103.35(3)°$, $\gamma = 102.83(3)°$. The calculated density is 1.90; Z = 2. Atom coordinates are given. The crystal structure consists of infinite chains of manganese and cadmium coordination polyhedra, running parallel to the x axis. The Mn atom displays octahedral coordination being linked to four water molecules and two O atoms of two ethylenediaminetetraacetato ligands. The Cd atom is seven-coordinate, being linked to four O and two N atoms of one ethylenediaminetetraacetato ligand and to an O atom of a second ligand. These ligands act as bridges between Mn···Cd and Cd···Cd metal ions. Each cadmium coordination polyhedron is bridged to a cadmium and two manganese coordination polyhedra, whereas the two Mn atoms act as bridges between two cadmium coordination polyhedra. The various chains are linked by hydrogen bonds [33].

X-ray structural studies for the compounds with M^{II} = Cu and Co are reported in [35]. The compounds are orthorhombic; space group $Pn2_1a$-C_{2v}^9 (No. 33). The lattice parameters of the isostructural compound with M^{II} = Zn are: a = 14.594(4), b = 13.294(4), c = 9.841(3) Å [35]. For $MnZn(edta)\cdot 6H_2O$, five absorptions bands were observed in the electronic spectrum (18000 to 30000 cm^{-1} region) which were assigned to the usual d-d transitions of the Mn^{2+} ion [32].

Partially dehydrated products of composition $MnM^{II}(edta)\cdot 2H_2O$ were obtained by heating the hexahydrates at 130 to 150°C [31, 32]. Dehydration temperatures and kinetic parameters of the dehydration reaction were determined by thermoanalytic studies [32, 36]. Treatment of the compounds with a 1:2 mixture of 2,2-dimethoxypropane and methanol yielded lower hydrates (for M^{II} = Cd or Co [34]) or mixed water-methanol solvates (for M^{II} = Ni, Cu, Zn [30]). Analyses and IR spectra of the products are given [30, 34].

$Mn_5(edta)_2(ClO_4)_2$ was obtained by heating $Mn_5(edta)_2(ClO_4)_2\cdot 20H_2O$ (see below) at ~120°C under P_{H_2O} < 0.2 atm. Its magnetic moment increases from 5.19 μ_B at 78 K to 5.72 μ_B at 300 K. The dependence of μ_{eff} on temperature indicates indirect exchange interactions between the Mn^{2+} ions. On dehydration, water molecules in the inner coordination sphere are probably substituted by bridging ClO_4^- ions. Coordination of ClO_4^- ions in the anhydrous compound is also suggested by the presence of bands at 930 and 460 cm^{-1} in the IR spectrum and by the splitting of the $\nu_3(ClO_4)$ and $\nu_4(ClO_4)$ bands at 1100 and 620 cm^{-1}. On heating $Mn_5(edta)_2(ClO_4)_2$ with 2.5°C/min, oxidation accompanied by explosion occurs at 265 to 270°C [27].

$Mn_5(edta)_2(ClO_4)_2\cdot nH_2O$. The compound with n = 20 occurs as a solid phase in the system $Mn_2(edta)$-$Mn(ClO_4)_2$-H_2O at 25°C in the concentration ranges 0.3 to 3.9 wt% $Mn_2(edta)$ and 5 to 19 wt% $Mn(ClO_4)_2$. It was synthesized by mixing concentrated aqueous solutions of $Mn_2(edta)\cdot 9H_2O$ and $Mn(ClO_4)_2\cdot 6H_2O$ or of $Mn_2(edta)\cdot 9H_2O$ and $NaClO_4$. The pink solid was washed with water and ethanol. The magnetic moment of the small needle-like crystals is 5.96 μ_B at 78 to 300 K.

When the complex is heated in the air (heating rate 2.5°C/min), the water splits off in one step in the temperature interval 50 to 120°C to give the anhydrous compound. When the complex is heated under quasi-isobaric and quasi-isothermal conditions (P_{H_2O} ~ 1 atm, t ≈ 100°C), the hexahydrate, $Mn_5(edta)_2(ClO_4)_2\cdot 6H_2O$ is formed which starts to split off its water at ~200°C [27].

References:

[1] Škramovský, S., Podlahová, J. (Collection Czech. Chem. Commun. **27** [1962] 1374/8).

[2] Richards, S., Pedersen, B., Silverton, J. V., Hoard, J. L. (Inorg. Chem. **3** [1964] 27/33, 27/8; Structure Reports, Vol. 29, 1964, pp. 544/8).

[3] Hoard, J. L., Pedersen, B., Richards, S., Silverton, J. V. (J. Am. Chem. Soc. **83** [1961] 3533/4; Structure Reports, Vol. 26, 1961, pp. 692).

[4] Pechurova, N. I., Martynenko, L. I., Snezhko, N. I. (Vestn. Mosk. Univ. Ser. II Khim. **19** [1978] 65/70; Moscow Univ. Chem. Bull. **33** No. 1 [1978] 48/52, 49, 51).

[5] Anan'eva, N. N., Polynova, T. N., Porai-Koshits, M. A. (Zh. Strukt. Khim. **15** [1974] 261/7; J. Struct. Chem. [USSR] **15** [1974] 239/43).

[6] Polynova, T. N., Anan'eva, N. N., Porai-Koshits, M. A., Martynenko, L. I., Pechurova, N. I. (Zh. Strukt. Khim. **12** [1971] 335; J. Struct. Chem. [USSR] **12** [1971] 310/1).
[7] Woodruff, W. H., Margerum, D. W. (Inorg. Chem. **13** [1974] 2578/85, 2578).
[8] Rad'ko, V. A., Yakimets, E. M. (Zh. Neorgan. Khim. **7** [1962] 683/6; Russ. J. Inorg. Chem. **7** [1962] 348/9).
[9] Sawyer, D. T., Paulsen, P. J. (J. Am. Chem. Soc. **81** [1959] 816/20, 817/8).
[10] Nikitin, V. D., Yakimets, E. M., Timakova, N. A., Rad'ko, V. A., Shabashova, N. V., Tribunskii, V. V. (Tr. Ural'sk. Politekhn. Inst. No. 130 [1963] 94/103, 100/1; C.A. **60** [1964] No. 11590).

[11] Yoshino, Y., Ouchi, A., Tsunoda, Y. (Sci. Papers Coll. Gen. Educ. Univ. Tokyo **13** [1963] 27/33, 32; C.A. **59** [1963] 12230).
[12] Sawyer, D. T. (Ann. N.Y. Acad. Sci. **88** [1960] 307/21, 308, 310, 312).
[13] Pechurova, N. I., Spitsyn, V. I., Bogdanovich, N. G., Martynenko, L. I. (Dokl. Akad. Nauk SSSR **198** [1971] 347/9; Dokl. Chem. Proc. Acad. Sci. USSR **196/201** [1971] 405/7).
[14] Stein, J., Fackler, J. P., McClune, G. J., Fee, J. A., Chan, L. T. (Inorg. Chem. **18** [1979] 3511/9, 3514).
[15] Wendlandt, W. W., Horton, G. R. (Nature **187** [1960] 769/70).
[16] Sýkora, J., Eybl, V. (Collection Czech. Chem. Commun. **32** [1967] 352/7).
[17] Myachina, L. I., Logvinenko, V. A., Knyazeva, N. N. (Izv. Sibirsk. Otd. Akad. Nauk SSSR Ser. Khim. Nauk **1974** No. 6, pp. 77/82; C.A. **82** [1975] No. 67575).
[18] Brintzinger, H., Munkelt, S. (Z. Anorg. Allgem. Chem. **256** [1948] 65/74, 70).
[19] Pozhidaev, A. I., Neronova, N. N., Polynova, T. N., Porai-Koshits, M. A., Logvinenko, V. A. (Zh. Strukt. Khim. **13** [1972] 344/5; J. Struct. Chem. [USSR] **13** [1972] 323).
[20] Logvinenko, V. A., Paulik, F., Paulik, J. (Dokl. Akad. Nauk SSSR **233** [1977] 129/32; Dokl. Chem. Proc. Acad. Sci. USSR **232/237** [1977] 118/21).

[21] Paulik, J., Paulik, F., Logvinenko, V. A. (J. Therm. Anal. **10** [1976] 123/32, 126/8, 130).
[22] Pozhidaev, A. I., Polynova, T. N., Porai-Koshits, M. A., Logvinenko, V. A. (Zh. Strukt. Khim. **14** [1973] 746/7; J. Struct. Chem. [USSR] **14** [1973] 696/7).
[23] Pozhidaev, A. I., Neronova, N. N., Polynova, T. N., Porai-Koshits, M. A., Logvinenko, V. A. (Zh. Strukt. Khim. **13** [1972] 738; J. Struct. Chem. [USSR] **13** [1972] 690).
[24] Nikolaev, A. V., Logvinenko, V. A., Shestak, Ya., Shkvara, F. (Dokl. Akad. Nauk SSSR **231** [1976] 146/9; Dokl. Phys. Chem. USSR **231** [1976] 1035/7).
[25] Logvinenko, V. A., Myachina, L. I. (J. Therm. Anal. **19** [1980] 45/50).
[26] Myachina, L. I., Logvinenko, V. A., Knyazeva, N. N. (Izv. Sibirsk. Otd. Akad. Nauk SSSR Ser. Khim. Nauk **1977** No. 5, pp. 34/8; C.A. **88** [1978] No. 28491).
[27] Myachina, L. I., Logvinenko, V. A., Grankina, S. A., Ikorskii, V. N., Sheludyakova, L. A. (Izv. Sibirsk. Otd. Akad. Nauk SSSR Ser. Khim. Nauk **1982** No. 6, pp. 112/8; C.A. **98** [1983] No. 78921).
[28] Charles, R. G. (J. Am. Chem. Soc. **78** [1956] 3946/50).
[29] Hovey, J. K., Tremaine, P. R. (J. Phys. Chem. **89** [1985] 5541/9, 5544).
[30] Vidal, P., Aroztegui, M., Roso, M. (Circ. Farm. **42** [1984] 235/40).

[31] Escriva, E., Beltran, D., Beltran, J. (Anales Quim. B **77** [1981] 330/4).
[32] Escriva, E., Fuertes, A., Beltran, D. (Transition Metal Chem. [Weinheim] **9** [1984] 184/90).
[33] Solans, X., Font-Altaba, M., Oliva, J., Herrera, J. (Acta Cryst. C **41** [1985] 1020/2).
[34] Vidal, P., Aroztegui, M., Moragas, M. R. (Circ. Farm. **43** [1985] 81/6).
[35] Solans, X., Font-Altaba, M. (Acta Cryst. C **39** [1983] 435/8).
[36] Nikolaev, A. V., Logvinenko, V. A., Gorbatchov, V. M., Myachina, L. I. (Therm. Anal. Proc. 4th Intern. Conf., Budapest 1974 [1975], pp. 95/103, 98, 103; C.A. **87** [1977] No. 44778).

22.10.3 Manganese(III) Complexes in Solution

Formation. Titrimetric [1 to 3], spectrophotometric [1, 2, 4, 5], and kinetic studies [6] of aqueous solutions of the complex salt $K[Mn(edta)]\cdot 2H_2O$ (p. 51) [1, 2, 6], of such solutions passed through the H^+ form of an ion exchanger [1], and of aqueous solutions containing Mn^{III} and the ligand in various ratios (from 2:1 to 1:9) [4, 5] at various pH's revealed the existence of three 1:1 complexes depending on the pH. The protonated **Mn(Hedta)** species (absorption maximum at 500 nm) is the main species in the 2 to 3 pH region, **$Mn(edta)^-$** (500 nm) in the 4 to 5 pH region, and the hydroxo species **$Mn(edta)(OH)^{2-}$** (450 nm) at pH > 5. The optical density at 500 nm, indicating complex formation, increases from pH ~1.5 [1, 5] to pH ~3 [1] or 3.5 [5], is nearly constant in the pH 3 to 4 [1, 7] or 3.5 to 4.5 region [5], and then decreases at higher pH values [1, 5].

The equilibrium constant of the reaction $Mn(edta)^- + H^+ \rightleftharpoons Mn(Hedta)$, log K = 2.7 at 4°C and I = 0.1 M, was determined by base titration [1]. The equilibrium constant of the reaction $Mn(edta)(OH)^{2-} + H^+ \rightleftharpoons Mn(edta)(H_2O)^-$, log K = 5.5 at 4°C and I = 0.1 M, was found by base titration, and log K = 5.3 was found spectrophotometrically [1, 2].

Aqueous solutions containing the $Mn(edta)^-$ ion, which has frequently been formulated as $Mn(edta)(H_2O)^-$ (see p. 46), were obtained (A) by dissolving the solid compound $K[Mn(edta)]\cdot 2H_2O$ (p. 51) in water [6, 8 to 11], (B) by reacting freshly prepared hydrated Mn^{III} oxide [14, 30], acetate [12], or pyrophosphate [4] with H_4edta, (C) by reacting a mixture of freshly prepared MnO_2 and KOH, obtained by the reduction of $KMnO_4$ with ethanol, or $KMnO_4$ itself with solid H_4edta or Na_2H_2edta [1, 2, 13] in acetic acid medium [14], or (D) by oxidation of the manganese(II) complex, $Mn(edta)^{2-}$, with an oxidizing agent of sufficiently high potential, e.g., O_2 [31], Br_2 [18, 19], MnO_2 [1], $KMnO_4$ [15], $NaBiO_3$ [7, 17], or PbO_2 [5, 7, 13, 16, 17], in acetic acid medium [5, 16, 17]. This method was particularly used for the colorimetric determination of manganese [7, 16, 17], oygen [31], or selenium [35]. The formation of the $Mn(edta)^-$ ion by oxidation of the $Mn(edta)^{2-}$ ion and thermodynamic and kinetic data of the oxidation reaction are described on pp. 34/5. In a procedure using this method, 2.2 mL of 0.1 M aqueous $MnSO_4$ solution and 5 mL of 0.11 M aqueous $NaH_2edta\cdot 2H_2O$ solution were mixed and shaken with 0.7 mL glacial acetic acid and 0.07 g PbO_2. After 2 min, the volume was made up to 50 mL, and the excess of lead salts was removed [16]; earlier work is reported in [7, 17].

The stability constant K_1 of the $Mn(edta)^-$ ion in aqueous solution was determined by spectrophotometry (sp) or by measurements of redox potentials (redox) of the $Mn(edta)^-/Mn(edta)^{2-}$ couple:

t in °C	I in mol/L	pH	method	log K_1	Ref.
18 to 20	variable	3.8	sp	17.35	[5]
?	1.0($NaClO_4$)	2	sp	26.99±0.28[a]	[4]
				27.26±0.2[b]	[4]
25	0.2($NaClO_4$)	not given ($CH_3COONa-CH_3COOH$)	redox	24.8[c]	[8]
25	0.2(KCl)	4.63 ($CH_3COONa-CH_3COOH$)	redox	24.9[d]	[9]

[a] By the molar ratio method. – [b] By the method of continuous variation. – [c] Using K_1 = 13.6 for $Mn(edta)^{2-}$. – [d] Using K_1 = 13.64 for $Mn(edta)^{2-}$.

The value log K_1 = 24.85 at 25° and I = 0.2 was evaluated as a "tentative" selected value in the "Critical Survey of Stability Constants of EDTA Complexes" [34] (categories of selected values: recommended, tentative, doubtful, and rejected).

A solution containing the Mn(Hedta) species was prepared by passing a solution of the potassium salt $K[Mn(edta)]\cdot 2H_2O$ (p. 51) through the hydrogen form of a cation-exchange resin [11]. The formation constant of Mn(Hedta) formed by the reaction of Mn^{3+} with $Hedta^{3-}$ ions was determined spectrophotometrically at 18 to 20°C as log K = 8.89 [5].

Structure and Absorption Spectra. Acid-base titration studies of aqueous solutions containing the $Mn(edta)^-$ or Mn(Hedta) species indicate that they are weak acids, $Mn(edta)^-$ being a monobasic, Mn(Hedta) a dibasic one. Coordination of one water molecule is therefore assumed, from which one proton is dissociated at higher pH to give the hydroxy species [1]. Subsequently, the $Mn(edta)^-$ ion was frequently formulated with one water molecule [10 to 12, 21, 32]. Splitting of bands in the electronic spectra of the complexes in aqueous [22] or dimethyl sulfoxide solution [23] indicates tetragonal distortion of the complex molecules. A seven-coordinate Mn atom is suggested in the $Mn(edta)(H_2O)^-$ complex on account of crystallographic evidence for solid Mn^{II}, Fe^{II}, and Fe^{III} ethylenediaminetetraacetato complexes and from consideration of the metal ion radii [32].

A high degree of covalent bonding in the complexes is indicated by the IR spectra and the kinetics of complex formation. This covalent character is due to significant overlap of the ligand orbitals with the unoccupied Mn^{III} orbitals [20].

Aqueous solutions containing the $Mn(edta)^-$ and/or Mn(Hedta) species show an absorption maximum at 500 nm [1, 4, 5, 22] with a constant absorptivity (log ε = 2.67) in the 2.5 to 4 pH range [1]. (Plots of absorptivity vs. pH at 500 nm are given in [1, 5, 7, 31]; plots of absorption spectrum at various pH in the 350 to 700 nm region are given in [1, 2, 5, 7, 31].) Slightly varied wavelengths are reported by other authors: 497 nm in the 1.5 to 2.3 pH range [10], the same at pH 4.5 with log ε = 2.54 [11], and 500 to 510 nm at pH 3 to 4 with log ε = 2.51 [7]. The molar extinction coefficient at 440 nm is log ε = 2.35 at pH 4.5 [11]. The maximum at 500 nm, which gives rise to a ruby red color of the solution, is shifted to lower wavelengths in solutions of pH > 5: λ_{max} = 470 nm at pH 6 to 8 [4] and $\lambda_{max} \approx$ 450 nm at pH ~9. The red color of the solution then changes to the yellow color of the hydroxo species, $Mn(edta)(OH)^{2-}$ [1, 2]. The maximum in the visible region was associated with transitions to the sublevels $^5B_{2g}$ and ^5E_g of the T_{2g} state split as a result of tetragonal distortion, while a rise in absorption in the UV region beginning with 400 nm without any pronounced maximum (shoulder at 280 nm) was assigned to a ligand → metal charge transfer transition [22]. A solution of $K[Mn(edta)]\cdot 2H_2O$ (p. 51) in dimethyl sulfoxide shows three absorption bands which were assigned as follows, assuming an idealized D_{4h} symmetry for the coordination geometry about the Mn^{III} atoms: UV region (no maximum given), $^5B_{1g} \rightarrow {}^5E_g$; 500 nm, $^5B_{1g} \rightarrow {}^5B_{2g}$; 950 nm (broad), $^5B_{1g} \rightarrow {}^5A_{1g}$. Addition of CaH_2 to remove water from the solution does not alter the spectrum, nor does the addition of water to the dimethyl sulfoxide solution [21].

The IR spectrum of a solution of $K[Mn(edta)]\cdot 2H_2O$ in D_2O shows two absorption bands, at 1650 and 1590 cm^{-1}, which were assigned to the $\nu_{as}(COO^-)$ vibration [20].

Chemical Reactions. Ligand Exchange. The $Mn(edta)^-$ ion reacts with the azide ion with exchange of the coordinated water molecule: $Mn(edta)(H_2O)^- + N_3^- \underset{k_b}{\overset{k_f}{\rightleftharpoons}} Mn(edta)(N_3)^{2-} + H_2O$ (I). Rate constants and activation parameters for the forward and back reactions at 25°C, I = 0.25 M ($NaClO_4$) in the pH range 3.5 to 5 are: $k_f = 0.103\ M^{-1}\cdot s^{-1}$, $\Delta H_f^* = 13.7$ kcal/mol, $\Delta S_f^* = -17.1\ cal\cdot mol^{-1}\cdot K^{-1}$; $k_b = 3.18 \times 10^{-3}\ s^{-1}$, $\Delta H_b^* = 13.0$ kcal/mol, $\Delta S_b^* = -26.3\ cal\cdot mol^{-1}\cdot K^{-1}$. Thermodynamic parameters for reaction I (ΔH and ΔG in kcal/mol, ΔS in $cal\cdot mol^{-1}\cdot K^{-1}$), calculated from the equilibrium constant (p. 53) at various temperatures or from the activation parameters (in parentheses) are: ΔG = 2.01 (2.0), ΔH = 1.28 (0.7), ΔS = 11.2 (9.2) [11].

Decomposition. Concentrated aqueous solutions of $K[Mn(edta)]\cdot 2H_2O$ are stable for several hours at −5°C in the dark. Frozen solutions could be preserved for weeks [13]. Maximum stability of the $Mn(edta)^-$ ion was observed at pH 3.7 to 4.3 in cold solutions, in the dark, and in the presence of an exess of Mn^{2+} ions (Mn^{2+}: ligand = 4:1) [31].

At room temperature under normal conditions, the solution is unstable. The complex is decomposed with discoloration, particularly by the effect of light und heat [6], because of an intramolecular redox reaction between the Mn^{III} atom and the coordinated ligand [8]. Reaction products are: the manganese(II) complex, $Mn(edta)^{2-}$ (p. 28), ethylenediaminetriacetate, formaldehyde, and carbon dioxide [6, 8]. The same products and also ethylenediaminediacetate were found during photolysis of aqueous K[Mn(edta)] [22]. The degree of decomposed $edta^{4-}$ (or $Hedta^{3-}$) per mol of complex is dependent on the initial pH of the solution. It attains a maximum of 0.5 mol per mol of complex at pH 2.2 to 3 and decreases to 0.1 mol at pH 8 (plot in the publication). Nearly 0.8 mol of carbon dioxide was evolved at pH 4 [6]. Intermediate products detected by an ESR and spectrophotometric study of a frozen solution irradiated by a mercury lamp at 77 K were found to be Mn^{II} complexes with radicals of the ligand and Mn^{III} complexes containing metal-C bonds produced by contraction of the chelate ring by elimination of a CO_2 molecule [22].

The rate of the redox reaction is strongly dependent on the pH of the solution [4, 6, 7]. Thus, the ruby red color of the solution disappeared in 110, 70, 30, 13, 2.3, or 0.85 h at pH 3.0, 3.5, 4.0, 4.5, 5.0, or 5.5, respectively [7]. A plot of the rate constant of the decomposition reaction vs. pH is shown in **Fig. 6** [6]. A pH dependence with somewhat greater rate constants at pH < 3 and lower rate constants at pH 4 to 8 was found by [4]. A steady increase of the rate of decomposition in the 4 to 5 pH range was found by [7] (plots in the papers).

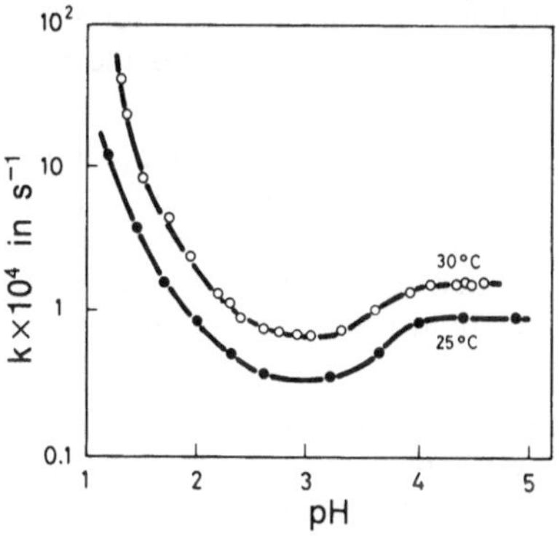

Fig. 6. Effect of pH on the rate of decomposition of manganese(III) ethylenediaminetetraacetate at 25 and 30°C [6].

The Mn(Hedta) species is less reactive than the $Mn(edta)^-$ species. At pH ranges < 2.5, dissociation of the chelate may take place resulting in the rapid reduction of the free Mn^{3+} ion [6]. On the other hand, the hydroxo species, $Mn(edta)(OH)^{2-}$, is more reactive than the $Mn(edta)^-$ species. It readily decomposes giving MnO_2 precipitation [1]. The enthalpy of activation, $\Delta H^{\neq}$, of the decomposition reaction is also pH-dependent. $\Delta H^{\neq}$ is 48.5, 28.9, 26.2, 19.8, and 18.2 kcal/mol at pH 1.4, 2.0, 3.0, 4.0, and 4.4. The values show that the decomposition reaction proceeds in a different manner in the lower pH ranges [6]. Kinetic parameters were found to be nearly constant in the pH ranges 3 to 6 in light [8] or 4 to 5 in the dark [6]. The listed first-order rate constants and enthalpies of activation were obtained following the concentrations by spectrophotometry at 25°C:

pH range	4 to 4.5	4 to 4.5	3 to 6	–
ionic strength	0.1 (KNO_3)	0.1 (KNO_3)	acetate buffer	0.12 (acetate buffer)
k in $10^{-4}s^{-1}$	0.856[a)]	1.17[b)]	0.120	0.48[c)]
$\Delta H^{\neq}$ in kcal/mol	18.2	16.2	23.5	–
Ref.	[6]	[6]	[8]	[12]

[a)] In the dark. – [b)] In the light of a 60 W tungsten lamp at a distance of 10 cm. – [c)] This rate constant corresponds to a half-life of about 4 h.

The following half-lives were determined spectrophotometrically for the $Mn(edta)^-$ or Mn(Hedta) complexes in the light or in the dark, respectively: 118 and 145 min at pH 4.5 (acetic acid), 114 and 207 min at pH 4.5 (sulfuric acid), 223 and 285 min, respectively, at pH 2.4 (sulfuric acid) [33].

The entropy of activation, $\Delta S^{\neq} = -0.6 \pm 1.1\ cal \cdot mol^{-1} \cdot K^{-1}$, was calculated [8]. Kinetic data from a polarographic study could be explained by assuming five parallel reactions: (1) $Mn(edta)(H_2O)^- \rightleftharpoons Mn(edta)(OH)^{2-} + H^+$, (2) $Mn(edta)(H_2O)^- \xrightarrow{k_1}$ dec. products, (3) $Mn(edta)(OH)^{2-} \xrightarrow{k_2}$ dec. products, (4) $Mn^{III}(edta)(H_2O)^- + Mn^{III}(edta)(OH)^{2-} \underset{k_b}{\overset{k_f}{\rightleftharpoons}} Mn^{IV}(edta)(OH)^- + Mn^{II}(edta)(H_2O)^{2-}$, (5) $Mn^{IV}(edta)(OH)^- \xrightarrow{k_4}$ dec. products. Rate constants at 25°C and enthalpies of activation for I = 0.2 (CH_3COONa) are: $k_1 = 0.55 \times 10^{-4} s^{-1}$, $\Delta H^{\neq} = 17.1$ kcal/mol; $k_2 = 1.21 \times 10^{-4} s^{-1}$, $\Delta H^{\neq} = 15.7$ kcal/mol; $k_f = 0.827\ M^{-1} \cdot s^{-1}$, $\Delta H^{\neq} = 22.3$ kcal/mol; $k_b/k_4 = 2.28 \times 10^3 M^{-1}$, $\Delta H^{\neq} = -0.6$ kcal/mol [24]. The presence of PbO_2 [7] or Br_2 [16] favors the decomposition of the complex, while varying the ionic strength between 0.001 and 1.0 M does not noticeably affect the rate of decomposition [6, 8]. Cetyltrimethylammonium bromide (CTAB) had a stabilizing effect on the complex, the absorbance remaining stable for at least 10 min instead of 3 min without CTAB [14]. In the presence of CTAB, the half-life of the complex at pH 4.07 (acetic acid) was 290 min in the light and 355 min in the dark [33].

The $Mn(edta)^-$ ion as well as the Mn(Hedta) and the $Mn(edta)(OH)^{2-}$ species become more stable and can be preserved for 5 h when the complex is extracted with a solution of tricaprylmethylammonium chloride (Aliquat-336 chloride) in 1,2-dichloroethane. The extractability of a 4×10^{-3} M complex solution containing a 10% excess of H_4edta by an equal volume of a 0.22 M solution of the extractant in 1,2-dichloroethane was 79, 70, and 30% at pH 2.8, 4, and 6.2, respectively [16].

In the presence of free ligand, decomposition products are the same as reported for the complex alone, i.e., $Mn(edta)^{2-}$, ethylenediaminetriacetate, formaldehyde, and carbon dioxide [8, 15, 25]. An analytic study at pH 6 with a three- to five-fold excess of free ligand shows that the main reactions can be written $Mn^{III}Y + Y \rightarrow Mn^{II}Y + Y^{\bullet} + CO_2$, $Mn^{III}Y + Y^{\bullet} + H_2O \rightarrow Mn^{II}Y + Y' + HCHO + H^+$, where Y is one of the deprotonated species of H_4edta, $Y^{\bullet}$ a radical which is formed by oxidation of Y, and Y′ ethylenediaminetriacetate ($Mn^{III}Y$ may be $Mn(edta)^-$ or $Mn(edta)(OH)^{2-}$). The occurrence of side reactions is suggested by the additional formation of N-methyl-ethylenediaminediacetic acid [25]. Kinetic data from a polarographic study at pH 4 to 9 with a ten-fold excess of free ligand could be explained by assuming 4 parallel reactions: reactions of the $Mn(edta)^-$ ion with H_2edta^{2-} or $Hedta^{3-}$ (k_1 and k_2 reactions, respectively) and reactions of the $Mn(edta)(OH)^{2-}$ ion with the same anions (k_3 and k_4 reactions, respectively). Rate constants at 20°C, I = 0.5 M, are between 6×10^{-3} and $7 \times 10^{-1}\ L \cdot mol^{-1} \cdot s^{-1}$. Enthalpies of activation are around 20 kcal/mol [21]. The data from another kinetic study at pH 2 to 5 followed by spectrophotometry could be explained by three parallel reactions: the k_1 and k_2 reactions of [21] and a third reaction independent of the ethylenediaminetetraacetate concentration (k_3 reaction). Second-order rate constants are in the same range as cited above, and the first-order rate constant is $k_3 \approx 2 \times 10^{-5}\ s^{-1}$, all at 20°C, I = 0.1 M. The enthalpy of activation of the k_3 reaction is 19 kcal/mol. A steady-state mechanism with formation of a reactive intermediate in a low concentration was assumed to explain the variation of the reaction rate with the concentration of the Mn^{III} complex [15]. The dependence of the rate of reduction of the Mn^{III} complexes on the ratio Mn:ligand = 1:1 to 1:9 at pH 2 was studied by spectrophotometry in [4].

Redox Reactions. The standard potential of the reaction $Mn(edta)^- + e^- \rightleftharpoons Mn(edta)^{2-}$ in aqueous solution is $E° = 0.82_5$ V vs. NHE at 25°C, I = 0.2 M ($NaClO_4$) [8]; $E° = 0.57_9$ V vs. SCE (= 0.82_3 V vs. NHE) at 25°C, I = 0.2 M (KCl), pH 4.63 [9]. Emf values of the system Mn^{III}–ligand in aqueous solution at pH 1 and 2 are reported and their relationship to the stability constant of the complex is discussed in [20]. The cyclic voltammogram of $K[Mn(edta)] \cdot 2H_2O$ in dimethyl sulfoxide shows a peak for the $Mn^{III} \rightarrow Mn^{II}$ reduction at $E_p = -0.25$ V vs. SCE at slow sweep rates. The couple is quasi-reversible [23].

The redox reactions of the Mn(Hedta), $Mn(edta)^-$, and $Mn(edta)(OH)^{2-}$ species with (excess) free ligand are described on p. 48; the intramolecular redox reaction of the Mn^{3+} ion with the coordinated ligand is described on p. 47.

The reduction of the $Mn(edta)(OH)^{2-}$ ion by a hydrated electron, e^-_{aq}, formed by pulse radiolysis, was investigated spectrophotometrically. For a solution with pH 11 to 12 and I = 0.2 M, the pseudo first-order rate constant is $k \leqq 2.2 \times 10^6$ s^{-1} as compared with 7.7×10^7 s^{-1} for the Mn^{3+} ion. The activation energy of the reaction is <4 kcal/mol [26]. The $Mn(edta)^-$ ion reacts with the hyperoxide ion, O_2^-, in DMSO or aqueous DMSO medium to form the Mn^{II} complex and O_2. The second-order rate constant is $5 \times 10^{-4} M^{-1} \cdot s^{-1}$ at 20°C as found by stopped-flow or rapid-scan spectrophotometry. In aqueous DMSO medium, the O_2^- ion disproportionates in a slow parallel reaction forming O_2 and H_2O_2. The latter also reacts in a very slow reaction with $Mn(edta)^-$ to form the Mn^{II} complex and O_2 [23]. Kinetics and mechanism of the redox reaction of $Mn(edta)^-$ with N_3^- to form the Mn^{II} complex and N_2, with intermediate formation of the azido complex, $Mn(edta)(N_3)^{2-}$, are reported on p. 46. Hydrazine is oxidized by the $Mn(edta)^-$ ion in aqueous solution with a nonintegral stoichiometry, giving N_2 and NH_3. The ratio of reacted $Mn(edta)^-$ to reacted hydrazine is pH-dependent (1.1 at pH 2.4 and 1.7 at pH 7.5). Possible mechanisms of the reaction are discussed [27]. Dihydroxybenzene compounds (H_2Q) are oxidized to give the corresponding quinones by two one-electron steps. Kinetic studies by stopped-flow spectrophotometry in the 2.5 to 6 pH range suggest a mechanism with intermediate formation of the complex $Mn(edta)(H_2Q)^-$ and subsequent electron transfer to give $Mn^{II}(edta)^{2-}$ and the semiquinone radical HQ•, the latter reaction as the rate determining step [13]. 3-Indoleacetic acid is oxidized to give carbon dioxide. The kinetics of the reactions of 3-indoleacetic acid or its anion with the complex were studied at various pH's by measuring the O_2 consumption of the intermediately formed radicals. The enthalpy of activation of the two reactions in phosphate-buffered solution is 13.7 and 16.0 kcal/mol, respectively. A mechanism involving five parallel reactions is proposed [28]. Oxovanadium(IV) complexes are oxidized to give dioxovanadium(V) complexes. The reactions of the $Mn(edta)^-$ ion with the VO^{2+} ion or with the complex of nitrilotriacetic acid, $VO(C_6H_6NO_6)^-$, proceed by parallel paths, one path acid-independent (k_o) and one path dependent inversely on the hydrogen ion concentration (k_1). The second-order rate constants at 25°C for I = 0.1 (acetate buffer) are: $k_o = 778$ $M^{-1} \cdot s^{-1}$ and $k_1 = 4.54$ s^{-1} for the VO^{2+} ion, and $k_o = 10.2$ $M^{-1} \cdot s^{-1}$ and $k_1 = 1.06 \times 10^{-2} s^{-1}$ for the $VO(C_6H_6NO_6)^-$ ion. The enthalpy of activation and entropy of activation for the k_1 path are: $\Delta H^{\neq} = 11.5 \pm 1.0$ kcal/mol and $\Delta S^{\neq} = -17 \pm 3$ $cal \cdot mol^{-1} \cdot K^{-1}$, respectively [12]. Rate constants for the 1 to 7°C temperature range are reported in [10]. The reaction of the $Mn(edta)^-$ ion with $VO(hedta)^-$ (H_3hedta = N-(2-hydroxyethyl)ethylenediamine-triacetic acid) or $VO(edta)^{2-}$ is first-order in $[H_3O^+]$ and in each of the reactant concentrations. The rate constant is $k[H_3O^+] = 5.09 \times 10^{-2} M^{-1} \cdot s^{-1}$ at pH 3.76 and $0.38 M^{-1} \cdot s^{-1}$ at pH 3 for the $VO(hedta)^-$ ion, and $k[H_3O^+] = 6.11 \times 10^{-2} M^{-1} \cdot s^{-1}$ at pH 3.76 for the $VO(edta)^{2-}$ ion, both reactions at 25°C, I = 0.1 (CH_3COONa) [12]. While an inner-sphere electron transfer mechanism was proposed for the reaction with VO^{2+} by [10], an outer-sphere mechanism was proposed for all the reactions by [12]. The oxidation of $Co(edta)^{2-}$ with $Mn(edta)^-$ to give the Mn^{II} complex and $Co(edta)^-$ is dependent on pH. While 37% $Co(edta)^-$ formed at pH 4.2, only 17% formed at pH 5. The second-order rate constant is 1.6 $M^{-1} \cdot s^{-1}$ at 25°C, I = 0.5, and pH ~ 5. An inner-sphere mechanism with an effective hydroxy-bridged path involving the hydroxy species $Mn(edta)(OH)^{2-}$ is proposed [29].

References:

[1] Yoshino, Y., Ouchi, A., Tsunoda, Y., Kojima, M. (Can. J. Chem. **40** [1962] 775/83, 776, 780, 782).
[2] Yoshino, Y., Tsunoda, Y., Ouchi, A. (Bull. Chem. Soc. Japan **34** [1961] 1194/5).
[3] Schwarzenbach, G. (Helv. Chim. Acta **32** [1949] 839/53, 840).
[4] Bogdanovich, N. G., Pechurova, N. I., Martynenko, L. I., Piunova, V. V. (Zh. Neorgan. Khim. **16** [1971] 2507/11; Russ. J. Inorg. Chem. **16** [1971] 1337/9).
[5] Mikhailova, T. V., Astakhov, K. V., Zhirnova, N. M. (Zh. Fiz. Khim. **45** [1971] 1106/9; Russ. J. Phys. Chem. **45** [1971] 618/20).
[6] Yoshino, Y., Ouchi, A., Tsunoda, Y. (Sci. Papers Coll. Gen. Educ. Univ. Tokyo **13** [1963] 27/33, 29).
[7] Rad'ko, V. A., Yakimets, E. M. (Tr. Ural'sk. Politekhn. Inst. No. 130 [1963] 62/9, 65; C.A. **60** [1964] 12653).
[8] Hamm, R. E., Suwyn, M. A. (Inorg. Chem. **6** [1967] 139/45, 141).
[9] Tanaka, N., Shirakashi, T., Ogino, H. (Bull. Chem Soc. Japan **38** [1965] 1515/7).
[10] Boone, D. J., Hamm, R. E., Hunt, J. P. (Inorg. Chem. **11** [1972] 1060/2).

[11] Suwyn, M. A., Hamm, R. E. (Inorg. Chem. **6** [1967] 2150/4).
[12] Nelson, J., Shepherd, R. E. (Inorg. Chem. **17** [1978] 1030/4).
[13] Giraudi, G., Mentasti, E. (Transition Metal Chem. [Weinheim] **6** [1981] 230/4).
[14] Rahim, S. A., Mohamed, S. H. (Talanta **25** [1978] 519/21).
[15] Schroeder, K. A., Hamm, R. E. (Inorg. Chem. **3** [1964] 391/5).
[16] Irving, H. M. N. H., Al-Jarrah, R. H. (Anal. Chim. Acta **74** [1975] 321/31, 324/5).
[17] Přibil, R., Hornychová, E. (Collection Czech. Chem. Commun. **15** [1950] 456/62, 457).
[18] Vierling, F. (Bull. Soc. Chim. France **1977** 404/9).
[19] Woodruff, W. H., Margerum, D. W. (Inorg. Chem. **13** [1974] 2578/85, 2580).
[20] Pechurova, N. I., Spitsyn, V. I., Bogdanovich, N. G., Martynenko, L. I. (Dokl. Akad. Nauk SSSR **198** [1971] 347/9; Dokl. Chem. Proc. Acad. Sci. USSR **196/201** [1971] 405/7).

[21] Shirakashi, T., Tanaka, N. (Nippon Kagaku Zasshi **91** [1970] 142/8, 144; C.A. **73** [1970] No. 18972).
[22] Stel'mashok, V. E., Poznyak, A. L. (Koord. Khim. **5** [1979] 1019/24; Soviet J. Coord. Chem. **5** [1979] 801/5).
[23] Stein, J., Fackler, J. P., McClune, G. J., Fee, J. A., Chan, L. T. (Inorg. Chem. **18** [1979] 3511/9, 3514).
[24] Tanaka, N., Shirakashi, T. (Nippon Kagaku Zasshi **90** [1969] 57/61; C.A. **70** [1969] No. 81371).
[25] Tanaka, N., Gomi, K., Shirakashi, T. (Nippon Kagaku Kaishi **1975** 444/8; C.A. **83** [1975] No. 21244).
[26] Anbar, M., Meyerstein, D. (Trans. Faraday Soc. **65** [1969] 1812/7).
[27] Brown, A., Higginson, W. C. E. (J. Chem. Soc. Dalton Trans. **1972** 166/70).
[28] Ricard, J., Nari, J. (Bull. Soc. Franc. Physiol. Veg. **9** [1963] 47/52; C.A. **60** [1964] 2878).

[29] Wilkins, R. G., Yelin, R. (J. Am. Chem. Soc. **89** [1967] 5496/7).
[30] Škramovský, S., Podlahová, J. (Collection Czech. Chem. Commun. **27** [1962] 1374/80).

[31] Herculano de Carvalho, A., Gonçalves Calado, J., Legrand de Moura, M. (Rev. Port. Quim. **5** [1963] 15/9).
[32] Shioyama, T. K. (Diss. Washington State Univ. 1978, pp. 5/6; Diss. Abstr. Intern. B **39** [1978] 695).
[33] Malaiyandi, M., Sastri, V. S. (Talanta **30** [1983] 983/5).
[34] Anderegg, G. (IUPAC Chem. Data Ser. No. 14 [1977] 23/4).
[35] Klochkovskii, S. P. (Zh. Analit. Khim. 29 [1974] 929/32; J. Anal. Chem. [USSR] **29** [1974] 793/6).

22.10.4 Isolated Manganese(III) Compounds

K[Mn(edta)]·2H_2O. A solution of the complex salt was prepared by reacting a suspension of freshly produced MnO_2 in aqueous KOH (obtained by reducing powdered $KMnO_4$ with aqueous ethanol) with H_4edta. After excess MnO_2 was filtered off from the deep cherry red solution, an equal volume of cold ethanol was added. The mixture was allowed to stand for 3 to 4 h in a cold, dark place. The separated crystals were washed with 90% ethanol, absolute ethanol, and ether, and air-dried in a cold, dark place. The yield was 45 to 60%, based on H_4edta [1, 2]. A similar method, which is described in detail for the preparation of the trans-1, 2-cyclohexanediaminetetraacetato complex K[Mn(cdta)]·2.5H_2O on p. 74, was used by [13]. K[Mn(edta)]·2H_2O was also prepared by adding powdered manganese(III) acetylacetonate (0.01 mol) to a thoroughly stirred mixture of KOH (0.01 mol) and H_4edta (0.01 mol) in cold water (15 mL). Cold ethanol was added, and the solution was kept in a refrigerator for 2 days. Well-defined crystals separated, which were washed with ethanol and sucked dry. Another crop was obtained by adding more ethanol and cooling again. The yield was 60% [3]. The compound was earlier formulated with 2.5 mol H_2O [1, 2, 13].

K[Mn(edta)]·2H_2O forms flat needles with a hexagonal cross section. It is orthorhombic with the lattice parameters (in Å): a = 6.576(4), b = 23.14(2), c = 10.038(8) [4]; a = 6.579(1), b = 23.161(7), c = 10.054(2) [10]; Z = 4 [4, 10]. Space group: $P2_12_12_1$-D_2^4 (No. 19). The structure was solved up to R = 0.050 [4] and R = 0.065 [10]. The crystals contain discrete octahedral $[Mn(edta)]^-$ anions with the Mn atom bonded to two N atoms and to four O atoms from four different carboxylate groups of the $edta^{4-}$ anion, **Fig. 7**, p. 52 [4, 10]. The octahedron is highly distorted; the N(1)-Mn-N(2) bond angle is 81°. The average Mn-N distance is 2.22(1) Å [10] or 2.20(3) Å [4]. There are two types of Mn-O bonds with average lengths of 1.90(1) and 2.04(1) Å. The potassium ion is octahedrally coordinated by two water oxygen atoms and four carboxylate oxygen atoms from neighboring $[Mn(edta)]^-$ units. The closest Mn-O_{H_2O} distance is 4.1 Å [10]. Other interatomic distances, bond angles, atomic positional [4, 10], and anisotropic thermal parameters [10] are given in the publications.

The measured density of K[Mn(edta)]·2H_2O at room temperature is 1.81 g/cm^3; the calculated density is 1.82 g/cm^3 [4, 10]. The compound is paramagnetic and has an effective magnetic moment of 4.89 μ_B at 21°C [1]. Absorptions in the IR spectrum (cm^{-1}) were assigned as follows (only selected data given): 2950, 2920, ν(CH); 1680, 1590, $\nu_{as}(COO^-)$; 1410, 1340, $\nu_s(COO^-)$; 1135, 1110, 1085, ν(CN) [5]. The band of the coordinated COO^- group at 1680 cm^{-1} was also observed by [8]. Earlier data are reported in [1]; spectra are given in [1, 6 to 8]. The IR spectrum indicates that the Mn-O and Mn-N bonds in the complex are highly covalent [5, 6]. The reflectance spectrum of K[Mn(edta)]·2H_2O shows a maximum at ~500 nm and another in the region below 330 nm; a spectrum is presented in [8].

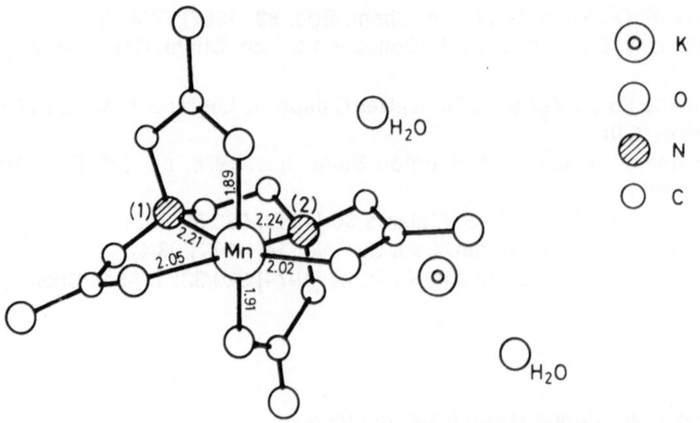

Fig. 7. Molecular structure of $K[Mn(edta)]\cdot 2H_2O$ (H atoms omitted) [10].

The crystalline compound is quite air-stable [4]. This is in contrast to the findings of earlier authors who found that it was stable only at 0°C in the dark [1]. Thermal decomposition starts at ~70°C [9]. At ~80°C, 0.5 mol of water separates; and in the temperature range 100 to 200°C, 12 to 13% of the total weight loss occurs, corresponding to the sum of the remaining water and decomposed ligand. The remaining ligand was completely decomposed at 260 to 300°C. The residue (~38% of the initial sample) was a mixture of K_2CO_3 and a product similar to $MnO\cdot MnCO_3$ [1]. Decomposition products include the unchanged ligand, ethylenediamine-triacetic acid, carbon oxide, carbon dioxide, and formaldehyde. The products indicate that during thermal decomposition Mn^{III} is reduced to Mn^{II}, and the coordinated ligand is oxidized [9].

$K[Mn(edta)]\cdot 2H_2O$ is readily soluble in water, slightly soluble in glacial acetic acid, and insoluble in most other organic solvents, such as absolute ethanol, ether, or acetone [1]. The complex in solution is treated on p. 45.

$Mn^{II}[Mn^{III}(edta)]_2\cdot 10H_2O$. Excess sodium hydroxide was added to an aqueous solution of $MnSO_4$ and H_2O_2. The precipitate, consisting of Mn^{III} oxide hydrates, was suspended in water and H_4edta was added. The resulting red-violet solution was allowed to stand at −5°C for some days, then the pink crystals which had separated were collected and washed with water. The density is 1.95 g/cm³. On standing in the air, $Mn[Mn(edta)]_2\cdot 10H_2O$ effloresces and gradually decomposes with reduction of Mn^{III}. The solubility in water is 0.6 g/100 g solution at 20°C [12].

References:

[1] Yoshino, Y., Ouchi, A., Tsunoda, Y., Kojima, M. (Can. J. Chem. **40** [1962] 775/83, 777/9).
[2] Yoshino, Y., Tsunoda, Y., Ouchi, A. (Bull. Chem. Soc. Japan **34** [1961] 1194/5).
[3] Thankarajan, N., Sen, D. N. (J. Indian Chem. Soc. **46** [1969] 959/60).
[4] Lis, T. (Acta Cryst. B **34** [1978] 1342/4).
[5] Pechurova, N. I., Spitsyn, V. I., Bogdanovich, N. G., Martynenko, L. I. (Dokl. Akad. Nauk SSSR **198** [1971] 347/9; Dokl. Chem. Proc. Acad. Sci. USSR **196/201** [1971] 405/7).
[6] Pechurova, N. I., Martynenko, L. I., Snezhko, N. I. (Vestn. Mosk. Univ. Ser. II Khim. **19** [1978] 65/70; Moscow Univ. Chem. Bull. **33** No. 1 [1978] 48/52).
[7] Yoshino, Y., Ouchi, A., Tsunoda, Y. (Sci. Papers Coll. Gen. Educ. Univ. Tokyo **13** [1963] 27/33, 32).

[8] Takeuchi, T., Ouchi, A. (Nippon Kagaku Kaishi **1974** 1486/8; C.A. **81** [1974] No. 113572).
[9] Shirakashi, T., Tanaka, N. (Nippon Kagaku Kaishi **1974** 1061/7, 1062; C.A. **82** [1975] No. 3650).
[10] Stein, J., Fackler, J. P., McClune, G. J., Fee, J. A., Chan, L. T. (Inorg. Chem. **18** [1979] 3511/9, 3513, 3515).

[11] Giraudi, G., Mentasti, E. (Transition Metal Chem. [Weinheim] **6** [1981] 230/4).
[12] Škramovský, S., Podlahová, J. (Collection Czech. Chem. Commun. **27** [1962] 1374/80, 1376, 1379).
[13] Hamm, R. E., Suwyn, M. A. (Inorg. Chem. **6** [1967] 139/42).

22.10.5 Mixed Ligand Manganese(II) and Manganese(III) Compounds

Complexes in Solution. Mixed ligand Mn^{II} complexes are formed by the reactions listed in Table 2, p. 54. The dinuclear species with $edta^{4-}$ and ethylenediamine(= en) is also formed by the reaction $2Mn(en)_3^{2+} + edta^{4-} \rightleftharpoons Mn_2(en)_4(edta) + 2en$ [2, 3].

Absorptions in the electronic spectra are observed at ~345, ~430, and ~515 nm (from graph) for both the $Mn(en)(edta)^{2-}$ and $Mn_2(en)_4(edta)$ species [2]. The latter reacts with the $edta^{4-}$ ion to give $Mn(en)(edta)^{2-}$ [2] and with complexes of various metals M to give $MnM(en)_4(edta)$ with M = Ni^{II}, Co^{II}, or Cu^{II} [3].

The Mn^{III} complex $Mn(edta)(N_3)^{2-}$ is formed by treating an aqueous solution containing the $Mn(edta)(H_2O)^-$ ion with an excess of azide ions. Rate constants as well as activation and thermodynamic parameters of the slow reaction $Mn(edta)(H_2O)^- + N_3^- \rightleftharpoons Mn(edta)(N_3)^{2-} + H_2O$ are given on p. 46. The equilibrium constant K = 32.1 L/mol was determined spectrophotometrically and K = 32.4 L/mol from the rate constants at 25°C in the pH range 3.5 to 5 and for I = 0.25 M ($NaClO_4$). K varies from 28.3 L/mol at 6.5°C to 33.5 at 30°C. The absorption spectrum of a solution with pH 4.5 has a maximum at 440 nm, which appears as a shoulder on a larger charge-transfer band extending into the visible region from the ultraviolet. Molar extinction coefficients at pH 5 are 1448 at 440 nm and 855 $M^{-1} \cdot cm^{-1}$ at 497 nm. Solutions of the ion are not stable and decompose with formation of $Mn(edta)^{2-}$ and N_2. The following mechanism has been proposed for the decomposition reaction: $Mn^{III}(edta)(N_3)^{2-} \rightarrow Mn^{II}(edta)^{2-} + N_3^0$ (I); $2N_3^0 \rightarrow 3N_2$ (II). The first-order rate constant of step I is 5.99×10^{-4} s^{-1} at the above conditions. The enthalpy and entropy of activation for reaction (I) are $\Delta H^* = 17.1 \pm 0.97$ kcal/mol and $\Delta S^* = -20.6 \pm 1.08$ cal $\cdot$ $mol^{-1} \cdot K^{-1}$, respectively [8].

$[Mn_2^{II}(H_2edta)_2(N_2H_4)_3]$ and **$[Mn_2^{II}(H_2edta)_2(N_2H_4)_3] \cdot 8H_2O$.** The octahydrate was prepared by mixing aqueous $Mn^{II}(H_2edta) \cdot 4H_2O$ (p. 38) with an excess of 25% aqueous $N_2H_4 \cdot H_2O$. After standing for 2 to 3 months at room temperature, white, faintly pink crystals separated. They were washed with ethanol and ether and dried over $CaCl_2$. The hydrate is soluble in water and inorganic acids, but insoluble in organic solvents. Dehydration occurs at 110 to 150°C to give the anhydrous compound, which decomposes to give MnO_2 on heating >315°C [1].

$K_2[Mn^{III}(edta)(N_3)] \cdot H_2O$. The manganese(III) complex was prepared by adding a stoichiometric amount of KN_3 to an aqueous slurry of a 1:1:2 MnO_2–$Mn(NO_3)_2$–H_4edta mixture at −10°C. The mixture was slowly neutralized with KOH and then allowed to stand at −10°C for 5 h. The excess MnO_2 was filtered off, and an equal volume of cold ethanol was added to the solution. The complex precipitated as dark brown crystals, which were washed with absolute ethanol and air-dried. The yield was ~65% based on KN_3 [8].

Table 2
Formation Constants of Mn^{II} Complexes with Ethylenediaminetetraacetic Acid and Other Ligands[a] at 25°C.

No.	reaction	log K	ionic strength	method[b]	Ref.
1	$Mn(edta)^{2-} + en \rightleftharpoons Mn(edta)(en)^{2-}$	0.91	1.5(en·HCl)	sp, pH	[2]
2	$Mn^{2+} + edta^{4-} + en \rightleftharpoons Mn(edta)(en)^{2-}$	14.38	1.5(en·HCl)	sp, pH	[2]
3	$Mn(en)(edta)^{2-} + Mn(en)_3^{2+} \rightleftharpoons Mn_2(en)_4(edta)$	3.62	1.5(en·HCl)	sp, pH	[2]
4	$Mn^{2+} + edta^{4-} + 4\,en \rightleftharpoons Mn_2(en)_4(edta)$	23.67	1.5(en·HCl)	sp, pH	[2]
5	$Mn(edta)^{2-} + C_2H_4NO_2^- \rightleftharpoons Mn(edta)(C_2H_4NO_2)^{3-}$	1.43	—	gl	[6]
6	$Mn(edta)^{2-} + C_6H_8N_3O_2^- \rightleftharpoons Mn(edta)(C_6H_8N_3O_2)^{3-}$	1.78	0.1($NaClO_4$)	gl	[5]
7	$Mn(edta)^{2-} + Mn(C_4H_5NO_4)_2^{2-} \rightleftharpoons Mn_2(C_4H_5NO_4)_2(edta)^{4-}$	2.95	1.5	sp	[9]
8	$2\,Mn^{2+} + edta^{4-} + 2\,C_4H_5NO_4^{2-} \rightleftharpoons Mn_2(C_4H_5NO_4)_2(edta)^{4-}$	23.1	1.5	sp	[9]
9	$Mn(edta)^{2-} + C_6H_6NO_6^{3-} \rightleftharpoons Mn(edta)(C_6H_6NO_6)^{5-}$	2.66	0.5($NaNO_3$)	gl	[7]
10	$Mn(edta)^{2-} + Mn(C_6H_6NO_6)_2^{4-} \rightleftharpoons Mn_2(C_6H_6NO_6)_2(edta)^{6-}$	3.01	1.5	sp	[9]
11	$2\,Mn^{2+} + edta^{4-} + 2\,C_6H_6NO_6^{3-} \rightleftharpoons Mn_2(C_6H_6NO_6)_2(edta)^{6-}$	27.6	1.5	sp	[9]
12	$Mn(edta)^{2-} + Fe(C_6H_6NO_6)_2^{4-} \rightleftharpoons MnFe(C_6H_6NO_6)_2(edta)^{6-}$	2.25	1.5	sp	[10]
13	$Mn^{2+} + Fe^{2+} + edta^{4-} + 2\,C_6H_6NO_6^{3-} \rightleftharpoons MnFe(C_6H_6NO_6)_2(edta)^{6-}$	39.6	1.5	sp	[10]

[a] $C_2H_5NO_2$ = glycine; $C_6H_9N_3O_2$ = histidine; $C_4H_7NO_4$ = iminodiacetic acid; $C_6H_9NO_6$ = nitrilotriacetic acid. – [b] sp = by spectrophotometry; pH = by the pH method, electrode not specified; gl = potentiometrically, with the glass electrode.

The strong IR absorption band in the 2070 to 2040 cm^{-1} region indicates that the azide ion is coordinated. The band was attributed to the antisymmetric stretching vibration. (Noncoordinated N_3^- should absorb in the 2140 cm^{-1} region.) Strong absorptions due to the antisymmetric stretching vibrations of the coordinated COO^- group are observed at 1670, 1640, and 1620 cm^{-1}. If one COO^- group were not coordinated, an absorption would be expected around 1750 cm^{-1}. The presence of a seven-coordinate manganese(III) ion in the complex is indicated by the IR spectrum [8].

References:

[1] Gogorishvili, P. V., Chkoniya, M. E., Akhobadze, D. A., Chkoniya, T. V. (Tr. Gruz. Politekhn. Inst. **1969** No. 1, pp. 1/13, 8; C.A. **72** [1970] No. 139140).
[2] Barkhanova, N. N., Fridman, A. Ya., Dyatlova, N. M. (Zh. Neorgan. Khim. **17** [1972] 2982/8; Russ. J. Inorg. Chem. **17** [1972] 1569/73, 1572).
[3] Barkhanova, N. N., Dyatlova, N. M., Fridman, A. Ya. (Zh. Neorgan. Khim. **18** [1973] 1489/94; Russ. J. Inorg. Chem. **18** [1973] 785/8).
[4] Hague, D. N., Martin, S. R. (J. Chem. Soc. Dalton Trans. **1974** 254/8, 256/7).
[5] Israéli, Y. J., Cecchetti, M. (J. Inorg. Nucl. Chem. **30** [1968] 2709/16, 2715).
[6] Israéli, Y. J. (Bull. Soc. Chim. France **1963** 1979).
[7] Israéli, Y. J. (Nature **201** [1964] 389/90).
[8] Suwyn, M. A., Hamm, R. E. (Inorg. Chem. **6** [1967] 2150/4).
[9] Polyakova, I. Ya., Fridman, A. Ya., Temkina, V. Ya. (Koord. Khim. **11** [1985] 467/71; Soviet J. Coord. Chem. **11** [1985] 264/8).
[10] Polyakova, I. Ya., Fridman, A. Ya., Dyatlova, N. M. (Koord. Khim. **11** [1985] 1468/72).

22.11 With the Tetramethyl Ester of Ethylenediamine-N,N,N',N'-tetraacetic Acid

$(CH_3OOCCH_2)_2NCH_2CH_2N(CH_2COOCH_3)_2$ ($=C_{14}H_{24}N_2O_8$)

$Mn(C_{14}H_{24}N_2O_8)Cl_2$ was prepared by adding a solution of the ligand (5 mmol) in dry methanol dropwise to a stirred solution of $MnCl_2$ (5 mmol), also in dry methanol. The complex crystallized from the solution on standing or after addition of ether.

The IR spectrum of the complex in Nujol shows ν(C=O) absorptions at 1740, 1725, and 1690 cm^{-1}, and absorptions of ν(CN) + ν(C–O) at 1250, 1235, and 1220 cm^{-1}. The absorption at 1740 cm^{-1} was assigned to a free ester C=O group, whereas the absorptions at 1725 and 1690 cm^{-1} are presumably due to axially and radially coordinated ester C=O groups, respectively. In the diffuse reflectance spectrum (visible region), very weak bands appear at 23020, 19000, and 17000 cm^{-1}. Susceptibility measurements yielded a magnetic moment of 6.21 μ_B at 20°C. This value and the reflectance spectral data are consistent with a high-spin complex of approximately octahedral structure. A 10^{-3} M solution of the complex in dry methanol has an electrical conductance of 128 $cm^2 \cdot \Omega^{-1} \cdot mol^{-1}$ at 25°C, corresponding to that of a 1:2 electrolyte. It seems likely that in the solid state, either one or both Cl atoms are coordinated, but in methanol undergo solvolytic displacement to give ionic chloride.

Reference:

Hay, R. W., Nolan, K. B., Shuaib, M. (Transition Metal Chem. [Weinheim] **5** [1980] 230/1).

22.12 With Ethylenediamine-N,N′,N′-triacetic-N-(alkylacetic) Acids

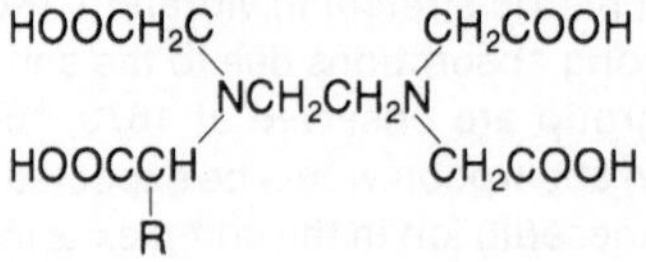

1) $R = C_{10}H_{21}$; $(= C_{20}H_{36}N_2O_8)$
2) $R = C_{14}H_{29}$; $(= C_{24}H_{44}N_2O_8)$

Surface-active chelate complexes were prepared by reacting manganese(II) sulfate with an excess of the tetrasodium salts of the ligands in aqueous solution [1, 2]. The 5 mM solution of the blackish brown complex with ligand 2 has a surface tension of 52.5 dyn/cm and an interfacial tension with kerosene of 9.5 dyn/cm at 21°C [1]. (Data for the complex with ligand 1 [2] are unclear.)

References:

[1] Takeshita, T., Maeda, S. (Yukagaku **19** [1970] 984/93, 989, 991; C. A. **74** [1971] No. 14393).
[2] Nakashima, H., Miyake, M. (Yukagaku **21** [1972] 416/21, 419; C. A. **77** [1972] No. 154290).

22.13 With Ethylenediamine-N,N′-diacetic-N,N′-bis(alkylacetic) Acids

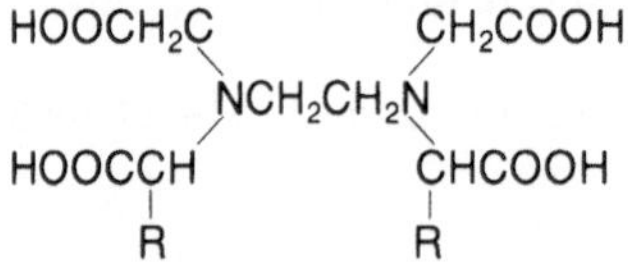

Stability constants of 1:1 manganese(II) complexes with the listed ligands in aqueous solution were determined potentiometrically at 20°C for I = 0.1 M (KNO_3) with the glass electrode (gl), polarographically (pol) by use of the exchange equilibrium with the Cd complex, or potentiometrically (pot) by use of the exchange equilibrium with the Cu complex and with tris(2-aminoethyl)amine (= tren):

R	CH_3	CH_3	CH_2CH_3	$CH_2CH_2CH_3$	$CH(CH_3)_2$
ligand H_4L	$C_{12}H_{20}N_2O_8$	$C_{12}H_{20}N_2O_8$	$C_{14}H_{24}N_2O_8$	$C_{16}H_{28}N_2O_8$	$C_{16}H_{28}N_2O_8$
log K_1	13.30±0.03	13.37±0.11	13.19±0.08	13.23±0.09	9.75±0.03
method	gl	pol	pol	pol	pot
Ref.	[1]	[2]	[3]	[3]	[3]

A surface-active grayish brown complex was prepared by reacting Mn^{II} sulfate with the tetrasodium salt of a ligand with $R = C_{10}H_{21}$ ($H_4L = C_{30}H_{56}N_2O_8$) in excess in aqueous solution of pH 11.6. The 3 mM solution of the complex has a surface tension of 37.3 dyn/cm and an interfacial tension with kerosene of 7.2 dyn/cm at 21°C [4].

References:

[1] Majer, J., Kotouček, M., Dvořáková, E. (Chem. Zvesti **20** [1966] 242/51, 248; C. A. **65** [1966] 3304).
[2] Novák, V., Kotouček, M., Lučanský, J., Majer, J. (Chem. Zvesti **21** [1967] 687/97, 692; C. A. **68** [1968] No. 8763).

[3] Novák, V., Dvořáková, E., Svičeková, M., Majer, J. (Chem. Zvesti **23** [1969] 861/8, 865).
[4] Takeshita, T., Maeda, S. (Yukagaku **19** [1970] 984/93, 990/1; C.A. **74** [1971] No. 14393).

22.14   With Ethylenediamine-N,N'-diacetic-N,N'-dipropanoic Acid

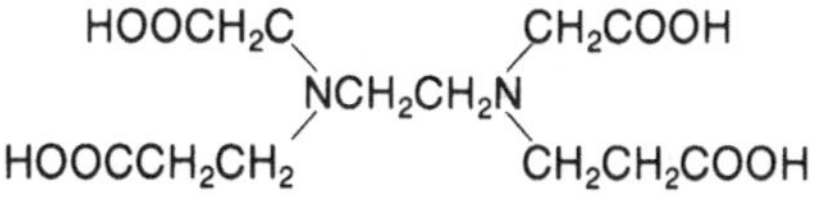

$$\begin{array}{ll} HOOCH_2C & CH_2COOH \\ & NCH_2CH_2N \qquad\qquad (=C_{12}H_{20}N_2O_8=H_4L) \\ HOOCCH_2CH_2 & CH_2CH_2COOH \end{array}$$

Complexes in Solution. The stability constant of the **$Mn^{II}(C_{12}H_{16}N_2O_8)^{2-}$** ion, log $K_1 = 7.41$, was determined potentiometrically (glass electrode) at 25°C in aqueous solution of $I = 0.1$ M KCl.

The stability constant of the manganese(III) complex ion, **$Mn^{III}(C_{12}H_{16}N_2O_8)^-$**, log $K_1 = 21.0$, was determined at 25°C for an aqueous solution of $I = 0.2$ M ($NaClO_4$) by measuring the emf of the $Mn^{III}(C_{12}H_{16}N_2O_8)^-/Mn^{II}(C_{12}H_{16}N_2O_8)^{2-}$ redox couple. The standard potential of the reaction $Mn^{III}(C_{12}H_{16}N_2O_8)^- + e^- \rightleftharpoons Mn^{II}(C_{12}H_{16}N_2O_8)^{2-}$ is $E° = 0.69$ V vs. NHE at 25°C and $I = 0.2$ M ($NaClO_4$).

The IR spectrum of the Mn^{III} complex in D_2O, showing the absorptions of the antisymmetric vibrations of the coordinated carboxylate group in the 1680 to 1600 cm^{-1} region, and the fact that the complex reacts with hydroxide and azide ions, indicate a seven-coordinate Mn atom in the complex with one H_2O molecule occupying the seventh coordination position. The absorption spectrum in the visible region of the red complex solution shows a major band at 490 nm (log $\varepsilon = 2.59$) and a shoulder at 463 nm of much smaller intensity, both of which are associated with d-d transitions.

The Mn^{III} complex ion decomposes slowly in aqueous solution. The rate of self-decomposition is pH-independent in the range of approximately pH 3 to 5. In this range, the first-order rate constant is $k = 1.47 \times 10^{-6}$ s^{-1} at room temperature and $I = 0.2$ M ($NaClO_4$). The activation parameters are $\Delta H^* = 3.58 \pm 1.07$ kcal/mol and $\Delta S^* = -9.43 \pm 3.63$ cal·mol^{-1}·K^{-1}. At pH values higher than 5, a steady increase in the rate of decomposition is observed. In the presence of free ligand, the rate increases linearly with the concentrations of the free ligand in all its forms. A two-part rate expression previously established for the Mn^{III} complex with cyclohexane-diaminetetraacetic acid (p. 70) is postulated where one part is first order in complex and the second is first order in complex and free ligand concentration. In basic solution, the equilibrium $Mn^{III}(C_{12}H_{16}N_2O_8)(H_2O)^- \rightleftharpoons Mn^{III}(C_{12}H_{16}N_2O_8)(OH)^{2-} + H^+$ is displaced to the right side. The equilibrium constant of this reaction, log $K = -6.22$, was determined spectrophotometrically. The **$Mn^{III}(C_{12}H_{16}O_8)(OH)^{2-}$** ion is the dominating species in the 5.8 to 6.8 pH range. Above pH 8.1, the ion decomposes readily to form $Mn(OH)_2$ and MnO_2. Solutions of the $Mn^{III}(C_{12}H_{16}N_2O_8)$-$(OH)^{2-}$ ion are yellow and show an absorption maximum at 480 nm. **$Mn^{III}(C_{12}H_{16}N_2O_8)(N_3)^{2-}$** forms on reaction of $Mn^{III}(C_{12}H_{16}N_2O_8)(H_2O)^-$ with azide ions in a 1:10 mole ratio. Solutions of this ion are yellow and show an absorption maximum at 480 nm as well. Their IR spectra show the bands of the four coordinated carboxylate groups in the 1680 to 1600 cm^{-1} region.

$K[Mn^{III}(C_{12}H_{16}N_2O_8)(H_2O)]\cdot 0.5H_2O$ was prepared by the procedure outlined for the Mn^{III} complex with ethylenediamine-N,N'-disuccinic acid (p. 59). The IR spectrum of the complex in KBr or Nujol shows the $\nu_{as}(COO^-)$ bands of the coordinated carboxylate groups with a complicated fine structure suggesting a seven-coordinate manganese atom.

Reference:

Shioyama, T. K. (Diss. Washington State Univ. 1978, pp. 1/92, 14/5, 28/45, 50/9; Diss. Abstr. Intern. B **59** [1978] 695).

22.15 With Ethylenediaminetetra-N,N,N',N'-propanoic Acid and Its Isomer
$R_2NCH_2CH_2NR_2$ $(= C_{14}H_{24}N_2O_8)$

ligand 1 with $R = CH_2CH_2COOH$ ligand 2 with $R = CH(CH_3)COOH$

Stability constants of the isomeric complex ions $Mn(C_{14}H_{20}N_2O_8)^{2-}$ in aqueous solution were determined potentiometrically (glass electrode): $\log K_1 = 4.7$ at 30°C and $I = 0.1\,M$ (KCl) for the complex with ligand 1 [1] and $\log K_1 = 6.06$ at 25°C and $I = 0.1\,M$ for the complex with ligand 2 [2].

References:

[1] Courtney, R. C., Chaberek, S., Martell, A. E. (J. Am. Chem. Soc. **75** [1953] 4814/8).
[2] Goetz, C. A., Debbrecht, F. J. (Iowa State Coll. J. Sci. **33** [1959] 267/77, 275).

22.16 With Ethylenediamine-N,N'-dimalonic, -disuccinic, or -diglutaric Acid

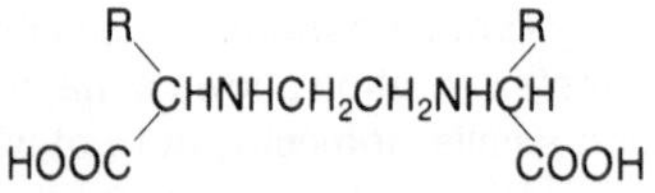

$$\underset{HOOC}{\overset{R}{\diagdown}}CHNHCH_2CH_2NH\underset{COOH}{\overset{R}{\diagup}}CH \qquad (= H_4L)$$

Complexes in Solution. Stability constants of the Mn^{II} complexes **MnL^{2-}** in aqueous solution with the deprotonated ligands for $I = 0.1\,M$ (KNO$_3$) [1 to 3] or 0.1 M (KCl) [4] were determined potentiometrically with an Hg electrode (Hg) or with a glass electrode (gl) and/or polarographically by use of the exchange equilibrium with the Cd complex (pol):

No.	R	H$_4$L	t in °C	method	log K$_1$	pH range of stability	Ref.
1	COOH	$C_8H_{12}N_2O_8$	25	Hg, pol	8.45	4 to 8	[1]
2	CH$_2$COOH	$C_{10}H_{16}N_2O_8$	20	gl	8.95	4 to 8	[2]
			25	gl	7.76		[4]
			25	gl, pol	8.50		[1]
			30	gl	5.11		[3]
3	CH$_2$CH$_2$COOH	$C_{12}H_{20}N_2O_8$	25	gl, pol	6.74	4 to 8	[1]
			30	gl	5.18		[3]

The complex **Mn^{III}(C$_{10}$H$_{12}$N$_2$O$_8$)$^-$** with ligand 2 exists in the pH range 3 to 6.8. Above pH 6.8, decomposition takes place to form $Mn(OH)_2$ and MnO_2. The electronic absorption spectrum of the complex in aqueous solution shows peaks at 465 nm ($\log \varepsilon = 2.59$) and 472 nm (2.60). The IR spectrum of the complex in D_2O shows the antisymmetric stretching vibrations of four coordinated carboxylate groups in the 1700 to 1600 cm^{-1} region, indicating hexadentate coordination of the ligand. This together with the fact that the complex does not react with OH^-, N_3^-, or I^- ions suggests an octahedral structure of the complex molecule (with no coordinated H_2O molecule).

The rate of self-decomposition of the complex in the pH range 3 to 5 is relatively low and practically pH-independent in this range. At higher pH values the rate steadily increases. The first-order decomposition rate constant has been determined spectrophotometrically as $k = 3.55 \times 10^{-8}\ s^{-1}$ at room temperature, pH 3 to 5, and I = 0.2 M($NaClO_4$). The activation parameters of the reaction are $\Delta H^* = 4.68 \pm 0.27$ kcal/mol and $\Delta S^* = 15.71 \pm 1.79\ cal \cdot mol^{-1} \cdot K^{-1}$. The greater stability of the $Mn^{III}(C_{10}H_{12}N_2O_8)^-$ ion compared to analogous complexes with ethylenediaminetetraacetic acid or similar ligands is attributed to its octahedral structure and the better "wrapping capability" of the ligand, which stabilizes the Mn^{3+} ion to a greater extent [4].

$K[Mn^{III}(C_{10}H_{12}N_2O_8)] \cdot nH_2O$. A compound with n = 1.5 was prepared by adding aqueous $KMnO_4$ (1 mmol) to a mixture of $MnCO_3$ (4 mmol), ligand 2 (5 mmol), and solid K_2CO_3 (4 mmol) at a rate of one drop every 15 s. After effervescence had ceased, the solution was stirred an additional 15 min, and then filtered and evaporated. The oil obtained was transformed into a glassy solid at 30 Torr over P_2O_5. Treatment with absolute ethanol gave a powder, which was dried in vacuum. The yield was 20 to 25% based on $MnCO_3$. When the complex was prepared using the (−)-isomer of ligand 2, it showed maxima in the optical rotary dispersion curve at 418 and 470 nm with rotations of −1.14 or 2.07 $deg \cdot M^{-1} \cdot cm^{-1}$, respectively. The IR spectrum of $K[Mn^{III}(C_{10}H_{12}N_2O_8)] \cdot 2H_2O$ in KBr shows bands of the antisymmetric vibrations of four coordinated carboxylate groups at 1650, 1638, and 1618 cm^{-1}. This suggests that the ligand is hexadentately coordinated involving the two N atoms and four carboxylate O atoms. (Analysis and IR spectrum is given for $K[Mn(C_{10}H_{12}N_2O_8)] \cdot 2H_2O$ [4].)

References:

[1] Samsonov, A. P., Gorelov, I. P. (Zh. Neorgan. Khim. **19** [1974] 2115/7; Russ. J. Inorg. Chem. **19** [1974] 1159/60).

[2] Majer, J., Jokl, V., Dvořáková, E., Jurčová, M. (Chem. Zvesti **22** [1968] 415/23, 421; C.A. **69** [1968] No. 70597).

[3] Tak, S. G., Sunar, O. P., Trivedi, C. P. (Indian J. Chem. **9** [1971] 1394/5).

[4] Shioyama, T. K. (Diss. Washington State Univ. 1978, pp. 1/92, 14/5, 24/5, 28/45, 50/9; Diss. Abstr. Intern. B **39** [1978] 695).

22.17 With Monoalkyl- or Monophenylethylenediamine-N,N,N',N'-tetraacetic Acids

$(HOOCCH_2)_2NCHRCH_2N(CH_2COOH)_2$ (= H_4L)

Complexes in Solution. Stability constants of manganese(II) complexes **MnL^{2-}** in aqueous solution were determined polarographically by use of the exchange equilibria with the Zn [1] and Cd complexes [2 to 6, 12]:

No.	R	H_4L	t in °C	I in mol/L	log K_1	Ref.
1	CH_3	$C_{11}H_{18}N_2O_8$	25	0.2(KNO_3)	14.85	[1]
			20	0.10(KNO_3)	15.28	[2]
2	C_2H_5	$C_{12}H_{20}N_2O_8$	20	0.10(KNO_3)	15.66 ± 0.09	[3]
3	$(CH_3)_2CH$	$C_{13}H_{22}N_2O_8$	20	0.10(KNO_3)	15.47 ± 0.12	[4]

No.	R	H_4L	t in °C	I in mol/L	log K_1	Ref.
4	$(CH_3)_2CHCH_2$	$C_{14}H_{24}N_2O_8$	20	0.10(KNO_3)	15.44±0.11	[4]
5	C_6H_{13}	$C_{16}H_{28}N_2O_8$	20	0.10(KNO_3)	15.51±0.10	[5]
6	C_6H_5	$C_{16}H_{20}N_2O_8$	20	0.10(KNO_3)	14.58±0.12	[6, 12]

The change in enthalpy for the complexation reaction of Mn^{2+} ions with the racemic form of ligand 1 is $\Delta H = -5.25$ kcal/mol at 25°C for the ionic strength 0.2 M(KNO_3) determined by the direct calorimetric method [7].

For the $(+)_{589}Mn^{II}\{(-)C_{11}H_{14}N_2O_8\}^{2-}$ ion, the complex with the optically active (−)-form of ligand 1, the molecular rotation in degree·mL·dm^{-1}·mol^{-1} of a 0.01 M solution of pH 4.80 was measured. It decreases from 730 at 365 nm to 199 at 589 nm at room temperature. The optical rotation is a linear function of complex concentration and also depends on the pH of the solution. The complex can be used for the spectropolarimetric titration of Mn^{II} with optimum conditions at 365 nm and pH 5.0 [8]. The time for half-exchange of the ligand in the $Mn^{II}\{(+)C_{11}H_{14}N_2O_8\}^{2-}$ ion by the (−)-form of free ligand 1 in buffered aqueous solution is 5 min at pH 6.30 or <1 min at pH 3.00, measured polarimetrically at 20°C [9].

The stability constant of the manganese(III) complex $\mathbf{Mn^{III}(C_{11}H_{14}N_2O_8)^-}$, with ligand 1, in aqueous solution of $I = 0.2$ M($NaClO_4$), log K = 26.5 at 25°C, was determined by measuring the emf of the $Mn^{II}(C_{11}H_{14}N_2O_8)^{2-}/Mn^{III}(C_{11}H_{14}N_2O_8)^-$ redox couple. The standard potential of the reaction $Mn^{III}(C_{11}H_{14}N_2O_8)^- + e^- \rightleftharpoons Mn^{II}(C_{11}H_{14}N_2O_8)^{2-}$ is $E° = 0.80$ V vs. NHE at 25°C and $I = 0.2$ M ($NaClO_4$). A seven-coordinate Mn^{III} atom in the complex molecule with one coordinated water molecule was suggested on account of the IR spectrum and substitution reactions with OH^- and N_3^- ions. The electronic spectrum shows maxima at 485 (log ε = 2.30) and 476 nm (2.27) [11].

The ion decomposes slowly in aqueous solution. The rate of self-decomposition is independent of pH within approximately pH 3 to 5. The first-order rate constant is $k = 3.91 \times 10^{-5}$ s^{-1} and $I = 0.2$ M ($NaClO_4$) in this range. The activation parameters are $\Delta H^* = 3.01 \pm 0.10$ kcal/mol and $\Delta S^* = -9.87 \pm 0.36$ cal·mol^{-1}·K^{-1}. At pH values higher than 5, a steady increase in the rate of decomposition is observed. For the decomposition in the presence of free ligand, the rate increases linearly with the concentration of free ligand in all its forms [11].

$\mathbf{Mn^{II}\{(+)C_{11}H_{16}N_2O_8\}}$. The solid compound was prepared by reaction of Mn^{II} acetate with the (+) form of ligand 1. It was purified by recrystallization [9].

$\mathbf{[Mn^{II}(C_{11}H_{16}N_2O_8)(H_2O)_2] \cdot 2H_2O}$ was prepared by treating an aqueous suspension of freshly prepared $MnCO_3$ with solid ligand 1 and evaporating the resulting solution to a fifth of its original volume. The white precipitate was washed with water and ether and dried over $CaCl_2$ [10]. The magnetic moment is $\mu_{eff} = 5.91$ μ_B at 299 K. The IR spectrum of the complex in Nujol shows the ν(OH) bands of water in the 3450 to 3100 cm^{-1} region. The ν(CO) band of the noncoordinated carboxyl groups appears at 1675 cm^{-1}, and the antisymmetric and symmetric stretching vibration bands of the coordinated carboxylate groups are observed at 1570 or 1410 cm^{-1}. The ν(CN) band at 1105 cm^{-1} indicates the coordination of the N atom. A band at 905 cm^{-1} is attributed to the coordinated water [10].

On heating, the separation of the water of crystallization is observed in the 60 to 100°C temperature range. In the 100 to 200°C range (DTA peak at ~145°C) the two coordinated water molecules are split off in an endothermic reaction, as shown by a thermoanalytic study. Above ~200°C, decarboxylation of the ligand takes place in two steps. The base titration curve of the complex in aqueous solution shows an inflection point at a base: complex ratio of two. The properties of the complex indicate that the ligand is tetradentate in this complex (coordination of two N and two carboxylate O atoms) with water O atoms occupying the remaining octahedral positions [10].

The manganese(III) compound **$K[Mn^{III}(C_{11}H_{14}N_2O_8)(H_2O)]\cdot 1.5H_2O$** was prepared by adding solid ligand 1 to a suspension of freshly prepared MnO_2 in aqueous KOH solution at 0°C. When the reaction was complete, excess MnO_2 was filtered off. To the dark red filtrate, an equal volume of 95% ethanol was added, and the solution placed in a dry ice-acetone bath until frozen. The frozen solution was allowed to come up to 10°C slowly. Upon filtration, a fine pink powder remained, which was washed with absolute ethanol, cooled to 0°C, and dried at 30 Torr. The yield was ~40% based on $KMnO_4$ [11].

The IR spectrum of the complex in KBr or Nujol shows the $\nu_{as}(COO^-)$ bands of four coordinated carboxylate groups in the 1700 to 1600 cm^{-1} region, suggesting hexadentate coordination of the ligand involving two N and four O atoms [11].

References:

[1] Ogino, H. (Bull. Chem. Soc. Japan **38** [1965] 771/7, 776).
[2] Novák, V., Lučanský, J., Svičeková, M., Majer, J. (Chem. Zvesti **32** [1978] 19/26, 24).
[3] Novák, V., Lučanský, J., Majer, J. (Chem. Zvesti **22** [1968] 721/32, 727).
[4] Novák, V., Lučanský, J., Majer, J. (Chem. Zvesti **22** [1968] 733/42, 739).
[5] Majer, J., Butvin, P., Novák, V., Svičeková, M., Füleová, E., Valášková, J., Novák, J. (Chem. Zvesti **33** [1979] 742/8, 744).
[6] Novák, V., Dvořáková, E., Svičeková, M., Majer, J. (Chem. Zvesti **23** [1969] 330/5, 333).
[7] Suarez Cardeso, J. M., Gonzales Garcia, S. (Anales Quim. [Madrid] B **71** [1975] 625/6).
[8] Palma, R. J., Reinbold, P. E., Pearson, K. H. (Anal. Chem. **42** [1970] 47/51).
[9] Bosnich, B., Dwyer, F. P., Sargeson, A. M. (Nature **186** [1960] 966).
[10] Suarez Cardeso, J. M., Gonzales Garcia, S. (Anales Quim. [Madrid] B **69** [1973] 491/7, 492, 494).

[11] Shioyama, T. K. (Diss. Washington State Univ. 1978, pp. 1/92, 16, 24, 41, 45, 51; Diss. Abstr. Intern. B **39** [1978] 695).
[12] Lučanský, J., Novák, V., Svičeková, M., et al. (Proc. 9th Conf. Coord. Chem., Bratislava, CSSR, 1983, pp. 265/9; C.A. **99** [1983] No. 94541).

22.18 With Dimethyl- or Diphenylethylenediamine-N,N,N′,N′-tetraacetic Acids

Stability constants of manganese(II) complexes **MnL^{2-}** with the ligands listed below in aqueous solution of ionic strength 0.1 M(KNO_3) at 20°C determined potentiometrically with an unspecified electrode (pot), potentiometrically with the Hg electrode (Hg), polarographically with the use of exchange equilibria systems (pol):

No.	ligand H_4L	formula	method	$\log K_1$	Ref.
1	$(HOOCCH_2)_2NC(CH_3)_2CH_2N(CH_2COOH)_2$	$C_{12}H_{20}N_2O_8$	pol	15.31	[1]
2	$(HOOCCH_2)_2NCH(CH_3)CH(CH_3)N(CH_2COOH)_2$	$C_{12}H_{20}N_2O_8$	Hg, pot	16.72	[2,3]
	(racemic form)		pol	16.30	[4]

No.	ligand H_4L	formula	method	log K_1	Ref.
3	$(HOOCCH_2)_2NCH(CH_3)CH(CH_3)N(CH_2COOH)_2$	$C_{12}H_{20}N_2O_8$	Hg, pot	14.10	[2, 3]
	(meso form)		pol	14.18	[4]
4	$(HOOCCH_2)_2NCH(C_6H_5)CH(C_6H_5)N(CH_2COOH)_2$ (racemic form)	$C_{22}H_{24}N_2O_8$	pot, pol	15.10	[5]

Protonation constants of the complexes $Mn(C_{12}H_{16}N_2O_8)^{2-}$, $K_H = [MnHL^-]/[MnL^{2-}][H^+]$, determined potentiometrically with the Hg electrode at 20°C for I = 0.1 M(KNO_3) are: log K_H = 2.68 for ligand 2 and log K_H = 3.46 for ligand 3 [1]. The *meso*-form of ligand 4 was used as a chelating agent in the analysis of multi-component mixtures for micro-amounts of Mn^{II} (or other di- and trivalent metals) by polarography [6].

References:

[1] Novák, V., Lučanský, J., Svičeková, M., Majer, J. (Chem. Zvesti **32** [1978] 19/26, 24).
[2] Irving, H. M. N. H., Sharpe, K. (J. Inorg. Nucl. Chem. **33** [1971] 203/15, 204, 209).
[3] Dvořáková, E., Majer, J. (Chem. Zvesti **20** [1966] 233/41, 239).
[4] Majer, J., Novák, V., Svičeková, M. (Chem. Zvesti **18** [1964] 481/92, 490).
[5] Lučanský, J., Novák, V., Svičeková, M., et al. (Proc. 9th Conf. Coord. Chem., Bratislava, CSSR, 1983, pp. 265/9; C.A. **99** [1983] No. 94541).
[6] Lučanský, J., Pikulikova, A. (Farm. Obzor **55** [1986] 3/10; C.A. **104** [1986] No. 198995).

22.19 With Carboxyethylenediamine-N,N,N′,N′-tetraacetic Acid

$(HOOCCH_2)_2NCH(COOH)CH_2N(CH_2COOH)_2$ (= $C_{11}H_{16}N_2O_{10}$)

The formation of a 1:1 Mn^{II} complex with the ligand in aqueous solution was revealed by conductometric measurements (continuous variation method). The stability constant ot the complex is log K = 11.34 at 20°C for I = 0.1 M(KNO_3), determined by the pH method. The optimum pH of formation is 6 to 8, Gonzales Garcia, S., Sanchez Santos, F. J., Morales Ayala, M. F. (Anales Quim. [Madrid] B **78** [1982] 22/6).

22.20 With Trimethylenediamine-N,N,N′,N′-tetraacetic Acid or Related Compounds

$(HOOCCH_2)_2N{-}CH(R)CH(R')CH(R'')\text{-}N(CH_2COOH)_2$

ligand 1	R = R′ = R″ = H;	(= $C_{11}H_{18}N_2O_8$ = H_4L)
ligand 2	R = R″ = CH_3; R′ = H;	(= $C_{13}H_{22}N_2O_8$ = H_4L)
ligand 3	R = R″ = H; R′ = OH;	(= $C_{11}H_{18}N_2O_9$ = H_4L)

22.20.1 Manganese(II) Complexes in Solution

The stability constant of the **$Mn(C_{11}H_{14}N_2O_8)^{2-}$ ion,** with ligand 1, log K_1 = 9.99 at 20°C for I = 0.1 M (KNO_3), was determined by pH titration (glass electrode) [1]; log $K_1 \leqq$ 10.8 at 25°C for I = 0.2 M (KNO_3) was determined polarographically by use of the exchange equilibrium with the Co complex [2]. Thermodynamic data of formation from Mn^{2+} and $(C_{11}H_{14}N_2O_8)^{4-}$, $\Delta H_1 = -0.72$ kcal/mol and $\Delta S_1 = 52.9$ cal·mol⁻¹·K⁻¹ at 20°C for I = 0.1 M (KNO_3), were determined calorimetrically [3]. Another value for ΔH_1, 2.1 kcal/mol, under the same conditions was calculated from literature data [4].

The stability constant of **$Mn(C_{11}H_{15}N_2O_8)^-$** formed by the reaction of Mn^{2+} with the anion HL^{3-} of ligand 1, log K = 4.82, was determined at 20°C for I = 0.1 M (KNO_3) [1]. The protonation constant $K_H = [Mn(C_{11}H_{15}N_2O_8)^-]/[Mn(C_{11}H_{14}N_2O_8)^{2-}]\cdot[H^+]$, log $K_H = 5.3$ at 20°C for I = 0.1 M, was calculated from literature data [4].

The stability constant of **$Mn(C_{13}H_{18}N_2O_8)^{2-}$** with the racemic form of ligand 2, log $K_1 = 10.97$ at 20°C for I = 0.1 M (KNO_3), was found by pH titration (glass electrode) [5].

Stability constants of **$Mn(C_{11}H_{14}N_2O_9)^{2-}$** with ligand 3 in aqueous solution were determined by emf measurements with a mercury electrode (Hg), with a hydrogen electrode (H), by pH potentiometry with a glass electrode (gl), by electrophoresis (el), or by polarography (pol):

t in °C	20	20	20	20	20	25
I in mol/L	0.1(KNO_3)	0.1(KNO_3)	0.1(KNO_3)	0.1(KCl)	0.1(KCl)	0.1(KNO_3)
log K_1	8.96	8.98	9	8.90	8.20	9.06 ± 0.04
method and Ref. ...	Hg [6]	gl [7]	el [8, 9]	H [10]	pol [11]	gl [12]

The complex is formed at pH 4.4 [11]. The fact that the stability constant is close to that for the complex with ligand 1 suggests that the hydroxy group is not involved in chelate formation [11, 12].

The protonation constant $K_H = [Mn(C_{11}H_{15}N_2O_9)^-]/[Mn(C_{11}H_{14}N_2O_9)^{2-}]\cdot[H^+]$ was determined by pH-potentiometry (glass electrode): log $K_H = 5.14 \pm 0.10$ at 25°C for I = 0.1 M (KNO_3) [12].

References:

[1] L'Eplattenier, F., Anderegg, G. (Helv. Chim. Acta **47** [1964] 1792/800, 1794).
[2] Ogino, H. (Bull. Chem. Soc. Japan **38** [1965] 771/7, 776).
[3] Anderegg, G. (Helv. Chim. Acta **47** [1964] 1801/15, 1808).
[4] Martell, A. E., Smith, R. M. (Critical Stability Constants, Vol. 1, Amino Acids, Plenum, New York 1974, p. 245).
[5] Novák, V., Svičeková, M., Dvořáková, E., Valášková, I., Majer, J. (Chem. Zvesti **35** [1981] 481/9, 484, 486).
[6] Dyatlova, N. M., Seliverstova, I. A., Dobrynina, N. A. (Tr. Vses. Nauchn. Issled. Inst. Khim. Reaktivov Osobo Chist. Khim. Veshchestv No. 28 [1966] 270/6, 275; C.A. **67** [1967] No. 15448).
[7] Majer, J., Dvořáková, E., Nagyová, M. (Chem. Zvesti **20** [1966] 313/20, 319).
[8] Jokl, V., Majer, J. (Chem. Zvesti **19** [1965] 249/58, 254).
[9] Jokl, V., Majer, J. (Acta Fac. Pharm. Bohemoslov. **11** [1965] 55/110, 91; C.A. **64** [1966] 14024).
[10] Podgornaya, I. V., Ivakin, A. A., Klyachina, K. N. (Zh. Obshch. Khim. **36** [1966] 2052/4; J. Gen. Chem. [USSR] **36** [1966] 2044/5).
[11] Dyatlova, N. M., Seliverstova, I. A., Yashunskii, V. G., Samoilova, O. I. (Zh. Obshch. Khim. **34** [1964] 4003/7; J. Gen. Chem. [USSR] **34** [1964] 4061/4).
[12] Thompson, L. C., Kundra, S. K. (J. Inorg. Nucl. Chem. **28** [1966] 2945/50).

22.20.2 Isolated Manganese(III) Complex

$K[Mn(C_{11}H_{14}N_2O_8)(H_2O)]\cdot 2.5H_2O$(?). (Analysis given for $K[Mn(C_{11}H_{14}N_2O_8)(H_2O)]$.) This compound with trimethylenediaminetetraacetic acid was prepared by the procedure outlined for the complex with the isomeric ligand methylethylenediaminetetraacetic acid, $K[Mn(C_{11}H_{14}N_2O_8)(H_2O)]\cdot 1.5H_2O$, on p. 61. The IR spectrum of the complex in KBr shows several absorptions due to the antisymmetric stretching vibrations of nonequivalent coordinated carboxylate groups. A seven-coordinate Mn atom surrounded by the hexadentate ligand and one water molecule is assumed.

The $Mn(C_{11}H_{14}N_2O_8)^-$ ion decomposes in a first-order reaction in aqueous solution. The first-order rate constant determined spectrophotometrically is $k=2.10\times10^{-4}\ s^{-1}$ for $I=0.2$ M ($NaClO_4$) in the 3 to 5 pH range. At pH > 5, the rate steadily increases. For the decomposition in the presence of free ligand, a linear increase of the rate with the free ligand concentration in all its forms was observed.

Reference:

Shioyama, T. K. (Diss. Washington State Univ. 1978, pp. 1/92, 16, 32/6, 50/5; Diss. Abstr. Intern. B **39** [1978] 695).

22.21 With N, N, N′, N′-Tetraacetic Acids Derived from Higher Aliphatic Diamines $(HOOCCH_2)_2N–R–N(CH_2COOH)_2$

Manganese(II) Complexes in Solution. The stability constants K_1 for MnL^{2-}, $K^{Mn}_{MnHL} = [MnHL^-]/[Mn^{2+}]\cdot[HL^{3-}]$, $K_H=[MnHL^-]/[MnL^{2-}][H^+]$, and $K_{Mn_2L}=[Mn_2L]/[Mn^{2+}][MnL^{2-}]$ of Mn^{II} complexes with the ligands 1 to 6 (= H_4L) in aqueous solution at 20°C for I = 0.1 M (KNO_3) [1, 2] or 0.1 M (KCl) [3, 4] were determined potentiometrically (glass electrode):

No.	R	ligand formula	log K_1	log K^{Mn}_{MnHL}	log K_H	log $K^{MnL}_{Mn_2L}$	Ref.
1	$(CH_2)_4$	$C_{12}H_{20}N_2O_8$	9.53	5.44	6.6	1.82	[1, 2]
2	$CH_2CH=CHCH_2$	$C_{12}H_{18}N_2O_8$	6.94	4.65	–	3.9	[3]
3	$CH_2C\equiv CCH_2$	$C_{12}H_{16}N_2O_8$	5.65	4.49	–	4.2	[4]
4	$(CH_2)_5$	$C_{13}H_{22}N_2O_8$	8.7	5.6	7.6	–	[2]
5	$(CH_2)_6$	$C_{14}H_{24}N_2O_8$	9.03	5.69	7.5	–	[2]
6	$(CH_2)_8$	$C_{16}H_{28}N_2O_8$	9.0	5.7	7.5	–	[2]

Thermodynamic quantities of formation for complexes MnL^{2-} from Mn^{2+} and L^{4-} were determined calorimetrically, ΔH_1 in kcal/mol and ΔS_1 in cal·mol^{-1}·K^{-1}, respectively, at 20°C, I = 0.1 M (KNO_3): 3.41 and 55.2 for ligand 1; 0.9 and 43 for ligand 4; 0.87 and 44.2 for ligand 5; 0.5 and 42.8 for ligand 6 [2]. The relatively low stabilities of the 1:1 complexes with ligands 2 and 3 and the relatively high stabilities of 2:1 complexes Mn_2L are accounted for by the rigid structure of the ligands, which facilitates the independent coordination of each iminodiacetate group to the Mn atoms [3, 4]. The complex with the deprotonated ligand 1 is oxidized by sodium bismuthate at pH 4 to give the Mn^{III} complex [5].

Manganese(III) Complex in Solution. Mn^{3+} ions form a violet 1:1 complex with the deprotonated ligand 1 in aqueous solution of pH 3.5 to 4.5. The complex was prepared as described above [5]. In the electronic absorption spectrum, the complex shows a maximum at 500 nm with $\varepsilon=83$ L·mol^{-1}·cm^{-1} [5, 6]. The complex is unstable because the coordinated ligand is readily oxidized by the Mn^{3+} ion. It is assumed that the Mn atom is seven-coordinate and has one water molecule coordinated in addition to the hexadentate ligand [5]. Upon addition of CN^- ions, the absorption maximum is shifted to 490 nm ($\varepsilon = 83\ L \cdot mol^{-1} \cdot cm^{-1}$) indicating the formation of a mixed ligand complex [6].

References:

[1] L'Eplattenier, F., Anderegg, G. (Helv. Chim. Acta **47** [1964] 1792/800, 1795).

[2] Anderegg, G. (Helv. Chim. Acta **47** [1964] 1801/15, 1805, 1808).

[3] Tikhonova, L. I., Tkacheva, G. I. (Zh. Neorgan. Khim. **21** [1976] 3264/9; Russ. J. Inorg. Chem. **21** [1976] 1799/802).

[4] Tikhonova, L. I., Samoilova, O. I., Yashunskii, V. G. (Zh. Neorgan. Khim. **24** [1979] 1237/42; Russ. J. Inorg. Chem. **24** [1979] 688/91).

[5] Vicente-Pérez, S., Cabrera-Martin, A., Gomis-Medina, F. (Quim. Anal. **31** [1977] 57/61).

[6] Vicente-Pérez, S., Cabrera-Martin, A., Gomis-Medina, F. (Quim. Anal. **31** [1977] 51/5, 52; C.A. **88** [1978] No. 96840).

22.22 With N,N,N',N'-Tetraacetic Acids Derived from Diaminoethers or a Diaminothioether

1) $(HOOCCH_2)_2NCH_2CH_2$–O–$CH_2CH_2N(CH_2COOH)_2$ ($=C_{12}H_{20}N_2O_9=H_4L$)
2) $(HOOCCH_2)_2NCH_2CH_2$–O–CH_2CH_2–O–$CH_2CH_2N(CH_2COOH)_2$ ($=C_{14}H_{24}N_2O_{10}=H_4L$)
3) $(HOOCCH_2)_2NCH_2CH_2$–S–$CH_2CH_2N(CH_2COOH)_2$ ($=C_{12}H_{20}N_2O_8S=H_4L$)

Manganese(II) Complexes in Solution. Stability constants for MnL^{2-} and $MnHL^-$ ions and protonation constants for MnL^{2-} ions ($K_H=[MnHL^-]/[MnL^{2-}][H^+]$) in aqueous solution of I = 0.1 M (KNO_3) determined potentiometrically with a glass electrode (gl) or with an Hg electrode (Hg), as well as thermodynamic data of formation for MnL^{2-} ions determined calorimetrically (cal) under the same conditions, are as follows:

No.	t in °C	method	log K_1	$-\Delta H_1$	ΔS_1	log K^{Mn}_{MnHL}	log K_H	Ref.
1	20	gl, cal	13.76	5.9	45.6	–	–	[1]
	20	Hg	13.2	–	–	–	–	[2]
	25	cal	–	5.6	41	–	–	[3]
2	20	gl, cal	12.28	8.16	21.5	7.02	4.2	[1]
	20	gl	12.11	–	–	6.59	3.94	[4]
	25	Hg	12.3	8.8	27	–	–	[2, 3]
3	20	gl	10.07	1.53	41.86	–	4.88	[1]
	20	gl	9.64	–	–	5.08	–	[5]

For the complex with ligand 2, $Mn(C_{14}H_{20}N_2O_{10})^{2-}$, so-called effective (or "arbitrary") stability constants (see p. 29) were calculated for the pH range 3 to 14 from literature values for log K_1, log K_H, and the pK values of the ligand. The protonation of the ligand anion, stepwise hydrolysis of metal ions, and formation of protonated and hydroxo complexes were taken into account [6].

Preliminary relaxation ^{17}O NMR measurements on aqueous solutions of $Mn(C_{14}H_{20}N_2O_{10})^{2-}$ ions suggest that the complex with ligand 2 is octahedral with the ligand occupying all coordination positions [7]. An Mn–proton distance of 3.70 Å, resulting from water proton relaxation studies, is like that for second shell water protons, thus indicating that no water molecule is in the first hydration shell and thus supporting the above structure [8]. ESR X-band and K-band spectra were taken from aqueous solutions of the $Mn(C_{14}H_{20}N_2O_{10})^{2-}$ ion at 25°C and from an aqueous sample frozen in a polydextran matrix [9].

Manganese(III) Complexes in Solution. Spectrophotometric measurements in the pH range 1 to 4 revealed the existence of the Mn^{III} complexes with ligand 2, $Mn(C_{14}H_{20}N_2O_{10})^-$ and $Mn(C_{14}H_{21}N_2O_{10})$, in aqueous solutions obtained by oxidizing Mn^{II} solutions with the stoichiometric amount of PbO_2 in the presence of the ligand. The complexes show absorption maxima at 480 and 560 nm. Increasing optical density in the pH range 1.9 to 3 and its constancy at pH 3 to 3.8 for both maxima shows increasing and maximal complex formation in these pH ranges. Stability constants are: $\log K_1 = 17.18$ and $\log K^{Mn}_{MnHL} = 10.04$, both values at 20°C. When the complex solution is allowed to stand, the violet color of the complexes gradually disappears indicating their decomposition evidently due to an oxidation-reduction process. The decomposition reaction obeys a first-order rate law, the rate depending markedly on the pH [10].

References:

[1] Anderegg, G. (Helv. Chim. Acta **47** [1964] 1801/15, 1811, 1812).
[2] Holloway, J. H., Reilley, C. N. (Anal. Chem. **32** [1960] 249/56; AD-226874 [1959] 1/28, 13; C.A. **1961** 16105).
[3] Wright, D. L., Holloway, J. H., Reilley, C. N. (Anal. Chem. **37** [1965] 884/92, 886).
[4] Fraústo da Silva, J. J. R., Gonçalves Calado, J. (Rev. Port. Quim. **5** [1963] 121/8; C.A. **61** [1964] 10298).
[5] Schwarzenbach, G., Senn, H., Willi, A. (unpublished data from Sillén, L. G., Martell, A. E., Stability Constants of Metal-Ion Complexes, Chem. Soc., [London] Spec. Publ. No. 17 [1964] p. 675).
[6] Amsheeva, A. A. (Zh. Analit. Khim. **35** [1980] 846/53; J. Anal. Chem. [USSR] **35** [1980] 553/9, 558).
[7] Zetter, M. S., Grant, M. W., Wood, E. J., Dodgen, H. W., Hunt, J. P. (Inorg. Chem. **11** [1972] 2701/6, 2706).
[8] Oakes, J., Smith, E. G. (J. Chem. Soc. Faraday Trans. II **77** [1981] 299/308, 302/6).
[9] Reed, G. H., Leigh, J. S., Pearson, J. E. (J. Chem. Phys. **55** [1971] 3311/6, 3314).
[10] Mikhailova, T. V., Astakhov, K. V., Zhirnova, N. M. (Zh. Fiz. Khim. **45** [1971] 2345/7; Russ. J. Phys. Chem. **45** [1971] 1326/7).

22.23 With *trans*-1,2-Cyclohexanediamine-N,N,N',N'-tetraacetic Acid ($= C_{14}H_{22}N_2O_8 = H_4cdta$)

22.23.1 Manganese(II) Complexes in Solution

Potentiometric pH titrations indicate the existence of the **$Mn(cdta)^{2-}$** and **$Mn(Hcdta)^-$** ions in aqueous solutions of pH ~2.3 containing Mn^{2+} ions and the ligand [1]. Complexation of 0, 10, 80, or 98% of the total manganese(II) at pH 2, 2.4, 3, or 4, respectively, was inferred from a sorption study of Mn^{2+} ions on a strongly acidic cation exchanger from an aqueous solution containing 7.2×10^{-3} M Mn^{II} and 1.8×10^{-2} M tetrasodium salt of the ligand [2]. The formation of a 1:1 complex of Mn^{II} and the ligand above pH 7 was observed during a polarographic study [3]. The formation of the $Mn(cdta)^{2-}$ ion by reduction of the Mn^{III} complex is described on p. 71.

Stability constants of $Mn(cdta)^{2-}$ were determined polarographically (pol), by distribution between two phases (dis), or potentiometrically with the Hg electrode (Hg). Thermodynamic quantities for the formation reaction from Mn^{2+} and $cdta^{4-}$ were obtained by calculation (calc), or by calorimetry (cal):

t in °C	I in M	methods	log K_1	$-\Delta H_1$ kcal/mol	$-\Delta G_1$ kcal/mol	ΔS_1 $cal \cdot mol^{-1} \cdot K^{-1}$	Ref.
20	0.1(KNO_3)	pol[a], cal	17.43[b]	4.14	23.37	65.6	[4, 5]
20	0.1($KClO_4$)	dis[c]	14.70	—	—	—	[6]
25	0.1(KNO_3)	Hg, calc	18.15	7.39	24.77	58.25	[1]
25	0.1(KNO_3)	calc[d], cal	17.38 [8]	7.1 [9]	22.7 [9]	52 [9]	[7, 8]

[a] By use of the exchange equilibria with the Cu and Cd complexes. – [b] The original value, log K_1 = 16.78 [5], was corrected by [4]. – [c] By extraction with a solution of 8-quinolinol in chloroform. – [d] The log K_1 value was calculated from that of [4, 5] at 20°C using the ΔH_1 value of [4].

The higher stability of the $Mn(cdta)^{2-}$ ion compared to the $Mn(edta)^{2-}$ ion (log $K_1 \approx 14$) is mainly due to the larger entropy change on formation of the former [4, 9]. With an effective stability constant K_{eff}, introduced by [10], the protonation of the $cdta^{4-}$ anions in the equilibrium solution is taken into account. $K_{eff} = [MnL]/[Mn][L]'_H$ where $[L]'_H$ is the total concentration of the noncomplexed ligand in all its forms at the given pH. $[L]'_H$ can be calculated from [L], the pK values of the ligand, and the pH [10]. A graph of log K_{eff} as a function of pH [2] is given on p. 30. By another effective stability constant, $K'_{eff} = [MnL]'/[Mn]' \cdot [L]'$, the formation of metal hydroxides and of protonated and hydroxo complexes is taken into account in addition to the protonated ligand species. [MnL]' and [Mn]' are calculated from [MnL] or [Mn], the pH, and the stability constants of the complexes occurring in addition to the MnL^{2-} or Mn^{2+} ions; log K'_{eff} is 1.78, 6.83, 10.70, 13.00, 14.93, 13.04, and 9.54 at pH 2, 4, 6, 8, 10, 12, and 14, respectively [11]. It deviates from log K_{eff} at pH<3 and >10. The effective stability constants are of importance in the analytical application of complex formation [2, 10, 11].

The kinetics of the reaction $Mn^{2+} + Hcdta^{3-} \underset{}{\overset{k_f}{\rightleftharpoons}} Mn(cdta)^{2-} + H^+$ was studied by amperometric titration using the dropping mercury electrode as indicator electrode. The second-order rate constant is $k_f = 1.2 \times 10^8\ M^{-1} \cdot s^{-1}$ at 25°C, I = 0.2 M [12]. Viscosimetric studies of solutions containing the $Mn(cdta)^{2-}$ ion suggest that the complex does not have a simple octahedral structure [13].

The equilibrium constant of the reaction $Mn(cdta)^{2-} + H^+ \rightleftharpoons Mn(Hcdta)^-$ at 20°C and I = 0.1 M (KNO_3) is log K = 3.19, determined potentiometrically (glass electrode) [1]. Other authors found log K = 2.8, polarographically [5]. For the decomposition reaction $Mn(cdta)^{2-} + H^+ \rightleftharpoons Mn^{2+} + (Hcdta)^{3-}$, the second-order rate constant at 25°C is $k = 317\ M^{-1} \cdot s^{-1}$ in the 4.7 to 6.1 pH range for I = 0.1 M ($NaClO_4$), determined spectrophotometrically [7], or $k = 220\ M^{-1} \cdot s^{-1}$ in the 4.8 to 5.7 pH range for I = 0.2 M, determined by amperometric titration [12]. The activation parameters for this reaction are: E_a = 17.0 kcal/mol, $\Delta H^{\neq}$ = 16.4 kcal/mol, $\Delta S^{\neq}$ = 10 $cal \cdot mol^{-1} \cdot K^{-1}$, and the preexponential factor is $A = 1.0 \times 10^{15}\ M^{-1} \cdot s^{-1}$. A mechanism is proposed with simultaneous break of the Mn-N bonds as the proton is added to one N atom [7].

The $Mn(cdta)^{2-}$ complex is oxidized by bromine in aqueous solution according to the parallel reactions (A) and (B) (see p. 68). The overall reaction is (C). A spectrophotometric study at 25°C, I = 1.1 M ($NaClO_4$), pH ≈ 3.5, in the presence of free manganese ions in excess (mole ratio complex : Mn^{2+} ions = 1:5) yielded the listed second-order rate constants k (in $M^{-1} \cdot s^{-1}$), the free energy changes ΔG and activation parameters (energies in kJ/mol, activation entropy in $J \cdot mol^{-1} \cdot K^{-1}$):

	k	ΔG	ΔH^{+}	ΔS^{+}
(A) $Mn(cdta)^{2-} + Br_2 \rightleftharpoons Mn(cdta)^{-} + Br_2^{-}$	2.5×10^{-3}	30.90	4.1 ± 1.3	-3.0 ± 1.6
(B) $Mn(cdta)^{2-} + Br_3^{-} \rightleftharpoons Mn(cdta)^{-} + Br_2^{-} + Br^{-}$	3.4×10^{-4}	37.93	4.6 ± 3.3	-2.5 ± 1.5
(C) $2Mn(cdta)^{2-} + Br_3^{-} \rightleftharpoons 2Mn(cdta)^{-} + 3Br^{-}$	—	—	—	—

The reactions are catalyzed by free manganese ions. The second-order rate constant of reaction (C) increases with increasing concentration of Mn^{2+} ions and with increasing content of Br_2 in the oxidant. The oxidation reaction proceeds by an inner-sphere mechanism [14, 15]. The Mn^{II} complex with $cdta^{4-}$ is oxidized by oxygen in basic solution to give the Mn^{III} complex. The reaction can be used for a spectrophotometric determination of oxygen dissolved in water [16, 17]. Exchange reactions of $Mn(cdta)^{2-}$ ions with Cu^{2+} or Co^{2+} ions in excess were studied spectrophotometrically [7] or by amperometric titration [12] at the conditions given on p. 67. The reactions proceed by dissociation of the $Mn(cdta)^{2-}$ ion according to $Mn(cdta)^{2-} + H^{+} \rightleftharpoons Mn^{2+} + (Hcdta)^{3-}$ and subsequent combination of the $Hcdta^{3-}$ ion with the Cu^{2+} or Co^{2+} ions [7, 12]. This behavior is a direct result of steric hindrance by the ligand molecule [7]. While in [7] was found that the dissociation reaction was rate determining in both cases, in [12] was observed that the combination reaction was rate determining for the exchange with Co^{2+} ions.

The analytical determination of manganese by use of complex formation with the ligand is described for example in [18]. The Mn complex (and also the Ca complex) with the ligand are effective in the decorporation of the radionuclide ^{54}Mn from mammals. The effect is accounted for by isotopic exchange [19].

References:

[1] Hseu, T.-M., Farng, L.-P., Siew, P.-Y., Tsai, I.-I. (J. Chinese Chem. Soc. [Taipei] [2] **25** [1978] 61/5).
[2] Povondra, P., Přibil, R., Šulcek, Z. (Talanta **5** [1960] 86/91).
[3] Jain, D. S., Gaur, J. N. (Trans. SAEST **3** [1968] 4/7).
[4] Anderegg, G. (Helv. Chim. Acta **46** [1963] 1833/42, 1836, 1839).
[5] Schwarzenbach, G., Gut, R., Anderegg, G. (Helv. Chim. Acta **37** [1954] 937/57, 947, 954).
[6] Starý, J. (Anal. Chim. Acta **28** [1963] 132/49, 139).
[7] Margerum, D. W., Menardi, P. J., Janes, D. L. (Inorg. Chem. **6** [1967] 283/9, 284, 286).
[8] Wright, D. L., Holloway, J. H., Reilley, C. N. (Anal. Chem. **37** [1965] 884/92, 886).
[9] Holloway, J. H., Reilley, C. N. (Anal. Chem. **32** [1960] 249/56, 253).
[10] Schwarzenbach, G., Flaschka, H. (Die Komplexometrische Titration, Enke, Stuttgart 1965, pp. 12/3).

[11] Amsheeva, A. A. (Zh. Analit. Khim. **35** [1980] 846/53; J. Anal. Chem. [USSR] **35** [1980] 553/9, 556).
[12] Kimura, M. (Nippon Kagaku Zasshi **90** [1969] 1246/50, 1249; C.A. **72** [1970] No. 71086).
[13] Yasuda, M. (Bull. Chem. Soc. Japan **42** [1969] 2547/9).
[14] Vierling, F., Wechsler, S. (Nouv. J. Chim. **2** [1978] 51/7, 55).
[15] Vierling, F. (Bull. Soc. Chim. France **1980** 144/8).
[16] Sastry, G. S., Hamm, R. E., Pool, K. H. (Anal. Chem. **41** [1969] 857/8).
[17] Malaiyandi, M., Sastri, V. S. (Talanta **30** [1983] 983/5).
[18] Parkash, R., Gupta, S. K., Singh, R. P., Singhal, R. L. (J. Indian Chem. Soc. **61** [1984] 731/5).
[19] Kuhn, A. (Strahlentherapie **137** [1969] 101/9).

22.23.2 Isolated Manganese(II) Compounds

Mn_2(cdta)·9H_2O and Mn_2(cdta). The hydrate was prepared similar to Mn_2(edta)·9H_2O (p. 41) by adding with heating $MnCO_3$ in small portions to a suspension of H_4cdta·H_2O in water until the acid was completely dissolved (pH 7 to 8). Crystals separated from the solution on standing in air at room temperature. These were washed with water and ethanol, and dried in air. Mn_2(cdta)·9H_2O is stable in air. On heating, dehydration begins at ~60 to 90°C. At ~170°C, the anhydrous compound, Mn_2(cdta), is formed, which decomposes at 300°C as shown by a thermoanalytic study (TG, DTG, and DTA) [1].

Mn(H_2cdta)·3H_2O and Na_2Mn(cdta)·nH_2O. Mn(H_2cdta)·3H_2O was prepared by reacting 0.02 mol of freshly precipitated $MnCO_3$ with 0.025 mol H_4cdta, both reactants suspended in water at about 100°C for 24 h. From the filtered solution, the compound separated as a white, microcrystalline solid, which was recrystallized from water. Na_2Mn(cdta)·nH_2O was obtained by neutralizing a solution of Mn(H_2cdta)·3H_2O with NaOH or $NaHCO_3$ and subsequently precipitating with acetone [2].

The magnetic moment determined for Mn(H_2cdta)·3H_2O corresponds to a high-spin d^5 complex. Bands (in cm^{-1}) in the IR spectra of Mn(H_2cdta)·3H_2O or Na_2Mn(cdta)·nH_2O were assigned as follows: 2920 and 2915, $\nu(CH)_{methylene}$; 1600 and 1580, $\nu_{as}(COO^-)$; 1350 and 1400, $\nu_s(COO^-)$; 1115 and 1115, respectively, ν(CN). Bands (in nm) in the UV and visible region observed for Mn(H_2cdta)·3H_2O were assigned to transitions from A_{1g}(S) to the given excited states: 360, $^4T_{2g}$(D); 414, $^4A_{1g}$, ^4E_g(G); 435, $^4T_{2g}$(G); 540, $^4T_{1g}$(G) [2].

A DTA curve for Mn(H_2cdta)·3H_2O indicates four endothermic effects. The peak at 120°C corresponds presumably to the separation of the water of crystallization. At 170°C, the coordinatively bound water is split off. The last two peaks, at 240 and 380°C, are attributed to the decarboxylation of the organic ligand, which takes place in two steps (weight losses in the TG curve at 250 and 350°C). On the basis of the DTA, TG, and IR results it is believed that the ligand is tetradentate in Mn(H_2cdta)·3H_2O, two water molecules occupying the remaining two positions of the coordination octahedron [2].

CaMn(cdta)·2.5H_2O. An aqueous solution of Ca_2(cdta)·7H_2O was added to a suspension of a stoichiometric amount of Mn_2(cdta)·9H_2O in a small volume of water, and the mixture heated until complete dissolution was attained. The solution was then filtered and evaporated until beginning of crystallization. The separated crystals were washed with water and ethanol and dried in air [1]. The IR spectrum of the complex in Vaseline, fluorine oils, or KBr tablets was recorded. Broad bands in the 3600 to 3000 cm^{-1} range are associated with the presence of water and different types of hydrogen bonds (from hydrate or crystal water molecules). The positions of the $\nu_{as}(COO^-)$ band (~1600 cm^{-1}) and $\nu_s(COO^-)$ band (~1400 cm^{-1}) indicate a prevalent ionic character of the Mn–O bond. The compound is stable in air [1].

MnM^{II}(cdta)·6H_2O (M^{II} = Ni, Cu) and **MnM^{II}(cdta)·9H_2O** (M^{II} = Zn, Co). The compounds were obtained at 5°C from aqueous or propanone-water solutions of the metal nitrates and Na_4cdta. Their structures are similar to those of MnM^{II}(edta)·nH_2O complexes, see p. 42 [3].

References:

[1] Myachina, L. I., Logvinenko, V. A., Grankina, Z. A., Sheludyakova, L. A. (Izv. Sibirsk. Otd. Akad. Nauk SSSR Ser. Khim. Nauk **1983** No. 4, pp. 75/83, 76; C.A. **98** [1983] No. 209083).

[2] Carmona Guzman, E., Gonzalez Garcia, F. (Anales Quim. **74** [1978] 1049/53).

[3] Fuertes, A., Miravitlles, C., Escrivá, E., Martínez-Tamayo, E., Beltrán, D. (Transition Metal Chem. [Weinheim] **10** [1985] 432/4).

22.23.3 Manganese(III) Complexes in Solution

Formation and Properties. Solutions containing the **Mn(cdta)$^-$** ion were prepared by reacting MnO_2 with H_4cdta [1 to 3] (see p. 74), by combining solutions containing the $Mn(H_2P_2O_7)^+$ ion and H_4cdta in a 1:1 mole ratio [4], or by oxidizing the Mn^{II} complex, $Mn(cdta)^{2-}$, with O_2 [5] or Br_2 (see p. 67). The protonated complex, **Mn(Hcdta)**, was obtained by passing a solution of the Mn(cdta)$^-$ complex through the H^+ form of a cation exchange resin [1]. In aqueous methanol (2.5 vol% H_2O), Mn(Hcdta) was obtained by addition of excess acid to a solution containing Mn(cdta)$^-$ ions [6, 7]. Titrimetric and spectrophotometric studies indicated the existence of the red Mn(Hcdta) and Mn(cdta)$^-$ species in the 2 to 7 pH range [1 to 5] and of a third species, the **Mn(cdta)(OH)$^{2-}$** ion, in the 9 to 11 pH range [1 to 3]. Values for the equilibrium constant of the reaction $Mn(cdta)(OH)^{2-} + H^+ \rightleftharpoons Mn(cdta)(H_2O)^-$ are: log K = 7.5 at 25°C, I = 0.1 [2, 3]; 7.74 at 25°C, I = 0.1 M (NaCl) [8]; and 8.11 at 25°C [1].

The stability constant of the Mn(cdta)$^-$ ion, log K_1 = 28.9, was determined at 25°C, pH 3 to 6, I = 0.2 ($NaClO_4$) by spectrophotometry and potentiometry [1], and log K_1 = 23.8 at pH 2, I = 1 ($NaClO_4$) by spectrophotometry [4].

Absorption spectra and titration studies indicate that in aqueous solutions one water molecule is coordinated to the Mn atom in Mn(cdta)$^-$, in contrast to the solid compound $K[Mn(cdta)] \cdot nH_2O$ (p. 74) [1 to 3]. The manganese atom is believed to be seven-coordinate [1, 2, 5, 8 to 11] as is the case for related aminecarboxylato complexes [1, 12] (p. 32). Solutions of the Mn(cdta)$^-$ ion show an absorption maximum at 500 nm [2 to 5] or 510 nm [1, 13] with an extinction coefficient of log $\varepsilon \approx 2.52$ [2, 3] or 2.54 [1] in the 3 to 5 [2, 3] or 3 to 7 pH range [1]. The maximum was assigned to the $^5E_g \rightarrow {}^5T_{2g}$ transition within the Mn^{3+} ion [2, 3]. The absorption maximum shifts to 448 nm with log $\varepsilon = 2.52$ for the $Mn(cdta)(OH)^{2-}$ ion [1].

The standard potential of the reaction $Mn^{III}(cdta)^- + e^- \rightleftharpoons Mn^{II}(cdta)^{2-}$ in aqueous solution, E° = 0.814 V vs. NHE at 25°C, pH 4, I = 0.2 M ($NaClO_4$), was determined by potentiometry using a Pt electrode and an SCE reference electrode [1]. A formal reduction potential of 0.76 V vs. NHE was obtained by cyclic voltammetry at 25°C, pH 7, 10^{-2} M phosphate, I = 0.1 M (NaCl) [8]. The cyclic voltammogram of $K[Mn(cdta)] \cdot H_2O$ in DMSO shows a peak for the $Mn^{III} \rightarrow Mn^{II}$ reduction at −0.30 V vs. SCE [14].

Chemical Reactions in Aqueous Solution. The chemical behavior of Mn^{III} complexes with the cdta^{4-} or Hcdta^{3-} anions is very similar to that described for the corresponding complexes with the edta^{4-} or Hedta^{3-} anions described on p. 46. However, the Mn(cdta)$^-$ ion is much less subject to redox decomposition than the Mn(edta)$^-$ complex [1, 4]. The former was therefore used for studies of the kinetics of various oxidation reactions. Only one ligand exchange reaction is reported. It is described at the end of this section.

In the intramolecular decomposition reaction of the Mn(cdta)$^-$ ion, products [1] and intermediate products during photolysis [15] correspond to those for the Mn(edta)$^-$ ion (p. 47). The first-order rate constant of this reaction is $k = 6.5 \times 10^{-6}\ s^{-1}$ at 25°C, pH 3 to 6 [1]. A similar value, $4.8 \times 10^{-6}\ s^{-1}$, was found at 20°C, pH 2 [4] for the rate constant of the decomposition reaction in the $Mn(H_2P_2O_7)^+ - H_4$cdta system [4]. Both values were determined by spectrophotometry [1, 4]. A plot of the rate constant vs. pH shows that the rate of decomposition sharply decreases above pH 1.5, reaches a minimum and constant value between pH 3 and 4, and sharply increases above pH 5 due to a higher rate of decomposition for the Mn(cdta)OH$^-$ ion than for the Mn(cdta)$^-$ ion [4]. The enthalpy and entropy of activation are: $\Delta H^{\ddagger} = 24.3$ kcal/mol and $\Delta S^{\ddagger} = -1.5\ cal \cdot mol^{-1} \cdot K^{-1}$ in the 3 to 6 pH range [1]; the activation energy is $E_a = 13.8$ kcal/mol at pH 2 [4]. The high sensitivity to light of the Mn(cdta)$^-$ complex is shown by the large difference of the half-lives in the dark (3240 min) and in the light (570 min) determined by spectrophotometry at pH 4.4 (H_2SO_4). Addition of cetyltrimethylammonium bromide had no marked effect on the stability of the complex (half-lives 3390 and 732 min) [5]. For the decomposition reaction of the $Mn(cdta)^-$ complex in the presence of free ligand, the rate of decomposition decreases sharply for ratios of ligand: Mn^{III} between 1 to 2, but remains approximately constant for ratios between 2 and 10, at 40°C, pH 2. The energy of activation for a ligand: Mn^{III} ratio of 10 was found to be 22.6 ± 0.2 kcal/mol at pH 2 [4]. An analytical study at pH 6 with a 3- to 5-fold free ligand excess gave the same or analogous products as was found for the $Mn(edta)^-$ ion in the presence of free ligand (p. 48). An analogous mechanism for the redox reaction was proposed [16]. Four parallel decomposition reactions were found in a kinetic study similar to that described on p. 48 for the ethylenediaminetetraacetato complexes with the same conditions except temperature. The second-order rate constants are $k_1 \approx 0$, $k_2 = 22.8$, $k_3 = 649$, and $k_4 = 0.63 \times 10^{-2}\ M^{-1} \cdot s^{-1}$ at 10.2°C. The energy of activation is 14.0 kcal/mol for both the k_2 and k_3 reactions [17].

The $Mn(cdta)^-$ ion is reduced to form the $Mn(cdta)^{2-}$ ion by a number of reducing agents. Complicated rate laws indicating several reaction steps were found in most cases. In several cases inverse acid dependence and inverse dependence on the Mn^{II} complex concentration were found. In most cases, a mechanism with inner-sphere electron transfer was proposed (for H_2O_2, N_2H_4, NH_2OH, benzene-1,2-diols, SCN^-, VO^{2+} as reducing agents). Kinetic studies were mostly performed by following the absorption of the $Mn(cdta)^-$ ion at 500 or 510 nm.

The reaction with H_2O_2 was studied in the following concentration ranges: $[H_2O_2]$, (5.6 to 80) $\times 10^{-3}$ M; $[Mn(cdta)^{2-}]$, (0 to 18.2) $\times 10^{-3}$ M; and $[H^+]$, 1.6×10^{-4} to 1.2×10^{-2} M. The reaction was found to be second-order in $[Mn(cdta)^{2-}]$, first-order in $[H_2O_2]$, and inverse first-order in $[H^+]$ and $[Mn(cdta)^{2-}]$. The kinetic results (rate constants and activation parameters given) indicate a mechanism which involves the formation of a peroxo complex, $Mn(cdta)HO_2^{2-}$, in a fast preequilibrium step. The peroxo complex decomposes to form the Mn^{II} complex and the $HO_2^{\bullet}$ radical which then reacts with a second $Mn(cdta)^-$ ion to form again an $Mn(cdta)^{2-}$ ion, oxygen, and protons [10, 18, 26].

The rate law for the reaction with hydrazine shows two terms, both first-order in $[Mn(cdta)^-]$ and $[N_2H_4]$. The k_1 term is independent of $[H^+]$, while the k_2 term is inverse first-order in $[H^+]$. Values for the second-order rate constants k_1 and k_2 at 20°C, I = 0.2 ($NaClO_4$), pH 2.5 to 4.5 are: 700 $M^{-1} \cdot s^{-1}$ and 0.90 s^{-1}, respectively. Activation parameters for the k_1 and k_2 reaction are $\Delta H^{\neq} = 7$ and 9 kcal/mol, $\Delta S^{\neq} = -22$ and -27 $cal \cdot mol^{-1} \cdot K^{-1}$, respectively. The mechanism proposed involves the formation of the $[Mn^{III}(cdta)(N_2H_5)]$ complex which then decomposes in two parallel reactions to form the Mn^{II} complex and the $N_2H_5^{\bullet 2+}$ and $N_2H_4^{\bullet +}$ radicals, the latter through intermediate formation of the $[Mn^{III}(cdta)(N_2H_4)]^-$ complex. The radicals then decompose to give N_2, NH_4^+, and H^+ ions [13]. For the reaction with NH_2OH (reacting species NH_3OH^+) a complex rate law was found. The rate is first-order in $[Mn(cdta)^-]$, higher than first-order in $[NH_3OH^+]$, and higher than inverse first-order in $[H^+]$. A mechanism similar to that for hydrazine was proposed with intermediate formation of the $[Mn^{III}(cdta)(NH_3OH)]$ complex, which decomposes after release of one proton to give the Mn^{II} complex and the $NH_2OH^{\bullet +}$ radical, the latter forming N_2, H_2O, and H^+ ions in a fast reaction [13]. The rate law for the reaction with dithionite, $S_2O_4^{2-}$, contains two terms, both first-order in $[Mn(cdta)^-]$, one first-order in $[S_2O_4^{2-}]$, the other half-order in $[S_2O_4^{2-}]$. The kinetic results indicate two competing reactions, one with the $S_2O_4^{2-}$ ion (k_1 reaction) the other with the sulfoxyl radical anion, $SO_2^{\bullet +}$ (k_2 reaction). Values for second-order rate constants are: $k_1 = 2.1 \times 10^6\ M^{-1} \cdot s^{-1}$, $k_2 \leqq 10^8\ M^{-1} \cdot s^{-1}$. An outer-sphere mechanism is proposed [19]. The oxidation of oxalate ions was studied over the pH 3 to 7 range. The rate law consists of the terms $2k_1[Mn(cdta)^-][HC_2O_4^-]$ and $2k_2[Mn(cdta)^-][C_2O_4^{2-}]$, indicating two parallel reactions with both $HC_2O_4^-$ (k_1 reaction) and $C_2O_4^{2-}$ ions (k_2 reaction), which both yield the Mn^{II} complex and the $C_2O_4^{\bullet -}$ radical anion. The latter reacts with another $Mn(cdta)^-$ ion to give the Mn^{II} complex again and CO_2 and H_2O. The second-order rate constants are:

$k_1 = 5.4 \times 10^{-3}\ M^{-1} \cdot s^{-1}$ and $k_2 = 7.1 \times 10^{-3}\ M^{-1} \cdot s^{-1}$ at 20°C, $I = 0.2(NaClO_4)$. Activation parameters are $\Delta H^{\neq} = 20.9$ kcal/mol and $\Delta S^{\neq} = 3.5\ cal \cdot mol^{-1} \cdot K^{-1}$ for the k_1 reaction and $\Delta H^{\neq} = 16$ kcal/mol and $\Delta S^{\neq} = -13\ cal \cdot mol^{-1} \cdot K^{-1}$ for the k_2 reaction. No indication could be found for formation of an intermediate $[Mn(cdta)(C_2O_4)]^{3-}$ complex [20]. Rate laws found for the oxidation of nitrite [18] and iodide [11] show that the reactions are inhibited by the Mn^{II} complex formed. In the oxidation of nitrite, the active species is the $N_2O_4^{2-}$ ion which is oxidized in two one-electron steps to form N_2O_4. The latter dismutes to give NO_3^- and NO_2^- [18]. A rate law was found for the oxidation of thiocyanate. The reaction appears to occur in two consecutive steps. Rate constants and activation parameters are given [10]. Oxidation of benzene-1,2-diols proceeds in an analogous manner as described on p. 49 for the $Mn(edta)^-$ complex [13].

VO^{2+} ions are oxidized to give VO_2^+ ions. The reaction proceeds in two parallel paths, one acid-independent (k_1) with VO^{2+} as the active species, and one with inverse $[H^+]$ dependence (k_2) with $VOOH^+$ as the active species. Second-order rate constants at 7°C, I = 1 M ($LiClO_4$) are: $k_1 = 65\ M^{-1} \cdot s^{-1}$ and $k_2 = 0.53\ s^{-1}$. Activation parameters are: $\Delta H^{\neq} = 13.1$ kcal/mol, $\Delta S^{\neq} = -3.4\ cal \cdot mol^{-1} \cdot K^{-1}$ for the k_1 reaction and $\Delta H^{\neq} = 11.3$ kcal/mol and $\Delta S^{\neq} = -20\ cal \cdot mol^{-1} \cdot K^{-1}$ for the k_2 reaction. The mechanism assumes the fast formation of complexes with the active species as intermediates which then decompose slowly to give the products [21]. $M(edta)^{2-}$ ions (M = Mn, Fe, Co) and $Co(cdta)^{2-}$ ions are oxidized to give the corresponding M^{III} complexes. The table lists second-order rate constants and activation parameters for the reactions:

No.	reducing agent	t in °C	pH	I in mol/L	k in $M^{-1} \cdot s^{-1}$	$\Delta H^{\neq}$ in kcal/mol	$\Delta S^{\neq}$ in $cal \cdot mol^{-1} \cdot K^{-1}$	Ref.
(1)	$Mn(edta)^{2-}$	25	4.5 to 6.5	(?)	1.2	7.1*)	−34	[22]
(2)	$Fe(edta)^{2-}$	5	4.5 to 6.5	0.05	$\sim 4 \times 10^5$	—	—	[22]
(3)	$Co(edta)^{2-}$	25	4.5 to 6.5	(?)	0.9	4.8	−42	[22, 27]
(4)	$Co(cdta)^{2-}$	25	2.1 to 4.8	0.5	0.45	—	—	[23]

*) From data at I = 0.25 M [22].

Reactions (3) and (4) are catalyzed by aqueous micellar cetyltrimethylammonium bromide. The reaction rates are enhanced by a factor of 600 or 160 at most, respectively [24].

The kinetics of the ligand exchange reaction between the $Mn(cdta)OH^{2-}$ ion and the CN^- ion could be accounted for by the mechanism $Mn(cdta)OH^{2-} + CN^- \underset{k_{-1}}{\overset{k_1}{\rightleftharpoons}} Mn(cdta)CN^{2-} + OH^-$ (1); $Mn(cdta)CN^{2-} + CN^- \xrightarrow{k_2} Mn(cdta)(CN)_2^{3-}$ (2); $Mn(cdta)(CN)_2^{3-} + 4CN^- \xrightarrow{fast} Mn(CN)_6^{3-} + cdta^{4-}$ (3). Rate constants at 25°C, pH 11, I = 0.25 M ($NaClO_4$), are: $k_1 = 0.015\ M^{-1} \cdot s^{-1}$ and $k_2 = 0.036\ M^{-1} \cdot s^{-1}$. Energies of activation are 11.3 kcal/mol for the k_1 reaction and 14.1 kcal/mol for the k_2 reaction. A crude estimate gave 4 M^{-1} for k_{-1}. The species $Mn(cdta)(CN)_2^{3-}$ is considered to be one in which one coordination site of the $cdta^{4-}$ ion has been replaced by a CN^- group [9].

Chemical Reactions in Other Solvents. A solution of $K[Mn(cdta)] \cdot H_2O$ in aqueous methanol (2.5 vol% H_2O) forms the same species, depending on H^+ ion concentration, as was shown for the aqueous solution on p. 70: the pink $Mn(cdta)^-$ complex (absorption maximum at 510 nm), the pink Mn(Hcdta) complex (520 nm), and the yellow $Mn(cdta)(OH)^{2-}$ complex (466 nm). In neutral methanol, the reduction of the $Mn(cdta)^-$ ion is extremely slow. The pseudo first-order rate constant is $2.24 \times 10^{-6}\ s^{-1}$. In the presence of excess acid, a redox reaction takes place to form the $Mn^{II}(Hcdta)^-$ ion and formaldehyde. The pseudo first-order rate constant is $2.1 \times 10^{-3}\ s^{-1}$ for $[H^+] = 2.05 \times 10^{-2}$ M at 25°C. The mechanism suggested involves as the rate-determining step, the transfer of an electron from CH_3OH to Mn^{III} to produce, after proton loss, a radical $CH_3O^•$ (or $CH_2OH^•$) which undergoes further oxidation by a second complex molecule to give formaldehyde. In alkaline methanol, the complex disproportionates to give $Mn(OH)_2$ and MnO_2. The reaction is first-order in $[Mn(cdta)^-]$ (up to about 50%) and also approximately first-order in base concentration. First-order rate constants increase from $1 \times 10^{-4}\ s^{-1}$ for 0.7×10^{-2} M base to $2.8 \times 10^{-4}\ s^{-1}$ for 2.7×10^{-2} M base. Activation parameters are $\Delta H^{\neq} = 16.1$ kcal/mol and $\Delta S^{\neq} = -15\ cal \cdot mol^{-1} \cdot K^{-1}$ [6]. The $Mn(cdta)^-$, Mn(Hcdta), or $Mn(cdta)(OH)^{2-}$ species react with 2,4,6-tri-tert-butylphenol in aqueous methanol (2.5 vol% H_2O) to give the corresponding phenoxy radical (which is stable). Pseudo-first order rate constants under neutral, acidic, or basic conditions per mol of phenol are 2.4 ± 10^{-4}, 3.4×10^{-2}, and $8.5 \times 10^{-2}\ M^{-1} \cdot s^{-1}$ [7]. Activation parameters are: $\Delta H^{\neq} = \sim 11$, ~ 16, and ~ 10 kcal/mol and $\Delta S^{\neq} = \sim -40$, ~ -10, and $\sim -30\ cal \cdot mol^{-1} \cdot s^{-1}$, respectively [25]. The rate-determining step involves the abstraction of one H atom from the phenol hydroxy group. Similar results were obtained with 2,6-di-tert-butylphenol and a series of 4-substituted 2,6-di-tert-butylphenols [7, 25]. For the oxidation of the hyperoxide ion, O_2^-, in DMSO, a second-order rate constant of $\sim 10^6\ M^{-1} \cdot s^{-1}$ was found. The rate enhancement compared with the ethylenediaminetetraacetato complex ($\sim 10^4\ M^{-1} \cdot s^{-1}$) is presumably related to steric factors [14].

References:

[1] Hamm, R. E., Suwyn, M. A. (Inorg. Chem. **6** [1967] 139/42).
[2] Takeuchi, T., Tsunoda, Y. (Nippon Kagaku Zasshi **88** [1967] 172/5; C.A. **66** [1967] No. 101204).
[3] Tsunoda, Y., Takeuchi, T. (Nippon Kagaku Zasshi **87** [1966] 626/7; C.A. **66** [1967] No. 34394).
[4] Lapkin, V. V., Pechurova, N. I. (Deposited Doc. VINITI-2663-77 [1977] 1/10, 6; C.A. **90** [1979] No. 61879).
[5] Malaiyandi, M., Sastri, V. S. (Talanta **30** [1983] 983/5).
[6] Poh, B. L., Stewart, R. (Can. J. Chem. **50** [1972] 3432/6).
[7] Stewart, R., Poh, B. L. (Can. J. Chem. **50** [1972] 3437/42).
[8] Adzamli, I. K., Davies, D. M., Stanley, C. S., Sykes, A. G. (J. Am. Chem. Soc. **103** [1981] 5543/7).
[9] Hamm, R. E., Templeton, J. C. (Inorg. Chem. **12** [1973] 755/7).
[10] Boone, D. J. (Diss. Washington State Univ. 1970 from Diss. Abstr. Intern. B **31** [1970] 3237/8).

[11] Shioyama, T. K. (Diss. Washington State Univ. 1978, pp. 1/92, 61, 66, 72, 78; Diss. Abstr. Intern. B **39** [1978] 695).
[12] Rettig, S. J., Trotter, J. (Can. J. Chem. **51** [1973] 1303/12).
[13] Arselli, P., Mentasti, E. (J. Chem. Soc. Dalton Trans. **1983** 689/96).
[14] Stein, J., Fackler, J. P., McClune, G. J., Fee, J. A., Chan, L. T. (Inorg. Chem. **18** [1979] 3511/9, 3514, 3516).
[15] Stel'mashok, V. E., Poznyak, A. L. (Koord. Khim. **5** [1979] 1019/24; Soviet J. Coord. Chem. **5** [1979] 801/5).
[16] Tanaka, N., Gomi, K., Shirakashi, T. (Nippon Kagaku Kaishi **1975** No. 3, pp. 444/8; C.A. **83** [1975] No. 21244).
[17] Shirakashi, T., Tanaka, N. (Nippon Kagaku Zasshi **91** [1970] 142/8; C.A. **73** [1970] No. 18972).
[18] Jones, T. E. (Diss. Washington State Univ. 1974; Diss. Abstr. Intern. B **35** [1974] 2642).
[19] Mehrotra, R. N., Wilkins, R. G. (Inorg. Chem. **19** [1980] 2177/8).
[20] Suwyn, M. A., Hamm, R. E. (Inorg. Chem. **6** [1967] 142/5).

[21] Boone, D. J., Hamm, R. E., Hunt, J. P. (Inorg. Chem. **11** [1972] 1060/2).
[22] Wilkins, R. G., Yelin, R. E. (Inorg. Chem. **7** [1968] 2667/9).
[23] Wilkins, R. G., Yelin, R. E. (J. Am. Chem. Soc. **92** [1970] 1191/4).
[24] Bhalekar, A. A., Engberts, J. B. F. N. (J. Am. Chem. Soc. **100** [1978] 5914/20, 5917).
[25] Poh, B. L. (Diss. Univ. British Columbia, Canada 1972; Diss. Abstr. Intern. B **33** [1972] 2506).
[26] Jones, T. E., Hamm, R. E. (Inorg. Chem. **13** [1974] 1940/3).
[27] Wilkins, R. G., Yelin, R. E. (J. Am. Chem. Soc. **89** [1967] 5496/7).

22.23.4 Isolated Manganese(III) Compound

K[Mn(cdta)] and **K[Mn(cdta)]·nH_2O** (n = 2.5, 1). The complex was obtained by adding H_4cdta to a suspension of freshly prepared MnO_2 in aqueous KOH solution at 10°C [1, 7, 8] or by adding the MnO_2 suspension to a slurry containing H_4cdta and $Mn(NO_3)_2$ in a 2:1 mole ratio at 0°C. After reacting for 0.5 to 1h, excess MnO_2 was filtered off and the dark red filtrate cooled to 0°C. The dark red crystals, which separated after addition of an equal volume of cold ethanol and several hours standing at 0°C, were washed with absolute ethanol [1, 2] and dried in vacuum at room temperature; the yield of K[Mn(cdta)]·2.5H_2O was 40 to 45% by the first or 80% by the second procedure, based on H_4cdta [1]. The same hydrate is reported in [5], while a trihydrate is reported in earlier work [7, 8]. The monohydrate was obtained by the same procedure but with final washing of the crystals with anhydrous ether and drying under reduced pressure at room temperature over silica gel [2]. It was recrystallized from a 3:2 mixture of water and methanol [3]. The monohydrate is also reported in [4, 5, 9].

K[Mn(cdta)]·H_2O forms irregularly developed crystals. An X-ray structure analysis (R = 0.067) shows that it is triclinic with the lattice parameters a = 5.60(3), b = 8.66(2), c = 7.36(1) Å, α = 103.8(1)°, β = 95.7(1)°, γ = 70.7(3)°; Z = 2; space group $P\bar{1}$-C_i^1 (No. 2). The structure is composed of layers of oxygen-bridged K^+ ions (figure in the paper) situated between layers of $[Mn(cdta)]^-$ anions, this basic structural unit repeating along the a axis (**Fig. 8**). The water molecule is two-fold disordered and is coordinated to the K^+ ion in both positions [3]. These positions are at a distance of 3.68 or 4.55 Å from the Mn^{III} atom (normal metal-water bond ~2.1 Å) [4]. The coordination polyhedron of the Mn^{III} atom is a very distorted octahedron. The Mn^{III} atom is coordinated by the two nitrogen and by four oxygen atoms of the ligand, which is in the chair form, showing only minor angular strain. The $[Mn(cdta)]^-$ anion has approximate C_2 symmetry with two classes of Mn-O bonds, the mean bond lengths of which are 1.887(7) and 2.015(22) Å; the mean Mn-N distance is 2.263(5) Å. The N-Mn-O or O-Mn-O angles are 77.1(2)° and 128.5(2)°, respectively. Individual bond lengths and bond angles, positional parameters, and thermal parameters are given in the publication. The K^+ ion is coordinated by eight or nine oxygen atoms depending on the position of the water molecule. Weak hydrogen bonds involving the water molecules are probably present. The calculated density is 1.69(2) g/cm³; the measured density is 1.656(6) g/cm³ at 20°C [3].

K[Mn(cdta)]·H_2O melts at 201 to 203°C [4]. The magnetic moment of K[Mn(cdta)]·2.5H_2O is 4.94 μ_B at 22.5°C, corresponding to a high-spin d^4 complex [7, 8]. The IR spectrum of the K[Mn(cdta)]·2.5H_2O in Nujol shows strong maxima near 3000 cm^{-1} (with a shoulder) and near 1600 cm^{-1} and two sharp maxima near 1400 cm^{-1} [7]. Similar but better resolved spectra were obtained from K[Mn(cdta)]·2.5H_2O and K[Mn(cdta)]·H_2O in KBr. Graphs are presented in [5].

K[Mn(cdta)]·2.5H_2O is stable at room temperature. A sample was kept unchanged for several months in the dark [1]. Both hydrates start dehydrating at 50 or 60°C [5, 7]. K[Mn(cdta)]·H_2O is completely dehydrated at 110 to 120°C [2, 5] (heating for 1 h [2]), forming the anhydrous compound K[Mn(cdta)]. Decomposition at 150 to 180°C gave *trans*-1,2-cyclohexanediaminetri- or tetraacetic acid (identified by the Co complexes) and HCHO, CO, and CO_2 (identified by gas chromatography). An intermolecular electron transfer forming Mn^{2+} ions is involved in thermal decomposition [5].

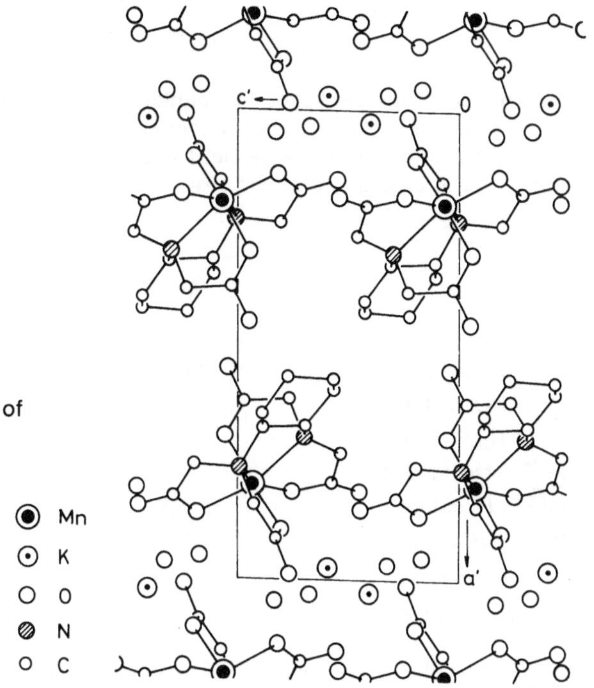

Fig. 8. Projection of the structure of K[Mn(cdta)]·H_2O down the b axis. (Hydrogen atoms are omitted.)

References:

[1] Hamm, R. E., Suwyn, M. A. (Inorg. Chem. **6** [1967] 139/42).
[2] Bhalekar, A. A., Engberts, J. B. F. N. (J. Inorg. Nucl. Chem. **40** [1978] 918/9).
[3] Rettig, S. J., Trotter, J. (Can. J. Chem. **51** [1973] 1303/12, 1304/5, 1310).
[4] Poh, B. L., Stewart, R. (Can. J. Chem. **50** [1972] 3432/6, 3435/6).
[5] Shirakashi, T., Tanaka, N. (Nippon Kagaku Kaishi **1974** No. 6, pp. 1061/7; C.A. **82** [1975] No. 3650).
[6] Stewart, R., Poh, B. L. (Can. J. Chem. **50** [1972] 3437/42).
[7] Takeuchi, T., Tsunoda, Y. (Nippon Kagaku Zasshi **88** [1967] 172/5; C.A. **66** [1967] No. 101204).
[8] Tsunoda, Y., Takeuchi, T. (Nippon Kagaku Zasshi **87** [1966] 626/7; C.A. **66** [1967] No. 34394).
[9] Stein, J., Fackler, J. P., McClune, G. J., Fee, J. A., Chan, L. T. (Inorg. Chem. **18** [1979] 3511/9).

22.24 With 1,4-Cyclohexanediamine-N,N,N′,N′-tetraacetic Acid (= $C_{14}H_{22}N_2O_8$ = H_4L)

A polarographic study shows that Mn^{II} forms an unstable 1:1 complex with the ligand in aqueous solutions of pH 7 to 9, Giuliani, A. M., Gattegno, D., Furlani, A. (Gazz. Chim. Ital. **97** [1967] 1076/88, 1079, 1088).

22.25 With Isomeric Phenylenediamine-N,N,N′,N′-tetraacetic Acids

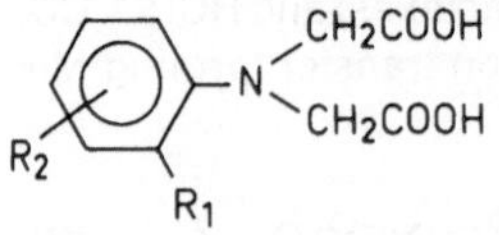

1) $R_1 = H$, $R_2 = 2\text{-}N(CH_2COOH)_2$; $(= C_{14}H_{16}N_2O_8)$
2) $R_1 = H$, $R_2 = 3\text{-}N(CH_2COOH)_2$; $(= C_{14}H_{16}N_2O_8)$
3) $R_1 = H$, $R_2 = 4\text{-}N(CH_2COOH)_2$; $(= C_{14}H_{16}N_2O_8)$
4) $R_1 = CH_3$, $R_2 = 4\text{-}N(CH_2COOH)_2$; $(= C_{15}H_{18}N_2O_8)$

Manganese(II) Complexes in Solution. Potentiometric (glass electrode) and conductometric measurements in aqueous solutions containing Mn^{2+} ions and the ligands [1 to 8] indicate the formation of the species listed. Stability constants or protonation constants were determined at 25°C and an ionic strength of 1 M$(NaClO_4)$ [1] or 0.1 M(KCl) [2 to 7].

equilibrium	log K for complexes with				
	ligand 1 [1]	ligand 2 [2, 3]	[4]	ligand 3 [5]	ligand 4 [6, 7]
$Mn^{2+} + L^{4-} \rightleftharpoons MnL^{2-}$	11.37	2.8_0*)	2.71	3.34	3.50
$Mn^{2+} + HL^{3-} \rightleftharpoons MnHL^-$	—	2.1_0	2.03	2.28	1.68
$Mn^{2+} + H_2L^{2-} \rightleftharpoons MnH_2L$	—	0.9*)	—	1.40	—
$MnL^{2-} + Mn^{2+} \rightleftharpoons Mn_2L$	—	2.2	1.63	1.25	—
$MnL^{2-} + H^+ \rightleftharpoons MnHL^-$	2.29	4.8	5.11	5.04	5.37
$MnHL^- + H^+ \rightleftharpoons MnH_2L$	1.7	4.4	—	4.00	—
$MnL(OH)^{3-} + H^+ \rightleftharpoons MnL(H_2O)^{2-}$	11.5	—	—	—	—

*) Corrected by a statistical factor.

Solutions of pH 6.3 containing the complex with ligand 1, $Mn(C_{14}H_{12}N_2O_8)^{2-}$ were studied by ^{17}O NMR spectroscopy under nitrogen atmosphere. Line broadenings and shifts in the spectrum of the light pink solution indicate the presence of one coordinated water molecule per Mn atom. Rate constant and activation parameters for single H_2O exchange are $k = 3.5 \times 10^8\ s^{-1}$ at 25°C, $\Delta H^{\neq} = 8.1$ kcal/mol, $\Delta S^{\neq} = 7.6$ cal·mol^{-1}·K^{-1}. The similarity of these values to those of the $Mn(edta)^{2-}$ complex (p. 32) suggests that the manganese atom in the $Mn(C_{14}H_{12}N_2O_8)^{2-}$ complex is seven-coordinate also. The ion is air-sensitive [9].

The additional formation of a dinuclear complex, $Mn_2(C_{14}H_{14}N_2O_8)_2$ in solutions containing Mn^{2+} ions and ligand 2 in a 1:1 mole ratio is assumed, based on potentiometric and conductometric studies. A solution of pH 4.6 is violet and shows an absorption maximum at 500 nm. At pH 6.5, which is the optimum pH value of complex formation, the maximum is shifted to 510 nm. At pH 8.8, the color vanishes [10].

[$Mn^{II}(H_2O)(C_{14}H_{12}N_2O_8)Mn^{II}(H_2O)_5$]. Well-formed, almost colorless and transparent crystals of the complex salt with ligand 1 crystallized from a solution which was prepared by neutralizing aqueous o-phenylenediaminetetraacetic acid with KOH, followed by addition of Mn^{II} acetate and then of ethanol [11].

The compound is triclinic, with the lattice parameters a = 9.893(1), b = 11.543(1), c = 9.844(1), α = 98.71(1)°, β = 104.59(1)°, γ = 97.70(1)°, Z = 2; space group $P\bar{1}$-C_i^1 (No. 2). The structure was solved up to R = 0.048. The experimental density is D_{exp} = 1.744 g/cm³, the calculated density is D_{calc} = 1.741 g/cm³ at 25°C. As shown in **Fig. 9**, the compound contains two non-equivalent Mn atoms. Mn(1) is coordinated by two N atoms and four O atoms of the hexadentate ligand and additionally by the O atom of one water molecule, O(6). Mn(2) is nearly octahedrally surrounded by one bridging carboxylate oxygen atom, O(32), and five water molecules. Pertinent atomic distances (in Å) are: Mn(1)–N(1) = 2.403(3), Mn(1)–N(2) = 2.428(4), Mn(1)–O(6) = 2.242(4), Mn(2)–O(32) = 2.137(3). The mean distance of Mn(1)–O(carboxylate) is 2.211 Å; the mean distance of Mn(2)–O(water) is 2.183 Å. Other atomic distances, pertinent angles, fractional atomic coordinates, and isotropic thermal parameters are given in the publication. All carboxylate oxygen atoms are receptors of hydrogen bonds from a water molecule, except for O(32), which acts as a bridging atom. The carboxylate oxygen atoms not bound to Mn(1) form two hydrogen bonds. O(1) and O(3) are also receptors. Thus, an intermolecular network of hydrogen bonds results [11].

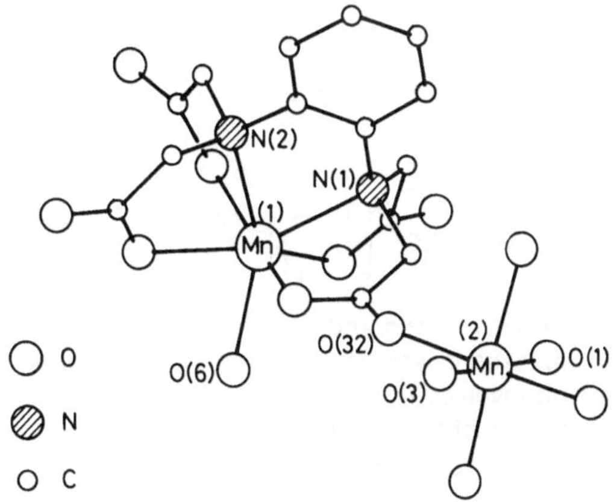

Fig. 9. Molecular structure of the Mn^{II} complex with o-phenylenediamine-N, N, N', N'-tetraacetic acid, $[Mn(H_2O)(C_{14}H_{12}N_2O_8)Mn(H_2O)_5]$. (Hydrogen atoms are omitted.)

$Mn^{II}_2(C_{15}H_{14}N_2O_8)\cdot nH_2O$. The dinuclear complex with ligand 4 (no analysis given) was obtained by adding a slight excess of a concentrated aqueous solution of $MnCl_2$ to a freshly prepared solution of the ligand of pH ~6. On addition of absolute ethanol, the complex precipitated. It was washed with a 9:1 water-ethanol mixture until free from Cl^- ions and dried at 70 to 110°C. The dried solid was yellowish, whereas the wet complex was pale rose [12].

The IR spectrum of the complex in Nujol shows characteristic bands at the following wave numbers (in cm^{-1}) which were assigned to vibrations of the groups given in parentheses: 3200 (OH); 1610 (sh), 1580 $(COO^-)_{coord}$; 1300 (H_2O); 1200 (CH_2COO^-); and 1100 (CN). The compound is strongly hygroscopic and insoluble in organic solvents. On heating, the water is split off at temperatures >200°C. Decompositon occurs at ~350°C [12].

$K[Mn^{III}(C_{14}H_{12}N_2O_8)(H_2O)]\cdot 1.5H_2O$. The complex with ligand 1 was prepared by the procedure given for the complex salt with methylethylenediamine-N, N, N', N'-tetraacetic acid on p. 61. Solutions of the complex show absorption maxima at 530 nm (log ε = 2.49) and 564 nm (2.51). The first-order rate constant of the decomposition reaction in the 3 to 5 pH range is $k = 1.37 \times 10^{-5}\ s^{-1}$ at room temperature and $I = 0.2\ M\ (NaClO_4)$ [13].

References:

[1] Nakasuka, N., Kunimatsu, M., Matsumura, K., Tanaka, M. (Inorg. Chem. **24** [1985] 10/5).
[2] Uhlig, E., Herrmann, D. (Z. Anorg. Allgem. Chem. **359** [1968] 135/46, 140).
[3] Uhlig, E., Herrmann, D. (Z. Anorg. Allgem. Chem. **365** [1969] 79/90, 81).
[4] Mederos, A., Rodríguez González, A., Rodríguez Ríos, B. (Anales Quim. **66** [1970] 531/42, 538).
[5] Rodríguez Ríos, B., Mederos, A. (Anales Quim. **65** [1969] 649/58, 656).

[6] Rodríguez Ríos, B., Mederos, A. (Anales Quim. **65** [1969] 557/66, 565).
[7] Rodríguez Ríos, B., Mederos, A. (Anales Quim. **65** [1969] 743/50, 748).
[8] Rodríguez Ríos, B., Mederos, A. (Anales Quim. **65** [1969] 751/62, 751).
[9] Liu, G., Dodgen, H. W., Hunt, J. P. (Inorg. Chem. **16** [1977] 2652/3).
[10] Gonzalez Garcia, S., Sanchez Santos, F. J. (Anales Quim. **71** [1975] 780/4).

[11] Nakasuka, N., Azuma, S., Katayama, C., et al. (Acta Cryst. C **41** [1985] 1176/9).
[12] Rodríguez Ríos, B., Mederos Pérez, A. (Anales Quim. **64** [1968] 47/54, 47/9, 52).
[13] Shioyama, T. K. (Diss. Washington State Univ. 1978, pp. 1/92, 17, 24, 51; Diss. Abstr. Intern. B **39** [1978] 695).

22.26 With 2,5-Dihydroxy-1,4-phenylenedimethylamine-N,N,N′,N′-tetraacetic Acid

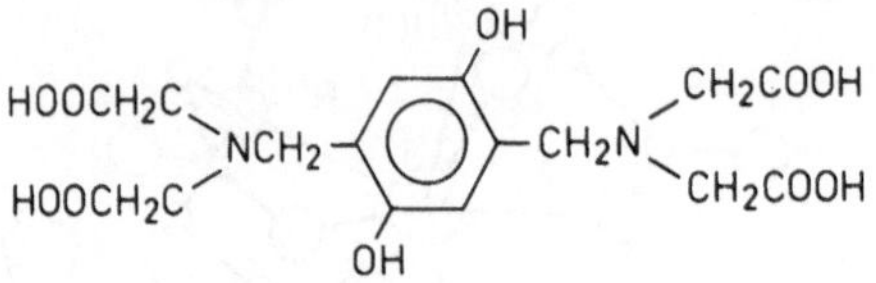

$(=C_{16}H_{20}N_2O_{10}=H_6L)$

Potentiometric and high-frequency titration studies revealed the formation of 1:1 and 2:1 complexes in solutions containing Mn^{2+} ions and the ligand with only one hydroxyl group being involved in complex formation. It may be coordinated in its protonized form. Stability constants of the various species were determined at 25°C potentiometrically (glass electrode) [1, 2]:

equilibrium	log K [1]	log K [2]
$Mn^{2+}+H_3L^{3-}\rightleftharpoons MnH_3L^-$	6.53	6.5
$Mn^{2+}+H_2L^{4-}\rightleftharpoons MnH_2L^{2-}$	8.11	8.1
$Mn^{2+}+HL^{5-}\rightleftharpoons MnHL^{3-}$	9.90	10.3
$2Mn^{2+}+HL^{5-}\rightleftharpoons Mn_2HL^-$	17.51	18.0

References:

[1] Tsirul'nikova, N. V., Temkina, V. Ya., Dyatlova, N. M., Rusina, M. N., Zhadanov, B. V., Lastovskii, R. P. (Zh. Analit. Khim. **25** [1970] 839/46; J. Anal. Chem. [USSR] **25** [1970] 724/30, 727/8).
[2] Dyatlova, N. M., Temkina, V. Y. (Koord. Khim. **1** [1975] 66/82; Soviet J. Coord. Chem. **1** [1975] 52/64, 54).

22.27 With 4,4′-Dicarboxy-2,2′-biphenylenediamine-N,N,N′,N′-tetraacetic Acid

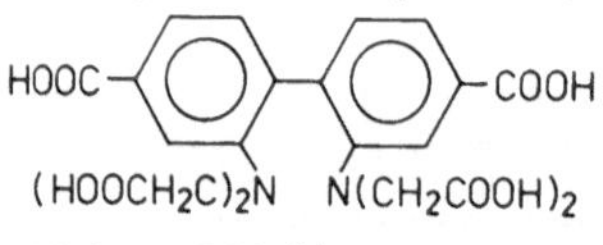

$(=C_{22}H_{20}N_2O_{12})$

By addition of Mn^{2+} ions to the fluorescent aqueous solution of the ligand at a pH value between 5 and 10, a non-fluorescent complex is formed, Kirkbright, G. F., Stephen, W. I. (Anal. Chim. Acta **32** [1965] 544/51, 548).

22.28 With N-Polyacetic Acids Derived from 4,4′-Methylenebis(phenylamine)

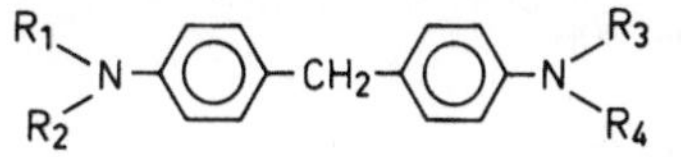

1) $R_1 = R_3 = H$, $R_2 = R_4 = CH_2COOH$; $(= C_{17}H_{18}N_2O_4)$
2) $R_1 = R_2 = R_3 = R_4 = CH_2COOH$; $(= C_{21}H_{22}N_2O_8)$

$[Mn(C_{17}H_{16}N_2O_4)]_2 \cdot 4H_2O$ and $Mn_2(C_{21}H_{18}N_2O_8) \cdot 4H_2O$. The complex with ligand 1 was prepared by reacting $MnCl_2$ with the stoichiometric amount of $Na_2(C_{17}H_{16}N_2O_4)$ in alcohol or aqueous alcohol. The white precipitate was washed with water until Cl^- free, then with methanol and ether, and dried in vacuum at room temperature. The melting point of the complex is >300°C [1]. The complex with ligand 2 was prepared by reacting an Mn^{II} salt with the stoichiometric amount of $Na_4(C_{21}H_{18}N_2O_8)$ in aqueous solution of pH 5 to 6. The precipitate was washed with water, then with thioether, and dried at 35 to 40°C [2].

Characteristic absorption maxima in the IR spectra (wave numbers in cm^{-1}) of the compounds in KBr were assigned as follows (maxima of the free ligands in parentheses): 2928 (2920), $\nu(CH_2)$; −(1715), $\nu(CH_2COOH)$; 1615 (1615), $\nu_{as}(COO^-)$; 1405 (1430), $\nu_s(COO^-)$; 1320, Ar–NH–CH_2 for the complex with ligand 1 [1]; and 2925 (2920), $\nu(CH_2)$; −(1720), $\nu(CH_2COOH)$; 1618 (1615), $\nu_{as}(COO^-)$; 1425 (1430), $\nu_s(COO^-)$; −(1380), $\nu(CH_2COOH)$; 1330, 1230 (1330, 1235), ν(Ar–N=) for the complex with ligand 2 (Ar = aryl) [2]. The IR data indicate coordination of the N atoms and of the carboxylate O atoms for both compounds [1, 2].

The complex with ligand 2 is unstable in air. It becomes blueish on standing and heating. It is sparingly soluble in water and organic solvents. The solubility in water is 0.27 g/L of solution [2]. The complex with ligand 1 is insoluble in water and alcohol. A dinuclear polymeric structure with tetracoordination by two ligand molecules in equatorial positions and by two water molecules in axial positions is proposed for this compound [1]. A dinuclear polymeric structure is also possible for the complex with ligand 2 [2].

References:

[1] Macarovici, C. G., Chis, E. (Rev. Roumaine Chim. **22** [1977] 657/64).
[2] Macarovici, C. G., Chis, E. (Rev. Roumaine Chim. **25** [1980] 95/103, 96, 98).

22.29 With N-Tetraacetic Acids Derived from Phthaleins or Sulfophthaleins

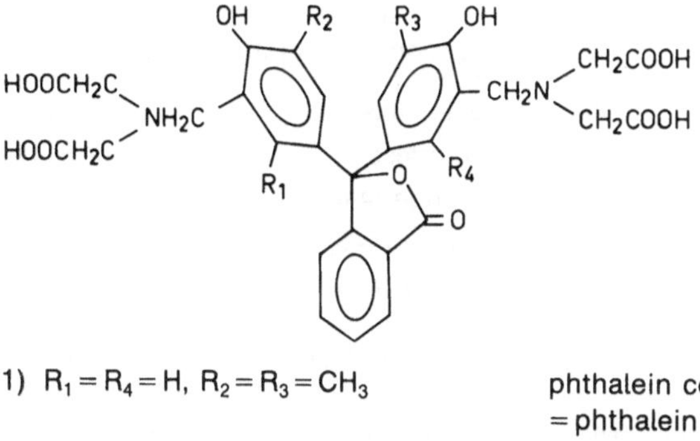

1) $R_1 = R_4 = H$, $R_2 = R_3 = CH_3$ — phthalein complexone (= metalphthalein = phthalein purple = $C_{32}H_{32}N_2O_{12}$)

2) $R_1 = R_4 = CH_3$, $R_2 = R_3 = CH(CH_3)_2$ — thymolphthalein complexone (= thymolphthalexon = $C_{38}H_{44}N_2O_{12}$)

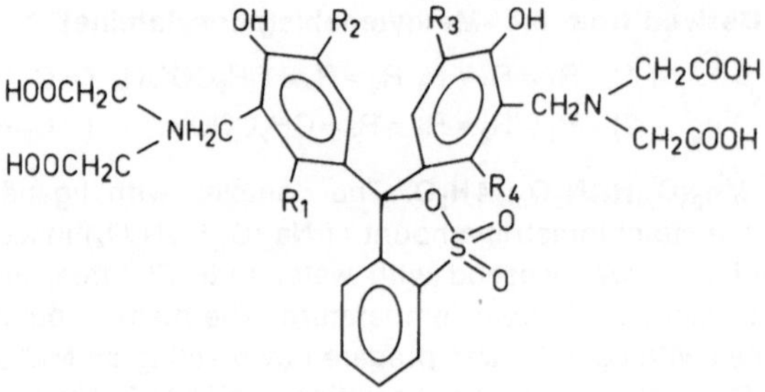

3) $R_1 = R_4 = H$, $R_2 = R_3 = CH_3$ xylenol orange ($= C_{31}H_{32}N_2O_{13}S$)
4) $R_1 = R_2 = R_3 = R_4 = CH_3$ methylxylenol blue ($= C_{33}H_{35}N_2NaO_{13}S$)
5) $R_1 = R_4 = CH_3$, $R_2 = R_3 = CH(CH_3)_2$ methylthymol blue ($= C_{37}H_{40}N_2Na_4O_{13}S$)

Only those complexes of Mn^{II} with metallochromic indicators for which recent publications exist are considered. Older data about complex formation of manganese with the above ligands can be found in the analytical literature, for instance [1 to 3]. In [3], the earlier literature, especially on Mn^{II} complexes with ligands 3 and 5, is reviewed.

Complex formation of Mn^{II} (or other divalent metals) is indicated: by color deepening at pH 10 for ligand 1 ($\lambda_{max} \approx 575$ nm) [4] or ligand 2 (blue [17], $\lambda_{max} \approx 600$ nm [4], ≈ 625 nm [17]); by a red-violet color reaction at pH 5 to 6 [3, p. 40] or >pH 6 [5] for ligand 3; and by a blue color reaction at pH 6 for ligand 4 [6] or 0 to 6.5 for ligand 5 [3, 7].

The logarithm of the stability constant for the protonated 1:1 complex with ligand 3, formed by the reaction $Mn^{2+} + C_{31}H_{28}N_2O_{13}S^{4-} \rightleftharpoons Mn(C_{31}H_{28}N_2O_{13}S)^{2-}$, is log K = 5.11 at 20°C, I = 0.1 M ($NaClO_4$, 0.06 M, and CH_3COONH_4, 0.04 M), pH 6.2 to 7.4 [8]; log K = 5.89 at pH 6.6 [9], both values determined by spectrophotometry. For an undefined 1:1 complex with ligand 5, stability constants were determined by pH-metric titration [10] or by spectrophotometry [9]. Thermodynamic parameters of formation for the complex with ligand 5 are reported [10].

Maxima in the electronic absorption spectra are: 580 nm (log ε = 4.58) [8] or 582 nm (4.35) [9] in the 6 to 9 pH range for the complex with ligand 3 [8, 9] and 615 nm (4.28) at pH 8.5 for the complex with ligand 5 [15]. Reversible thermochromic changes were observed for aqueous solutions of the ligands 1 through 5 containing a 10- to 100-fold excess of Mn^{2+} ions in a narrow pH range characteristic of the individual system [11, 12]:

No.	ligand	pH at 25°C	color of complexes (λ_{max} in nm) at room temperature	at 60°C
1	phthalein complexone	5.27	pale violet	violet (578)
2	thymolphthalein complexone	6.41	colorless	blue (614)
3	xylenol orange	4.92	orange (440)	reddish violet (583)
4	methylxylenol blue	5.04	yellow (446)	blue (615)
5	methylthymol blue	5.85	yellow (438)	green (605)

The high-temperature color and absorption maxima of the complexes strongly resemble the color or absorption maxima of the parent ligands (and also of the complexes) in alkaline medium [11, 12]. The change in color with temperature is explained by an equilibrium between a species with an uncoordinated phenolic hydroxy group and a species with a deprotonated coordinated phenolate group [11, 12]. Chelate formation involving the two nitrogen atoms and two carboxylate groups, but no phenolate oxygen atom, is concluded from another spectrophotometric study of Mn^{II} complexes with ligand 3. In the latter case, however, equimolar concentrations of Mn^{II} and the ligand were used [8].

Time and temperature had no effect on the complex stability in the pH range 6 to 9 for ligand 3 [8, 9] and at pH 7 for ligand 5 [9]. Above pH 9, the intensity of the absorption band at 580 nm for the complex with ligand 3 decreases as a result of formation of another complex species [8]. The use of the ligands as complexometric indicators for the determination of manganese was proposed, for ligand 1 [13], for ligand 2 [16], for ligand 4 [14], and for ligand 5 [7].

References:

[1] Přibil, R. (Komplexometrie, Vol. 1, Prinzipien und Grundbestimmungen, Leipzig 1963, pp. 45, 47, 50, 51).
[2] Schwarzenbach, G., Flaschka, H. (Die Komplexometrische Titration, Enke, Stuttgart 1965, pp. 33/5).
[3] Buděšinský, B. (in: Flaschka, H. A., Barnard, A. J., Chelates in Analytical Chemistry, Vol. 1, London – New York 1967, pp. 15/47).
[4] Körbl, J., Přibil, R. (Collection Czech. Chem. Commun. **23** [1958] 1213/8).
[5] Körbl, J., Přibil, R., Emr, A. (Collection Czech. Chem. Commun. **22** [1957] 961/6).
[6] Vytřas, K., Vytřasová, J. (Chem. Zvesti **28** [1974] 779/88, 787).
[7] Körbl, J., Přibil, R. (Collection Czech. Chem. Commun. **23** [1958] 873/80, 877).
[8] Bogachuk, L. G., Sheka, I. A. (Ukr. Khim. Zh. **42** [1976] 899/905; Soviet Progr. Chem. **42** No. 9 [1977] 1/7, 1/2, 6).
[9] Tataev, O. A., Anisimova, L. G. (Zh. Analit. Khim. **26** [1971] 184/7; J. Anal. Chem. [USSR] **26** [1971] 166/8).
[10] Saraswat, I. P., Sharma, C. L., Sharma, A. (J. Indian Chem. Soc. **55** [1978] 757/8).

[11] Nakada, S., Yamada, M., Ito, T., Fujimoto, M. (Bull. Chem. Soc. Japan **52** [1979] 766/71).
[12] Nakada, S., Yamada, M., Ito, T., Fujimoto, M. (Chem. Letters **1977** 1243/6).
[13] Belcher, R., Leonard, M. A., West, T. S. (Chem. Ind. [London] **1958** 128/9).
[14] Vytřas, K., Mach, V., Kotrlý, S. (Chem. Zvesti **29** [1975], 61/7, 65).
[15] Tikhonov, V. N. (Zh. Analit. Khim. **22** [1967] 658/63; J. Anal. Chem. [USSR] **22** [1967] 571/4).
[16] Přibil, R., Kopanica, M. (Chemist-Analyst **48** [1959] 35/6).
[17] Parkash, R., Gupta, S. K., Singh, R. P., Singhal, R. L. (J. Indian Chem. Soc. **61** [1984] 731/5).

22.30 With N-Ethyldiethylenetriamine-N,N′,N″-triacetic Acid

$C_2H_5N(CH_2COOH)CH_2CH_2N(CH_2COOH)CH_2CH_2NH(CH_2COOH)$ ($= C_{12}H_{23}N_3O_6$)

The stability constant of the $\mathbf{Mn^{II}(C_{12}H_{20}N_3O_6)^-}$ complex at 25°C for I = 0.1 M (KNO_3) is log K_1 = 14.3. It was polarographically determined using the exchange equilibrium with the Cd complex, Hama, H., Takamoto, S. (Nippon Kagaku Kaishi **1975** No. 7, pp. 1182/5; C.A. **83** [1975] No. 121751).

22.31 With Derivatives of Diethylenetriamine-N,N,N″,N″-tetraacetic Acid

$$R{-}N\begin{matrix} CH_2CH_2N(CH_2COOH)_2 \\ CH_2CH_2N(CH_2COOH)_2 \end{matrix}$$

1) $R = CH_2{=}CHCH_2OCH_2CH_2$; $(= C_{17}H_{29}N_3O_9 = H_4L)$

2) $R = HOOCCH_2CH_2$; $(= C_{15}H_{25}N_3O_{10})$

Potentiometric pH titration studies of solutions containing Mn^{2+} ions and ligand 1 showed the existence of the MnL^{2-} and $MnHL^{-}$ ions. Stability constants at 20°C for I = 0.1 M (KCl) determined with a glass electrode using tris(2-aminoethyl)amine (tren) as an auxiliary ligand are: log K_1 = 12.00 and log K^{Mn}_{MnHL} = 8.24. The log K values suggest that the ether oxygen atom is involved in coordination [1]. Complex formation of Mn^{II} and ligand 2 in the 2.5 to 4.2 pH range and at pH 9.2 was indicated by polarographic studies [2].

References:

[1] Tikhonova, L. I., Samoilova, O. I., Lyubchanskii, E. R., et al. (Khim. Farm. Zh. **18** [1984] 166/70; Pharm. Chem. J. **18** [1984] 101/5).

[2] Vasil'eva, V. F., Lavrova, O. Yu., Dyatlova, N. M., Yashunskii, V. G. (Zh. Obshch. Khim. **36** [1966] 674/9; J. Gen. Chem. [USSR] **36** [1966] 688/92).

22.32 With Diethylenetriamine-N,N,N′,N″,N″-pentaacetic Acid

$(HOOCCH_2)_2NCH_2CH_2N(CH_2COOH)CH_2CH_2N(CH_2COOH)_2$ $(= C_{14}H_{23}N_3O_{10} = H_5L)$

22.32.1 Manganese(II) Complexes in Solution

The existence of the species MnL^{3-}, $MnHL^{2-}$, MnH_2L^{-}, and Mn_2L^{-} has been established by spectrophotometric and potentiometric studies [1 to 5]. The MnL^{3-} ion is the predominant species at pH 6 to 9 [10]. Stability constants of the various species and the protonation constant of the MnL^{3-} ion were determined potentiometrically with a glass electrode (gl) [1, 3], with an Hg electrode (Hg) [2, 4, 5], or polarographically (pol) [6] by the exchange method [1, 3, 6] using the stability constant of the Cd complex as a reference [3] and tris(2-aminoethyl) amine (= tren) as an auxiliary ligand [1, 3]:

equilibrium	log K[a)]	log K[b)]	log K[c)]	log K[d)]	log K[e)]
$Mn^{2+} + L^{5-} \rightleftharpoons MnL^{3-}$	15.60	15.13	15.5	15.1	15.6
$Mn^{2+} + HL^{4-} \rightleftharpoons MnHL^{2-}$	8.63	9.00	—	9.1	—
$MnL^{3-} + H^{+} \rightleftharpoons MnHL^{2-}$	4.64	4.42	4.5	4.40	—
$MnL^{3-} + Mn^{2+} \rightleftharpoons Mn_2L^{-}$	2.09	—	—	—	—

a) At 20°C (Hg), I = 0.1M ($NaNO_3$) [2]. – b) At 20°C (gl), I = 0.1M (KCl) [3]. – c) At 25°C (Hg), I = 0.1M (KNO_3) [4, 5]. – d) At 25°C (gl), I = 0.1M (KNO_3) [1]. – e) At 25°C (pol), I = 0.1M (KCl) [6].

Effective (or "arbitrary" [7]) stability constants defined on p. 29 taking into consideration side reactions of the Mn^{2+} ion (hydrolysis), of the ligand (formation of protonated anions), or of the complex (formation of hydroxy or protonated species) were calculated and are presented in [7]. Thermodynamic data of formation, ΔH_1, ΔG_1, and ΔS_1, for the MnL^{3-} species (ΔH_1 determined calorimetrically in aqueous solution for I = 0.1M (KNO_3), ΔG and ΔH in kcal/mol, ΔS in $cal \cdot mol^{-1} \cdot K^{-1}$) are: −7.18, −20.92, and 47.0, respectively, at 20°C [8]; −7.5, −21.1, and 46, respectively, at 25°C [9].

The half-wave potential of MnL^{3-} at pH 9.3 and room temperature is −1.04 V against SCE (−1.58 V for the free Mn^{2+} ion) [6]. The complex is effective in supplying Mn as a micronutrient to plants [11]. The release of Mn from the chelate in the mammalian body was observed by [12]. The ligand applied as the Ca chelate [12 to 14] or Mn chelate [13, 15] is effective in removing Mn^{2+} ions (labeled by ^{52}Mn) [13], the radionuclides ^{46}Sc, ^{65}Zn, ^{144}Ce [15], and ^{54}Mn [13, 14] from mammalian bodies. The Mn chelate was more effective than the Ca chelate [13]. The preparation and water relaxation properties of proteins labeled with Mn chelates of the ligand were studied in [10].

References:

[1] Chaberek, S., Frost, A. E., Doran, M. A., Bicknell, N. J. (J. Inorg. Nucl. Chem. **11** [1959] 184/96, 184, 190).
[2] Anderegg, G., Nägeli, P., Müller, F., Schwarzenbach, G. (Helv. Chim. Acta **42** [1959] 827/36, 834).
[3] Durham, E. J., Ryskiewich, D. P. (J. Am. Chem. Soc. **80** [1958] 4812/7).
[4] Holloway, J. H., Reilley, C. N. (Anal. Chem. **32** [1960] 249/56, 253).
[5] Wänninen, E. (Acta Acad. Aboensis B **21** No. 17 [1960] 1/110, 31, 34; C.A. **1960** 19263).
[6] Dyatlova, N. M., Temkina, V. Ya., Seliverstova, I. A. (Kompleksony Sredstvo Izvestkovogo Khloroza Rast. **1965** 39/46, 40; C.A. **64** [1966] 1604).
[7] Amsheeva, A. A. (Zh. Analit. Khim. **35** [1980] 846/53; J. Anal. Chem. [USSR] **35** [1980] 553/9, 557).
[8] Anderegg, G. (Helv. Chim. Acta **48** [1965] 1722/5).
[9] Wright, D. L., Holloway, J. H., Reilley, C. N. (Anal. Chem. **37** [1965] 884/92, 886).
[10] Lauffer, R. B., Brady, T. J. (Magn. Reson. Imaging **3** [1985] 11/6 from C.A. **103** [1985] No. 101218).

[11] Wallace, A., Wallace, G. A. (J. Plant Nutr. **6** [1983] 451/60).
[12] Koutenský, J., Jonáková, M., Eybl, V., Sýkora, J., Mertl, F. (Prac. Lek. **19** [1967] 52/6; C.A. **67** [1967] No. 10005).
[13] Zablotna, R., Geisler, J., Szot, Z., Żylicz, E. (Nukleonika **19** [1974] 905/15, 908; C.A. **82** [1975] No. 108258).
[14] Kuhn, A. (Strahlentherapie **137** [1969] 101/9).
[15] Szot, Z., Żylicz, E., Zablotna, R., Geisler, J., Czechowska, Z. (Nukleonika **19** [1974] 917/31, 923/4; C.A. **82** [1975] No. 120975).

22.32.2 Manganese(III) Complexes in Solution

The formation of red 1:1 complexes of Mn^{3+} ions with the ligand was revealed by visual [1] or spectrophotometric studies [2, 3]. Complex solutions were prepared by treating solutions containing Mn^{II} complexes with an oxidant, for instance $NaBiO_3$, $(NH_4)_2S_2O_8$, PbO_2 [1, 2], or by reacting a solution containing $MnH_2P_2O_7^+$ ions with the ligand [3]. Complex formation begins at pH 1.45. Maximum formation of the MnL^{2-}, MnH_2L, and MnH_3L^+ species is observed in the 3.4 to 6 pH range (λ_{max} = 530 nm). The breakdown of the complexes occurs at pH > 6. The stability constants log K_1 = 19.35, log $K^{Mn}_{MnH_2L}$ = 5.36, and log $K^{Mn}_{MnH_3L}$ = 3.70 at 18 to 22°C, pH 3.4 to 6 and $I \leqq 0.01$ M were determined [2]. However, formation of only two species, the MnL^{2-} ion (λ_{max} = 505 nm) at pH ~1 to 2 and of a hydroxy complex not specified (λ_{max} = 480 nm) at pH 2.5 to 7 was observed by other authors. The stability constant log K_1 = 31.06 at pH 2, I = 1.0 M ($NaClO_4$), is reported [3]. The solutions are unstable due to an intramolecular redox reaction [2, 3], which is first order with respect to the Mn^{III} concentration [3].

References:

[1] Hernández Méndez, J., Medina Escriche, J., Corberán Martínez, M. A. (Acta Salmanticensia Cienc. No. 54 [1975] 61/78, 73; C.A. **86** [1977] No. 128299).
[2] Mikhailova, T. V., Astakhov, K. V., Zhirnova, N. M. (Zh. Fiz. Khim. **45** [1971] 1773/6; Russ. J. Phys. Chem. **45** [1971] 1003/5).
[3] Bogdanovich, N. G., Pechurova, N. I., Martynenko, L. I., Piunova, V. V. (Zh. Neorgan. Khim. **16** [1971] 2507/11; Russ. J. Inorg. Chem. **16** [1971] 1337/9).

22.33 With Polycarboxylic Acids Derived from Triethylenetetraamine or Tetraethylenepentaamine

$(HOOCCH_2)_2NCH_2CH_2[N(CH_2COOH)CH_2CH_2]_nN(CH_2COOH)_2$

ligand 1 with n = 2 $(= C_{18}H_{30}N_4O_{12} = H_6L)$

ligand 2 with n = 3 $(= C_{22}H_{37}N_5O_{14} = H_7L)$

Potentiometric (glass and Hg electrode) [1], polarographic (exchange method) [2], and titrimetric studies [3, 4] indicate that 1:1 [1 to 3] and 2:1 complexes (Mn:ligand) [1, 4] form in solutions containing Mn^{2+} ions and either of the ligands. Some stability and protonation constants for the complexes with ligand 1 are: $\log K_1 = 14.65$ (glass electrode) and 14.30 (Hg electrode) at 25°C and I = 0.1M (KNO_3) [1]; 14.62 (polarographically) at 25°C and I = 0.1 M (KCl) [2]. Log K = 8.74 was found for $MnL^{4-} + H^+ \rightleftharpoons MnHL^{3-}$; log K = 3.75 for $MnHL^{3-} + H^+ \rightleftharpoons MnH_2L^{2-}$; and log K = 6.54 for $MnL^{4-} + Mn^{2+} \rightleftharpoons Mn_2L^{2-}$ under the above conditions [1]. The stability constant for the 1:1 complex with ligand 2 is $\log K_1 = 14.0$ at 25°C and I = 0.1 M (KCl) [2]. Plots of the pH dependence of effective or "conditional" [5] stability constants (defined on p. 29) for 1:1 and 1:2 complexes with ligand 1 taking into account hydrolysis of the metal ion, formation of protonated ligand anions, and formation of hydroxy or protonated complexes are presented in [5]. The titrimetric determination of manganese by use of ligand 1 taking into consideration the effective stability constants of the 1:1 and 1:2 complexes, is described in [5].

References:

[1] Harju, L. (Anal. Chim. Acta **50** [1970] 475/89, 484).
[2] Dyatlova, N. M., Temkina, V. Ya., Seliverstova, I. A. (Kompleksony Sredstvo Izvestkovogo Khloroza Rast. **1965** 39/46, 40; C.A. **64** [1966] 1604).
[3] Přibil, R., Veselý, V. (Talanta **9** [1962] 939/43).
[4] Přibil, R., Veselý, V. (Talanta **12** [1965] 191/2).
[5] Harju, L., Ringbom, A. (Anal. Chim. Acta **49** [1970] 205/19, 210).

22.34 With Polyazacycloalkane-N-polyacetic Acids

No.	ligand	X	Z	formula
1	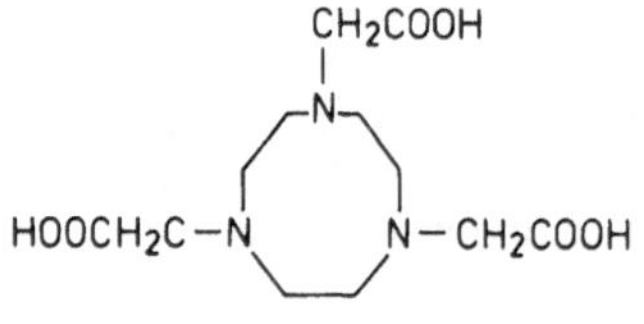	—	—	$C_{12}H_{21}N_3O_6 = H_3L$

No.	ligand	X	Z	formula
2	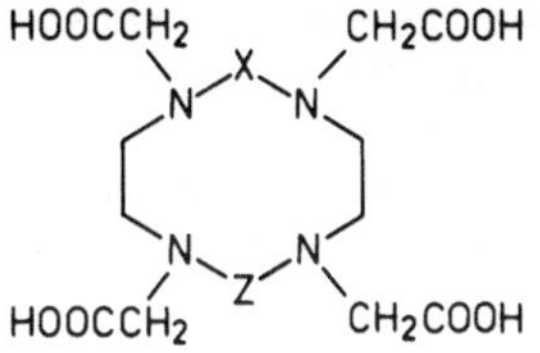	CH_2CH_2	CH_2CH_2	$C_{16}H_{28}N_4O_8$ $= H_4L$
3		CH_2CH_2	$CH_2CH_2CH_2$	$C_{17}H_{30}N_4O_8$ $= H_4L$
4		CH_2CH_2	$CH_2CH(CH_3)CH_2$	$C_{18}H_{32}N_4O_8$ $= H_4L$
5		CH_2CH_2	$CH_2CH(C_2H_5)CH_2$	$C_{19}H_{34}N_4O_8$ $= H_4L$
6		$CH_2CH_2CH_2$	$CH_2CH_2CH_2$	$C_{18}H_{32}N_4O_8$ $= H_4L$
7	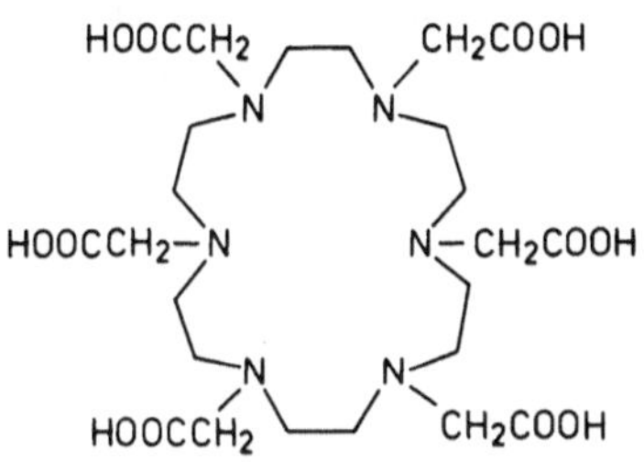	—	—	$C_{24}H_{42}N_6O_{12}$ $= H_6L$

Stability constants of 1:1 manganese(II) complexes with the ligands 2 to 6 in aqueous solution were determined potentiometrically with a glass electrode at 20°C and I = 0.1 M KCl [1]:

ligand	2	3	4	5	6	7
log K_1	17.8	14.9	16.4	9.2	11.2	14.2

The solvent water proton longitudinal relaxation rate $1/T_1$ for solutions of complexes with ligand 1 and 2 was studied as a function of the magnetic field and the pH. Competition experiments for these complexes with those of Mn^{II} complexes of ethylenediaminetetraacetic acid (p. 28) and diethylenetriaminepentaacetic acid (p. 82) were also performed. The relevance of the results is discussed in view to the use of these complexes in NMR imaging [3].

$Na[Mn^{II}(C_{12}H_{18}N_3O_6)]\cdot 6H_2O$. The complex with ligand 1 was prepared by reacting a solution of the ligand adjusted to pH 7 with concentrated aqueous $Mn(ClO_4)_2\cdot 6H_2O$. After another adjustment to pH 7 and addition of ethanol, the resulting solution was kept at room temperature for three days. The colorless crystals which appeared were collected, washed with ethanol and ether, dried in the air, and recrystallized from water-ethanol (1:1). Magnetic susceptibility measurements showed that the Curie-Weiss law is obeyed in the 100 to 300 K temperature range. The effective magnetic moment is 5.7 μ_B. The IR spectrum shows the ν(CO) band at 1580 cm^{-1} [2].

$[Mn^{III}(C_{12}H_{18}N_3O_6)]$ was prepared by oxidizing $Na[Mn(C_{12}H_{18}N_3O_6)]\cdot 6H_2O$ in aqueous solution with $K_2S_2O_8$ at 70°C. On cooling to 0°C, red, needle-shaped crystals precipitated, which were washed with ethanol and ether and dried in air [2].

The compound crystallizes in the monoclinic space group $P2_1/c-C^5_{2h}$ (No. 14) with a = 8.94(1), b = 13.88(1), c = 11.72(1) Å, β = 104.3(1)°; Z = 4. It is isomorphous with $M(C_{12}H_{18}N_3O_6)$, M = Al, Cr^{III}, Co^{III}. The structure consists of $[Mn(C_{12}H_{18}N_3O_6)]$ molecules. The trigonal prismatic ligand is coordinated distorted-octahedrally through three nitrogen and three oxygen atoms. The magnetic moment, $\mu_{eff} = 4.95\ \mu_B$, indicates a high-spin complex and is constant in the 100 to 300 K temperature range. In the IR spectrum, two ν(CO) vibration bands, at 1700 and 1650 cm^{-1}, indicate the presence of two differently coordinated carboxylate groups with two different Mn–O distances [2].

The compound is only slightly soluble in water. It shows in aqueous solution absorption maxima at 481 and 471 nm (log ε = 2.85 for both). The invariance of the spectrum indicates that the complex is stable for at least 24 h at 25°C in the 1 to 10 pH range. A formal potential at 20°C, pH 7, I = 0.05 M KCl, for the reaction $Mn^{III}(C_{12}H_{18}N_3O_6)^{\circ} + e^- \rightleftharpoons Mn^{II}(C_{12}H_{18}N_3O_6)^-$ in aqueous solution of E° = 0.80 V vs. NHE was determined by cyclovoltammetry [2].

References:

[1] Stetter, H., Frank, W., Mertens, R. (Tetrahedron **37** [1981] 767/72).
[2] Wieghardt, K., Bossek, U., Chaudhuri, P., et al. (Inorg. Chem. **21** [1982] 4308/14).
[3] Geraldes, C. F. G. C., Sherry, A. D., Brown, III, R. D., Koenig, S. H. (Magn. Resonance Med. **3** [1986] 242/50 from C.A. **105** [1986] No. 93923).

23 Complexes with Hydrazinecarboxylic Acid and Derivatives

23.1 With Hydrazinecarboxylic Acid $H_2NNHCOOH$ ($=N_2H_3COOH=CH_4N_2O_2$)

$Mn(N_2H_3COO)_2$. The crystalline compound was prepared by reacting the calculated amount of $MnCl_2$ with dry, liquid $N_2H_3COON_2H_5$ (obtained from $(NH_4)_2CO_3$ and hydrazine hydrate) and concentrating the resulting solution [1]. $Mn(N_2H_3COO)_2$ was also obtained by dehydration of its dihydrate [8, 9]. The characteristic IR absorptions (in cm^{-1}) of the compound in KBr were assigned as follows: 1608, $\nu_{as}(COO^-)$; 1375, $\nu_s(COO^-)$; 1590, 1484, $\delta(N_2H_3)$; 1208, $\varrho(N_2H_3)$; 1000, $\nu(NN)$; 820, $\nu(OCO)$ or $\varrho(NH_2)$; 660, $\varrho(COO^-)$; 620, $\varrho(NH_2)$ or $\delta(COO^-)$; and 430, $\nu(Mn–N)$ [1]. Thermoanalytical studies (TG, DTG, DTA; heating rate 20°C/min) indicate the exothermic decomposition of the compound in the 170 to 300°C temperature range (DTA peaks at 170, 180, and 200°C) [11] to give MnO as the final product [10, 11] with $MnC_2O_4 \cdot N_2H_4$ as the intermediate [11]. Previous thermoanalytical studies are reported in [1, 8, 9]. $Mn(N_2H_3COO)_2$ (and other $M^{II}(N_2H_3COO)_2$ complexes) are of interest as precursors of finely divided metals and metal oxides [10, 11].

$Mn(N_2H_3COO)_2 \cdot 2H_2O$ was prepared by treating an aqueous solution containing Mn^{2+} ions (25 ml; 0.4 M) with a 5% solution of N_2H_3COOH in $N_2H_4 \cdot H_2O$ until the precipitate initially formed just dissolved. The clear solution thus obtained was kept open to the atmosphere. A crystalline solid separated from the solution within a couple of days [11]. The crystals, which were found to be light blue [8, 9] or colorless thick plates [2 to 4], were washed with ethanol and stored under hexane [9]. Previous preparations employed neat liquid N_2H_3COOH [9] or an aqueuos solution of N_2H_3COOH as the reactants with the Mn salt [5]. Or, carbon dioxide was bubbled through an aqueous solution containing an Mn^{II} salt and $N_2H_4 \cdot H_2O$ [2 to 4, 8, 12]. $Mn(N_2H_3COO)_2 \cdot H_2O$ also forms when the compound $[Mn(N_2H_4)_2](CH_3COO)_2$ ("Manganese" D3 [1982], p. 68) in its mother liquor is exposed to the air [4].

$Mn(N_2H_3COO)_2 \cdot 2H_2O$ is orthorhombic; the lattice constants are: $a=11.052 \pm 0.010$, $b=9.862 \pm 0.020$, $c=7.847 \pm 0.005$ Å; Z=4. The space group is Pba2-C_{2v}^8 (No. 32) [2 to 4]. The $N_2H_3COO^-$ anions are coordinated by one N atom and one O atom forming two five-membered chelate rings. The remaining octahedral positions are occupied by water O atoms. As shown in **Fig. 10**, p. 88, the complex has a chain structure. The chains are formed by two alternating types of coordination octahedra which both have trans (O,O), cis (O,O), cis (N,N) structures. Half of the water molecules are directly coordinated to the metal (in the type II coordination octahedra); the other half appear as water of crystallization (type I octahedra). The chains are held together by strong hydrogen bonds either directly or via the water of crystallization. The Mn–N bond lengths are 2.208 ± 0.016 (type I) or 2.189 ± 0.016 (type II). In octahedron I there are two different $Mn–O_{carboxylate}$ bond lengths, 2.138 ± 0.014 and 2.248 ± 0.014 Å. In octahedron II, the $Mn–O_{carboxylate}$ bond length is 2.145 ± 0.018 Å and the $Mn–O_{H_2O}$ bond length is 2.213 ± 0.016 Å. Other bond lengths, bond angles, atomic coordinates, structure factors, and anisotropic thermal parameters are given in [2]. The measured and calculated densities are 1.893 and 1.872 g/cm^3 [2 to 4]. Characteristic IR absorptions (in cm^{-1}) of the compound in KBr pellets were assigned as follows: $\nu(OH)$, 3500; $\nu(NH)_{H\text{-}bonded}$, 3340, 3280, 3180, 2960; $\nu_{as}(COO^-)+\delta(H_2O)$, 1655; $\delta(N_2H_3)$, 1590, 1560; $\nu_s(COO^-)$, 1510; $\varrho(NH_2)$, 1210, 1200, 1105; $\nu(N–N)$, 995 [9], 998 [12]; $\nu(Mn–N)$, 415 [9]. The compound is piezoelectric [2, 3].

$Mn(N_2H_3COO)_2 \cdot 2H_2O$ is very stable at room temperature [2]. TG, DTG, and DTA analyses (heating rate 20°C/min [9, 11]) indicate dehydration in the 110 to 175°C temperature range (DTA peak at 120°C [9], 125°C [8], or 130°C [11]) to give the anhydrous compound [8, 9, 11].

$Mn(N_2H_3COO)_2 \cdot 2N_2H_4$ was prepared by treating a fine powder of $Mn(N_2H_3COO)_2 \cdot 2H_2O$ with $N_2H_4 \cdot H_2O$ [11]. Previous authors obtained the colorless monoclinic crystals by free evaporation of an ammoniacal solution of an Mn^{II} halide and $N_2H_4 \cdot H_2O$ [6, 7]. A single-crystal

X-ray determination yielded the lattice parameters a = 8.217 ± 0.01, b = 7.449 ± 0.005, c = 8.567 ± 0.01 Å, β = 110°56′ ± 6′ and Z = 2. The space group is $P2_1/c-C_{2h}^5$ (No. 14). The manganese complex is isostructural with the corresponding zinc and cobalt complexes [7]. The X-ray data suggest an octahedral molecular structure with N and O atoms of chelating hydrazinecarboxylate anions in the equatorial positions and N atoms of monodentate hydrazine molecules in the axial positions [6, 7]. The calculated density is 1.819 g/cm³, and the experimental density at room temperature is 1.800 g/cm³. The complex is paramagnetic. The IR spectrum shows the ν(N–N) band of the unidentately coordinated N_2H_4 at ~925 cm⁻¹ in addition to the ν(N–N) band of the coordinated $N_2H_3COO^-$ anion at ~1000 cm⁻¹ [11]. $Mn(N_2H_3COO)_2 \cdot 2N_2H_4$ decomposes slowly in air [7]. Thermoanalytic studies (TG, DTG, and DTA; heating rate 20°C/min) indicate the exothermic decomposition of the compound in the 130 to 290°C temperature range (DTA peaks at 175 and 195°C) with MnO as the final product and probably $Mn(N_2H_3COO)_2$ as the intermediate [11]. $Mn(N_2H_3COO) \cdot 2N_2H_4$ is soluble in water. It decomposes in dilute acids, evolving CO_2 [7].

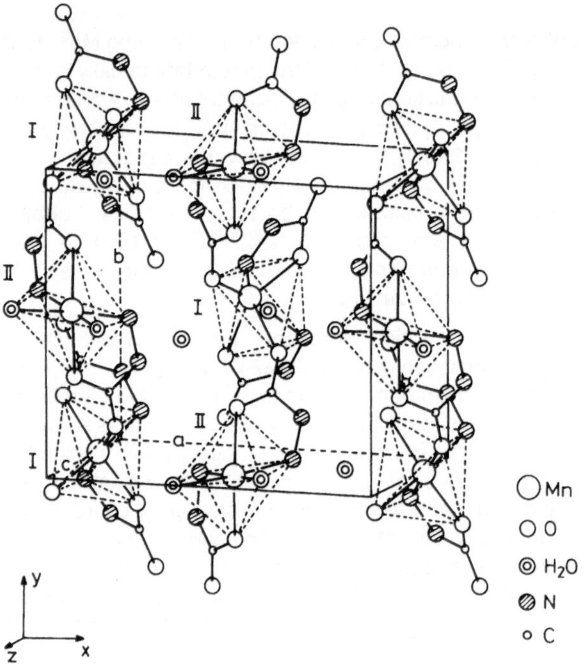

Fig. 10. Clinographic projection of the structure of $Mn(N_2H_3COO)_2 \cdot 2H_2O$ [2].

References:

[1] Patil, K. C., Budkuley, J. S., Pai Verneker, V. R. (J. Inorg. Nucl. Chem. **41** [1979] 953/5).
[2] Braibanti, A., Tiripicchio, A., Manotti-Lanfredi, A. M., Camellini, M. (Acta Cryst. **23** [1967] 248/54, 249, 253).
[3] Braibanti, A., Tiripicchio, A., Manotti-Lanfredi, A. M., Dallavalle, F. (Ric. Sci. **36** [1966] 1210/3).
[4] Braibanti, A., Bigliardi, G., Manotti-Lanfredi, A. M. (Ateneo Parmense II **1** [1965] 81/6, 82; C.A. **65** [1966] 16174).
[5] Slivnik, J., Rihar, A., Sedej, B. (Monatsh. Chem. **98** [1967] 200/3).
[6] Braibanti, A., Bigliardi, G., Manotti-Lanfredi, A. M., Tiripicchio, A. (Nature **211** [1966] 1174/5).

[7] Braibanti, A., Bigliardi, G., Canali Padovani, R. (Gazz. Chim. Ital. **95** [1965] 877/84, 878, 880).
[8] Gogorishvili, P. V., Chkoniya, M. E., Akhobadze, D. A., Chkoniya, T. V. (Tr. Gruz. Politekhn. Inst. **1969** No. 1, pp. 7/13, 10/2; C.A. **72** [1970] No. 139140).
[9] Patil, K. C., Soundararajan, R., Goldberg, E. P. (Syn. React. Inorg. Metal-Org. Chem. **13** [1983] 29/43).
[10] Macek, J., Rahten, A., Slivnik, J. (Proc. 1st Eur. Symp. Therm. Anal., London 1976, pp. 161/3; C.A. **87** [1977] No. 47531).
[11] Ravindranathan, P., Patil, K. C. (Proc. Indian Acad. Sci. Chem. Sci. **95** [1985] 345/56, 346, 350/2).
[12] Braibanti, A., Dallavalle, F., Pellinghelli, M. A., Leporati, E. (Inorg. Chem. **7** [1968] 1430/3).

23.2 With the 2-(1-Phthalazinyl)ethyl Ester of Hydrazinecarboxylic Acid

$H_2NNHCOOCH_2CH_2$

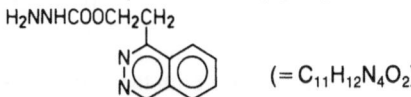

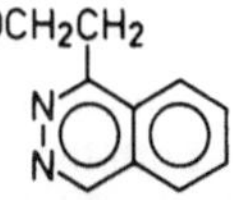

($=C_{11}H_{12}N_4O_2$)

$Mn(C_{11}H_{12}N_4O_2)_3(OH)_2$ was prepared by adding 0.1 M aqueous binazine (monohydrochloride of the ligand) to 0.1 M aqueous $MnSO_4$ and then adding 1 M aqueous NaOH in small portions. The yellow precipitate was dried over solid NaOH. The compound melts at 185°C. It can be extracted with chloroform, amyl alcohol, ethyl acetate, butyl acetate, and benzene. A chloroform solution shows an absorption maximum at 380 nm. The absorbance is at its optimum and is most stable at pH 8 to 11, where the extinction coefficient is 11000 $L \cdot mol^{-1} \cdot cm^{-1}$. This sensitive reaction can be used for the determination of manganese in deep sea waters, Sikorska-Tomicka, H. (Chem. Anal. [Warsaw] **22** [1977] 761/5; C.A. **88** [1978] No. 176880).

23.3 With 2-(2-Phenylhydrazino)propanoic Acid and Derivatives

p-$XC_6H_4NHNHCH(CH_3)COOH$ (= HL)

1) X = H; ($=C_9H_{12}N_2O_6$) 2) X = CH_3; ($=C_{10}H_{14}N_2O_4$) 3) X = NO_2; ($=C_9H_{11}N_3O_4$)

Stability constants of the species MnL^+ and MnL_2 in 40:60 vol% ethanol-water were determined potentiometrically (glass electrode) at various temperatures and I = 0.1 M. Below are listed log K_n values at 30°C and the thermodynamic parameters of formation, ΔG and ΔH in kcal/mol and ΔS in $cal \cdot mol^{-1} \cdot K^{-1}$. (The log K_n values for other temperatures are not given in the paper):

No.	log K_1	log K_2	$-\Delta G_1$	$-\Delta G_2$	$-\Delta H_1$	$-\Delta H_2$	ΔS_1	ΔS_2
1	2.92	2.54	4.05	3.52	3.27	3.16	3.05	1.19
2	3.03	2.66	4.22	3.69	3.71	3.22	1.69	1.55
3	2.63	—	3.65	—	3.17	—	1.54	—

The stability constants become higher (data not given) with increasing solvent ethanol content. Separation of the ΔG and ΔH values into their electrostatic ($\Delta G_e \approx -2.15$ kcal/mol and $\Delta H_e \approx 0.8$ kcal/mol) and cratic components ($\Delta G_c \approx \Delta H_c \approx -4.2$ kcal/mol) shows the prevailing covalent nature of bonding in these complexes, Kasi Vishwanatham, R. N., Ram, Kashi, Ram Reddy, M. G. (Indian J. Chem. A **22** [1983] 270/2).

24 Complexes with Amides and Related Compounds

General

In the first two sections of this chapter are described complexes with amides of monocarboxylic acids or of di- and tetracarboxylic acids and related compounds (phthalimide, phthalanilic acids). Complexes with N-carbamoylpyrazole and -imidazoles as well as with carbamates are treated in Section 24.3, those with urea, its derivatives, and related compounds (biuret, azo- and hydrazinedicarboxamides) in Section 24.4. Complexes with derivatives of guanidine and biguanide are found in Section 24.5.

Complexes with carboxamides are prepared by direct reaction of the manganese salt with excess ligand, e.g., complexes with dimethylformamide, or by reaction of the components in water or alcohol. Complexes with urea exist as stable phases in the ternary systems Mn^{II} salt-urea-water. The infrared and Raman spectra of the complexes with amides of carboxylic acids or urea are characterized in general by displacement of the CO stretching mode of the amide to lower wave numbers, indicating that ligation occurs through the carbonyl oxygen atom. The formation of the Mn-O bond causes an increase in the C-N bond strength shown by a shift of ν(CN) to higher wave numbers. The δ(OCN) bands also undergo significant changes. Amides derived from pyridinecarboxylic acids may be coordinated through both the carbonyl oxygen and heterocyclic nitrogen atoms as shown by the X-ray study of the isothiocyanato complex with N,N-diethyl-3-pyridinecarboxamide, $Mn(C_{10}H_{14}N_2O)_2(NCS)_2$, (see p. 118). The X-ray study of the chloro complex with N,N-diethyl-3-pyridinecarboxamide, $Mn(C_{10}H_{14}N_2O)_4Cl_2$ (see p. 117), shows the Mn atom to be surrounded by two Cl^- ions and four pyridine ring nitrogens of the organic ligands. For many complexes, a polymeric structure is discussed, where the organic ligand or the coordinated anions have a bridging function. The octahedral arrangement of the coordinating atoms around the Mn atom was confirmed by magnetic measurements.

Manganese(III) complexes with deprotonated biguanides are prepared from the acid sulfates of the ligands and a manganese(III) salt in strongly alkaline solution. Manganese(IV) compounds are obtained by reaction with $KMnO_4$. Reaction of these compounds with the appropriate alkali salts in dilute acid yields the complex salts with the nondeprotonated ligands. Two biguanide units are bidentately coordinated to manganese, each by two imino nitrogen atoms forming two six-membered chelate rings. The remaining positions of the coordination octahedron are occupied by oxygen atoms. Dimeric structures with O bridges are indicated by magnetic measurements.

24.1 Complexes with Amides of Monocarboxylic Acids

24.1.1 With Formamide $HC(O)NH_2$ ($=CH_3NO$)

$Mn(CH_3NO)_4Cl_2$ was detected as a stable solid phase in the solubility isotherm of the ternary system $MnCl_2$–CH_3NO–H_2O at 25°C together with the species $Mn(CH_3NO)_2Cl_2$ (see p. 91), and $MnCl_2 \cdot 4H_2O$ [1]. For preparation the finely ground $MnCl_2 \cdot 4H_2O$ was dissolved in the minimum of ethanol and the amide added dropwise with stirring at room temperature until the ratio of salt : amide = 1:10. The solution was allowed to stand for 24 h, by which time a precipitate had formed [10]. It crystallizes also from aqueous [2], or aqueous methanolic [7] solution. According to [3] a mixture of equal volumes of acetone and formamide was used for preparation. The crystals, pale pink leaflets [7], were washed first with acetone [3] or formamide [10] then with ether [3, 10] and dried in a vacuum over P_2O_5 and paraffin wax for 2 to 3 d [10]. The heat of formation from anhydrous $MnCl_2$ and $4CH_3NO$, $\Delta H = -23.77 \pm 0.05$ kcal/mol, was calculated from the heat of solution of the components in water at 25°C [3]. A value of $\Delta H = -26.03 \pm 0.04$ kcal/mol was recalculated with the solution enthalpy of $MnCl_2$ being equal to 16.02 kcal/mol [4].

X-ray studies (Weissenberg and crystal rotation) reveal a monoclinic lattice, space group $C_{2/c}$-C_{2h}^6 (No. 15) with lattice constants a = 7.25, b = 14.13, c = 14.60 Å and β = 126.6°; Z = 4, V = 1216 Å^3. The compound is isostructural with $Co(CH_3NO)_4Cl_2$. The density D = 1.65 g/cm^3 was measured by flotation ($CHBr_3$ + benzene) or 1.67 g/cm^3, calculated from X-ray data [2]. The IR spectrum of the complex in fluorinated hydrocarbon, liquid paraffin, or hexachlorobutadiene mulls shows in the 4000 to 700 cm^{-1} region characteristic absorption bands. Their wave numbers (in cm^{-1}) are given with the observed shift Δν (in parentheses) relative to the free ligand in 0.02 M $CHCl_3$ solution and with the assignments: ν_{as}(NH) 3330(−175), ν_s(NH) 3257(−138) and 3200(−195), ν(CO) 1697(−25) and 1655(−67), δ(NH_2) 1587(−3), ν(C–N) 1330(+40) and 1320(+30). The observed decrease of the ν(CO) and increase of the ν(C–N) frequencies indicate coordination of formamide to manganese through oxygen. The splitting of these bands into two components of equal intensity suggests nonequivalent (Mn–O) coordination bonds where two formamide molecules are joined less strongly to the metal than the other ligands. The appearance of three ν(NH) bands (instead of two) in the 3400 to 3200 cm^{-1} region can be attributed to changes in the strength of the NH···Cl hydrogen bonds as a consequence of coordination [3]; also see [7]. Two bands in the far IR at 234 and 204 cm^1 were assigned to ν(Mn–O) and one band at 148 cm^{-1} to ν(Mn–Cl) vibration modes. In the Raman spectrum the corresponding bands occur at 232 and 136 cm^{-1} and can be associated with a monomeric structure. The IR band at 204 cm^{-1} may also be a ligand mode [10]. The electronic solid-state spectrum, obtained from KCl-KBr discs, shows five absorption bands in the λ = 320 to 530 nm region [10].

The complex is hygroscopic [2] and incongruently soluble in water [1]. The solution enthalpy ΔH_{sol} = 10.86 ± 0.03 kcal/mol was determined calorimetrically at 25°C [3].

$Mn(CH_3NO)_2Cl_2$ was detected as the second formamide-containing solid phase in the system $MnCl_2$-formamide-H_2O at 25°C; see above. It is congruently soluble in water [1].

$Mn(CH_3NO)_4Br_2$ and **$Mn(CH_3NO)_4I_2$** were prepared following the procedure described for $Mn(CH_3NO)_4Cl_2$ at 0°C under N_2. The Raman spectrum of the bromide reveals bands assigned to ν(Mn–O) at 236 cm^{-1} and to ν(Mn–Br) at 114 cm^{-1} which suggest a monomeric structure. The electronic spectrum shows five absorption bands between λ = 315 and 570 nm [10].

$Mn(CH_3NO)_4SO_4$ crystallizes from an aqueous solution containing $MnSO_4$ and formamide in a 1:10 mole ratio at room temperature. The density D = 1.81 g/cm^3 was measured by the flotation ($CHBr_3$ + benzene) method [2].

$Mn(CH_3NO)_3SO_4$ and **$Mn(CH_3NO)SO_4 \cdot 2H_2O$.** The compounds exist as stable solid phases in the solubility isotherm of the system $MnSO_4$-formamide-H_2O at 25°C together with the species $MnSO_4 \cdot 4H_2O$. The anhydrous complex $Mn(CH_3NO)_3SO_4$ is incongruently soluble in water [5].

$Mn(CH_3NO)_4(NCS)_2$. To a methanolic solution of $Mn(NCS)_2$, excess formamide (50% by volume) was added and the mixture was allowed to crystallize. The complex forms transparent plate-like crystals and yields X-ray powder diagrams which show it to be isostructural with the related complexes of cobalt and nickel [6]. The IR spectrum (KBr pellets) recorded in the 2200 to 400 cm^{-1} region shows characteristic absorption bands at 1680 cm^{-1} assigned to ν(CO), at 2125 and 2065 cm^{-1} assigned to ν(CN), at 792 to ν(CS), and at 487, 474, 447 cm^{-1} to δ(NCS) of the thiocyanate groups which are coordinated to Mn through terminal nitrogen [6, 7]. $Mn(CH_3NO)_4(NCS)_2$ decomposes in air; it is soluble in water, alcohol, and dimethylformamide, sparingly soluble in formamide [6].

$Mn(CH_3NO)_4X(NCS)$ (X = NO_3, Cl). Methanolic solutions of manganese(II)nitrate or chloride and KNCS were mixed in a 1:1 mole ratio, the precipitated KNO_3 or KCl was filtered off and the stoichiometric amount of formamide was added to the filtrate. The precipitates were filtered off and dried. They are pale brownish needles with greenish tint in the case of $Mn(CH_3NO)_4NO_3(NCS)$, whereas the crystals of $Mn(CH_3NO)_4Cl(NCS)$ are pinkish and brittle. The IR spectrum (vaseline mulls) shows in the 4000 to 400 cm^{-1} region characteristic absorption bands which were assigned in a similar way as for $Mn(CH_3NO)_4(NCS)_2$ [7, 8]. The assignment of IR bands to vibrations of coordinated monodentate nitrato groups in $Mn(CH_3NO)_4NO_3(NCS)$ is not unambiguous since bands assignable to coordinated formamide exist in that region. However, the absence of bands due to the free NO_3^- ion (around 1390 to 1380 cm^{-1}) enables one to make such assignment. The complexes are readily soluble in water but $Mn(CH_3NO)_4Cl(NCS)$ is only sparingly soluble in $CHCl_3$ and acetone [7].

$Mn(CH_3NO)_n(CH_3COO)_2$ (n = 1 to 4). $Mn(CH_3NO)_4(CH_3COO)_2$ was prepared by dissolving $Mn(CH_3COO)_2 \cdot 4H_2O$ (12.2 g) in formamide (18 g) with stirring and heating at 100 to 110°C until a clear solution had formed which then was allowed to crystallize. After 2 to 3 d the pale pink crystals were collected, washed with acetone, and dried in vacuum over P_2O_5. The compound is hygroscopic and melts at 95°C. X-ray diffraction patterns show it to be apparently isostructural with the corresponding complexes of cobalt and nickel. The IR spectrum obtained from KBr pellets and mineral oil mulls in the 4000 to 400 cm^{-1} region shows characteristic absorption bands which were assigned as follows (bands of free ligand in cm^{-1} are in parentheses): $\nu(NH_2)$ 3400 broad (3400), $\nu(CO)$ 1695(1700), $\nu_{as}(COO) + \delta(NH_2)$ 1590 broad (1615), $\nu(COO)$ 1430, $\nu(CN)$ 1345(1310), $\nu(CC)$ 940, 790, $\delta(COO)$ 660, $\delta(OCN)$ 625(604). The acetate groups are monodentate. It is concluded that formamide is coordinated to Mn through the carbonyl oxygen, although only an upward shift of the $\nu(C–N)$ band, but only a slight change of the $\nu(CO)$ ligand band, was observed [9].

$Mn(CH_3NO)_4(CH_3COO)_2$ loses the formamide molecules in successive steps upon heating at 155 to 190, 190 to 210, 210 to 224, and 224 to 241°C yielding $Mn(CH_3NO)_3(CH_3COO)_2$, $Mn(CH_3NO)_2(CH_3COO)_2$, $Mn(CH_3NO)(CH_3COO)_2$, and $Mn(CH_3COO)_2$, respectively. $Mn(CH_3NO)_4(CH_3COO)_2$ is soluble in water, methanol, and ethanol, but less soluble in acetone. Conductivity measurements on aqueous and ethanol solutions at 25°C and a dilution rate V = 1000 L/mol yield the molar conductance $\Lambda = 238$ and $19.2\,cm^2 \cdot \Omega^{-1} \cdot mol^{-1}$, respectively, indicating a non-electrolyte in ethanol and a 1:2 electrolyte in water which replaces the organic ligands [9].

References:

[1] Baidinov, T. B., Imanakunov, B. J. (Izv. Akad. Nauk Kirg. SSR **1977** No. 3, pp. 61/3; C.A. **87** [1977] No. 77732).
[2] Nardelli, M., Coghi, L. (Ric. Sci. **29** [1959] 134/8; C.A. **1960** 1151).
[3] Barvinok, M. S., Mashkov, L. V., Obozova, L. A. (Zh. Neorgan. Khim. **20** [1975] 429/32; Russ. J. Inorg. Chem. **20** [1975] 237/8).
[4] Barvinok, M. S., Mashkov, L. V. (Zh. Neorgan. Khim. **25** [1980] 2846/8; Russ. J. Inorg. Chem. **25** [1980] 1570/2).
[5] Baichalova, S., Imanakunov, B. (Izv. Akad. Nauk Kirg. SSR **1970** No. 6, pp. 48/51, 50; C.A. **75** [1971] No. 26007).
[6] Eristavi, D. I., Tsintsadze, G. V., Kereselidze, L. B. (Tr. Gruz. Politekhn. Inst. **1968** No. 7, pp. 39/44, 41; C.A. **74** [1971] No. 60328).
[7] Eristavi, D. I., Tsintsadze, G. V., Kereselidze, L. B. (Tr. Gruz. Politekhn. Inst. **1970** No. 1, pp. 37/44, 40; C.A. **76** [1972] No. 135214).
[8] Eristavi, D. I., Tsintsadze, G. V., Kereselidze, L. B. (Tr. Gruz. Politekhn. Inst. **1971** No. 5, pp. 33/7, 34; C.A. **80** [1974] No. 150540).
[9] Khodzhaev, O. F., Azizov, T. A., Parpiev, N. A. (Koord. Khim. **3** [1977] 1710/7; Soviet J. Coord. Chem. **3** [1977] 1340/7).
[10] Powell, D. B., Woollins, A. (Spectrochim. Acta A **41** [1985] 1023/33, 1024, 1027).

24.1.2 With N-Methylformamide $HC(O)NHCH_3$ (= C_2H_5NO)

$Mn(C_2H_5I{\bullet}Ͻ)_6(ClO_4)_2$. For preparation, $Mn(ClO_4)_2 \cdot 6H_2O$ was dissolved in excess triethyl orthoformate for dehydration and a small excess of N-methylformamide was added followed by ether. The resulting oil was extracted two or three times with ether to give a white hygroscopic solid which melts at 82 to 84°C. The IR spectrum of the complex in Nujol mulls, recorded in the range 4000 to 250 cm^{-1}, shows an absorption band at 1653 cm^{-1} assigned to ν(CO) which on complexation had shifted to lower frequencies by 14 cm^{-1}, thus indicating coordination of the ligand to Mn through the carbonyl oxygen. Susceptibility measurements at 302 K (corrected for diamagnetism) yield the magnetic moment $\mu_{eff} = 5.90 \pm 0.1 \mu_B$ in nitromethane (or $5.8 \pm 0.1 \mu_B$ in $CH_2Cl_2 + C_2H_5NO$) indicating a high-spin-only octahedral configuration. The complex hydrolyzes in water. It is soluble in highly polar solvents such as nitromethane, alcohols, N-methylformamide and insoluble in less polar solvents such as $CHCl_3$, CH_2Cl_2, ether, and ethyl acetate. Conductivity measurements in 10^{-3} to 10^{-4} M nitromethane solutions at 25°C yield the molar electrical conductance $\Lambda = 154\ cm^2 \cdot \Omega^{-1} \cdot mol^{-1}$ of a 1:2 electrolyte [1].

$Mn(C_2H_5NO)_4X_2$ ($X = NO_3$, Cl, NCS, NCSe). The nitrato and chloro complexes were prepared by adding N-methylformamide to the solutions of manganese nitrate or chloride in aqueous acetone in a mole ratio of 1 Mn : 2 to 5 ligand. To prepare the isothiocyanato complex the ligand was added to a solution of $Mn(NCS)_2$ obtained by reacting a manganese(II) salt, with KSCN (mole ratio 1:2) in acetone [2]. $Mn(C_2H_5NO)_4(NCSe)_2$ was prepared from $Mn(NCSe)_2$ and N-methylformamide (mole ratio 1:4) in acetone in the same way as for the isothiocyanato complex [3]. The precipitated crystals were washed with acetone and dried [2]. The selenocyanate crystallized when the acetone solution was placed in a desiccator over P_2O_5 for 4 to 5 d [3].

$Mn(C_2H_5NO)_4(NO_3)_2$ and $Mn(C_2H_5NO)_4(NCS)_2$ form brittle pale pink crystals; those of $Mn(C_2H_5NO)_4Cl_2$ are pink and tiny [2] and those of $Mn(C_2H_5NO)_4(NCSe)_2$ are colorless [3]. The identity of the complexes was established by crystallooptic methods and X-ray diagrams, which show the nitrato complex to be isostructural with those of cobalt, nickel, and zinc and the chloro and isothiocyanato complexes with the corresponding ones of cobalt and nickel [2]. The IR spectra of the isothio- and isoselenocyanato complexes show absorption bands (in cm^{-1}) due to vibrations of the ligand [3] and of the NCS^- or NCSe groups as follows [4]:

complex	ν(CN)	ν(CO) [3]	ν(CS, CSe)	δ(NCS, NCSe)
$Mn(C_2H_5NO)_4(NCS)_2$	2080	—	—	476
$Mn(C_3H_5NO)_4(NCSe)_2$	2090	1665 to 1655	620	422

The shift of the band assigned to ν(CO) vibrations of N-methylformamide in $Mn(CH_3NO)_4$-$(NCSe)_2$ to lower frequencies by about 20 cm^{-1} suggests its coordination to Mn through the carbonyl oxygen [3]. The position of the bands assigned to NCS or NCSe vibration modes in the isothio- or isoselenocyanato complexes indicates bonding of these groups to Mn by terminal nitrogen, thus completing the six-coordinate structure of the complexes [3, 4].

The complexes are air-stable [2] except for the complex with X = NCSe which decomposes in air with separation of selenium [3]. They are readily soluble in water [2, 3] and dimethylformamide, slightly soluble in acetone, methanol, and benzene, and insoluble in $CHCl_3$, ether, and toluene [2]; the selenocyanate is readily soluble in acetone [3].

$Mn(C_2H_5NO)_2Cl_2$. A solution of $MnCl_2 \cdot 4H_2O$ in N-methylformamide (6 mL) and triethyl orthoformate (5 mL) was stirred for 1 h; ether (50 mL) was then added and the mixture was stirred for another 1 h, then decanted off. The solid was digested with ether (50 mL) for 1 h, absolute ethanol (5 mL) and, after 30 min of stirring, ether (10 mL) was added. The resulting solid was scraped off, crushed, and stirred until it formed a fine powder which was washed first with a mixture of alcohol and ether (1:2), then with ether and dried in vacuum. The light pink hygroscopic complex decomposes at 129°C without melting. Its IR spectrum (Nujol mulls) shows a ν(CO) band at 1654 cm^{-1}. The shift of $\Delta\nu = -18$ cm^{-1} and the insolubility of the complex in water and organic solvents suggest a polymeric octahedral structure where manganese is coordinated by the carbonyl oxygen of the organic ligands and bridging Cl^- anions [5].

References:

[1] Mackay, R. A., Poziomek, E. J. (Inorg. Chem. **7** [1968] 1454/7).
[2] Eristavi, D. I., Tsintsadze, G. V., Kereselidze, L. B. (Soobshch. Akad. Nauk Gruz. SSR **59** [1970] 57/9; C.A. **73** [1970] No. 136860).
[3] Tsintsadze, G. V., Skopenko, V. V., Kereselidze, L. B. (Soobshch. Akad. Nauk Gruz. SSR **61** [1971] 53/5; C.A. **74** [1971] No. 150532).
[4] Eristavi, D. I., Tsintsadze, G. V., Kereselidze, L. B. (Tr. Gruz. Politekhn. Inst. **1971** No. 5, pp. 33/7, 34; C.A. **80** [1974] No. 150540).
[5] Mackay, R. A., Poziomek, E. J. (J. Chem. Eng. Data **14** [1969] 271/2).

24.1.3 With N,N-Dimethylformamide $HC(O)N(CH_3)_2$ (= C_3H_7NO = dmf)

24.1.3.1 Manganese(II) Complex in Solution

The ^{14}N NMR spectrum of $Mn(ClO_4)_2$ in dimethylformamide solution studied at 25 and 50°C with field frequencies of 2.3 and 2.8 MHz reveals a downfield shift of the resonant nucleus frequency in the solvent indicating the presence of a complex cation assumed to be $[Mn(dmf)_6]^{2+}$. The observed ^{14}N spin relaxation in this complex is dominated by isotropic spin exchange whereas the ^{14}N quadrupol interaction is less significant, Chen, Tzeng-ming (J. Phys. Chem. **76** [1972] 1968/72).

24.1.3.2 Isolated Manganese(II) Compounds

24.1.3.2.1 $Mn(dmf)_nX_2$ Complexes

$Mn(dmf)_nCl_2$ (n = 1, 2) and **$Mn(dmf)_nBr_2$** (n = 1, 2, 3). The complexes $Mn(dmf)_2Cl_2$ and $Mn(dmf)_3Br_2$ were prepared by dissolution of the anhydrous manganese(II) salt in a large excess of dimethylformamide. After removal of the excess solvent by distillation in vacuum at 60°C, the compounds crystallize from the saturated solution at room temperature [1]. The chloride was also precipitated by addition of ether to the saturated solution. The fine crystalline pink precipitate was washed with ether, reprecipitated several times from dimethylformamide [2] and dried in vacuum over silica gel [1]. Susceptibility measurements at 20°C yield $\chi_{mol} = 13830$ cm^3/mol. The magnetic moment $\mu_{eff} = 5.71$ μ_B indicates a high-spin octahedral Mn^{II} complex [3]. Thermal analysis (DTA, TG, heating rate 5°C/min) reveals three endothermic effects at 110, 205, and 345°C which indicate formation of $Mn(dmf)Cl_2$, boiling of its solution in dimethylformamide, and final decomposition of the complex, respectively. The TG data suggest formation of $Mn(dmf)Cl_2$ between 115 and 135°C and of $Mn(dmf)_{0.5}Cl_2$ between 160 and 230°C; $MnCl_2$ is left above 280°C [1]. $Mn(dmf)_3Br_2$ reveals three endothermic effects at 75, 250, and 360°C and an exothermic one at 395°C (decomposition). The DTG data suggest formation of $Mn(dmf)_2Br_2$ in the range from 150 to 190°C and of $Mn(dmf)Br_2$ between 220 and 290°C; the latter leaves $MnBr_2$ above 340°C [1].

$Mn(dmf)_2X_2 \cdot 2H_2O$ (X = Cl, Br). The chloride appears as a stable solid phase in the solubility isotherm of the ternary system $MnCl_2$–dmf–H_2O at 25°C together with the solid species $MnCl_2 \cdot 4H_2O$. The pure complex resulted from solutions containing ≦41.2 wt% $MnCl_2$ and ≧48.4 wt% dimethylformamide. It is pale pink, not hygroscopic and has a (measured) density of 1.58 g/cm³ [3]. $Mn(dmf)_2Br_2 \cdot 2H_2O$ was prepared in a manner similar to that for the chloride [13]. IR absorption bands of the complexes in liquid paraffin mulls are assigned as follows:

complex	$\nu(OH)_{H_2O}$	ν(CO)	δ(HOH)	$\nu_a(NCH_3)$	ν(C–N)	δ(OCN)	ν(Mn–O)	ν(Mn–X)
X = Cl	3500	1660	1625(sh)	1260	1121	682	386	365
X = Br	3470	1660	1620(sh)	1255	1120	682	383	224
ligand	—	1685	—	1265	1099	660	—	—

The shift of the ν(CO) band toward lower and of the ν(C–N) (amide III) band toward higher wave numbers indicates coordination of dimethylformamide to Mn through the carbonyl oxygen. This causes a redistribution of electron density between the C=O and C–N bonds tending toward the state $^-O–\acute{C}=\overset{+}{N}<$ which is associated with a weakening of the CO and strengthening of the CN bond [4]. As already shown by [5] for the anhydrous complexes $Mn(dmf)_2X_2$ with X = Cl, Br, and $Mn(dmf)_4I_2$ the δ(OCN) vibration modes also undergo significant changes. Coordination of the water molecules is indicated by bands assignable to ν(OH) and δ(HOH) vibration modes of water. Thus, it can be assumed for the $Mn(dmf)_2X_2 \cdot 2H_2O$ complexes that the metal is surrounded octahedrally by the dmf ligands, water and halide anions [4] (also see [3]).

$Mn(dmf)_2Cl_2 \cdot 2H_2O$ is readily soluble in alcohol but insoluble in carbon tetrachloride or benzene. Thermal analysis (DTA, DTG) of the chloro complex at a heating rate of 10°C/min up to 800°C reveals endothermic effects at 90, 140, 195, and 320°C, the first of them being due to melting of the complex and loss of adsorbed moisture. The effects between 120 and 320 indicate dehydration and decomposition of the complex which is complete at 320°C leaving $MnCl_2$ [3].

$Mn(dmf)_6(ClO_4)_2$. A concentrated aqueous solution of manganese(II) perchlorate (slightly acidified by $HClO_4$) was added dropwise to cooled dimethylformamide. Then water was pumped off in a rotary evaporator at 40°C. The complex crystallizing upon cooling was recrystallized from anhydrous dimethylformamide at about 80°C, separated without washing and dried in vacuum over silica gel or in a high vacuum. If crystallization occurred below room temperature a solvate containing one additional dmf molecule was obtained which in vacuum was lost again. Upon heating, the complex explodes with high shattering power [6].

The electronic spectrum of $Mn(dmf)_6(ClO_4)_2$ in dimethylformamide solution shows absorption maxima at 537, 437, 406, 360, 341, and 302 nm; ligand field parameter $\Delta \approx 8500\ cm^{-1}$, nephelauxetic parameter $\beta_{35} = 0.92$. These data are indicative of an octahedral arrangement of the six dmf molecules coordinated through oxygen around the Mn^{2+} ion as suggested also by lowering of the ν(CO) frequency of dimethylformamide in the IR spectrum and its high dipole moment [6].

$Mn(dmf)(CN)_2$. Preparation occurred by passing a calculated amount of NH_4CN (obtained by thermal decomposition of $(NH_4)_4Fe(CN)_6$ under N_2) into a solution of anhydrous $MnCl_2$ in dimethylformamide under inert conditions. The light brown precipitate was washed with ethanol and ether and dried in vacuum or air at 80°C. X-ray powder patterns (for interplanar distances and intensities see the paper) show the complex to be crystalline with an individual structure. The IR spectrum (petrolatum mulls) in the 4000 to 400 cm^{-1} region shows characteristic absorption bands at 2160 and 1668 cm^{-1} assigned to ν(CN) of the cyanide group and to ν(CO) of dimethylformamide, respectively, together with other basic vibration modes at 1260, 1120, 1070, 680, 510, and 432 cm^{-1}. A shift of the ν(CO) band of the ligand to lower wave numbers (17 cm^{-1}), and of the ν(C–N) and δ(NCO) bands to higher wave numbers (26 and

15 cm^{-1}, respectively) was observed on complexation. The position of the ν(CN) band suggests a structure with bridging cyanide groups strongly bonded to manganese. The complex hydrolyzes in water, and is soluble in dimethyl sulfoxide but insoluble in chloroform. dimethylformamide, and dimethylacetamide [7].

$Mn(dmf)_4X_2$ (X = NCS, NCSe). Ethanol solutions of $Mn(NO_3)_2$ or $MnCl_2$ and KNCS or KNCSe were reacted in a 1:2 mole ratio and the precipitates of KNO_3 or KCl were filtered off; then dimethylformamide was added and the mixture was allowed to crystallize. The isothiocyanato complex forms transparent, well-shaped crystals which can be stored in a sealed vessel for a long time without change. They melt at 86°C without decomposition. $Mn(dmf)_4(NCSe)_2$ forms transparent colorless crystals which partly decompose on prolonged storage and melt at about 91°C with slight decomposition [8].

The compounds are isostructural and isomorphous with the corresponding ones of Co and Ni; the crystal habits (faces) and absence of a piezoelectric effect suggest a centrosymmetric crystal structure [8, 9]. X-ray studies show the complexes to be triclinic, space group $P\bar{1}$-C_i^1 (No. 2) with Z = 2 [8]. Lattice constants (a, b, c in Å), densities (D in g/cm^3), and refractive indices are shown below:

complex	a	b	c	$\alpha=\beta$	γ	D	n_α	n_β	n_γ
$Mn(dmf)_4(NCS)_2$	12.06	11.70	7.10	90°	92°30′	1.25	1.571	1.585	1.597
$Mn(dmf)_4(NCSe)_2$	12.70	12.30	7.35	90°	92°30′	1.63	1.587	1.610	1.615

Magnetic moments of μ_{eff} = 5.78 and 6.04 μ_B were observed (temperature not given) [8]. The IR spectrum (recorded from liquid paraffin or fluorocarbon mulls in the 4000 to 400 cm^{-1} region) shows downward shifts (in parentheses) of the ν(CO) and upward shifts of the δ(NCO) band relative to free dimethylformamide:

complex	ν(CO)	δ(NCO)	ν(CN)	ν(CS, Se)	δ(NCS, Se)
$Mn(dmf)_4(NCS)_2$	1649(−20)	674(+17)	2076, 2030	788	477
$Mn(dmf)_4(NCSe)_2$	1650(−19)	674(+17)	2084, 2037	606	413?

The observed splitting of the ν(CN) anion bands may be ascribed either to the presence of two NC(S,Se) groups or to the influence of the crystal state on these vibrations. Since the most intense ν(CN) and ν(CS,CSe) bands occur at higher frequencies than in the uncomplexed anions, coordination of the thiocyanate and selenocyanate groups through terminal nitrogen is assumed. The approximate force constants of the Mn–N bond (~1.3 and ~0.9 mdyn/Å) estimated from the positions of the ν(CN) and ν(CS,CSe) bands in the complexes suggest a stronger bonding of Mn to NCS^- than to $NCSe^-$ [10]. Thus, considering the spectral, crystallographic, and other physical data, a transoctahedral configuration of the complexes with the dmf ligands in equatorial and NCS or NCSe groups in axial positions is proposed [8 to 10]. The complexes decompose slowly in air and the $Mn(dmf)_4(NCS)_2$ may deliquesce [8]. They are readily soluble in water. $Mn(dmf)_4(NCS)_2$ dissolves also in dimethylformamide [8].

References:

[1] Glavas, M., Ribar, J. (J. Inorg. Nucl. Chem. **31** [1969] 291/5, 293).

[2] Haberditzl, W., Friebe, R., Havemann, R. (Z. Physik. Chem. [Leipzig] **228** [1965] 73/80, 75/7).

[3] Kim, T. P., Kazybaev, S. A., Imanakunov, B. I., Dzhunusov, A. (Zh. Neorgan. Khim. **26** [1981] 3129/31; Russ. J. Inorg. Chem. **26** [1981] 1672/4).

[4] Kim, T. P., Imanakunov, B. J., Kazybaev, S. A. (Zh. Neorgan. Khim. **30** [1985] 2817/21; Russ. J. Inorg. Chem. **30** [1985] 1604/7).
[5] Jungbauer, M. A. J., Curran, C. (Nature **202** [1964] 290).
[6] Schneider, W. (Helv. Chim. Acta **46** [1963] 1842/8, 1844, 1847).
[7] Kuntyi, O. I., Mikhalevich, K. N., Semenishin, D. I. (Koord. Khim. **5** [1979] 685/8; Soviet J. Coord. Chem. **5** [1979] 539/42).
[8] Skopenko, V. V., Tsintsadze, G. V. (Zh. Neorgan. Khim. **9** [1964] 2675/7; Russ. J. Inorg. Chem. **9** [1964] 1442/3).
[9] Tsintsadze, G. V., Porai-Koshits, M. A., Antsyshkina, A. S. (Zh. Strukt. Khim. **8** [1967] 296/302; J. Struct. Chem. [USSR] **8** [1967] 253/8, 254).
[10] Kharitonov, Yu. Ya., Tsintsadze, G. V. (Zh. Neorgan. Khim. **10** [1965] 35/40; Russ. J. Inorg. Chem. **10** [1965] 18/23, 22).

24.1.3.2.2 Mixed Ligand Compounds

Manganese diketonato complexes containing dimethylformamide have already been described in "Mangan" D 1, 1979, pp. 83, 114, and 118, the solvate $K_2Mn(CN)_6 \cdot 3dmf$ in "Mangan" D 2, 1980, p. 249. Dimethylformamide adducts of $MnbpyCl_2$ and $Mnphen_2F_2 \cdot 6H_2O$ were reported in "Manganese" D 3, 1982, pp. 211/2 and 248, respectively, and a phthalocyanine complex $[MnPc(dmf)_2]^+$ in "Manganese" D 4, 1985, pp. 206/7.

With 8-Hydroxyquinoline (C_9H_7NO)

$Mn(dmf)_2(C_9H_6NO)_2$ was prepared by addition of excess dimethylformamide to the 8-hydroxyquinolato complex $Mn(C_9H_6NO)_2$. After 3 h the precipitate was filtered off, washed with water and cold n-pentane and dried in air for 3 d. The product then was recrystallized from a mixture of $CHCl_3$ and hexane and dried first in air and then in a desiccator over P_2O_5 and paraffin. The IR spectrum of the complex (Vaseline mulls) shows shifts of the ν(Mn–O) and of the ν(C–O) vibration modes of $Mn(C_9H_6NO)_2$ indicating coordination of the dmf donor molecules. Bands of ν(Mn–O) were observed at 505, 565, and 590 cm^{-1}, whereas $Mn(C_9H_6NO)_2$ shows bands at 510 and 570 cm^{-1}. The thermal stability of the mixed ligand complex in air was studied between 25 and 500°C with a derivatograph at a heating rate of 10°C/min. $Mn(dmf)_2(C_9H_6NO)_2$ loses both dmf molecules at 125°C within 30 min of isothermal heating, to leave $Mn(C_9H_6NO)_2$ which decomposes at 340±10°C. A comparison with other 8-hydroxyquinolato complexes containing dmf shows that the stability increases in the order Cu< Mn≪Co<Ni [1].

With Acridine N-Oxides

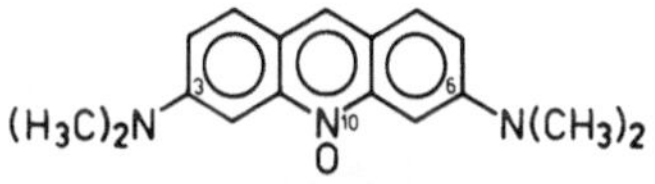

ligand 1 (=$C_{17}H_{19}N_3O$)

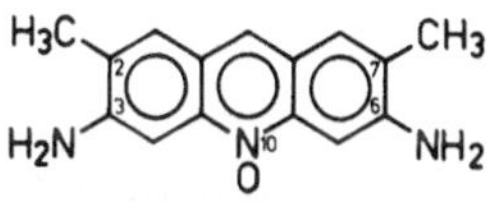

ligand 2 (=$C_{15}H_{15}N_3O$)

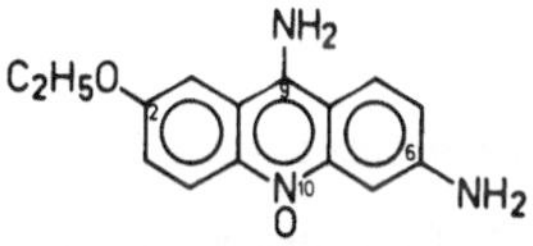

ligand 3 (=$C_{15}H_{15}N_3O_2$)

Complexes of composition **$[Mn(C_{17}H_{19}N_3O)_3(dmf)(NCS)_2]$** with ligand 1, **$[Mn(C_{15}H_{15}N_3O)_2(NCS)_2] \cdot dmf$** with ligand 2, and **$[Mn(C_{15}H_{15}N_3O_2)(dmf)(NCS)_2]$** with ligand 3 were prepared by reacting $Mn(NCS)_2$ with the corresponding ligand (mole ratio 1:1 up to 1:4) in dimethylformamide. The complexes crystallize upon cooling to −10°C or removing some of the solvent by distillation. Coordination of Mn through the N-oxidic oxygen and the carbonyl oxygen is suggested. Shifts of ν(CO) and ν(NO) to lower frequencies were observed in the IR spectra of the complexes in Nujol mulls or KBr disks (Δν(NO)≈130 cm^{-1}). The black complex $[Mn(C_{17}H_{19}N_3O)_3(dmf)(NCS)_2]$ reveals two absorption bands assignable to ν(Mn–O) vibrations at 400 and 300 cm^{-1}. Characteristic absorption bands (in cm^{-1}) of the brown compound

$[Mn(C_{15}H_{15}N_3O)_2(NCS)_2]\cdot$dmf were assigned as follows: 2060 to ν(CN); 1340, 1280, 1220 to ν(NO); 785, 765, 707 to ν(CS), and 475, 425, 410 to δ(NCS) vibrations of the N,S bridging thiocyanate group. The electronic spectrum of $[Mn(C_{15}H_{15}N_3O)_2(NCS)_2]\cdot$dmf in methanol exhibits an absorption band at 325 nm which is characteristic of a hexacoordinated configuration. The molar electrical conductance $\Lambda = 17.3$ $cm^2\cdot\Omega^{-1}\cdot mol^{-1}$ of a 0.001 M solution in methanol at 25°C indicates a nonelectrolyte. The properties of the brown complex with ligand 3 are similar [2].

References:

[1] Bublik, Zh. N., Mazurenko, E. A., Volkov, S. V., Gerasimenko, N. V. (Ukr. Khim. Zh. **51** [1985] 1123/7; Soviet. Progr. Chem. **51** No. 11 [1985] 1/5).

[2] Müller, R., Brunn, J., Böhland, H. (Z. Anorg. Allgem. Chem. **529** [1985] 209/15).

24.1.3.3 Isolated Manganese(III) Compound

$Mn(dmf)_6(ClO_4)_3$. To an ice-cold solution of $Mn(CH_3COO)_3\cdot 2H_2O$ in a minimum amount of dimethylformamide an ice-cold solution of $HClO_4$ was added slowly in slight excess ($Mn:HClO_4 > 1:3$). The resulting reddish violet syrupy liquid was washed several times with benzene to remove excess dimethylformamide. Further concentration in vacuum yields violet crystals of the complex which were washed with dry ether and dried over P_2O_5. The compound is stable in the solid state if moisture is excluded [1]. In the IR spectrum (KBr disks) the absorption band assigned to ν(CO) vibrations at 1650 cm^{-1} is shifted to lower frequencies by 32 cm^{-1} upon complexation [1, 2]. Bands assigned to vibrations of the uncoordinated ClO_4^- ion appear at 1090(ν_3) and 625(ν_4) cm^{-1} [3]. The electronic reflectance spectrum (solid state) shows three absorption bands at 20830, 17860, and 15380 cm^{-1}, assigned to the electron transitions: $e_{1g}\rightarrow b_{1g}$, $b_{2g}\rightarrow b_{1g}$, and $a_{1g}\rightarrow b_{1g}$, respectively, from which the field-splitting parameters $D_q = 1786$, $D_s = 2621$, and $D_t = 865$ cm^{-1} were derived. The presence of three bands indicates a lowering of the complex symmetry from O_h to the tetragonally distorted D_{4h} symmetry caused by the Jahn-Teller effect [2]. A solution of the complex in dimethylformamide shows only one absorption maximum at 20620 cm^{-1} ascribed to the $^5E_g\rightarrow {}^5T_{2g}$ electron transition which is the only spin-allowed one in a high-spin octahedral $Mn^{III}(d^4)$ complex, in agreement with the magnetic moment $\mu_{eff} = 5.10\,\mu_B$ resulting from susceptibility measurements at 298 K. The molar electrical conductance $\Lambda = 390\,cm^2\cdot\Omega^{-1}\cdot mol^{-1}$ of a 0.001 M solution in acetonitrile measured at 25°C indicates a 1:3 electrolyte. The complex is soluble in dichloroethane, acetonitrile, and acetone but insoluble in solvents like CCl_4, $CHCl_3$, and benzene. In dilute solutions the complex decomposes slowly but its stability can be increased by the addition of perchloric acid [1].

References:

[1] Prabhakaran, C. B., Patel, C. C. (J. Inorg. Nucl. Chem. **30** [1968] 867/9).

[2] Prabhakaran, C. B., Patel, C. C. (J. Inorg. Nucl. Chem. **34** [1972] 2371/4, 2373).

24.1.4 With Acetamide $CH_3C(O)NH_2$ (= C_2H_5NO)

$Mn(C_2H_5NO)_6(NO_3)_2$ crystallizes from an aqueous solution containing $Mn(NO_3)_2$ and acetamide in a 1:10 mole ratio at room temperature. X-ray studies reveal a triclinic lattice with lattice constants a = 7.29, b = 9.28, c = 10.78 Å, and α = 80.0°, β = 74.2°, γ = 72.2°, V = 665 $Å^3$; calculated density $D_x = 1.33$ g/cm³. A value of 1.41 g/cm³ was measured by flotation ($CHBr_3$ + benzene) [1].

$Mn(C_2H_5NO)_nCl_2$ (n = 2, 4). The complexes were obtained from ethanol solutions of $MnCl_2$ and acetamide in the corresponding mole ratio. The mixtures were kept at 0 to 2°C for crystallization. After 24 h the crystals were washed with acetone and dried between sheets of filter paper [2]. $Mn(C_2H_5NO)_2Cl_2$ also was prepared by refluxing stoichiometric amounts of anhydrous $MnCl_2$ and acetamide in CCl_4 or benzene [4]. The enthalpy of formation from solid $MnCl_2$ and two mol acetamide (solid) was found to be $\Delta H = -44.9 \pm 0.1$ kJ/mol by calorimetric measurements in 2N HCl solution at 25°C, and the standard enthalpy of formation from the elements was calculated to be $\Delta H^{\circ}_{298} = -1159.0 \pm 1.2$ kJ/mol [9].

The light pink crystals of $Mn(C_2H_5NO)_2Cl_2$ melt at 188°C [4]. They form flat tablets with the refraction indices $n_\alpha = 1.501$ and $n_\gamma = 1.561$ [2]. The IR and Raman (Ar laser excitation) spectra of $Mn(C_2H_5NO)_2Cl_2$ and of the N-deuterated complex **$Mn(C_2H_3D_2NO)_2Cl_2$** (obtained by recrystallization of $Mn(C_2H_5NO)_2Cl_2$ from D_2O) are recorded in the 4000 to 30 cm^{-1} region: the intense IR band around 1650 cm^{-1} assigned to ν(CO) vibrations had shifted upon coordination to lower frequencies by about 50 to 100 cm^{-1} relative to gaseous, and by about 15 to 20 cm^{-1} relative to liquid acetamide. In the far IR a split band assignable to ν(Mn–O) occurs at 424 and 408 cm^{-1}. Coordination of the Cl^- ion is suggested by a single band at 248 cm^{-1} assigned to ν(Mn–Cl) and two bands at 229 and 183 cm^{-1} assignable to δ(Cl–Mn–Cl) or ν(O–Mn–O). The single ν(Mn–Cl) band is consistent with the assumption of a trans-octahedral structure of the complex, since the IR and Raman spectra also do not coincide in that region, as is required by the mutual exclusion rule for a trans-arrangement. The bands observed below 100 cm^{-1} are ascribed mainly to lattice vibrations [5]. The enthalpy of solution in aqueous 2N HCl solution was determined calorimetrically at 25°C: $\Delta H = -4.82 \pm 0.02$ kJ/mol [9].

$Mn(C_2H_5NO)_4Cl_2$ forms elongated colorless crystal rods with the refractive indices $n_\alpha = 1.627$ and $n_\gamma = 1.654$; its identity was confirmed by X-ray diffraction studies [2].

Both complexes are readily soluble in water but insoluble in acetone, ether, and benzene. Thermal analysis (heating rate 3°C/mm) suggests a complicated decomposition. Breaking of the C–N bonds is assumed in the 135 to 230°C region, and of the C–C and C=O bonds in the 230 to 330°C range [2].

$Mn(C_2H_5NO)_nX_2 \cdot H_2O$ and **$Mn(C_2H_5NO)_4I_2$**. Compounds with n = 2, 4 for X = Cl, n = 4 for X = Br and the anhydrous iodide were detected upon studying the reaction of manganese halides (X = Cl, Br, I) with acetamide in aqueous solution at 25°C by the isothermal solubility method. They could be crystallized from equilibrium solutions and their properties were determined by methods of physicochemical analysis (no data presented) [3].

$Mn(C_2H_5NO)_6(ClO_4)_2$ and **$Mn(C_2H_5NO)_4(ClO_4)_2 \cdot 2H_2O$**. The compounds appear as stable solids in the solubility isotherm of the system $Mn(ClO_4)_2$–C_2H_5NO–H_2O at 25°C together with the solid species $Mn(ClO_4)_2 \cdot 6H_2O$ and acetamide. $Mn(C_2H_5NO)_6(ClO_4)_2$ forms in solutions containing >46.1 wt% of acetamide via replacement of all six aquo ligands in $Mn(ClO_4)_2 \cdot 6H_2O$ by acetamide. $Mn(C_2H_5NO)_4(ClO_4)_2 \cdot 2H_2O$ crystallizes from solutions containing 27.6 up to 46.0 wt% of acetamide. The two water molecules are probably also coordinated to Mn. Both complexes apparently involve acetamide as a monodentate ligand bonded to Mn through oxygen and are congruently soluble in water [6].

$Mn(C_2H_5NO)_2(CH_3COO)_2$ and **$Mn(C_2H_5NO)(CH_3COO)_2 \cdot H_2O$**. The complexes were prepared by dissolving $Mn(CH_3COO)_2 \cdot 4H_2O$ with stirring in molten acetamide at 80 to 85°C, the Mn-to-ligand mole ratio being 1:4 and 1:2, respectively. Stirring was continued until a clear viscous melt resulted which was cooled to room temperature in a desiccator over H_2SO_4 for crystallization requiring 7 d for the anhydrous compound. The crystals were washed repeatedly with cold water and dried in vacuum [7]. The identity of the complexes was determined by X-ray powder diagrams. Interplanar distances and relative intensities are given in the paper. $Mn(C_2H_5NO)_2$-$(CH_3COO)_2$ melts at 60°C, $Mn(C_2H_5NO)(CH_3COO)_2 \cdot H_2O$ at 55°C. The IR spectra are recorded in the 4000 to 400 cm^{-1} region. The ν(CO) band is shifted to lower, and the ν(C–N) band to higher frequencies by about 10 cm^{-1} in the complex. From a comparison of the spectra of the complexes and free acetamide and from the shift of the ν(C–C) band of the acetate ion it is concluded that the acetate group is bidentate. Thus an octahedral structure involving two bidentate acetato groups and monodentate acetamide ligands is assumed for $Mn(C_2H_5NO)_2$-$(CH_3COO)_2$. The structure of $Mn(C_2H_5NO)(CH_3COO)_2 \cdot H_2O$ is apparently similar. One of the acetamide ligands is replaced by water, which is split off upon heating at 130°C. $Mn(C_2H_5NO)_2(CH_3COO)_2$ decomposes at 190 to 200°C. The intermediate thermolysis products, $Mn(C_2H_5NO)(CH_3COO)_2$ at 180 to 210°C and $Mn_2(C_2H_5NO)(CH_3COO)_4$ at 210 to 228°C, were identified by elementary analysis [8].

The complexes, $Mn(C_2H_5NO)_2(CH_3COO)_2$ and $Mn(C_2H_5NO)(CH_3COO)_2 \cdot H_2O$, are nonelectrolytes in ethanol and decompose in water. Measurements at 25°C in ethanol yield molar electrical conductivities of $\Lambda = 36.6$ and 31.9 $cm^2 \cdot \Omega^{-1} \cdot mol^{-1}$. In water both compounds behave as 1:2 electrolytes due to solvolysis [8].

References:

[1] Nardelli, M., Coghi, L. (Ric. Sci. **29** [1959] 134/8; C.A. **1960** 1151).

[2] Smol'nikov, Yu. P., Dzashiashvili, T. K., Mal'donato, S. E. (Tr. Molodykh Nauchn. Sotrudn. Aspirantov Inst. Neorgan. Khim. Elektrokhim. Akad. Nauk Gruz.SSR **1974** 75/8; Ref. Zh. Khim. **1974** No. 23V131; C.A. **83** [1975] No. 70828).

[3] Imanakunov, B., Baichalova, S., Alymkulova, K. (Mater. Nauchn. Konf. Posvyashch. 100th Letiyu Period Zakona L. I. Mendeleeva, Frunze, USSR, 1969 [1970], p. 143; C.A. **76** [1972] No. 28318).

[4] Paul, R. C., Dev, R. (Indian J. Chem. **5** [1967] 267/71, 269; C.A. **68** [1968] No. 81839).

[5] Tsivadze, A. Yu., Kharitonov, Yu. Ya., Tsintsadze, G. V., Smirnov, A. N., Tevzadze, M. N. (Zh. Neorgan. Khim. **19** [1974] 3321/6; Russ. J. Inorg. Chem. **19** [1974] 1818/22, 1819).

[6] Goryunov, Yu. A. (Sb. Nauchn. Tr. Yaroslav. Pedagog. Inst. **1977** No. 164, pp. 13/6; C.A. **89** [1978] No. 49626).

[7] Azizov, T. A., Khodzhaev, O. F., Parpiev, N. A. (Uzb. Khim. Zh. **20** No. 5 [1976] 6/7; C.A. **86** [1977] No. 71869).

[8] Khodzhaev, O. F., Azizov, T. A., Parpiev, N. A. (Koord. Khim. **3** [1977] 1495/502; Soviet J. Coord. Chem. **3** [1977] 1163/9, 1166/8).

[9] Barvinok, M. S., Mashkov, L. V. (Zh. Neorgan. Khim. **30** [1985] 2972/3; Russ. J. Inorg. Chem. **30** [1985] 1693/4).

24.1.5 With N,N-Dialkylacetamides $CH_3C(O)N(R)_2$ with $R = CH_3$ ($= C_4H_9NO$) and $R = C_2H_5$ ($= C_6H_{13}NO$)

The phthalocyanine complex $[MnPc(C_4H_9NO)_2]$ containing dimethyl acetamide was reported in "Manganese" D 4, 1985, pp. 198 and 206/7.

Complexes in Solution. The electron spin resonance spectra of $Mn(NO_3)_2 \cdot 4H_2O$ solutions in N,N-diethylacetamide containing a small amount of water were taken at X and Q bands at room temperature. They reveal the presence of manganese ions being solvated by diethylacetamide. These species have altered hyperfine splitting and g factor due to nitrate ions in an outer coordination sphere, so that they may be considered as loose ion pairs. In addition to these species, the existence of tight ion pairs and a small fraction of manganese ions with mixed diethylacetamide-water solvation is discussed. Presumably there are no well-defined chemical species present in such solutions but constituents of fluctuating aggregations [1].

$Mn(C_4H_9NO)_2Cl_2 \cdot nH_2O$ (n=1, 6). The hexahydrate was detected as a solid phase in the solubility isotherm of the ternary system $MnCl_2$–dimethylacetamide–H_2O at 25°C together with the species $MnCl_2 \cdot 4H_2O$. It can be crystallized from solutions containing $\lesssim$40 wt% $MnCl_2$ and $\gtrsim$7 wt% dimethylacetamide [2]. $Mn(C_4H_9NO)_2Cl_2 \cdot 6H_2O$ melts at 48°C. The density D=1.1784 g/cm^3 was determined at 20°C by pycnometry with $CHCl_3$, CCl_4, or toluene. The IR spectrum in KBr disks or liquid paraffin mulls was recorded in the 4000 to 400 and 400 to 200 cm^{-1} regions: The broad bands in the 3430 to 3200 cm^{-1} region are apparently due to ν(OH) modes of H_2O hydrogen bonded to the organic ligand. The overlapping of the ν(CO) and δ(H_2O) bands prevents determination of the positions of the H_2O molecules. A downward shift of the ν(CO) band at 1615 cm^{-1} by 40 cm^{-1} and an upward shift of the ν(C–N) band at 1205 cm^{-1} and δ(NCO) band at 615 cm^{-1} by 30 cm^{-1} were observed in comparison to the free amide. A far IR band at 448 cm^{-1} was ascribed to ν(Mn–O). The ν(Mn–Cl) band at 334 cm^{-1} indicates bonding of the Cl^- ions in the inner coordination sphere. This is supported by a rather low electrical conductance Λ=18.74 cm$^2 \cdot \Omega^{-1} \cdot$mol^{-1} measured at 25°C in dimethylacetamide solution which indicates an only partial ionization. Thus, the coordination sphere of Mn must contain the organic ligand, the Cl^- ions and some of the water molecules. Thermal analysis shows that the complex loses all water at 132°C, one molecule of dimethylacetamide at 195°C and the last one at 319°C, to leave $MnCl_2$ [3].

$Mn(C_4H_9NO)_2Cl_2 \cdot H_2O$ was obtained by reacting $MnCl_2$ and dimethylacetamide in a 1:4, 1:6, and 1:10 mole ratio. The IR spectrum was recorded as mineral oil, fluorinated mineral oil, or hexachlorobutadiene mulls in the 4000 to 400 cm^{-1} region and the Raman spectrum (excited by ionized Ar, λ=4880 Å) in the 3600 to 20 cm^{-1} region. Shifts of the ν(CO) band to lower frequencies (30 cm^{-1} relative to liquid and 70 cm^{-1} relative to gaseous dimethylacetamide) and upward shifts of the ν(CC) and δ(NCO) bands were observed. The bands at 3340 and 550 cm^{-1} suggest the presence of coordinated water. The observed Raman frequencies of ν(Mn–Cl) at 228 and 260 cm^{-1} and δ(Cl–Mn–Cl) at 184 cm^{-1} are characteristic of octahedral Mn complexes [4].

$Mn(C_4H_9NO)Cl_2 \cdot H_2O$. Crystallization of the complex from a solution of $MnCl_2 \cdot 4H_2O$ in excess dimethylacetamide was accomplished by removing the excess ligand in vacuum (p=0.5 Torr for 48 to 60 h) or addition of anhydrous ether. The oily products were washed repeatedly with ether for purification and dried in vacuum over P_4O_{10} at room temperature. The IR spectrum of the pale yellow complex shows an absorption band assigned to ν(CO) vibrations at 1630 cm^{-1} (Δν=−32 cm^{-1}). The molar electrical conductance Λ=24 cm$^2 \cdot \Omega^{-1} \cdot$mol^{-1}, measured at 25°C on a 0.003 M solution in dimethylacetamide, indicates a nonelectrolyte in that solvent and suggests coordination of the Cl^- ions to manganese. The complex decomposes on heating without melting [5].

$Mn(C_4H_9NO)_4(ClO_4)_2$ was prepared from $Mn(ClO_4) \cdot 6H_2O$ and dimethylacetamide in about the same way as $Mn(C_4H_9NO)Cl_2 \cdot H_2O$, see above. The light pink complex melts at 73 to 75°C. Its IR spectrum shows a ν(CO) band at 1629 cm^{-1}. The molar electrical conductance Λ=141 cm$^2 \cdot \Omega^{-1} \cdot$mol^{-1} of a 0.003 M solution in dimethylacetamide at 25°C is that of a 1:2 electrolyte containing ionic perchlorate [5].

$Mn(C_4H_9NO)_2Br_2 \cdot 4H_2O$ was prepared from $MnBr_2 \cdot 4H_2O$ and dimethylacetamide in the same way as $Mn(C_4H_9NO)Cl_2 \cdot H_2O$; see above. The light yellow complex shows an IR absorption band of ν(CO) at 1612 cm^{-1}. Susceptibility measurements at 300 K (corrected for diamagnetism) yield the magnetic moment μ_{eff} = 5.90 μ_B indicative of a spin-free, probably octahedral, $Mn^{II}(d^5)$ complex. The compound decomposes on heating [5].

$Mn(C_4H_9NO)(CN)_2$. Preparation is carried out in the same way as described for $Mn(dmf)(CN)_2$ (see p. 95), replacing dimethylformamide by dimethylacetamide. X-ray powder diagrams show the light brown complex to be crystalline with an individual structure. The IR spectrum shows absorption bands of ν(CN) at 2160 cm^{-1} and of ν(CO) at 1620 cm^{-1}. The position of the ν(CN) band suggests a structure with bridging cyanide groups strongly bonded to Mn. The complex hydrolyzes in water; its solubility in organic solvents resembles that of $Mn(dmf)(CN)_2$ [6].

$Mn(C_4H_9NO)_4X_2$ (X = NCS, NCSe). Stoichiometric amounts of $Mn(NO_3)_2 \cdot 6H_2O$ and KNCS or KNCSe in acetone solution were mixed, the precipitate of KNO_3 was filtered off and a large excess (6- to 10-fold mole ratio) of dimethylacetamide was added to the filtrate. The mixture was stored in vacuum over $CaCl_2$ for crystallization. The stable crystals are colorless (X = NCS) or pale yellow (X = NCSe). X-ray powder diagrams (for spacings d see paper) suggest isomorphism with the corresponding complexes of cobalt and nickel [7]. $Mn(C_4H_9NO)_4(NCS)_2$ was found to be monoclinic, space group $P2_1/a\text{-}C_{2h}^5$ (No. 14), with the lattice constants a = 13.878(3), b = 9.509(2), c = 11.299(2) Å and γ = 111.47(2)°; Z = 2, V = 1381.6(5) $Å^3$. The structure was solved from 1180 independent reflections and refined to a final R = 0.030. It is shown in **Fig. 11** together with the bond lengths (in Å) and the bond angles around the Mn atom. The metal atoms are placed in the symmetry center of slightly distorted octahedra consisting of the four oxygens of the monodentate organic ligands and the two nitrogens of the thiocyanate groups in trans positions. The carbonyl groups have planar configurations indicating an sp^2 bond hybridization. The density of 1.25 g/cm^3 was calculated from X-ray data [11]. The IR spectrum was recorded on Vaseline or hexachlorobutadiene mulls, and its most relevant absorption bands (in cm^{-1}) were assigned as follows:

X	ν(CO)	ν(C–N) + δ(NCH_3)	$ν_{as}$(CNC)	ν(C–C)
NCS	1619	1521, 1413	1271	974
NCSe	1624, 1614	1520, 1413, 1404	1270	971

X	$ν_s$(CNC)	δ(NCO)	δ(CNC)	ν(Mn–O)
NCS	757	614, 601	487, 435	411
NCSe	757, 732	615, 600	488, 404	407

Bands attributable to ν(CN) vibrations of the thiocyanate or selenocyanate anions appear between 2190 and 2170 cm^{-1}, the ν(CS) band of the thiocyanate occurs at 800 to 790 cm^{-1}, and a band around 470 cm^{-1} was ascribed to δ(NCS). A band assignable to ν(CSe) was not observed but two bands at 425 and 405 cm^{-1} were ascribed to δ(NCSe) vibrations of the selenocyanate groups [8]. The complexes are readily soluble in water, methanol, ethanol, $CHCl_3$, acetone, dimethylformamide, dioxane, and dimethyl sulfoxide but insoluble in ether and benzene [7].

$Mn(C_4H_9NO)_6(BF_4)_2$. A solution of $Mn(BF_4)_2 \cdot 6H_2O$ in excess anhydrous dimethylacetamide was heated and the excess solvent removed by distillation under reduced pressure. The resulting solid was recrystallized twice from anhydrous dimethylacetamide and dried in vacuum [9]. The IR spectrum shows an absorption band assigned to ν(C=O) at 1617 to 1618 cm^{-1}. Conductivity measurements on solutions in dimethylacetamide at 25°C yield an (extrapolated) limiting molar conductance $Λ = 82.3\ cm^2 \cdot Ω^{-1} \cdot mol^{-1}$ of a 1:2 electrolyte containing the (possibly slightly distorted) octahedral complex cation $[Mn(C_4H_9NO)_6]^{2+}$ [10].

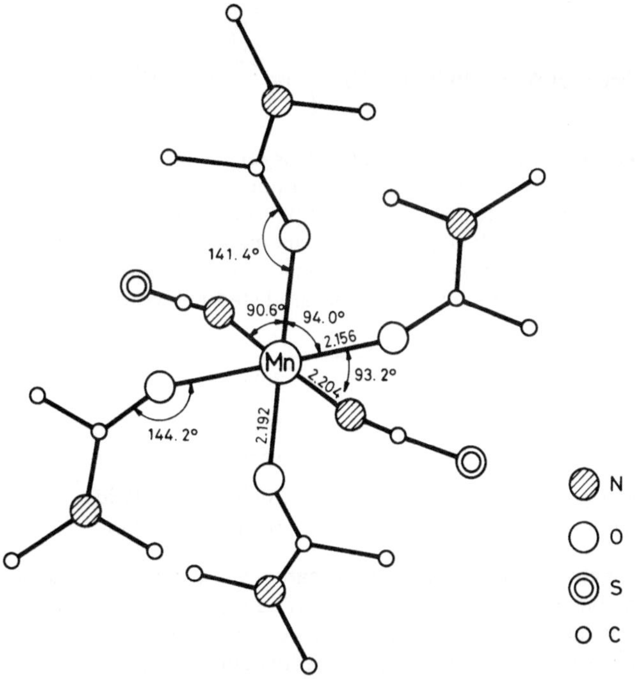

Fig. 11. Structure of $Mn(C_4H_9NO)_4(NCS)_2$ [11].

References:

[1] Kleinmann, V., Stockhausen, M. (Z. Naturforsch. **36a** [1981] 751/8, 752).

[2] Berdiev, A., Imanakunov, B. I., Yun, P. T., Lukina, L. I. (Izv. Akad. Nauk Kirg.SSR **1981** No. 2, pp. 41/3; C.A. **96** [1982] No. 169555).

[3] Berdiev, A., Imanakunov, B. I., Kazybaev, S. A., Abdybakirova, B. (Zh. Neorgan. Khim. **30** [1985] 2304/8; Russ. J. Inorg. Chem. **30** [1985] 1309/11).

[4] Tsivadze, A. Yu., Tsintsadze, G. V., Khugashvili, Ts. L., Kharitonov, Yu. Ya. (Koord. Khim. **3** [1977] 1839/45; Soviet J. Coord. Chem. **3** [1977] 1442/7, 1446).

[5] Bull, W. E., Madan, S. K., Willis, J. E. (Inorg. Chem. **2** [1963] 303/6).

[6] Kuntyi, O. I., Mikhalevich, K. N., Semenishin, D. I. (Koord. Khim. **5** [1979] 685/8; Soviet J. Coord. Chem. **5** [1979] 539/42).

[7] Khugashvili, T. G. (Trudy Gruz. Politekhn. Inst. im. V. I. Lenina **1975** No. 4, pp. 20/2; C. A. **87** [1977] No. 160960).

[8] Tsintsadze, G. V., Tsivadze, A. Yu., Khugashvili, Ts. L., Smirnov, A. N. (Soobshch. Akad. Nauk Gruz.SSR **75** [1974] 337/40; C.A. **82** [1975] No. 36927).

[9] Kamieńska, E., Uruska, B. (Bull. Acad. Polon. Sci. Ser. Sci. Chim. **21** [1973] 587/92, 587).

[10] Kamieńska, E., Uruska, B. (Electrochim. Acta **22** [1977] 181/3).

[11] Tsintsadze, G. V., Tsintsivadze, T. I., Bel'skii, V. K., Sobolev, A. N., Turiashvili, T. N., Iashvili, E. A., Elerdashvili, M. A. (Soobshch. Akad. Nauk Gruz.SSR **116** [1984] 517/20; C.A. **103** [1985] No. 46223).

24.1.6 With N-(2-Pyridyl)acetamide 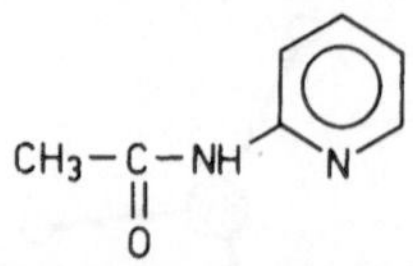 (= $C_7H_8N_2O$)

$[Mn(C_7H_8N_2O)_2(H_2O)_2](NO_3)_2$ and **$Mn(C_7H_8N_2O)_2X_2$** (X = Cl, Br). The complexes were prepared by reacting saturated ethanolic solutions of manganese nitrate, chloride or bromide with excess N-(2-pyridyl)acetamide. The pale pink precipitates were washed with ethanol and dried in vacuum over silica gel. Magnetic moments resulting from susceptibility measurements at room temperature and the positions of the IR absorptions (in cm^{-1}) assigned to the amide I band and to the amide II band are listed below together with those of the free ligand:

complex	μ_{eff} in μ_B	amide I	amide II
$[Mn(C_7H_8N_2O)_2(H_2O)_2](NO_3)_2$	6.00	1673	1538
$[Mn(C_7H_8N_2O)_2Cl_2]$	5.96	1668	1538
$[Mn(C_7H_8N_2O)_2Br_2]$	5.98	1667	1538
ligand $C_7H_8N_2O$	—	1694	1535

Coordination of the ligand to Mn through carbonyl oxygen and pyridine ring nitrogen with formation of a chelate ring is suggested. The nitrato complex shows absorption bands due to free NO_3^- ions at 1380, 830, and 720 cm^{-1}. The color and magnetic moments allow to assume the compounds as high-spin Mn^{II} complexes with octahedral structure. The water molecules of the nitrato complex appear to be coordinated in the inner sphere, Z. Warnke (Roczniki Chem. **49** [1975] 263/71, 264/6; C.A. **83** [1975] No. 21138).

24.1.7 With Chloroacetamide $ClCH_2C(O)NH_2$ (= C_2H_4ClNO)

$Mn(C_2H_4ClNO)_6(SbCl_6)_2$ was prepared, as described for the ethyl acetato complex in [1], by stirring manganese (II) chloride with a slight excess of $SbCl_5 \cdot CH_3NO_2$ in chloroacetamide or its mixture with nitromethane [2]. The yellow complex is hygroscopic and melts at 260°C with decomposition. The IR spectrum shows a shift of the ν(CO) absorption band toward lower frequencies. The rather strong band at about 345 cm^{-1} is apparently due to the characteristic ν_3 mode of the octahedral $SbCl_6^-$ anion [2].

References:

[1] Driessen, W. L., Groeneveld, W. L., Van der Wey, F. W. (Rec. Trav. Chim. **89** [1970] 353/67, 354).

[2] De Bolster, M. W. G., Driessen, W. L., Groeneveld, W. L., Van Kerkwijk, C. J. (Inorg. Chim. Acta **7** [1973] 439/44, 441).

24.1.8 With Diacetamide $CH_3C(O)NHC(O)CH_3$ (= $C_4H_7NO_2$)

The nitrato and chloro complexes **$[Mn(C_4H_7NO_4)_2(H_2O)(NO_3)]NO_3$** and **$[Mn(C_4H_7NO_2)(H_2O)Cl_2]$** were prepared by grinding together stoichiometric amounts of $Mn(NO_3)_2 \cdot 6H_2O$ or $MnCl_2 \cdot 4H_2O$ and diacetamide (7 g) in a mortar for 15 min and drying the resulting paste in vacuum for 10 min. The dry colorless (nitrate) or faintly pink (chloride) powders were washed thoroughly with ether to remove excess ligand and dried in vacuum. **[Mn($C_4H_7NO_2$)$_3$](ClO_4)$_2$** was obtained from ethyl acetate solutions of $Mn(ClO_4)_2 \cdot 6H_2O$ and diacetamide (mole ratio 1:4); the white precipitate was washed with ether to remove unreacted diacetamide and dried in vacuum (at 0.01 Torr) and room temperature for at least 24 h [2]. It was prepared also by grinding together $Mn(ClO_4)_2 \cdot 6H_2O$ and diacetamide in stoichiometric proportions [1]. For the preparation of **[Mn($C_4H_7NO_2$)$_2$(NCS)$_2$]·H_2O** slightly heated aqueous solutions of $Mn(NCS)_2$ and diacetamide were mixed in a 1:1 up to 10:1 mole ratio, evaporated with heating (no boiling) and left in the air for further evaporation at room temperature. The crystals which precipitated after 2 to 3 d were washed with ether and dried. They are readily soluble in water, ethanol, acetone, and acetonitrile [3].

Infrared studies demonstrate that diacetamide assumes the trans-trans configuration upon complexation to the manganese ion with coordination occurring through oxygen. Shifts of the ν(CO) bands in the 1739 to 1732 and 1705 to 1695 cm^{-1} region to lower wave numbers were observed [1 to 3]. The downward shift of the ν(NH) bands in the 3400 to 3000 region suggests the presence of hydrogen bonding in the complex perchlorate [2]. Coordination of the thiocyanate groups to Mn through the nitrogen atoms is indicated by two bands occurring at 2093 and 902(?) cm^{-1} which are assigned to ν(CN) and ν(CS) vibrations, respectively, of NCS. Thus an octahedral structure of the complex with bidentate acetamide and monodentate thiocyanate ligands is assumed [3]. Susceptibility measurements reported for the perchlorate at 299 K yield the magnetic moment $\mu_{eff} = 6.06 \pm 0.04\,\mu_B$, indicative of a high-spin, probably octahedral, $Mn^{II}(d^5)$ complex. The molar electrical conductance $\Lambda = 152\ cm^2 \cdot \Omega^{-1} \cdot mol^{-1}$ of a 0.001 M solution in acetone is that of a 1:2 electrolyte and indicates occurrence of uncomplexed ClO_4^- ions [2].

References:

[1] Gentile, P. S., Shankoff, T. A. (J. Inorg. Nucl. Chem. **28** [1966] 1283/9, 1285).
[2] Kraihanzel, C. S., Grenda, S. C. (Inorg. Chem. **4** [1965] 1037/42).
[3] Kharitonov, Yu. Ya., Tsintsadze, G. V., Tsivadze, A. Yu., Tabidze, E. I. (Koord. Khim. **4** [1978] 1609/10; Soviet J. Coord. Chem. **4** [1978] 1229).

24.1.9 With N,N'-(2,6-Pyridinediyl)bis(acetamide)

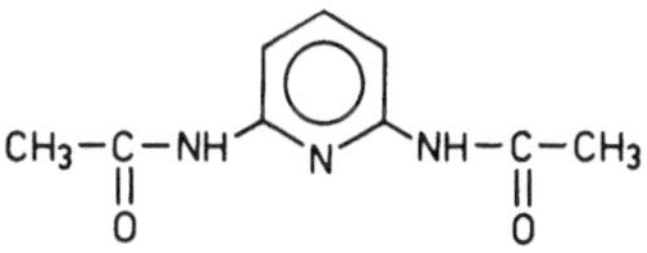

($=C_9H_{11}N_3O_2$)

Mn($C_9H_{11}N_3O_2$)X_2 (X = NO_3, Cl, Br, NCS). A solution of the appropriate manganese salt (0.01 mol) in a 1:1 mixture of acetone and ethanol was mixed with a hot acetone solution of the ligand (0.02 mol) and refluxed for 2 h on a water bath. The crystalline solid was washed with acetone and dried. IR absorption bands of the complexes in KBr and Nujol are shown below in comparison to bands of the free ligand (wave numbers in cm^{-1}):

assignment ...	ν(CO)	ν(CN) + δ(NH)	δ(NH)	$\delta(CO)_{in\ plane}$	$\delta(CO)_{out\ of\ plane}$
complex	1640 to 1650	1550 to 1560	1250 to 1265	655	580 to 585
ligand	1690	1520	1230	650	550

The observed shifts are compatible with coordination of the ligand to Mn through carbonyl oxygen, which is supported by a band assignable to ν(Mn–O) at about 370 cm^{-1}. The broad band in the 3250 to 3000 cm^{-1} region ascribed to ν(NH) confirms the ketone conformation of the ligand in the complex. The bands referred to deformations of the pyridine ring appear at about 1605, 630, and 425 cm^{-1} with shifts to higher frequencies by about 20 cm^{-1} in the complexes, and suggest coordination of the pyridine ring nitrogen to Mn, as concluded also from a band assignable to ν(Mn–N) at 240 cm^{-1}. The nitrato complex shows several new bands assignable to different ν(NO) vibrations around 1740, 1475 to 1460, 1260 to 1255, and 900 to 890 cm^{-1} and one weak band assignable to $\nu(Mn–ONO_2)$ at about 265 cm^{-1}, suggesting monodentate coordination of the nitrato group. The chloro and bromo complexes exhibit bands ascribed to ν(Mn–Cl) at 260 cm^{-1} and to ν(Mn–Br) at 210 cm^{-1}, indicating coordination of the halogenide. The thiocyanato complex shows characteristic bands at 2050, 815, and 485 cm^{-1} assigned to ν(CN), ν(CS), and δ(NCS) vibrations of N-bonded thiocyanate groups.

The electronic spectrum of the complexes in Nujol mulls or solutions in dimethyl sulfoxide shows three weak absorption bands around 16500, 19200, and 22000 cm^{-1} which suggest spin-forbidden electron transitions. These are compatible with five-coordinate complexes of a trigonal bipyramidal geometry where the tridentate O, O, N-bonded organic ligand will occupy the equatorial plane and the anions are at the axial positions. The magnetic moments resulting from susceptibility measurements at room temperature are μ_{eff} = 5.80, 5.84, 5.88, and 5.86 μ_B for the nitrato, chloro, bromo, and isothiocyanato complexes, respectively, thus indicative of their high-spin nature. Conductivity measurements show the complexes to be nonelectrolytes in dimethylformamide solution. The complexes are readily soluble in dimethylformamide but only partially soluble in common organic solvents.

Reference:

Sangal, S. K., Sahni, S. K., Rana, V. B. (Acta Chim. Acad. Sci. Hung. **110** [1982] 19/23).

24.1.10 With Derivatives of Acetoacetamide and -anilide

$CH_3C(O)CH_2C(O)NHR \rightleftharpoons CH_3C(OH)=CHC(O)NHR$

Complexes in Solution

The formation constants of the complexes with acetoacetanilide $CH_3C(O)CH_2C(O)NHC_6H_5$ (= $C_{10}H_{11}NO_2$) in aqueous ethanol (70 vol%), $Mn(C_{10}H_{10}NO_2)^+$ and $Mn(C_{10}H_{10}NO_2)_2$, were determined potentiometrically (glass electrode) at 25°C and ionic strength I = 0.1 M (NaCl): log K_1 = 4.78, log K_2 = 4.18 [1]. A value of log K_1 = 3.38 for the complex with N-(2-pyridyl)acetoacetamide, $CH_3C(O)CH_2C(O)NHC_5H_4N$ (= $C_9H_{10}N_2O_2$) in aqueous solution, $Mn(C_9H_9N_2O_2)^+$, was determined by pH-potentiometric titrations at 25 ± 0.05°C and I = 0.10 M (KNO_3); log K_2 = 2.9 was approximated by extrapolation [2]. The comparison with other metal complexes of acetoacetanilide and N-(2-pyridyl)acetoacetamide reveals that the stability constants are in accordance with the Irving-Williams order $Zn^{2+} < Cu^{2+} > Ni^{2+} > Co^{2+} > Fe^{2+} > Mn^{2+}$ [1, 2].

Isolated Compounds

$Mn(C_{10}H_9BrNO_2)_2$ was prepared by reaction of $Mn(NO_3)_2$ in aqueous solution with α-bromoacetoacetanilide, $CH_3C(OH)=CBrC(O)NHC_6H_5$ (= $C_{10}H_{10}BrNO_2$) in aqueous 6 N ammonia, as described for other acetoacetanilide complexes; see e.g. [3]. The brownish black precipitate was washed repeatedly with warm acetone or ethanol. The compound does not melt or decompose visibly upon heating up to 360°C. The IR spectrum (Nujol mulls), recorded in the 4000 to 650 cm^{-1} region, shows a ν(NH) absorption band at 3350 cm^{-1}. The shift to higher wave numbers in comparison to the free ligand ($\Delta\nu = 150\ cm^{-1}$) indicates disengagement of the NH group from intramolecular hydrogen bonding and nonbonding of the metal. It is assumed that both carbonyl groups (one of them enolized and deprotonated) are coordinated to Mn as in the corresponding complex of chromium. $Mn(C_{10}H_9BrNO_2)_2$ is insoluble in all common organic solvents [4].

$Mn_2(C_{15}H_{15}NO_4)_2 \cdot 6H_2O$. The dinuclear complex with N-acetoacetyl-N-2-tolylacetoacetamide $CH_3C_6H_4N[C(O)CH_2C(O)CH_3]_2$ ($= C_{15}H_{17}NO_4$) was prepared by slow addition of manganese(II) acetate (1 mmol) in 100 mL ethanol to an equimolar ethanolic solution of the ligand and NaOH. The colored mixture was kept at 25°C for several hours and then at room temperature for another 2 h. The precipitate was filtered off, washed with ethanol and dried in a vacuum. The IR spectrum (KBr disks) shows an absorption band around 1590 cm^{-1} ascribed to ν(CO) and a further band around 1525 cm^{-1} assigned to ν(C=C) vibration modes. The broad band in the 3400 to 3200 cm^{-1} region is apparently due to the stretching modes of water. The magnetic moment of the complex is that for Mn^{2+} ions with no magnetic coupling. The complex is sparingly soluble in common organic solvents and more soluble in coordinating solvents such as dimethyl sulfoxide or pyridine [5].

$Mn_3(C_{15}H_{15}NO_4)_2(OH)ClO_4$. The trinuclear complex was prepared by two different routes: To a suspension of $Mn_2(C_{15}H_{15}NO_4)_2 \cdot 6H_2O$ in ethanol, an ethanolic solution of $Mn(ClO_4)_2$ was added in a 1:1 mole ratio or an ethanolic solution of $Mn(ClO_4)_2$ was added slowly to that of N-acetoacetyl-N-2-tolylacetoacetamide in a 3:2 mole ratio. The residue was digested with water for 2 d. The solid was filtered off, washed with water and dried in vacuum. After stirring at room temperature for 2 h, the solution was evaporated to dryness. The residue was digested with water and the precipitate was washed and dried in vacuum. The IR spectrum resembles that of $Mn_2(C_{15}H_{15}NO_4)_2 \cdot 6H_2O$ (see above). The observed ClO_4^- bands indicate noncoordination. Magnetic moments lower than the spin-only values are indicative of interaction between the metal ions [5].

References:

[1] Shoukry, M. M. (Rev. Roumaine Chim. **29** [1984] 283/8, 285).
[2] Harries, H. J. (J. Inorg. Nucl. Chem. **29** [1967] 2484/6).
[3] Sen, D. N., Umapathy, P. (Indian J. Chem. **6** [1968] 516/20, 516).
[4] Thankarajan, N., Sreeman, P. (Current Sci. [India] **44** [1975] 420/1).
[5] Fenton, D. E., Tate, J. R., Casellato, U., Tamburini, S., Vigato, P. A., Vidali, M. (Inorg. Chim. Acta **83** [1984] 23/31, 24/6).

24.1.11 With Propionamide $C_2H_5C(O)NH_2$ ($= C_3H_7NO$)

$Mn(C_3H_7NO)_3Cl_2$. The complex was prepared from manganese chloride and propionamide by a method similar to that applied to the synthesis of the related acetamide complexes $Mn(C_2H_5NO)_nCl_2$, see p. 99. The IR spectrum of Vaseline and fluorinated oil mulls in the 4000 to 400 and 400 to 30 cm^{-1} regions and the Raman spectrum of crystalline samples are recorded and compared to the spectrum of the free ligand. Coordination of propionamide to Mn through carbonyl oxygen is postulated. Bands observed at 419 and 411 cm^{-1} were assigned to ν(Mn–O). The three bands assignable to ν(Mn–Cl) vibrations at 262, 249, and 234 cm^{-1} suggest a five-coordinate structure of the complex with a trans-arrangement of the Cl atoms. The bands in the 210 to 140 cm^{-1} region can be ascribed to skeleton deformation of the complex [1].

$Mn(C_3H_7NO)(CH_3COO)_2$. For preparation, $Mn(CH_3COO)_2$ $4H_2O$ was added to molten propionamide at 100 to 110°C with intense stirring until a clear cinnamon-colored melt had formed. Finally a solid appeared which was cooled to room temperature. The product was triturated until it became a fine powder which was washed with ethanol and acetone and dried in vacuum over P_2O_5 [2].

The X-ray diagram reveals the strongest reflex at d = 9.19 Å. The IR spectrum (KBr disks and Vaseline mulls) and the conductance data suggest the presence of five-coordinate manganese(II) with bidentate O, O-bonded acetato groups and monodentate carbonyl oxygen-bonded propionamide. The bands at 945 cm^{-1} and 680 cm^{-1} (broad) are ascribed to ν(C–N) modes and to δ(CCO) modes of the acetato groups. An intense band at 470 cm^{-1} is assigned to δ(CCN) of propionamide. Coordination of the acetato groups to Mn is suggested by a shift of the ν(C–C) band to higher frequency by 22 cm^{-1} relative to ionic acetates. The bands in the 430 to 412 cm^{-1} region can be associated with vibrations of the Mn–O (propionamide or acetate) bonds. Thermal analysis reveals two endothermic effects at 170 and 195°C, indicating stepwise release of the propionamide ligand, which is confirmed by thermogravimetry.

The low electrical conductance $\Lambda = 28.9\ cm^2 \cdot \Omega^{-1} \cdot mol^{-1}$ of a 0.001 M ethanol solution measured at 25°C indicates a nonelectrolyte. In water the complex dissociates as shown by its conductance characteristic of a 1:2 electrolyte [2].

References:

[1] Tsivadze, A. Ya., Smirnov, A. N., Kharitonov, Yu. Ya., Tsintsadze, G. V., Tevzadze, M. N. (Koord. Khim. **3** [1977] 514/23; Soviet J. Coord. Chem. **3** [1977] 393/400, 394).

[2] Azizov, T. A., Khodzhaev, O. F., Zuparov, M. Z., Parpiev, N. A. (Koord. Khim. **8** [1982] 916/21, 917, 920; C.A. **97** [1982] No. 119518).

24.1.12 With Acrylamide $CH_2{=}CHC(O)NH_2$ (= C_3H_5NO)

$Mn(C_3H_5NO)_4X_2$ (X = NO_3, Cl) and **$Mn(C_3H_5NO)_4Cl_2 \cdot 2H_2O$.** The complex $Mn(C_3H_5NO)_4(NO_3)_2$ was obtained by reacting a solution of $Mn(NO_3)_2 \cdot 6H_2O$ in a mixture of absolute ethanol and triethyl orthoformate with excess ligand (mole ratio 1:7). Some diethyl ether was added to precipitate the white complex which was washed with dry diethyl ether, and dried in vacuum. It melts at 88 to 89°C [1]. $Mn(C_3H_5NO)_4Cl_2$ was prepared by treating finely ground $MnCl_2 \cdot 4H_2O$ with an acetone solution of acrylamide (mole ratio 1:5). The precipitate was washed with hot benzene followed by ether, and dried in vacuum [2]. The enthalpy of formation from $MnCl_2$ and acrylamide, $\Delta H_f = -10.62 \pm 0.06$ kcal/mol, was calculated from the solution enthalpy of the complex (-7.24 ± 0.02 kcal/mol), of anhydrous $MnCl_2$ (-16.02 kcal/mol), and of the ligand in aqueous 2n HCl (-3.16 ± 0.03 kcal/mol), determined calorimetrically at 25°C. Using $MnCl_2 \cdot 4H_2O$ as the manganese salt according to $MnCl_2 \cdot 4H_2O + 4C_3H_5NO \rightarrow Mn(C_3H_5NO)_4Cl_2 + 4H_2O$, a formation enthalpy of the complex $\Delta H_f = 4.4 \pm 0.6$ kcal/mol, results [3].

$Mn(C_3H_5NO)_4Cl_2 \cdot 2H_2O$ was detected as a stable solid phase in the solubility isotherm of the ternary system $MnCl_2$-acrylamide-H_2O studied at 20°C. Its identity was confirmed by chemical analysis. Physical properties (molecular weight, heat of formation, melting and decomposition points, pH of saturated aqueous solution) were studied but no data are given in the paper [4].

The IR spectrum of the nitrato complex (Nujol mulls) recorded in the 4000 to 200 cm^{-1} region shows the ν(NH) absorption bands in positions which suggest strong hydrogen bonding between the nitrato groups and acrylamide. The bands at 1670 ± 5, 1627 ± 3 and 1585

(±10) cm^{-1}, assigned to ν(CO), ν(CN), and NH_2 bending vibrations, are shifted to lower frequencies by 5, 23, and 25 cm^{-1}, respectively, compared to the free ligand, and the broad δ(NH) ligand band at 700 cm^{-1} is shifted to 650 cm^{-1}. Several bands below 400 cm^{-1} may be associated with ν(Mn–O) vibrations. The broadened anion bands suggest slightly coordinated NO_3^- ions. The spectrum of the chloro complex reveals bands of ν_{as}(NH) at 3385 cm^{-1}, ν_s(NH) at 3175 and 3145 cm^{-1}, ν(CO) at 1680 cm^{-1} and δ(NH) at 1600 cm^{-1} [2].

$Mn(C_3H_5NO_2)_2X_2$ (X = Cl, Br). Absolute ethanol solutions of manganese chloride or bromide and an ethanolic solution of acrylamide were combined in a 1:6 or 1:8 mole ratio, refluxed for 1 h and allowed to stand 1 to 2 d for crystallization. The light pink crystals were washed with ethanol and ether, and dried in vacuum. The melting points, magnetic moments (from susceptibility measurements at room temperature), absorption maxima ν_{max} (in cm^{-1}) observed in the electronic spectrum of 0.01 M solutions in $CHCl_3$ and acrylamide with molar extinction coefficient ε (in $L \cdot mol^{-1} \cdot cm^{-1}$) and molar electrical conductance Λ of 0.001 M acetone solutions of the complexes (in $cm^2 \cdot \Omega^{-1} \cdot mol^{-1}$) are given below:

complex	m.p. in °C	μ_{eff} in μ_B	ν_{max}	ε	ν_{max}	ε	Λ
$Mn(C_3H_5NO)_2Cl_2$	188	5.9	27500	0.35	24600	0.2	11.2
$Mn(C_3H_5NO)_2Br_2$	208	5.92	27200	0.3	24500	0.2	12.0

In the IR spectra of the complexes the ν(NH) absorption bands of the ligand (at 3350 and 3180 cm^{-1}) are almost unaltered whereas the bands assigned to ν(CO) and ν(CN) + δ(NH_2) shifted from 1680 cm^{-1} to about 1660 cm^{-1} and from 1440 to 1460 cm^{-1}, respectively. A band assignable to ν(Mn–O) was observed in the 450 to 420 cm^{-1} region. The magnetic data and electronic spectra support the assumption of a spin-free octahedral configuration of the complexes, which are nonelectrolytes in acetone solution [5].

$Mn(C_3H_5NO)_6X_2$ (X = ClO_4, BF_4). The white complexes were prepared from manganese perchlorate or tetrafluoroborate and acrylamide in absolute ethanol containing triethyl orthoformate in the same way as was $Mn(C_3H_5NO)_4(NO_3)_2$ (see p. 108). They are apparently isomorphous with each other and with the corresponding complexes of cadmium as suggested by the similar X-ray diffraction patterns. The perchlorate melts at 110 to 111°C, the tetrafluoroborate at 100 to 102°C. The IR spectra in the 1670 to 1585 cm^{-1} and 400 to 200 cm^{-1} regions are similar to that of $Mn(C_3H_5NO)_4(NO_3)_2$ and are interpreted alike. However, the shift of the δ(NH) ligand absorption band near 700 cm^{-1} to 600 cm^{-1} and the unsplit anion bands indicate less hydrogen bonding and an ionic nature of the complexes [1].

References:

[1] Reedijk, J. (Inorg. Chim. Acta **5** [1971] 687/90, 688).

[2] Barvinok, M. S., Mashkov, L. V. (Zh. Neorgan. Khim. **19** [1974] 571/3; Russ. J. Inorg. Chem. **19** [1974] 310/1).

[3] Barvinok, M. S., Mashkov, L. V. (Zh. Neorgan. Khim. **25** [1980] 2846/8; Russ. J. Inorg. Chem. **25** [1980] 1570/2).

[4] Kalaeva, M. I., Zaruba, N. V., Gusak, N. N., Mirko, S. A., Mamaeva, B. A. (Zh. Neorgan. Khim. **24** [1979] 553/4; Russ. J. Inorg. Chem. **24** [1979] 308/9).

[5] Samantaray, A., Panda, P. K., Mohapatra, B. K. (J. Indian Chem. Soc. **57** [1980] 430/2).

24.1.13 With Derivatives of Benzamide

ligand 1 $C_6H_5C(O)NRC(O)CH_3$ with $R = 3\text{-}NO_2C_6H_4$ $(= C_{15}H_{12}N_2O_4)$

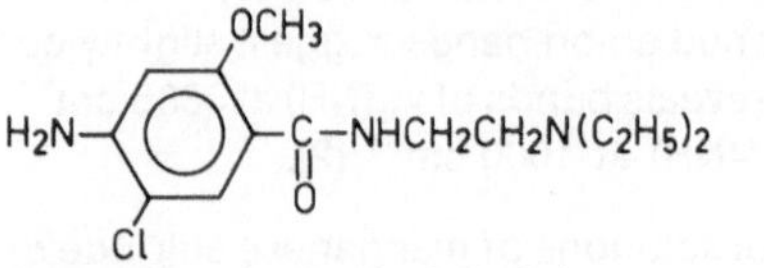

ligand 2 $(= C_{14}H_{22}ClN_3O_2)$

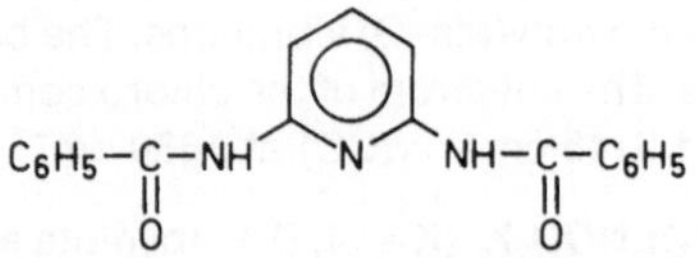

ligand 3 $(= C_{19}H_{15}N_3O_2)$

$Mn(C_{15}H_{11}N_2O_4)_2$. An aqueous solution of $MnCl_2$, adjusted to pH of 4.0 to 4.5 with sodium acetate, was boiled and a solution of ligand 1 in aqueous ethanol added slowly with stirring. The pale yellow complex precipitates on addition of aqueous 2 N ammonia at pH 6.5 to 8.0. It was digested and washed with hot water, then with ethanol, and dried at 120°C. Susceptibility measurements (temperature not given) yield the magnetic moment $\mu_{eff} = 5.61\ \mu_B$. As shown by thermal analysis (TG, DTA), the complex is stable up to 220°C. It is insoluble in methanol, soluble in chloroform, acetone, isobutyl ketone and tributyl phosphate [1].

$Mn(C_{14}H_{22}ClN_3O_2)_3Cl_2$, prepared from $MnCl_2$ and ligand 2 in water, reveals a magnetic moment of 5.9 μ_B and a molar conductance $\Lambda = 192.45\ cm^2 \cdot \Omega^{-1} \cdot mol^{-1}$. IR data indicate bonding through the amide nitrogen and the N atom of the $N(C_2H_5)_2$ group. The electronic spectrum was investigated [2].

$Mn(C_{19}H_{15}N_3O_2)X_2$ ($X = NO_3$, Cl, Br, NCS). The complexes with ligand 3 were prepared from the appropriate manganese salt and the ligand in the same way as the corresponding complexes with N,N'-(2,6-pyridinediyl)bis(acetamide), $Mn(C_9H_{11}N_3O_2)X_2$; see p. 105. The IR and electronic spectra are similar to the spectra of those complexes and are also interpreted in terms of a trigonal bipyramidal structure with the pyridinediamine derivative acting as tridentate O,O,N-bonded ligand and with coordinated anions. The magnetic moments at room temperature, $\mu_{eff} = 5.83$, 582, 584, and 5.87 μ_B, for $X = NO_3$, Cl, Br, and NCS, respectively, are indicative of high-spin Mn^{II} compounds. The complexes are partly soluble in common organic solvents, but highly soluble in dimethylformamide, where they behave as nonelectrolytes [3].

References:

[1] Mandal, S. K., Das, J. (J. Indian Chem. Soc. **54** [1977] 951/3).
[2] Mahto, C. B. (J. Indian Chem. Soc. **62** [1985] 731/3).
[3] Sangal, S. K., Sahni, S. K., Rana, V. B. (Acta Chim. Acad. Sci. Hung. **110** [1982] 12/23).

24.1.14 With Salicylamide and N-Acetylsalicylamide

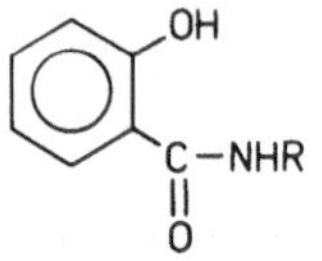

$R = H$; $(= C_7H_7NO_2)$
$R = CH_3CO$; $(= C_9H_9NO_3)$

$[Mn(C_7H_7NO_2)(H_2O)Cl_2]$ was prepared by refluxing a methanolic solution of manganese(II)chloride and salicylamide for about 6 h, after which the solvent was removed completely on a water bath. The residue was washed repeatedly with $CHCl_3$ and finally with diethyl ether and then dried in vacuum. The white complex melts above 300°C. The IR spectrum (Nujol mulls) recorded in the 4000 to 200 cm^{-1} region shows characteristic absorption bands. Their wave numbers (in cm^{-1}) are given with the assignment and shift relative to free salicylamide or water (in parentheses): $\nu(OH)$ 3456 (−169); $\nu_{as}(NH)$ 3424 (−109); $\nu_s(NH)$ 3240 (−185);

ν(CO), amide I, 1644 (−31); β(NH_2), amide II, 1620 (−10); ν(C–O) of salicylamide 1242 (−13). The observed band shifts allow the assumption of a polymeric octahedral structure where six-coordinate manganese is bonded to the carbonyl and phenolic oxygens of the same salicylamide molecule and to the amide nitrogen of a second one as well as to one oxygen of water and two chloride ions. From the shift of the ν(CO) band and the ionization potential (8.048 eV) a stability constant of the complex, log K = 5.86, was calculated. Susceptibility measurements at room temperature yield the magnetic moment μ_{eff} = 6.01 μ_B indicating a spin-free Mn^{II} complex of octahedral geometry. [$Mn(C_7H_7NO_2)(H_2O)Cl_2$] loses one molecule of water upon heating at 130 to 160°C. It decomposes in water but it is soluble in ethanol, DMF, and DMSO. The molar electrical conductance of a 0.001 M solution in DMF, Λ = 3.5 $cm^2 \cdot \Omega^{-1} \cdot mol^{-1}$, is that of a nonelectrolyte whereas the value measured in DMSO, Λ = 7.5 $cm^2 \cdot \Omega^{-1} \cdot mol^{-1}$, indicates a 1:2 electrolyte in agreement with the apparent molecular weight of 96.4 obtained by cryoscopy in the same solvent, which is about a third of the calculated value (281.0) [1].

$Mn(C_7H_6NO_2)_2$. Aqueous solutions of a manganese(II) salt and salicylamide were mixed in a 1:2 mole ratio and NaOH solution was added to attain pH≈8 [1] or ethanolic solutions of $MnCl_2 \cdot 4H_2O$ and salicylamide in ethanolic NaOH (4 g/100 mL) were mixed in a 1:2 mole ratio and refluxed on a water bath for about 12 h and then left overnight. The resulting black powdery precipitate was washed with ethanol, hot water, ethanol again, and dried at 110°C [2]. The complex is stable up to 360°C [2]; also see [1]. A stability constant, log K = 5.269, was calculated as for [$MnCl_2(C_7H_7NO)(H_2O)$] (see above [1]). The positions of the IR absorption bands are similar to those of [$MnCl_2(C_7H_7NO_2)(H_2O)$]. A polymeric octahedral structure where six-coordinate manganese is bonded chelate-like to the carbonyl and phenolic oxygens of two salicylamide molecules and to two nitrogens of two other molecules is assumed [1, 2]. The magnetic moment μ_{eff} = 6.1 μ_B indicates a spin-free Mn^{II} complex of octahedral geometry. $Mn(C_7H_6NO_2)_2$ is extremely insoluble in water and common organic solvents [1, 2], but slightly soluble in DMF and dimethyl sulfoxide [1]. It is resistant to the action of mineral acids and alkalies [2].

$Mn(C_9H_8NO_3)_2 \cdot H_2O$. An aqueous solution of $MnCl_2$ was added with stirring to an aqueous suspension of the Na salt of N-acetylsalicylamide (mole ratio 1:2). The precipitate was washed with water, alcohol, and ether. The IR spectrum shows a ν(CO) absorption at 1670 cm^{-1} ($\Delta\nu$(CO) = −20 cm^{-1}). No bands assignable to ν(OH) modes were found in the region above 3000 cm^{-1}. This indicates deprotonation of the phenolic OH groups and formation of Mn–O bonds. A polymeric structure with the ligand being tetradentate bridging is assumed. Thermal analysis (DTA) reveals dehydration at 70°C. The dehydrated complex is stable up to 275°C and decomposes at 290°C. Decomposition proceeds rapidly at 370°C and is complete at 650°C [3].

References:

[1] Aggarwal, R. C., Singh, B., Singh, T. B. (Gazz. Chim. Ital. **111** [1981] 425/8).
[2] Maurya, P. L., Agarwala, B. V., Dey, A. K. (Inorg. Nucl. Chem. Letters **13** [1977] 145/8).
[3] Khudaiberdiev, E. Kh., Faizieva, S. S., Azizov, T. A. (Uzb. Khim. Zh. **1985** No. 1, pp. 12/4; C.A. **102** [1985] No. 196876).

24.1.15 With 2-Pyridinecarboxamide (= Picolinamide = $C_6H_6N_2O$) and Its 1-Oxide (= $C_6H_6N_2O_2$)

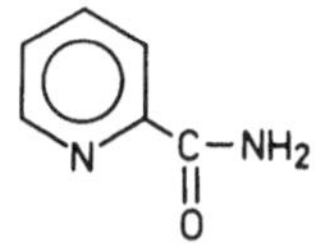

$Mn(C_6H_6N_2O)_2(H_2O)_2X_2$. Complexes with X = Cl or Br were prepared by adding an aqueous solution of $MnX_2 \cdot 4H_2O$ (0.01 mol) to that of picolinamide (0.02 mol) and evaporating the mixture until crystallization occurred. The complexes were recrystallized from water, washed with ether, and dried in air. As shown by thermal analysis (TG) under streaming N_2 at a heating rate of 3°C/min, the complexes lose the water at 132°C (X=Cl) and 143°C (X=Br). Color changes from yellow to white and from pale green to pale yellow, respectively, were observed. Dehydration enthalpies, resulting from DSC studies (heating rate 16°C/min) were: $\Delta H = 86.3$ kJ/mol (chloro complex), $\Delta H = 101.3$ kJ/mol (bromo complex). Estimated enthalpy values for the loss of the first ligand molecule were $\Delta H = 40.1$, and 43.9 kJ/mol, respectively [1].

The manganese(II) picolinamide complex in aqueous 0.2 M citric acid solution (buffered with Na_2HPO_4) was studied by polarography. At pH=2.75, two reduction waves with half-wave potentials $E_{1/2} = -0.918$ and -1.392 V (vs. SCE) were observed which upon raising the pH become more negative. The first wave, corresponds to a reversible one-electron reaction, the second wave indicates electrochemical reduction of the Mn^{II} complex ion in acidic solution. The first wave increases in strength up to pH 6; it then becomes weaker and disappears at pH 9.60 [2].

$Mn(C_6H_6N_2O_2)Cl_2$ was prepared by reacting $MnCl_2 \cdot 4H_2O$ with the 1-oxide of 2-pyridinecarboxamide in boiling ethanol. The yellow precipitate was dried in vacuum over anhydrous $CaCl_2$. The IR spectrum (Nujol mull) shows the following characteristic absorption bands (in cm^{-1}), with shifts relative to the free ligand in parentheses: ν(CO) 1672 (−3); ν(CN) 1391 (+17); ν(NO) 1225 (−8); δ(NO) 853 (+2). The observed shifts of ν(CO) and ν(NO) to lower and of ν(CN) to higher frequencies in the complex indicate coordination of the bidentate ligand to Mn through the oxygens of the carbonyl and N-oxide groups. Thus, a halogen-bridged structure of the complex is proposed since the ν(NO) frequency is not lower than in other monomeric metal complexes of the ligand. Susceptibility measurements at 89 and 294 K yield the magnetic moment $\mu_{eff} = 5.75\,\mu_B$. The complex obeys the Curie-Weiss law with the Curie constant $\Theta = 0$ K [3].

References:

[1] Kennedy, T., MacSween, D. R. (Therm. Anal. Proc. 3rd Intern. Conf., Davos, Switz., 1971 [1972], pp. 689/96, 690, 694; C.A. **78** [1973] No. 131573).

[2] Tuichiev, E. T., Korneva, L. E., Murtazaev, A. M. (Mater. Yubileinoi Resp. Nauchn. Konf. Farmat. Posvyashch 50th Letiyu Obraz. SSSR, Tashkent 1972, pp. 159/61; C.A. **82** [1975] No. 161780).

[3] Sanders, A. E., Philips, D. K. (Inorg. Chim. Acta **59** [1982] 125/32, 126/8, 131).

24.1.16 With N,N′-Ethylene- or Arylenebis(2-pyridinecarboxamides)

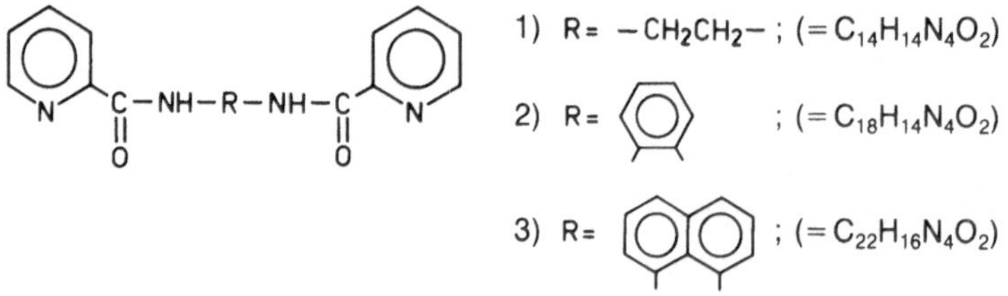

Chloro complexes of composition **$Mn(C_{14}H_{14}N_4O_2)Cl_2 \cdot 0.5H_2O$** and **$Mn(C_{18}H_{14}N_4O_2)Cl$** were prepared by reacting $MnCl_2$ with the corresponding ligand in hot ethanol. The pale yellow microcrystalline complex with ligand 1 was washed with a minimum of water and ethanol [1].

The large yellow crystals of $Mn(C_{18}H_{14}N_4O_2)Cl_2$ were washed only with ethanol; they changed to a cream-white powder after drying [2].

$Mn(C_{22}H_{16}N_4O_2)Cl_2 \cdot 2H_2O$ is reported to form on addition of anhydrous(?) $MnCl_2$ to a warm (70°C) solution of ligand 3 (mole ratio 1:1) in absolute(?) ethanol under vigorous stirring. After cooling to room temperature the precipitate was washed with acetone and ether and dried in vacuum over P_4O_{10} [3].

Characteristic absorption bands observed in the IR spectra of the complexes (KBr disks and Nujol mulls) are given below together with the observed shifts (in parentheses) in comparison to the free ligand (wave numbers in cm^{-1}). For additional data see the papers:

complex	ν(OH)	ν(NH)	amide I ν(CO)	amide II ν(CN) + δ(NH_2)	amide III ν(CN) + δ(NH_2)	Ref.
$Mn(C_{14}H_{14}N_4O_2)Cl_2 \cdot 0.5H_2O$	—	—	1645, 1630 (−15, −30)	1550 (+15)	1340 (+10)	[1]
$Mn(C_{18}H_{14}N_4O_2)Cl_2$	—	3360 (+40)	1660 (−15)	—	1335 (+3)	[2]
$Mn(C_{22}H_{16}N_4O_2)Cl_2 \cdot 2H_2O$	3440	3180	1633 (−36)	1537 (+18)	1280 to 1010	[3]

The ν(NH) band of ligand 1 at 3300 cm^{-1} is masked in the complex by broad bands due to coordinated water. Coordination of the ligand to Mn through the oxygen of the $CONH_2$ group and the pyridine ring nitrogens of the apparently tetradentate ligand is indicated by the shift of the ν(CO) band to lower wave numbers and the shift of the amide II, amide III, and the skeletal vibrations to higher wave numbers on complexation. The amide protons may be involved in hydrogen bonding between the amide nitrogen and the oxygen atoms [2, 3]. For $Mn(C_{14}H_{14}N_4O_2)Cl_2 \cdot 0.5H_2O$ a trans-arrangement involving bis-bidentate bridging behavior of the ligand and coordination of the Cl^- ions is assumed [1]. A polymeric structure with coordinated Cl^- ions is suggested for $Mn(C_{18}H_{14}N_4O_2)Cl_2$ by the insolubility of this complex and the decomposition pattern in its thermogravimetric analyses [2]. In the complex $Mn(C_{22}H_{16}N_4O_2)Cl_2 \cdot 2H_2O$, the ligand is assumed to occupy the equatorial plane and the water molecules are at the apical sites, whereas the Cl^- ions are not coordinated [3].

The magnetic moments μ_{eff} = 5.98 [1], 5.65 [2], and 5.87 μ_B [3] (from susceptibility measurements at room temperature) are indicative of high-spin octahedral Mn^{II} complexes. The diffuse reflectance spectrum of $Mn(C_{22}H_{16}N_4O_2)Cl_2 \cdot 2H_2O$ shows a weak band at ~490 nm assignable to the electron transition $^6A_{1g} \rightarrow ^4T_{1g}$ (^4G) which is also characteristic for an octahedral configuration [3].

$Mn(C_{14}H_{14}N_4O_2)Cl_2 \cdot 0.5H_2O$ is dehydrated between 110 and 170°C. Further decomposition with loss of ligand and Cl_2 occurs between 300 and 800°C [1]. $Mn(C_{18}H_{14}N_4O_2)Cl_2$ is insoluble in common solvents and melts with decomposition in the range 350 to 790°C [2]. $Mn(C_{22}H_{16}N_4O_2)Cl_2 \cdot 2H_2O$ darkens upon heating between 280 and 310°C and decomposes above 310°C. It is readily soluble in methanol, dimethylformamide, and dimethyl sulfoxide, slightly soluble in cold water and nitromethane, and insoluble in acetone, nitrobenzene, and nonpolar solvents. It is decomposed by dilute mineral acids [3].

$Mn_2(C_{14}H_{14}N_4O_2)_3(ClO_4)_4 \cdot 2H_2O$ was prepared by adding a hot aqueous solution of $Mn(NO_3)_2 \cdot 6H_2O$ to a suspension/solution of the ligand in hot water and stirring until all ligand had dissolved. Precipitation of manganese hydroxide was prevented by adding a few drops of $HClO_4$. After filtration, a warm saturated aqueous solution of $NaClO_4$ was added dropwise. The fine crystalline complex was washed well with water and ethanol and dried. The IR spectrum

(KBr disks) reveals bands of ν(CO) at 1655 and 1625 cm^{-1}, of ν(CN) + δ(NH_2) at 1555, 1530 (amide II), and 1360 cm^{-1} (amide III). The position of the perchlorate bands is indicative of uncoordinated ClO_4^- ions. The compound thus may be considered as a dimeric complex with bridging ligands, between the Mn atoms. They are only coordinated by the carbonyl oxygen and pyridine nitrogen atoms. An octahedral N_3O_3 configuration is in agreement with the observed high-spin magnetic moment (μ_{eff} = 5.86 μ_B). A complex hydrogen bonding system involving water, amide protons, and ClO_4^- ions could bind such dimeric anions strongly in the crystal lattice. The coordinated water is lost below 200°C. To prevent possible explosion, the compound was not heated above 200°C [1].

References:

[1] Barnes, D. J., Chapman, R. L., Stephens, F. S., Vagg, R. S. (Inorg. Chim. Acta **51** [1981] 155/62).
[2] Chapman, R. L., Vagg, R. S. (Inorg. Chim. Acta **33** [1979] 227/34).
[3] Zafiyropoulos, T. F., Perlepes, S. P., Ioannou, P. V., Tsangaris, I. M., Galinos, A. G. (Z. Naturforsch. **36b** [1981] 87/93).

24.1.17 With 3-Pyridinecarboxamide

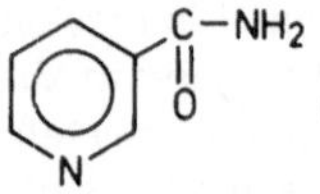

(= Nicotinamide = $C_6H_6N_2O$)

A mixed ligand complex [$Mnaca_2(C_6H_6N_2O)_2$] was described in "Mangan" D1, 1979, p. 84.

$Mn(C_6H_6N_2O)_2X_2$ (X = Cl, Br, I). The chloro and bromo complexes were prepared by mixing saturated ethanolic solutions of $MnCl_2$ or $MnBr_2$ and nicotinamide in a 1:2 mole ratio with continuous stirring [1, 2]. The resulting light pink precipitates were washed with alcohol and ether. The iodo complex was obtained by reacting saturated solutions of anhydrous MnI_2 (free from I_2) in ether and of nicotinamide in alcohol in a 1:2 mole ratio with protection from light. The light yellow precipitate was washed several times with ether [1]. The chloro complex was dried at about 70°C in an air oven [8].

The IR spectrum of the chloro complex (Nujol mull) recorded in the 4000 to 400 cm^{-1} region, shows almost unaltered absorption bands assigned to ν(NH) vibrations at 3330 and 3120 cm^{-1}, to ν(CO) at 1680 and to ν(CN) of amide at 1120 cm^{-1}. Bands assigned to ν(C=C) and ν(C=N) modes of the pyridine ring occur at 1580, 1570, and 1480 cm^{-1} together with other ring modes at 990, 605, and 405 cm^{-1}. The shift of these ring modes to higher frequencies in the complex by 18 to 30 cm^{-1} suggests bonding of nicotinamide to Mn only through the pyridine ring nitrogen without participation of the $CONH_2$ group [2, 8]. A far IR band at 220 cm^{-1} assigned to a ν(Mn–Cl) mode involves bridging Cl^- ions in a chlorine-bridged polymeric octahedral structure proposed for the complex. The electronic spectrum (Nujol mull) recorded in the 33000 to 6000 cm^{-1} region shows three absorption bands at 18690, 23530, and 26310 cm^{-1} ascribed to the electron transitions $^6A_{1g} \rightarrow {}^4T_{1g}(^4G)$, $^6A_{1g} \rightarrow {}^4E_g$, $^4A_{1g}(^4G)$, and $^6A_{1g} \rightarrow {}^4E_g(^4D)$, respectively. The ligand field-splitting parameter D_q = 974 cm^{-1} derived from the $^6A_{1g} \rightarrow {}^4E_g(^4G)$ transition is consistent with an octahedral environment of the Mn^{2+} ion [8].

Susceptibility measurements at room temperature yield the magnetic moment μ_{eff} = 6.10 μ_B, indicative of a high-spin Mn^{II} (d^5) complex [8]. In the following table are listed the melting points (m.p.) of the complexes, their solubility S in water at 25°C, and molar electrical conductance Λ in 0.001 M aqueous solution:

complex	m.p. in °C	S in g/100 mL H_2O	Λ in $cm^2 \cdot \Omega^{-1} \cdot mol^{-1}$
$Mn(C_6H_6N_2O)_2Cl_2$	320	12	248
$Mn(C_6H_6N_2O)_2Br_2$	312	66	240
$Mn(C_6H_6N_2O)_2I_2$	280	108	247

The complexes behave as 1:2 electrolytes in water due to solvolysis [1]. The chloro complex was found to be stable and insoluble in most organic solvents except nitrobenzene where it is slightly soluble and a nonelectrolyte as shown by its low conductivity (Λ=11 to 3 $cm^2 \cdot \Omega^{-1} \cdot mol^{-1}$) [2].

$Mn(C_6H_6N_2O)_4(NCO)_2$ was obtained from $Mn(NCO)_2$ and excess ligand in ethanolic solution. Stirring with the equimolar amount of $Hg(SCN)_2$ in acetone yields the complex $Mn(C_6H_6N_2O)_2[Hg(SCN)_2(OCN)_2]$; see p. 116 [9].

$Mn(C_6H_6N_2O)_4(NCS)_2$ and **$Mn(C_6H_6N_2O)_2(NCS)_2 \cdot 2H_2O$.** The anhydrous complex was prepared by the method recommended in [3] for the synthesis of the complex $Mn(C_6H_4N_2)_4(NCO)_2$ with 3-cyanopyridine: an aqueous solution containing $Mn(NO_3)_2 \cdot 6H_2O$ and NaSCN (mole ratio 1:2) was added to an aqueous ethanolic solution of nicotinamide (Mn to ligand ratio = 1:5). The crystals which separated on standing were washed with ethanol containing some ligand, and recrystallized from ethanol containing a 5% excess of ligand [3]. $Mn(C_6H_6N_2O)_2(NCS)_2 \cdot 2H_2O$, obtained from nicotinamide in aqueous ethanol and $Mn(NCS)_2$ (Mn to ligand ratio = 1:2), was washed with ethanol and ether and dried in vacuum over $CaCl_2$. It is described as a white powder [5] consisting of elongated crystals [6]. IR data are reported for the hydrate $Mn(C_6H_6N_2O)_2(NCS)_2 \cdot H_2O$ (mineral oil or hexachlorobutadiene mulls) in the range 4000 to 400 cm^{-1}. The observed shift of the bands assignable to the pyridine ring toward higher frequencies (by 15 to 40 cm^{-1}) and the only slight shift of the ν(NH) and ν(CO) bands in the complex clearly indicate coordination of nicotinamide to Mn only through the pyridine ring nitrogen with no participation of the $CONH_2$ group. The disappearance or intensity changes of ligand bands in the 800 to 400 cm^{-1} region and new bands around 565 (broad, intense) and 475 cm^{-1} may be associated with the formation of metal-to-ligand bonds. The positions and intensity of the ν(CN) and ν(CS) bands of the thiocyanate groups indicate their coordination to Mn through the nitrogen atoms [5].

frequency	ν(CN)	ν(CS)	δ(NCS)	ν(Mn–NCS)	Ref.
in cm^{-1}	2078	798	475	259	[4][a]
in cm^{-1}	2116(2063)	795	473	—	[5]

[a] Data reported for the anhydrous complex (CsI optics).

The hydrate is sparingly soluble in water and its characteristic crystals may serve to detect nicotinamide by microanalysis [6].

$Mn(C_6H_6N_2O)_2(CH_3COO)_2$. Finely ground $Mn(CH_3COO)_2 \cdot 4H_2O$ (0.022 mol) was added in small portions to molten nicotinamide (0.1 mol) over 20 to 25 min. After cooling, the solidified mass was dissolved in hot methanol. The precipitate formed upon cooling was recrystallized from methanol, washed with ethanol, and dried in vacuum. The complex, whose identity was established by X-ray diffraction studies, melts at 125°C. The IR spectrum shows similar absorption bands of ν(NH), ν(CO), and ν(ring) as reported for the other complexes with nicotinamide. Bands of $\nu_{as}(COO)$ and $\nu_s(COO)$ were observed at 1544 and 1455 cm^{-1}. The position of the ν(C–C) band of the acetato group at 951 cm^{-1} differs only slightly (+6 cm^{-1}) from that of anhydrous $Mn(CH_3COO)_2$. Therefore bidentate coordination of the acetato groups was suggested. The assumed octahedral environment of Mn is in agreement with the low electrical conductance $\Lambda = 32\ cm^2 \cdot \Omega^{-1} \cdot mol^{-1}$ of a 0.001 M solution in ethanol, indicating a nonelectrolyte while in water a value of $\Lambda = 217\ cm^2 \cdot \Omega^{-1} \cdot mol^{-1}$ indicates a 1:2 electrolyte. Thermal analysis (DTA, TG) shows that the complex starts to decompose at 245°C with loss of nicotinamide which is complete at 330°C when $Mn(CH_3COO)_2$ is left [7].

$Mn(C_6H_6N_2O)_2[Hg(SCN)_2(NCO)_2]$ was obtained by reaction of $MnHg(SCN)_2(NCO)_2$ (prepared from $MnCl_2 \cdot 4H_2O$, KNCO, and $Hg(SCN)_2$) with excess ligand (five-fold) in acetone solution, or from $Mn(C_6H_6N_2O)_4(NCO)_2$ and the equimolar amount of $Hg(SCN)_2$ in a mixture of acetone and alcohol. The absorption bands in the IR spectrum (polyethylene disks) at 2228, 2153, and 2132(sh) cm^{-1} were ascribed to ν(CN), at 1232 to ν(CO) at 778 cm^{-1} to ν(CS), at 647(br) to δ(NCO), at 458, and 424 cm^{-1} to δ(NCS), at 258 cm^{-1} (br) to ν(Mn–N), at 199 cm^{-1} to ν(Hg–S), and at 155 cm^{-1} to δ(N–Mn–N). The absorption regions due to the NCS and NCO groups are almost unchanged with respect to those in $Mn[Hg(SCN)_2(NCO)_2]$, where they indicate the presence of bridging NCS and NCO groups in a polymeric complex structure. The ligand nicotinamide is coordinated to Mn through the pyridine ring nitrogen as suggested by the upward shift of the lowest ring vibration modes at 604 and 405 cm^{-1}. Thus, Mn appears to be surrounded by six nitrogens of the nicotinamide, thiocyanate, and cyanate groups in an octahedral ligand field [9].

References:

[1] Azizov, M. A., Khamraev, A. D., Khakimov, K. K. (Uzb. Khim. Zh. **7** No. 4 [1963] 32/4; C. A. **60** [1964] 183).
[2] Paul, R. C., Arora, H., Chadha, S. L. (Inorg. Nucl. Chem. Letters **6** [1970] 469/73, 470).
[3] Nelson, J., Nelson, S. M. (J. Chem. Soc. **1969** 1597/603, 1603).
[4] Singh, P. P., Khan, S. A. (Indian J. Chem. A **14** [1976] 176/8).
[5] Tsivadze, A. Yu., Tsintsadze, G. V., Gongadze, N. P., Kharitonov, Yu. Ya. (Koord. Khim. **1** [1975] 1221/8; Soviet J. Coord. Chem. **1** [1975] 1021/7, 1025).
[6] Sandri, G. (Atti Accad. Sci. Ferrara **35** [1957/58] 25/33, 27; C.A. **1961** 13777).
[7] Khodzhaev, O. F., Azizov, T. A., Parpiev, N. A. (Zh. Neorgan. Khim. **23** [1978] 2942/8; Russ. J. Inorg. Chem. **23** [1978] 1633/7, 1635).
[8] Ahuja, I. S. (Indian J. Chem. A **22** [1983] 262/4).
[9] Ojha, T. N., Sharma, S. B. (Chem. Era **19** No. 2 [1983] 34/7; C.A. **101** [1984] No. 16211).

24.1.18 With N-Methyl- and N-(Hydroxymethyl)-3-pyridinecarboxamide

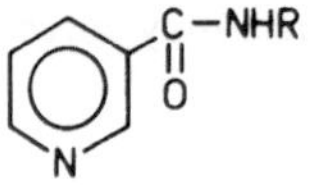

R = CH_3 (= $C_7H_8N_2O$)
R = CH_2OH (= $C_7H_8N_2O_2$)

$Mn(C_7H_8N_2O)_2Cl_2$ and **$Mn(C_7H_8N_2O_2)X_2$** complexes with X = Cl, Br were obtained from MnX_2 compounds and the ligands in ethanol. $Mn(C_7H_8N_2O)_2Cl_2$ was washed with ethanol and dried at ~70°C in an air oven [1]; the $Mn(C_7H_8N_2O_2)X_2$ complexes were washed with ethanol containing a little of the ligand, then with ether and dried over silica gel [2].

The IR spectrum of $Mn(C_7H_8N_2O)_2Cl_2$ (Nujol mulls) shows significant upward shifts of the pyridine ring modes but almost unchanged ν(NH), ν(CO), and ν(CN) bands which indicate coordination of the ligand to Mn only through the pyridine ring nitrogen. A far-IR band at 220 cm^{-1} assigned to a ν(Mn–Cl) mode involves bridging Cl^- ions in a chlorine-bridged polymeric octahedral structure proposed for the complex [1]. Chelate formation with the hydroxymethyl derivative seems improbable, but coordination of the carbonyl oxygen is indicated by a shift of the ν(CO) band [2].

The electronic spectrum of $Mn(C_7H_8N_2O)_2Cl_2$ (Nujol mulls) recorded in the 33000 to 6000 cm^{-1} region shows two absorption bands at 18690 and 23800 cm^{-1} ascribed to the electron transitions $^6A_{1g} \rightarrow {}^4T_{1g}(^4G)$ and $^6A_{1g} \rightarrow {}^4E_g, {}^4A_{1g}(^4G)$, respectively. A ligand field-splitting parameter $D_q = 957\ cm^{-1}$ derived from the latter transition is consistent with an octahedral environment of the Mn^{2+} ion. Susceptibility measurements yield the magnetic moment $\mu_{eff} = 6.12\ \mu_B$ indicative of a high-spin Mn^{II} (d^5) complex [1].

References:

[1] Ahuja, I. S. (Indian J. Chem. A **22** [1983] 262/4).
[2] Warnke, Z. (Roczniki Chem. **48** [1974] 2119/28, 2121, 2126; C.A. **83** [1975] No. 21067).

24.1.19 With N,N-Diethyl-3-pyridinecarboxamide

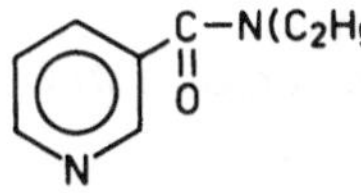

($= C_{10}H_{14}N_2O$)

$Mn(C_{10}H_{14}N_2O)_nCl_2$. Complexes with n = 2, 3, 4 were prepared by reacting aqueous ethanolic solutions of manganese chloride and ligand in a 1:n mole ratio [1]. For $Mn(C_{10}H_{14}N_2O)_4Cl_2$ X-ray diffractometer studies (MoKα radiation) reveal a triclinic lattice, space group $P\bar{1}$-C_i^1 (No. 2) with the lattice constants a = 15.549(7), b = 12.508(6), c = 7.696(4) Å and α = 111.1(1)°, β = 125.2(1)°, γ = 92.8(1)°; Z = 1, V = 1047.5 Å. The structure was solved by the heavy atom method (from 2121 independent reflections) and refined to R = 0.112. The resulting molecular structure is shown in **Fig. 12** below, together with the bond lengths and some bond angles involving the coordination polyhedron around the Mn atom which lies in an inversion center. The metal atom is surrounded by two Cl^- ions and four pyridine ring nitrogens of the organic ligand at the corners of a slightly distorted octahedron where the almost planar pyridine rings of two independent ligands form an angle of 84°. Conjugation between the π-electron systems of the pyridine ring and carboxamide groups does not occur [2].

The IR spectra of the complexes $Mn(C_{10}H_{14}N_2O)_nCl_2$ were investigated as mineral oil or hexachlorobutadiene mulls in the 4000 to 400 cm^{-1} region. The complexes with n = 2 and 3 also show bands in the 3590 to 3460 cm^{-1} region which tentatively are ascribed to the presence of water [1]. The almost unshifted ν(CO) bands at ~1630 cm^{-1} and the shift of the ligand ν(ring) band at 1588 cm^{-1} to higher frequency in the complexes by 11 cm^{-1} (n = 3, 4) or

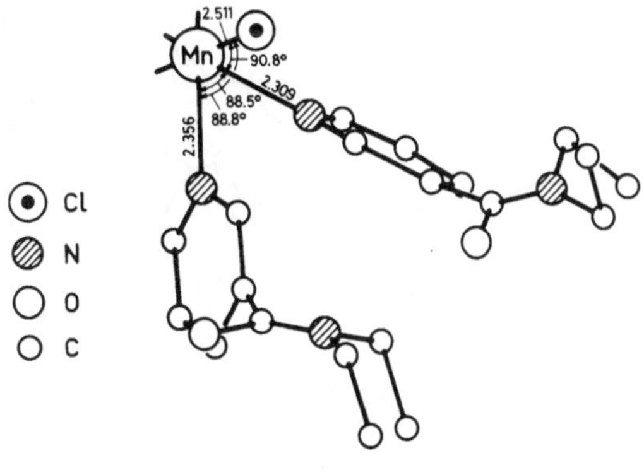

Fig. 12. Projection of the independent part of the molecule $Mn(C_{10}H_{14}N_2O)_4Cl_2$ on the (001) plane.

24 cm^{-1} (n = 2) as well as similar shifts of bands in the 1510, 1050, 635, and 405 cm^{-1} regions suggest coordination of manganese only to the pyridine ring nitrogen of the ligand and non-coordination of the carboxamide group [1]; also see [3]. Bands assignable to ν(Mn–N) and ν(Mn–Cl) modes are expected below 400 cm^{-1} [1]. Thus, a six-coordinate structure including the Cl^- ions was proposed for $Mn(C_{10}H_{14}N_2O)_4Cl_2$ and a four-coordinate one for $Mn(C_{10}H_{14}N_2O)_2Cl_2$ [3]. However, the Raman spectrum of concentrated aqueous solutions of $Mn(C_{10}H_{14}N_2O)_2Cl_2$ shows shifts of the ν(CO) and pyridine ring bands which suggest a structure containing the amide as a bridging ligand bonded also through the carboxamide oxygen as in $Mn(C_{10}H_{14}N_2O)_2(NCS)_2$ (see below) but the latter bond will disrupt upon dilution [4].

$Mn(C_{10}H_{14}N_2O)_2(NCX)_2$ (X = S, Se). The isothiocyanato complex was prepared by first reacting ethanolic solutions of $Mn(NO_3)_2 \cdot 6H_2O$ and KSCN in a 1:2 mole ratio; KNO_3 was removed and the stoichiometric amount of nicotinic acid diethylamide in aqueous solution was added. The white crystals which deposited after 10 to 15 d of standing were washed with ethanol and dried [1, 5]. The compound may be obtained also from $Mn(NCS)_2$ and ligand in a concentrated aqueous solution [6]. The $Mn(C_{10}H_{14}N_2O)_2(NCSe)_2$ complex was prepared in the same way with KSeCN instead of KSCN and acetone as the solvent [1]. The mixture was allowed to crystallize in vacuum over $CaCl_2$ for 5 to 10 d and the white crystals were washed with acetone and dried. $Mn(C_{10}H_{14}N_2O)_2(NCS)_2$ is air-stable $Mn(C_{10}H_{14}N_2O)_2(NCSe)_2$ is stable, only in a closed vessel [5].

X-ray diffraction data of single crystals (MoKα radiation) reveal for $Mn(C_{10}H_{14}N_2O)_2(NCS)_2$ a triclinic lattice, space group P1-C_i^1 (No. 2) with the lattice constants a = 9.332(8), b = 7.193(10), c = 11.159(8) Å and α = 113.9(1)°, β = 96.8(1)°, γ = 105.7(1)°; Z = 1, V = 637.1 $Å^3$. The structure was solved from 1797 independent reflections and refined up to R = 0.042. The crystal structure consists of octahedral centrosymmetric units shown in **Fig. 13** with bond lengths and some bonding angles around Mn which are aligned parallel to [010] and held together by pairs of bridging ligand molecules bound to Mn through the pyridinic nitrogen and the carbonyl oxygen atoms. The thiocyanate ions are coordinated to manganese through their nitrogen atoms. The metal atom in the center of the only slightly distorted octahedron is surrounded by two carboxamide oxygens, two pyridine and two thiocyanate nitrogens, each pair of equal atoms being in trans positions. The carboxamide moiety of the ligand appears to be bonded more strongly to Mn than the pyridine moiety as concluded from the Mn↔O(2.177 Å) and Mn↔N_{py} (2.324 Å) bond lengths. Density D = 1.375 g/cm^3 (from X-ray data), D = 1.365 g/cm^3 (measured) [6].

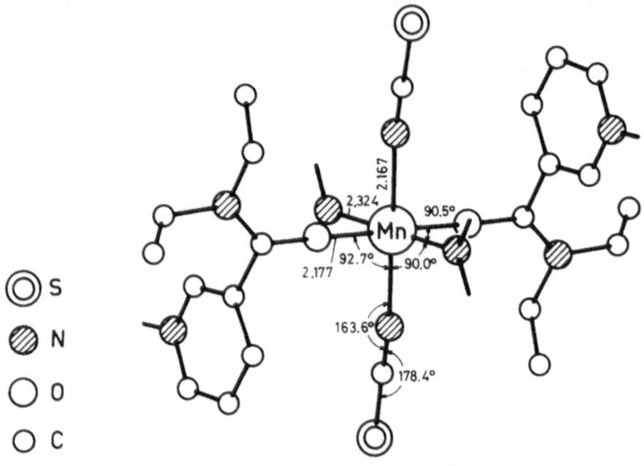

Fig. 13. Molecular structure of $Mn(C_{10}H_{14}N_2O)_2(NCS)_2$ with bond lengths (in Å) and some bond angles [6].

$Mn(C_{10}H_{14}N_2O)_2(NCSe)_2$ is also triclinic, space group P1-C_1^1 (No. 1) with the lattice constants a = 9.420(4), b = 7.187(6), c = 11.258(5) Å and α = 114.18(3)°, β = 83.92(5)°, γ = 73.88(7)°; Z = 2. The structure was solved by the heavy atom method and refined up to R = 0.093. The atomic parameters, bond lengths, and bond angles are given in the paper. The Mn atom is in octahedral coordination with 2 N atoms from the NCSe ligands in trans position, 2 N atoms of the heterocycle, and 2 carbonyl oxygens from the two 3-pyridinecarboxamide ligands. The density of 1.48 g/cm³ was calculated from the X-ray data [9].

The IR spectrum of the complexes (mineral oil or hexachlorobutadiene mulls) was recorded in the range 4000 to 400 cm^{-1}. A ν(CO) band was observed at 1610 cm^{-1}. The shift of this band toward lower frequencies by 20 cm^{-1} and of some bands assignable to ring vibrations toward higher frequencies by about 10 to 15 cm^{-1} in comparison to the free ligand suggest coordination to Mn through the carboxamide oxygen and pyridine ring nitrogen [1] (also see [4]). A bidentate bridging nature of the ligand in the complexes with bonding of the anions through terminal nitrogen thus is derived from the IR spectrum [1] in agreement with the results of X-ray structural analysis of $Mn(C_{10}H_{14}N_2O)_2(NCS)_2$, (see above). Bands of $\nu(CN)_{NCS}$ were observed at 2080 and 2028 cm^{-1}, of $\nu(CN)_{NCSe}$ at 2087 and 2035 cm^{-1}, of ν(CS) at 803 and 780 cm^{-1}, of ν(CSe) at 801 and 596 cm^{-1} [6].

$Mn(C_{10}H_{14}N_2O)_2(NCS)_2$ is readily soluble in dimethylformamide, soluble in water, ethanol, acetone, dioxane, and benzene, insoluble in ether. $Mn(C_{10}H_{14}N_2O)_2(NCSe)_2$ is soluble in water, dimethylformamide, and dimethyl sulfoxide, rather slightly soluble in methanol and ethanol, insoluble in ether, acetone, and benzene [5].

$Mn(C_{10}H_{14}N_2O)_2(NCS)_2 \cdot 2H_2O$ was obtained from concentrated aqueous solutions of $Mn(NCS)_2$ and the ligand as short prismatic, slightly pinkish crystals [7]. The compound is apparently not identical with $Mn(C_{10}H_{14}N_2O)_2(NCS)_2$ [6].

$Mn(C_{10}H_{14}N_2O)_2(CH_3COO)_2 \cdot 2H_2O$. Preparation from $Mn(CH_3COO)_2 \cdot 4H_2O$ and the ligand in aqueous solution (mole ratio 1:2). The mixture was allowed to crystallize in air. After 1 month the resulting crystals were washed with water and dried in air. The IR absorption spectrum of the complex (liquid paraffin and fluorinated hydrocarbon mulls) was recorded in the 4000 to 400 cm^{-1} region and the Raman spectrum (Ar laser) between 4000 and 20 cm^{-1}. The shift of the ν(CO) ligand band to lower frequencies by 15 cm^{-1} and the occurrence of a broad ν(OH) band around 3200 cm^{-1} suggest strong hydrogen bonding between water in the inner coordination sphere and the carboxamide group of the unidentate ligand which is coordinated to manganese only through the pyridine ring nitrogen as demonstrated by the raising of the ν(ring) frequency in the 1600 cm^{-1} region. The assignment of $\nu_{as}(COO)$ is difficult since the intense amide I band is observed in the range 1600 to 1640 cm^{-1}. The $\nu_s(COO)$ vibration was observed at ~1480 cm^{-1} together with other vibration modes. Bands at 441 and 404 cm^{-1} in the IR spectrum (441 and 406 cm^{-1} in the Raman spectrum) were assigned to ν(Mn–O), absorptions at 268 (IR) and 271 cm^{-1} (Raman) to ν(Mn–N). It is assumed that the complex is octahedral containing six-coordinate manganese which is bonded to two pyridine ring nitrogens of the nicotinic acid diethylamide ligands, two oxygens of the unidentate acetato groups and two oxygens of coordinated water [8].

References:

[1] Tsivadze, A. Yu., Tsintsadze, G. V., Gongadze, N. P., Kharitonov, Yu. Ya. (Koord. Khim. **1** [1975] 1084/94; Soviet J. Coord. Chem. **1** [1975] 912/21, 918).

[2] Rubinchik, B. Ya., Ionov, V. M., Rybakov, V. B., et al. (Zh. Strukt. Khim. **18** [1977] 209/11; J. Struct. Chem. [USSR] **18** [1977] 178/9).

[3] Khakimov, Kh. Kh., Azizov, M. A., Kantsepol'skaya, K. M., Khamraev, A. D. (Zh. Strukt. Khim. **10** [1969] 1036/40; J. Struct. Chem. [USSR] **10** [1969] 918/21).
[4] Tsintsadze, G. V., Tsivadze, A. Yu., Gongadze, N. P. (Tr. Gruz. Politekhn. Inst. **1979** No. 2, pp. 5/8, 7; C.A. **92** [1980] No. 68843).
[5] Tsintsadze, G. V., Tsulukidze, L. A., Tsivadze, A. Yu., Gongadze, N. P. (Tr. Gruz. Politekhn. Inst. **1975** No. 4, pp. 23/6; C.A. **86** [1977] No. 25390).
[6] Bigoli, F., Braibanti, A., Pellinghelli, M. A., Tiripicchio, A. (Acta Cryst. B **29** [1973] 39/43, 41).
[7] Sandri-Cavicchi, S. (Farmaco Ed. Practica **14** [1959] 241/500, 244).
[8] Tsivadze, A. Yu., Kharitonov, Yu. Ya., Smirnov, A. N., et. al (Zh. Neorgan. Khim. **24** [1979] 1269/78; Russ. J. Inorg. Chem. **24** [1979] 705/12, 706, 710).
[9] Tsintsadze, G. V., Tsivtsivadze, T. I., Orbeladze, P. V., Mikelashvili, Z. V., Ibragimov, B. T. (Nauchn. Tr. Gruz. Politekhn. Inst. im. V. I. Lenina **1985** No. 1, pp. 66/9 from C.A. **104** [1986] No. 139747).

24.1.20 With 4-Pyridinecarboxamide

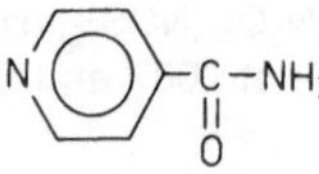

(= Isonicotinamide = $C_6H_6N_2O$)

$Mn(C_6H_6N_2O)_2Cl_2$ obtained from hot ethanolic solutions of manganese(II) chloride and isonicotinamide [1], was washed with ethanol and dried at about 70°C in an air oven [2]. The white solid is quite stable and melts sharply [1]. The IR spectrum (Nujol mulls) clearly indicates coordination of isonicotinamide to Mn only through the pyridine ring nitrogen without participation of the $CONH_2$ group. The frequency of the ν(Mn–Cl) band at 228 cm^{-1} suggests bridging chloride ions in a halogen-bridged polymeric octahedral chain structure [1, 2]. Its electronic spectrum recorded on Nujol mulls shows two absorption bands at 18870 and 23500 cm^{-1} ascribed to the electron transitions $^6A_{1g} \rightarrow {}^4T_{1g}(^4G)$ and $^6A_{1g} \rightarrow {}^4E_g$, $^4A_{1g}(^4G)$, respectively. The ligand field-splitting parameter $D_q = 927\ cm^{-1}$ derived from the latter transition is consistent with an octahedral environment of the Mn^{2+} ion [2]. The magnetic moment $\mu_{eff} = 6.10\ \mu_B$, resulting from susceptibility measurements at room temperature indicates a high-spin Mn^{II} (d^5) complex [2]. The low electrical conductivity of 0.001M solutions in dimethylformamide is that of a nonelectrolyte. Except for dimethylformamide the complex is insoluble in all common organic solvents [1].

References:

[1] Ahuja, S., Prasad, I. (Inorg. Nucl. Chem. Letters **12** [1976] 777/84, 778, 782).
[2] Ahuja, I. S. (Indian J. Chem. A **22** [1983] 262/4).

24.1.21 With 2-Pyrazinecarboxamide

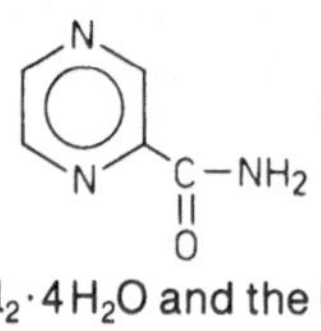

(= $C_5H_5N_3O$)

$Mn(C_5H_5N_3O)_2Cl_2$. A mixture of $MnCl_2 \cdot 4H_2O$ and the ligand (mole ratio 1:3), dissolved in tert-butanol was refluxed and cooled to room temperature. The resulting light yellow precipitate was washed with hot tert-butanol and dried in vacuum. The IR spectrum of KBr disks recorded in the 4000 to 400 cm^{-1} region and of Nujol mulls in the 650 to 200 cm^{-1} region shows only small shifts of the ligand bands assigned to ν_s(OCN) at 1720 to 1710 cm^{-1} and to ν_{as}(OCN) at 1734 cm^{-1} upon complexation, but nevertheless coordination of the amide group is assumed. The shift of the band at 450 cm^{-1}, ascribed to a pyrazine ring out-of-plane deformation to higher frequency in the complex by 15 cm^{-1} evidently indicates coordination of ring nitrogen to Mn. A strong band at 235 cm^{-1} tentatively assigned to ν(Mn–Cl) may indicate bridging chloride bonds. Terminal coordination of the ligand is suggested by a band of medium intensity at 950 cm^{-1}. The electronic reflectance spectrum exhibits weak bands at λ = 340, 390, and 550 nm, which are typical of octahedral arrangement. Susceptibility measurements at 27°C yield the magnetic moment $\mu_{eff} = 5.97\,\mu_B$, indicative of a high-spin $Mn^{II}(d^5)$ complex [1].

$Mn(C_5H_5N_3O)_2(CH_3COO)_2$. Saturated alcoholic solutions of $Mn(CH_3COO)_2$ and 2-pyrazinecarboxamide were mixed in a 1:2 mole ratio and stirred vigorously at 50 to 60°C until the mixture became clear. Upon cooling to room temperature in a desiccator a crystalline precipitate separated which was washed repeatedly with benzene and acetone and dried to constant weight in a desiccator at room temperature. The IR spectrum (KBr disks or Nujol mulls) in the 4000 to 400 cm^{-1} region shows characteristic absorption bands (in cm^{-1}) which are given with the shift relative to the free ligand (in parentheses) and assignment as follows: 3438 (+14) to $\nu_{as}(NH_2)$; 3298 (0) to $\nu_s(NH_2)$; 3180 (+5) to $2\delta(NH_2)$; 1685 (−35) to ν(CO); 1608 (sh, +21); 1560 (+29); 1487 (+3) to ν(ring); 1580 to $\nu_{as}(COO)$, and 1438 to $\nu_s(COO)$ of acetate; 1380 (0) to ν(CN), 1350, 1320 to $\delta(CH_3)$. The almost unaffected ν(NH) bands and the shift of the ν(CO) band to lower frequency indicate bonding of the carbonyl oxygen of the $CONH_2$ group to Mn. The bands assignable to the pyrazine ring are shifted to higher frequencies and suggest coodination also of a ring nitrogen to form a chelate ring with the bidentate N,O-bonded ligand. The ν_{as}(C–C) acetate band occurs in the same region (960 to 940 cm^{-1}) as in $Mn(CH_3COO)_2$, thus indicating unidentate coordination of the acetato groups to Mn through oxygen. Thus, considering bidentate coordination of two 2-pyrazinecarboxamide ligands and two acetate ligands, an octahedral configuration of the complex can be assumed. Thermal analysis (in a derivatograph) reveals an endothermic effect at 160°C due to melting and further effects in the 190 to 210°C region associated with a weight loss of 28.9% indicate decomposition of the complex by loss of one ligand molecule and formation of a more stable intermediate. An endothermic effect in the 250 to 320°C range suggests simultaneous decomposition of $Mn(CH_3COO)_2$ and pyrazinecarboxamide, and the exothermic effect at 390 to 500°C indicates combustion of the decomposition products leaving the metal oxide [2]. Conductivity measurements in alcoholic solution (dilution rate V = 1000) at 25°C yield the molar conductance $\Lambda = 16.7\ cm^2 \cdot \Omega^{-1} \cdot mol^{-1}$ indicating a nonelectrolyte but in water the complex dissociates forming a 1:2 electrolyte [2].

References:

[1] Sanyal, G. S., Modak, A. B., Mudi, A. K. (Indian J. Chem. A **21** [1982] 1044/8, 1045).

[2] Azizov, T. A., Khodzhaev, O. F., Parpiev, N. A. (Koord. Khim. **4** [1978] 1234/8; Soviet J. Coord. Chem. **4** [1978] 929/32).

24.1.22 With Tetrazole-5-carboxamide

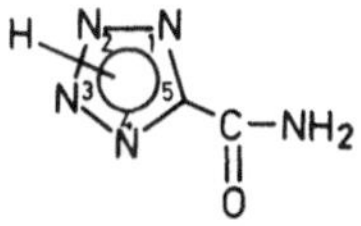

(= $C_2H_3N_5O$ = HL)

$Mn(C_2H_2N_5O)_2$ and **$Mn(C_2H_2N_5O)_2 \cdot 2H_2O$.** The anhydrous complex was obtained by a hydrolytic conversion of tetrazole-5-carbonitrile to the carboxamide. Acetone solutions of excess carbonitrile were refluxed in the presence of manganese(II) nitrate hydrate for 30 min and cooled to room temperature. The resulting fine precipitate was washed with ethanol and ether and dried in vacuum at 90°C for several hours. $Mn(C_2H_2N_5O)_2 \cdot 2H_2O$ was prepared by mixing aqueous solutions of the Mn salt and tetrazole-5-carboxamide. The product, which separated after standing for a few hours, was washed and dried in vacuum as above.

The IR spectra (KBr pellets or Nujol mulls) of both the anhydrous and hydrated complexes are almost identical, as probable OH and HOH vibrations are obscured by the νNH vibration bands. The band at 2284 cm^{-1} due to ν(CN) of the initially present carbonitrile has completely disappeared in the spectrum of the anhydrous complex. Characteristic absorption bands at 3450, 3350, 3250, 1690, and 1635 cm^{-1} are assigned to the amide $\nu_{as}(NH)$, $\nu_s(NH)$, and ν(NH) caused by hydrogen bond formation, to ν(CO) of the amide I band, and to δ(NH) and ν(CN) of the amide II band, respectively. Except for the ν(CO), which is unshifted, all amide bands upon coordination are displaced to higher frequencies by 82, 91, 59, and 26 cm^{-1}, respectively. A ligand band at 3150 cm^{-1}, owing to ν(NH) of the ring, has disappeared in the complex spectra, thus confirming ligand deprotonation. The occurrence of different absorption bands in the far IR spectrum at 239, 214, and 198 cm^{-1} indicates coordination of different nitrogen atoms. A three-dimensional polynuclear structure of the poorly soluble complexes is assumed with hexa-coordinate Mn^{2+} ions chelated by the amide and two ring nitrogen atoms of each ligand anion.

Reference:

Franke, P. L., Groeneveld, W. L. (Inorg. Chim. Acta **40** [1980] 111/4).

24.2 Complexes with Amides of Di- and Tetracarboxylic Acids and Related Compounds

24.2.1 With Oxalic Acid Monoamide $H_2NC(O)C(O)OH$ (= Oxamic Acid = $C_2H_3NO_3$)

$K_2[Mn(C_2HNO_3)_2(H_2O)_2]$. Anhydrous $MnCl_2$ (0.015 mol) was added to a solution of oxamic acid (0.03 mol) in dimethyl sulfoxide (40 mL). The mixture was heated at 40 to 50°C for 1 h and a concentrated solution of KOH in deionized water (10 mL) was added dropwise. The solution was heated again for 30 min and cooled for crystallization at pH 8.0. The pale pink crystals were washed with acetone and refluxed again with dimethyl sulfoxide for 1 h to remove unreacted oxamic acid, then washed several times with anhydrous acetone and dried in vacuum over P_4O_{10} at 80°C for 24 h.

The IR spectrum (KBr pellets) shows a single, strong absorption band at 3220 cm^{-1} whereas oxamic acid reveals two bands at 3300 and 3200 cm^{-1} ascribed to antisymmetric and symmetric ν(NH) vibrations of the amido group. A broad band at 1660 to 1630 cm^{-1} and a medium one at 1690 cm^{-1} replace the $\nu_{as}(CO)$ and amide bands of the free acid at 1730 and 1670 cm^{-1}, respectively, whereas the amide band at 1590 cm^{-1} is almost unaffected. The δ(OH) band of the COOH group at 1328 cm^{-1} is much weaker in the complex. It is concluded that Mn is coordinated to the carboxylate oxygen and to the nitrogen atoms, after ionization of one amidic hydrogen. The electronic diffuse reflectance spectrum shows four bands at 29000, 20300, 15100, and 10700 cm^{-1} which (except for the latter band) are due to electron transitions of manganese(II) in an octahedral crystal field. The magnetic moment $\mu_{eff} = 1.81\ \mu_B$ resulting from susceptibility measurements at 20°C indicates a low-spin complex.

Reference:

Koumis, J. K., Tsangaris, J. M., Galimos, A. G. (Z. Naturforsch. **33b** [1978] 987/9).

24.2.2 With Malonic Acid Diamide $H_2NC(O)CH_2C(O)NH_2$ (= Malonamide = $C_3H_6N_2O_2$)

$Mn(C_3H_6N_2O_2)_2(NO_3)_2$. To a hot ethanolic solution of manganese nitrate a hot ethanolic solution of malonamide was added dropwise with stirring up to a mole ratio of 1:2 [1] or solutions of $Mn(NO_3)_2$ and malonamide (mole ratio 1:2) in ethanol or methanol were refluxed for 8 h [2]. The precipitate was washed after 2 h with ethanol, and dried in a vacuum over $CaCl_2$

[1]. The colorless (white) crystals [2] according to X-ray diffractometer studies (MoKα radiation) are triclinic in space group $P\bar{1}$-C_i^1 (No. 2) with the lattice constants a = 7.132(1), b = 7.365(1), c = 7.674(1) Å and α = 91.45(1)°, β = 117.01(1)°, γ = 104.44(1)°; Z = 1. The structure was solved from 2165 reflections by the heavy atom method and refined up to R = 0.032. For atomic positions see the paper. A projection of the structure along (010) is shown in **Fig. 14**. The molecules of malonamide enter into the complex as bidentate oxygen bonded ligands occupying the equatorial planes of the octahedral coordination polyhedron and the nitrate ions occupy the (trans) apical sites as monodentate ones. The organic ligands assume an intermediary flat-chair conformation and are linked together by manganese as the inversion center of the six-membered metallocycles. The complex thus can be considered as a tetragonal bipyramid with a slight distortion from octahedral symmetry (< O–Mn–O = 85.2° to 94.8°, Mn ↔ O = 2.128 ± 0.009 Å) and weakened axial bonds. The nitrato groups assume a flat structure with (N–O) bond lengths between 1.217 and 1.261 Å and there exist intermolecular hydrogen bonds between these groups and the malonamide ligands with a length of 2.92 Å. The density D = 1.84 g/cm³ was calculated from X-ray data [3].

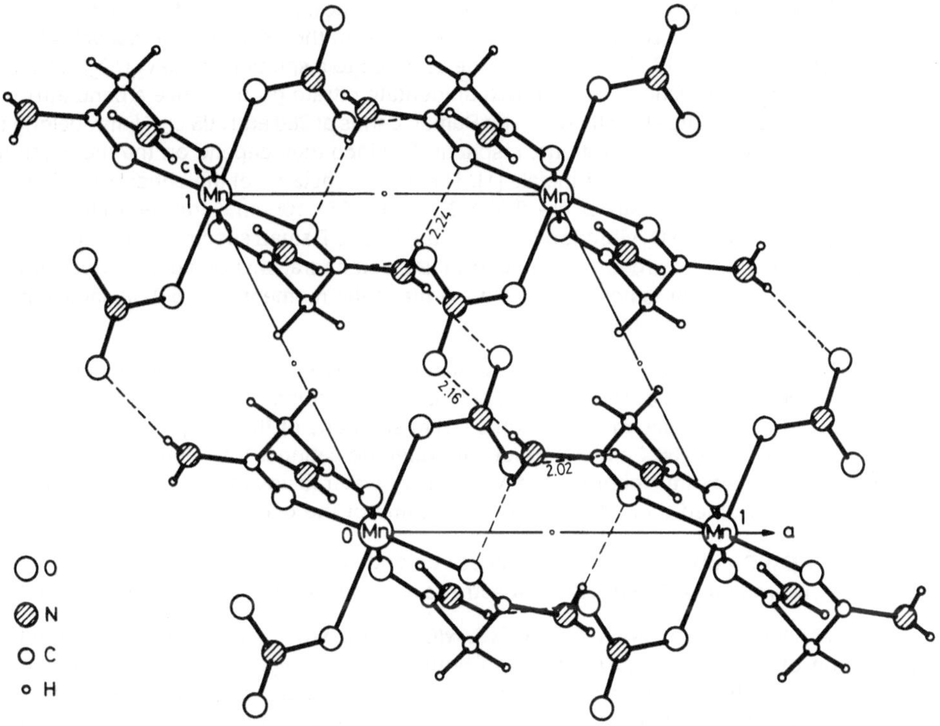

Fig. 14. Structure of $Mn(C_3H_6N_2O_2)_2(NO_3)_2$ projected along (010) [3].

Characteristic IR absorption bands of the complex in liquid paraffin, fluorinated oil, or hexachlorobutadiene mulls observed between 4000 and 400 cm^{-1}, and the bands of the Raman spectrum (Ar laser excitation with λ = 5145 Å) in the 4000 to 100 cm^{-1} region are shown on p. 124, together with the assignments [1]:

assignment	ν(NH)	amide I	amide II, δ(NH)	ν(C–N) + $\nu_3(NO_3)$
IR	3385, 3305, 3220	1670	1620	1470, 1440
Raman	3360, 3270, 3200	1665	1605	1465, 1439

assignment	$\nu_3(NO_3)$, δ(CCH)	ν(CC)	$\delta(NO_3)$	$\nu_1(NO_3)$	$\nu(Mn{-}ONO_2)$	$\nu(Mn{-}O_{ligand})$
IR	1330	905	830, 802	1050	—	—
Raman	1326, 1295	900	766	1045	388	308

A semiempirical analysis of the normal vibrations was performed for free and coordinated malonic acid diamide and their C- and N-deuterated derivatives. The experimental vibrational frequencies are interpreted [4]. In the IR and Raman spectra the high frequency component of the ligand ν(NH) vibrations is almost unshifted in the complex whereas the low frequency component had shifted to higher frequencies by about 40 to 70 cm^{-1} thus excluding direct linking of the NH_2 groups to Mn. The shift of the ν(CO) band to lower frequencies by 10 to 15 cm^{-1} and of the ν(C–N) bands to higher frequencies by about 20 cm^{-1} indicates coordination of malonamide to Mn through the carbonyl oxygens [1, 2] as shown by the X-ray structure analysis [3] and changing of the bonding force constants (lowering of k_{CO} from 16.50 to 15.30 and raising of k_{CN} from 13.20 to 13.60 was calculated from the IR data given above) [4]. The strong bands at 1470 and 1330 cm^{-1} (IR) can be ascribed to a splitting of the $\nu_3(NO_3)$ vibration mode which is characteristic of coordinated unidentate nitrato groups since a band attributable to $\nu(NO_3)$ also appears at 1050 cm^{-1}. The Raman bands at 388 and 308 cm^{-1} may belong to $\nu(Mn{-}O)_{NO_3}$ and $\nu(Mn{-}O)_{ligand}$ vibrations, respectively, since they depend on the metal ion as shown by a comparison with other metals [1]. The force constants of the bonds Mn–O and of the angles Mn–O–C and O–Mn–O are 2.9, 1.25, and 1.35×10^6 cm^{-2}, respectively [4]. The differing positions of the NO_3 vibration modes in the IR and Raman spectra support the trans-arrangement of the nitrato groups in a six-coordinate octahedral environment [1]. Susceptibility measurements at room temperature yield the magnetic moment $\mu_{eff} = 6.01\ \mu_B$ indicating a spin-free octahedral $Mn^{II}(d^5)$ complex.

$Mn(C_3H_6N_2O_2)_2(NO_3)_2$ decomposes upon heating to 215°C; it is soluble in water and coordinating solvents such as DMF, DMSO, but insoluble in common organic solvents [2]. The molar electrical conductance measured on 0.001 M solutions of the complex is $\Lambda = 76.9\ cm^2 \cdot \Omega^{-1} \cdot mol^{-1}$ in DMF and $238.4\ cm^2 \cdot \Omega^{-1} \cdot mol^{-1}$ in water indicating a 1:1 electrolyte in DMF and complete replacement of the coordinated NO_3^- ions by water in aqueous solution to give a 1:2 electrolyte. It is assumed that DMF may replace some of the coordinated NO_3^- ions.

$Mn(C_3H_6N_2O_2)_2Cl_2$ obtained from manganese chloride and malonamide (mole ratio 1:2) in hot ethanol shows similar IR and Raman spectra as recorded for the nitrato complex above [2].

The observed shifts of the ν(NH), ν(CO), ν(CN), and ν(C–C) bands suggest bidentate coordination of malonamide to Mn through both carbonyl oxygens forming six-membered chelate rings [5]. This is apparently supported by a Raman band at 288 cm^{-1} assignable to $\nu(Mn{-}O_{ligand})$. The Raman band at 265 cm^{-1} is ascribed to ν(Mn–Cl) of coordinated chloride ions [5]; for far IR bands of similar assignment (335 to 260 cm^{-1} region) see [2]. Susceptibility measurements at room temperature yield $\mu_{eff} = 5.9\ \mu_B$ [2].

The white compound decomposes upon heating at 245°C, it is soluble in water and hot coordinating solvents (DMF, DMSO) but insoluble in common organic solvents [2]. The molar conductance $\Lambda = 5.3\ cm^2 \cdot \Omega^{-1} \cdot mol^{-1}$ of a 0.001 M solution in DMF is that of a nonelectrolyte. In water the complex behaves as a 1:2 electrolyte ($\Lambda = 236.7\ cm^2 \cdot \Omega^{-1} \cdot mol^{-1}$) due to complete replacement of the coordinated Cl^- ions by water [2].

$MnCl_2\cdot2C_3H_6N_2O_2\cdot2H_2O$ and **$2MnCl_2\cdot3C_3H_6N_2O_2\cdot4H_2O$** were detected as stable solid phases of the quaternary system $MnCl_2$–$CoCl_2$-malonamide-water at 25°C [6].

$Mn(C_3H_6N_2O_2)_2SO_4$ was prepared from manganese sulfate and malonamide (mole ratio 1:2) in aqueous solution by precipitation with ethanol. The white precipitate was washed with ethanol, tetrahydrofuran, and ether, and dried in a vacuum. The IR spectrum resembling that of the nitrato and chloro complexes (except for the anion) suggests coordination of malonamide to Mn through both carbonyl oxygens. Coordination of the sulfato group as a chelating ligand is indicated by three bands assignable to $\nu_3(SO_4)$ in the 1235 to 1030 cm^{-1} region and one band assigned to $\nu_1(SO_4)$ in the 940 to 964 cm^{-1} region. Susceptibility measurements yield $\mu_{eff}=6.1\,\mu_B$ [2]. $Mn(C_3H_6N_2O_2)_2SO_4$ decomposes upon heating above 300°C, it is soluble in water but insoluble in all common organic solvents. The molar conductivity $\Lambda=248.6$ $cm^2\cdot\Omega^{-1}\cdot mol^{-1}$ of an aqueous 0.001 M solution is that of a 1:2 complex and demonstrates complete replacement of the coordinated sulfato groups by water.

References:

[1] Tsivadze, A. Yu., Kharitonov, Yu. Ya., Tsintsadze, G. V., Tsintsadze, E. V., Smirnov, A. N. (Zh. Neorgan. Khim. **23** [1978] 1572/8; Russ. J. Inorg. Chem. **23** [1978] 866/70, 868).
[2] Aggarwal, R. C., Singh, B., Singh, T. B. (Indian J. Chem. A **19** [1980] 137/40, 138).
[3] Porai-Koshits, M. A., Nikolaev, V. P., Butman, L., Tsintsadze, G. V. (Koord. Khim. **6** [1980] 793/804, 797, 800; C.A. **93** [1980] No. 58745).
[4] Smirnov, A. N., Tsivadze, A. Yu., Kharitonov, Yu. Ya., Tselebrovskaya, B. G. (Koord. Khim. **6** [1980] 1688/706; Soviet J. Coord. Chem. **6** [1980] 831/48, 837/9, 843).
[5] Tsivadze, A. Yu., Tsintsadze, G. V., Tsintsadze, E. V., Kharitonov, Yu. Ya. (Zh. Neorgan. Khim. **23** [1978] 1006/13; Russ. J. Inorg. Chem. **23** [1978] 556/60, 557).
[6] Kozlova, L. T. (Sb. Nauchn. Tr. Yaroslav. Gos. Ped. Inst. No. 205 [1984] 31/5; C.A. **103** [1985] No. 28075).

24.2.3 With Tetramethyldiamides of Dicarboxylic Acids
$(CH_3)_2NC(O)[CH_2]_nC(O)N(CH_3)_2$ (= L)

No.	n	pertinent acid	ligand formula
1	0	oxalic acid	$C_6H_{12}N_2O_2$
2	1	malonic acid	$C_7H_{14}N_2O_2$
3	2	succinic acid	$C_8H_{16}N_2O_2$

$MnL_3(ClO_4)_2$. The complexes were prepared by treating first $Mn(ClO_4)_2\cdot6H_2O$ with 2,2-dimethoxypropane (80 to 100% excess) at room temperature for 2 h and then with the required ligand in a 1:3 mole ratio. The initially resulting oil was washed and stirred with diethyl ether until it solidified. The white complexes which decomposed upon attempted recrystallization are crystalline and slightly hygroscopic [1, 2], $Mn(C_8H_{16}N_2O_2)_3(ClO_4)_2$ shows a faint pinkish tint [2].

In the following table are listed the melting point (d = with decomposition), molecular weight from osmometric measurements in nitromethane with calculated value (in parentheses), magnetic moment μ_{eff} from susceptibility measurements at 23°C, the IR absorption band (in cm^{-1}) tentatively assigned to ν(CO) vibrations and its shift $\Delta\nu$ upon complexation [1, 2]:

complex	m.p. in °C	molecular weight	μ_{eff} in μ_B	$\nu(CO)$	$\Delta\nu$ in cm^{-1}	Ref.
$Mn(C_6H_{12}N_2O_2)_3(ClO_4)_2$	325 to 330(d)	690(686)	6.02	1635	$-20^{2)}$	[2]
$Mn(C_7H_{14}N_2O_2)_3(ClO_4)_2$	204 to 207(d)	—	6.03	1616	-29	[1]
$Mn(C_8H_{16}N_2O_2)_3(ClO_4)_2$	160	$660(771)^{1)}$	5.66	1605	-22	[2]

[1] Some tendency of dissociation in the solvent. – [2] Referred to the high frequency component of the $\nu(CO)$ doublet at 1655 cm^{-1} which disappears in the complex [2].

Coordination of the ligand to Mn through both the carbonyl oxygens in each case and a probably octahedral environment of manganese is assumed [1, 2]. The complexes are monomeric in nitromethane [2]. The molar electrical conductivity $\Lambda = 159$ $cm^2 \cdot \Omega^{-1} \cdot mol^{-1}$ of $[Mn(C_7H_{14}N_2O_2)_3](ClO_4)_2$ in nitromethane solution indicates a 1:2 electrolyte and confirms the formula with noncoordinated ClO_4^- ions. The hygroscopic complex decomposes in the presence of moisture and air [1].

References:

[1] Bull, W. E., Ziegler, R. G. (Inorg. Chem. **5** [1966] 689/92, 691).
[2] Good, M. L., Siddal III, T. H. (J. Inorg. Nucl. Chem. **30** [1968] 2679/87, 2682, 2685/7).

24.2.4 With Tetrapropyldiamides of Dicarboxylic Acids

ligand 1 $(C_3H_7)_2N-\underset{\underset{O}{\|}}{C}-CH_2-\underset{\underset{CH_3}{|}}{N}-CH_2-CH_2-\underset{\underset{CH_3}{|}}{N}-CH_2-\underset{\underset{O}{\|}}{C}-N(C_3H_7)_2$ $(C = C_{20}H_{42}N_4O_2)$

ligand 2 to 5 $(C_3H_7)_2N-\underset{\underset{O}{\|}}{C}-CH_2-O-R-O-CH_2-\underset{\underset{O}{\|}}{C}-N(C_3H_7)_2$

ligand No.	2	3	4	5
R	(benzene ring)	(cyclohexane ring)	(naphthalene)	(structure)
formula	$C_{22}H_{36}N_2O_4$	$C_{22}H_{42}N_2O_4$	$C_{26}H_{38}N_2O_4$	$C_{28}H_{40}N_2O_4$

$Mn(C_{20}H_{42}N_4O_2)Br_2$ was prepared in the same way as the other tetrapropyldiamide complexes, see below, by reaction of ligand 1 with manganese(II) bromide in methanol containing 2,3-dimethoxypropane. After evaporation of the solvent the residue was redissolved in 2-propanol and allowed to crystallize. The violet-pink crystals dried in vacuum at 25°C and recrystallized from 2-propanol melt at 218 to 220°C [1].

$[Mn(C_{22}H_{36}N_2O_4)_2](ClO_4)_2$. A solution of manganese(II) perchlorate (0.5 to 1 mmol) in methanol (20 mL), containing 2,3-dimethoxypropane (1 mL) for dehydration, was heated with stirring for 30 min. The required amount of ligand was added and the solvent evaporated to a small volume. The residue dissolved in $CHCl_3$ was filtered and allowed to crystallize. The colorless crystals, dried in vacuum at 25°C for 24 h and recrystallized from $CHCl_3$, melt at 224 to 225°C. The electronic spectrum (KBr pellets) shows a weak absorption band at 406 cm^{-1} indicative of eight-coordinate Mn in a dodecahedral arrangement. Coordination of the ligand to Mn through the carboxamide group oxygens is suggested. The molar electrical conductivity of a 5×10^{-4} M solution in nitromethane, $\Lambda = 150$ $cm^2 \cdot \Omega^{-1} \cdot mol^{-1}$, corresponds to a 1:2 electrolyte [1].

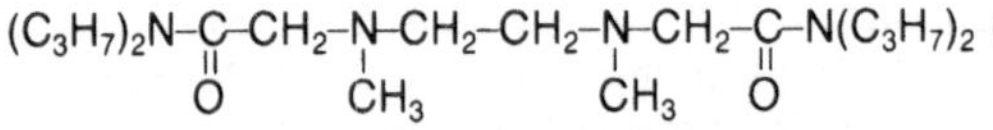

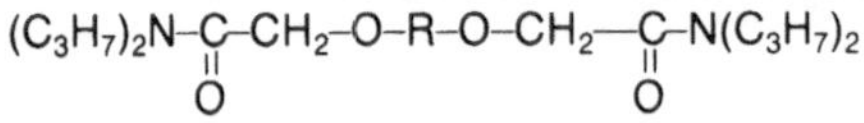

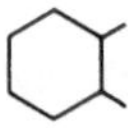

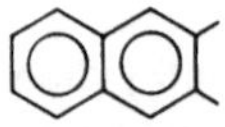

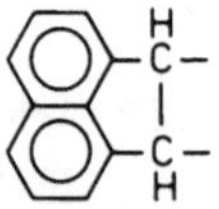

[Mn($C_{22}H_{36}N_2O_4$)$_2$]MnBr$_4$ and **[Mn($C_{22}H_{42}N_2O_4$)$_2$]MnBr$_4$** complexes were prepared as follows: A methanol or ethanol solution containing $MnBr_2$ and the corresponding ligand in a 1:1 mole ratio was evaporated to a small volume. The residue of [Mn($C_{22}H_{36}N_2O_4$)$_2$]MnBr$_4$ was redissolved in nitromethane, that of [Mn($C_{22}H_{42}N_2O_4$)$_2$]MnBr$_4$ in methanol, and the filtered solutions were allowed to crystallize. The crystals were dried in vacuum at 25°C for 24 h [1]. [Mn($C_{22}H_{36}N_2O_4$)$_2$][MnBr$_4$] forms yellow cubes melting at 297 to 299°C. The light green [Mn($C_{22}H_{42}N_2O_4$)$_2$]MnBr$_4$ melts at 255 to 256°C. The complexes are unstable in water [1].

X-ray diffractometer studies with MoKα radiation of single crystals reveal a tetragonal lattice for [Mn($C_{22}H_{36}N_2O_4$)$_2$]MnBr$_4$, space group $I\bar{4}$-S_4^2 (No. 82) with lattice constants a = b = 13.661(5) and c = 14.994(4) Å; Z = 2. The structure was solved from 660 unique reflections and refined up to R = 0.051; for atomic positions see the paper. Both Mn^{2+} ions occupy sites of local $\bar{4}(S_4)$ symmetry. From the loci of the Mn^{2+} ions in the position (½, 0, ¼; etc.) each manganese is coordinated to all eight oxygens of two ligand molecules forming the complex cation [Mn($C_{22}H_{36}N_2O_4$)$_2$]$^{2+}$; thus the asymmetric unit consists of one half ligand molecule with a crystallographic twofold rotation axis connecting the two halves as shown in **Fig. 15**. The resulting coordination geometry can be described in terms of the two tetrahedra of 42m (D_{2d}) symmetry formed by these eight atoms with four 2.168(13) and four 2.380(11) Å Mn–O distances. Except for the propyl side chains, the ligand molecules are nearly planar. The atoms show large thermal vibration parameters, especially those of the propyl side chains. The shape of the complete complex unit (see the paper), is reminiscent of that of natural ionophores where the cation is completely enclosed by a hydrophilic cavity lined with oxygen atoms. From the other set of Mn^{2+} ions at (0, 0, 0; etc.) each manganese is surrounded by four Br^- ions at a distance of 2.514 Å. The tetrahedron of Br^- ions has edges of 4.26 and 4.03 Å indicative of their mutual contacting. In the unit cell packing the [MnBr$_4$]$^{2-}$ anions occupy holes between the complex cations but show no interactions with them. Density D = 1.42 g/cm^3 (measured) or 1.44 g/cm^3 (calculated from X-ray data) [2].

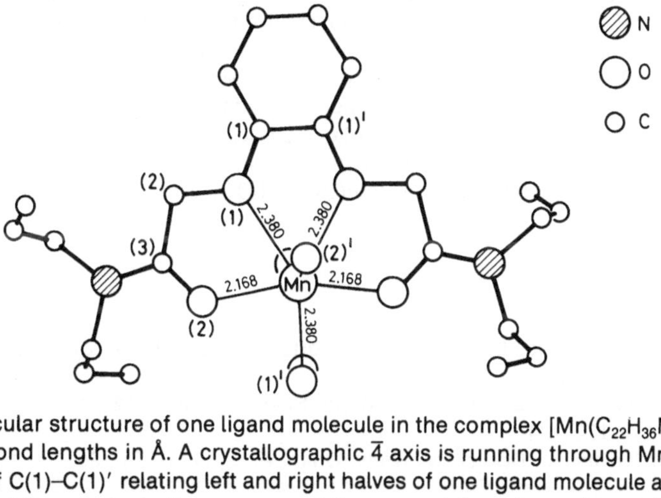

Fig. 15. Molecular structure of one ligand molecule in the complex [Mn($C_{22}H_{36}N_2O_4$)$_2$]-[MnBr$_4$] with bond lengths in Å. A crystallographic $\bar{4}$ axis is running through Mn^{2+} and the midpoint of C(1)–C(1)′ relating left and right halves of one ligand molecule and first and second ligands (only the oxygens of the second ligand are shown) [2].

[Mn($C_{22}H_{42}N_2O_4$)$_2$]MnBr$_4$ also has a tetragonal lattice, space group $I\bar{4}$-S_4^2 (No. 82), with lattice constants a = b = 11.695(4) and c = 21.299(5) Å; Z = 2. The structure was solved from 1642 unique reflections and refined to R = 0.041. For atomic positions see the paper. The elementary cell shows two types of Mn^{2+} ions, the ones being complexed by two ligand molecules lie on the $\bar{4}$ axis at the special positions (0, ½, ¼; etc.). Stereoscopic views of the complex looking down the $\bar{4}$ (crystallographic c) axis and down the a axis and packing of the complex in the unit cell are shown in [3]. The molecular geometry of the complex cations $[Mn(C_{22}H_{42}N_2O_4)_2]^{2+}$ demonstrates that the ligand parts with the complexing oxygen atoms are in essentially the same conformation whereas the flexible propyl side chains have different conformations. The individual tetrahedra of the complex cation (site symmetry $\bar{4}$) have a D_{2d} symmetry requiring only one distance and two angles for each one: (Mn–$O_{carbonyl}$ = 2.185 Å, O_{CO}–Mn–O_{CO} = 92.3° and 156.8°; Mn–O_{ether} = 2.370 Å, O–Mn–O = 67.3° and 133.9°). The Mn atoms of the second set (at 0, 0, 0; etc.), also with site symmetry $\bar{4}$ are coordinated by 4 Br^- ions. The Mn–Br distances in the tetrahedron are 2.50 Å and the bond angles Br–Mn–Br are 110.1° or 108.2°. The $Br^- \leftrightarrow Br^-$ edge distances of 4.06 and 4.10 Å show that the Br^- ions at the tetrahedron apices are contacting. The density D = 1.40 g/cm³ was both measured and calculated from X-ray data [3].

The IR spectrum of $[Mn(C_{22}H_{36}N_2O_4)_2]_2MnBr_4$ (KBr pellets) shows the ν(CO) absorption band at 1621 cm^{-1}. The shift of 39 cm^{-1} toward lower frequency in the complex confirms the coordination of the ligand through carbonyl oxygen. The electronic spectrum of the solid complex (KBr pellets) and its solution in acetonitrile shows nine absorption bands in the 360 to 730 nm region. The strong bands of the $[MnBr_4]^{2-}$ anion overshadow the weaker ones of the eight-coordinate $[Mn(C_{22}H_{36}N_2O_4)_2]^{2+}$ cation. The colorless solution in methanol shows absorption bands at 360, 404 (sharp), 430 and 500 to 600 nm (broad) suggesting the conversion of $[Mn(C_{22}H_{36}N_2O_4)_2]^{2+}$ and $[MnBr_4]^{2-}$ into octahedral complexes with six-coordinate Mn^{II} [1].

The molar electrical conductance data of $[Mn(C_{22}H_{36}N_2O_4)_2]_2MnBr_4$ and $[Mn(C_{22}H_{42}N_2O_4)_2]$-$MnBr_4$ in nitromethane solution (~0.001 M) are in the range of 1:1 electrolytes: Λ = 54 and 74 $cm^2 \cdot \Omega^{-1} \cdot mol^{-1}$, respectively [1].

$Mn(C_{26}H_{38}N_2O_4)Br_2 \cdot H_2O$ and **$Mn(C_{28}H_{40}N_2O_4)Br_2$** complexes were prepared from manganese dibromide and ligand 4 or 5 (see p. 126) in the same way as for $[Mn(C_{22}H_{42}N_2O_4)_2][MnBr_4]$. However, $Mn(C_{26}H_{38}N_2O_4)Br_2 \cdot H_2O$ was recrystallized from a mixture of ethyl acetate and $CHCl_3$ (added to turbidity) and $Mn(C_{28}H_{40}N_2O_4)Br_2$ from a mixture of CH_2Cl_2 and methanol, $Mn(C_{26}H_{38}N_2O_4)Br_2 \cdot H_2O$ melts at 297 to 299°C, $Mn(C_{28}H_{40}N_2O_4)Br_2$ at 258 to 260°C. The IR spectrum of the complexes (not reported) suggests coordination of the ligands to Mn through carboxamide oxygen. The molar electrical conductance Λ = 75 $cm^2 \cdot \Omega^{-1} \cdot mol^{-1}$ of $Mn(C_{28}H_{40}N_2O_4)Br_2$ measured in nitromethane solution (0.0001 M) indicates a 1:1 electrolyte with a structure probably similar to $[Mn(C_{22}H_{36}N_2O_4)_2[MnBr_4]$, see above. The complexes are unstable to water [1].

References:

[1] Readio, J., Borowitz, I. J., Poblack, N., Porter, J., Weiss, L., Borowitz, G. B. (J. Coord. Chem. **11** [1981] 135/42).
[2] Neupert-Laves, K., Dobler, M. (J. Cryst. Spectrosc. Res. **12** [1982] 271/86, 273/6; C.A. **97** [1982] No. 172784).
[3] Neupert-Laves, K., Dobler, M. (Helv. Chim. Acta **60** [1977] 1861/71, 1866/8).

24.2.5 With Phthalimide

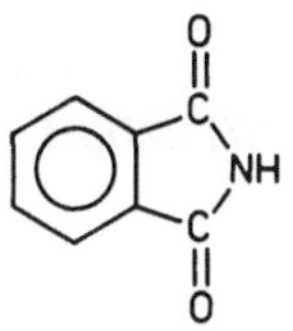

($= C_8H_5NO_2$)

$Mn(C_8H_4NO_2)_2$ was obtained by treating $MnCl_2 \cdot 4H_2O$ with potassium phthalimide in a 1:2 mole ratio. The brown precipitate was washed repeatedly with water and dried over P_4O_{10}. The IR spectrum recorded as Nujol mulls shows an absorption band assigned to ν(CO) vibrations which upon complexation had shifted to lower frequencies by 15 cm^{-1}. Such shift appears to be caused by a change of the electron density at the (deprotonated) nitrogen. It is assumed that the bond between Mn and the ligand has some covalent character since the solubility in polar solvents is low. The visible spectrum of a 0.001 M solution in nitrobenzene shows absorption bands at 340, 350, 400, 450, and 550 nm.

$Mn(C_8H_4NO_2)_2 \cdot 4RR'NH$. The mixed complexes with primary and secondary aliphatic amines were formed when calculated amounts of $Mn(C_8H_4NO_2)_2$ and amine were shaken in acetone medium for 6 d. The resulting black semicrystalline products which were washed several times with acetone could not be recrystallized. The substituents R and R′ of the amines are tabulated below together with IR absorption bands of the complexes and their shift Δν relative to the free ligands, the visible absorption bands and molar electrical conductance data of 0.001 M solutions in nitrobenzene at 25°C:

ligand			IR absorption of complex ν in cm^{-1}				visible spectrum λ in nm				Λ in $cm^2 \cdot \Omega^{-1} \cdot mol^{-1}$
R	R′	RR′NH	$\nu(NH_{amine})$	Δν(NH)	ν(CO)	Δν(CO)					
H	H	NH_3	3250	−150	1710	−30	355	400	—	550	0.88
CH_3	H	CH_5N	3265	−135	1700	−40	350	405	450	555	1.0
C_2H_5	H	C_2H_7N	3265	−135	1700	−40	350	405	455	550	0.82
C_3H_7	H	C_3H_9N	3270	−130	1705	−35	340	405	450	555	0.93
i-C_3H_7	H	C_3H_9N	—	—	—	—	—	—	—	—	0.87
C_4H_9	H	$C_4H_{11}N$	3300	−100	1695	−45	348	400	455	—	0.66
i-C_4H_9	H	$C_4H_{11}N$	—	—	—	—	—	—	—	—	0.69
C_5H_{11}	H	$C_5H_{13}N$	3310	−90	1700	−40	355	405	450	550	0.14
i-C_5H_{11}	H	$C_5H_{13}N$	—	—	—	—	—	—	—	—	0.31
CH_3	CH_3	C_2H_7N	3310	−90	1700	−40	352	405	—	550	0.55
C_2H_5	C_2H_5	$C_4H_{11}N$	3340	−60	1700	−40	345	400	450	555	1.0

The shift of the ν(NH) bands to lower frequencies indicates coordination of the amines to Mn forming rather strong bonds between manganese and nitrogen. The decrease of Δν(NH) with increasing molecular weight of the amine suggests lowering of the complex stability from the ammonia to the diethylamine complex. The lowering of the ν(CO) frequency by about 40 cm^{-1} indicates a further strengthening of the covalency nature of the bond between Mn and phthalimide. Probably the shift is caused by the formation of H bonds between the carbonyl oxygens and amine nitrogens and hindrance of the normal vibrations of the CO groups, which apparently are not coordinated. The weak bands in the visible spectrum are consistent with the assumed octahedral structure where the phthalimide ligands occupy the apical and the amines occupy the equatorial sites of the octahedron. The evident nonelectrolyte nature of the complexes in nitrobenzene, in which they are sparingly soluble, supports such assumption.

Reference:

Narain, G., Shukla, P. (Australian J. Chem. **20** [1967] 227/31).

24.2.6 With Phthalanilic Acids

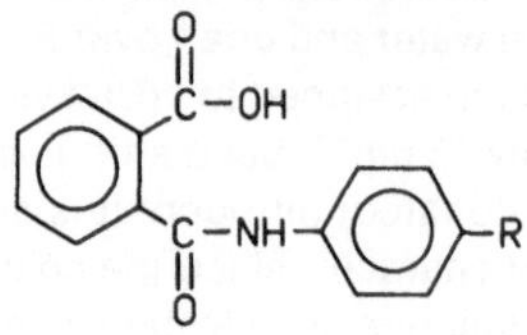

No.	R	formula
1	H	$C_{14}H_{11}NO_3$
2	NO_2	$C_{14}H_{10}N_2O_5$
3	Cl	$C_{14}H_{10}ClNO_3$
4	Br	$C_{14}H_{10}BrNO_3$
5	OCH_3	$C_{15}H_{13}NO_4$
6	CH_3	$C_{15}H_{13}NO_3$

Formation constants of 1:1 and 1:2 complexes with ligand 1 to 6 (= H_2L) in aqueous methanol (20 vol%) were determined by potentiometric pH titrations at different temperatures and ionic strength I = 0.1 M ($NaClO_4$). Values of log K_1 and log K_2 are tabulated below together with thermodynamic parameters, ΔH and ΔG in kcal/mol, ΔS in cal·mol^{-1}·K^{-1}:

ligand	t	22°C	30°C	40°C	−ΔG°	−ΔH	ΔS
1	log K_1	5.30	5.25	5.15	29.50	10.54	64.22
	log K_2	4.55	4.50	4.45	26.44	9.58	57.04
2	log K_1	6.60	6.55	6.45	36.82	15.31	72.93
	log K_2	5.30	5.25	5.20	30.21	7.99	75.27
3	log K_1	6.40	6.35	6.25	35.86	15.52	68.78
	log K_2	4.45	4.40	4.35	26.11	9.58	55.94
4	log K_1	6.45	6.35	6.25	36.11	14.73	72.34
	log K_2	4.80	4.75	4.70	27.49	8.49	64.48
5	log K_1	7.65	7.55	7.45	42.59	18.37	82.01
	log K_2	5.50	5.45	5.40	31.34	11.59	86.82
6	log K_1	6.40	6.30	6.20	35.61	17.95	59.79
	log K_2	5.15	5.10	5.05	29.37	13.39	53.97
accuracy	log K	±0.04	±0.04	±0.04	±0.56	±0.56	±2.4

Depending on the substituents R, the complex stabilities decrease in the order R = OCH_3 > NO_2 > CH_3 > Br > Cl > H. A comparison with other metal complexes of the same ligand reveals the order $Fe^{2+} > Cu^{2+} > Ni^{2+} > Co^{2+} > Mn^{2+}$. The exothermic values of ΔG and ΔH suggest coordination of the ligands to Mn through both the deprotonated carboxyl oxygen and amide nitrogen. The extreme stability of the complexes with ligand 5 (R = OCH_3) is apparently due to the increased electron density in the reaction center of the chelate.

Reference:

Sharma, C. L., Arya, R. S., Narvi, S. S., Mishra, V. (J. Indian Chem. Soc. **61** [1984] 677/9).

24.2.7 With Ethylenediamine-N,N,N',N'-tetraacetamide

$[H_2NC(O)CH_2]_2NCH_2CH_2N[CH_2C(O)NH_2]_2 = C_{10}H_{20}N_6O_4$

$Mn(C_{10}H_{20}N_6O_4)Cl_2 \cdot 0.5H_2O$. A solution of the ligand (5 mmol) in a minimum volume of hot water was added dropwise to a solution of $MnCl_2$ (5 mmol) in hot methanol. The colorless complex crystallized upon cooling or addition of ether. The N-deuterated complex $Mn(C_{10}H_{12}D_8N_6O_4)Cl_2 \cdot 0.5D_2O$ was obtained by evaporating to dryness a solution of the parent complex in hot D_2O under reduced pressure.

The IR spectrum of the complex in Nujol shows a number of characteristic absorptions (in cm^{-1}), which were assigned as follows: 3250 ($v_{as}NH_2$); 3120 (v_sNH_2); 1660, 1605, 1560, $v(CO)$, $\delta(NH_2)$; 1340 $v(CN)$. The broad bands in the 3500 to 3000 cm^{-1} region include bands due to coordinated water and lattice vibrations. Susceptibility measurements at 20°C yielded a magnetic moment of 6.02 μ_B, indicating a high-spin octahedral complex. The ligand is believed to be hexadentate, since the IR spectrum of the complex in the 1700 to 1600 cm^{-1} region closely resembles that of $Fe(C_{10}H_{20}N_6O_4)Cl_2$, which was characterized by reflectance spectral and conductivity data. The manganese complex is insoluble in methanol.

Reference:

Hay, R. W., Nolan, K. B., Shuaib, M. (Transition Metals Chem. [Weinheim] **4** [1979] 142/6).

24.2.8 With Alkyl- or Dialkylethylenediamine-N,N,N',N'-tetraacetic Acid Diamides (= H_2L)

$$
\begin{array}{c}
HO(O)CCH_2 \\
 \Big\rangle N-CHR-CHR'-N \Big\langle \\
H_2N(O)CCH_2
\end{array}
\begin{array}{c}
CH_2C(O)OH \\
\\
CH_2C(O)NH_2
\end{array}
$$

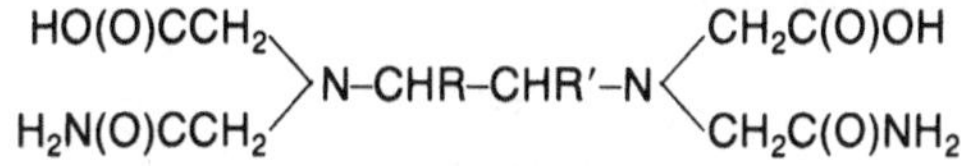

1) $R = CH_3$, $R' = H$; (= $C_{11}H_{20}N_4O_6$)
2) $R = C_2H_5$, $R' = H$; (= $C_{12}H_{22}N_4O_6$)
3) $R = R' = CH_3$; (= $meso$-$C_{12}H_{22}N_4O_6$)
4) $R = R' = CH_3$; (= ($\pm$)-$C_{12}H_{22}N_4O_6$)

Formation constants of the species MnL, $MnHL^+$, and Mn_2L^{2+} were determined potentiometrically (pH method) at 37°C and $I = 0.15$ M (NaCl). Logarithmic values of the overall formation constants $\beta_{pqr} = [Mn_pL_qH_r^{(2p+r-2q)+}]/[Mn^{2+}]^p \cdot [L^{2-}]^q \cdot [H^+]^r$ are tabulated below:

ligand	complex	log β_{pqr} [1]	ligand	complex	log β_{pqr} [2]
1	MnL	9.762 ± 0.006	3	MnL	7.615 ± 0.004
	$MnHL^+$	10.980 ± 0.056		$MnHL^+$	9.452 ± 0.046
	MnL_2^{2-}	11.599 ± 0.026	4	MnL	10.626 ± 0.008
2	MnL	9.382 ± 0.004		$MnHL^+$	12.225 ± 0.060
				$MnH_{-1}L_2^{3-}$	1.274 ± 0.023

It is assumed that the two carboxylate groups and the two diaminoalkane nitrogens are the coordination sites of the ligands [1, 2].

References:

[1] Huang, Z.-X., May, P. M., Quinlan, K. M., Williams, D. R., Creighton, A. M. (Agents Actions **12** [1982] 536/42; C.A. **97** [1982] No. 156072).
[2] Huang, Z.-X., Creighton, A. M., Williams, D. R. (Inorg. Chim. Acta **107** [1985] L29/L32).

24.3 Complexes with N-Carbamoylpyrazole and -imidazoles and with Carbamates

24.3.1 With N-Carbamoylpyrazole

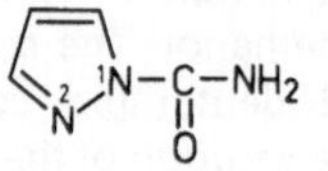

($=C_4H_5N_3O$)

$Mn(C_4H_5N_3O)Cl_2$. Solutions of manganese(II) chloride and N-carbamoylpyrazole in acetone were mixed in a 1:2 mole ratio. The resulting white powdery precipitate was washed with acetone and dried with diethyl ether; it melts at 250°C. Most relevant absorption bands (in cm^{-1}) observed in the IR spectrum (Nujol mulls and KBr pellets) in the 4000 to 180 cm^{-1} region were assigned as follows: 3318, 3260 to ν(NH); 1708 to ν(CO); 1601 to δ(NH); 1538 to ν_{ring}; 1202 to $\nu(C\text{–}N)_{amide}$; 1100 to δ(NH); 965, 918 to δ_{ring}; 710 to γ(NH); 640 to γ_{ring}; 588 to γ(NH) out-of-plane, and 543 to δ(CO). It is concluded from the shift of the ν(CO) and γ(NH) out-of-plane vibrations to lower frequencies by 7 and 17 cm^{-1}, respectively, and from displacement of the ν(NH) vibration from 3235 to 3260 cm^{-1}, that the ligand is coordinated to Mn mainly through carbonyl oxygen and the N(2) atom of the ring. The only small shift of the ν(CO) band can be explained by superposition of two reverse effects caused by coordination, Terheijden, J., Driessen, W. L., Groeneveld, W. L. (Transition Metal Chem. [Weinheim] **5** [1980] 346/50, 348).

24.3.2 With Derivatives of N-Carbamoylimidazole

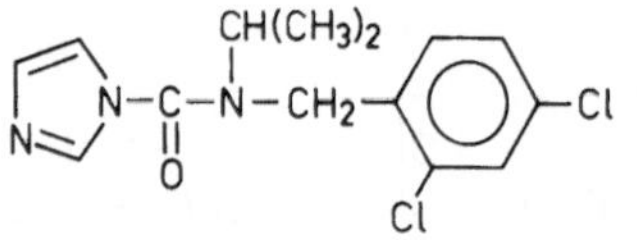

ligand 1 ($=C_{14}H_{15}Cl_2N_3O$)

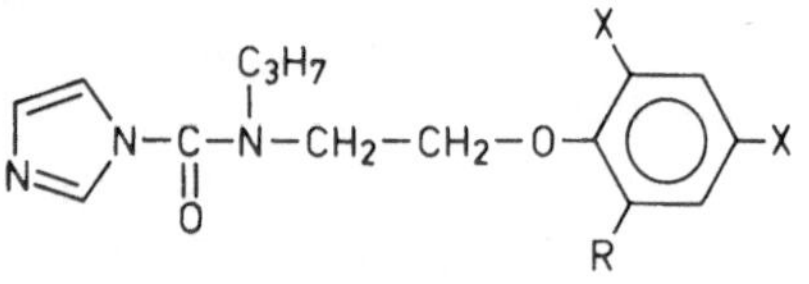

ligand	X	R	formula
2	H	H	$C_{15}H_{19}N_3O_2$
3	Cl	Cl	$C_{15}H_{16}Cl_3N_3O_2$
4	Cl	CH_3	$C_{16}H_{19}Cl_2N_3O_2$

Complexes of composition **$Mn(C_{14}H_{15}Cl_2N_3O_2)_4Cl_2$** and **$Mn(C_{15}H_{19}N_3O_2)_4Cl_2$** were prepared from $MnCl_2 \cdot 4H_2O$ with the stoichiometric amounts of ligand 1 or 2, respectively, in methanol. After stirring for 3 h, the solvent was removed and the residue treated with ether. The complexes, which were dried in vacuum at room temperature, melt at 167 to 169°C or at 86 to 88°C, respectively.

$Mn(C_{15}H_{16}Cl_3N_3O_2)_2Cl_2$ was obtained from ethanolic solutions of $MnCl_2 \cdot 4H_2O$ and ligand 3. The mixture was stirred for 18 h. After 2 d the precipitate was filtered off, washed with water and dried in vacuum at 50°C. It melts at 137 to 139°C.

$Mn(C_{16}H_{19}Cl_2N_3O_2)_2Cl_2$, precipitated by addition of an aqueous solution of $MnCl_2 \cdot 4H_2O$ to an ethanolic solution of ligand 4 in the required amount, was redissolved in a small volume of CH_2Cl_2. To the filtrate the same volume of ether was added. The solid was dried in vacuum at room temperature. It melts at 132 to 134°C.

The complexes may serve as fungicides for seeds of cereals and as additives for fertilizers.

Reference:

Birchmore, R. J., Brookes, R. F., Copping, L. G., Wells, W. H. (Ger. Offen. 2812662 [1978] 1/28, 18, 21/3; C.A. **90** [1979] No. 89566).

24.3.3 With Phenylcarbamates $C_6H_5OC(O)NHR$

ligand 1 R = C_6H_5; (= $C_{13}H_{11}NO_2$)
ligand 2 R = $(CH_2)_6NHC(O)OC_6H_5$; (= $C_{20}H_{24}N_2O_4$)

Complexes in Solution. The inhibiting activity of manganese(II) complexes with both ligands was studied by determining the rate of oxidation of ethylbenzene by O_2 in glacial acetic acid in the presence of azoisobutyronitrile (inhibitor) and manganese compounds as catalysts. Complex formation affects the reaction rate. The 1:1 complexes form in each case by reaction of $Mn(CH_3COO)_2 + L \rightleftharpoons MnL(CH_3COO)_2$ with L = ligand 1 or 2. The formation constants K = 18000 and 44000 L/mol, respectively, were calculated from the rate constant of the reaction $MnL + RO_2^{\bullet} \rightarrow$ inactive products. In pure hydrocarbons the urethanes are inhibitors, whose efficiency is increased by complexation with Mn (18.8 times with ligand 1, 5.8 times with ligand 2).

Reference:

Tochina, E. M., Postnikov, L. M., Antipova, V. F., Melamed, V. I (Izv. Akad. Nauk SSSR Ser. Khim. **1970** 479/82; Bull. Acad. Sci. USSR Div. Chem. Sci. **1970** 438/40).

24.3.4 With Other Carbamates

Remark. Formation constants of complexes with the monoethyl ester of ethylenediamine-N,N',N'-triacetic acid are reported on p. 26.

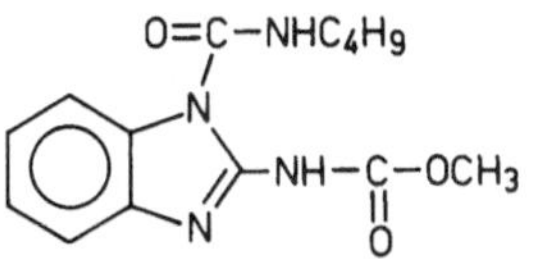

ligand 1 (= $C_{14}H_{18}N_4O_3$)

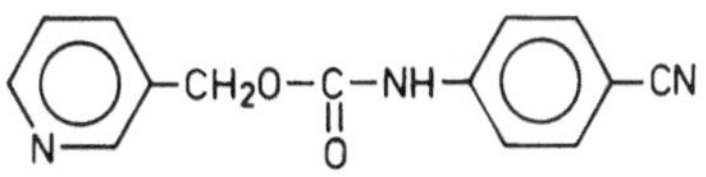

ligand 2 (= $C_{14}H_{11}N_3O_2$)

$Mn(C_{14}H_{18}N_4O_3)_2Cl_2$. Solutions of $MnCl_2$ and ligand 1 in a mixture of alcohol and $CHCl_3$ (50 vol%) were combined in a 1:2 mole ratio with continuous stirring and concentrated. Then benzene was added and the mixture was refluxed for 1 h and concentrated again. A mixture of alcohol and $CHCl_3$ was added to precipitate the white complex, which was dried in a vacuum. Characteristic absorption bands observed in the IR spectrum (Nujol mulls) were assigned as follows (wave numbers in cm^{-1}, bands of the ligand in parentheses: 1625 (1642) to ν(CO) of the NC(O)NH moiety; 1700 (1742) to ν(CO) amide I; 1590 (1608) to δ(NH) amide II, 1325 (1340) to ν(CN) amide III vibrations of the $NHC(O)OCH_3$ moiety. The downward shift of both ν(CO) bands suggests coordination of the carbonyl oxygens to different Mn atoms in a polymeric complex. Susceptibility measurements by the Gouy method at 24.5°C yield the magnetic moment $\mu_{eff} = 5.55\ \mu_B$, indicative of a high-spin Mn^{II} (d^5) complex [1].

$Mn(C_{14}H_{11}N_3O_2)_2Cl_2$ precipitates upon adding a solution of anhydrous $MnCl_2$ in monomethyl glycol to ligand 2 in the same solvent on cooling. The compound, which melts at 200 to 205°C, is useful as a rodenticide [2].

References:

[1] Savindra Randhawa, Pannu, R. S., Chopra, S. L. (J. Indian Chem. Soc. **61** [1984] 474/5).
[2] Weiler, E. D., Röhm & Haas Co. (U.S. 4172893 [1975/79] 1/5; C.A. **92** [1980] No. 105858).

24.4 Complexes with Urea and Related Compounds

24.4.1 With Urea $OC(NH_2)_2$ (= CH_4N_2O = ur)

General. Complexes of composition $Mn(ur)_nX_2$ with n = 2, 3, 4, 6, 8, and 10 are described. The highest urea content was found for the bromides, iodides (n = 10) and isothiocyanates (n = 8). The IR spectra of the compounds show shifts of the ν(CO) band toward lower and of the ν(C–N) band to higher frequencies in comparison to the free ligand. These shifts and the observed bands in the far-IR, assigned to ν(Mn–O) frequencies, indicate bonding of urea to manganese through the oxygen of the carbonyl group.

The complexes with X = Cl, Br, I, and ClO_4 have been studied by ESR, far-IR, and measurements of the low-temperature magnetic susceptibility. The $Mn(ur)_2X_2$ species with X = Cl, Br are halogen-bridged linear-chain compounds with oxygen-bonded urea in the axial positions. The Mn atoms are antiferromagnetically coupled. The $Mn(ur)_4X_2$ compounds appear to be tetragonally coordinated with the anions in the axial positions. The zero-field splittings of the ground state have been determined as 0.11 and 0.18 cm^{-1} for X = Cl and Br, respectively. The octahedral cations $[Mn(ur)_6]^{2+}$ were found to be slightly distorted. Mn–O vibrations occur in the 200 to 235 cm^{-1} region.

An urea adduct $[Mnaca_3]\cdot 3CH_4N_2O$ was described in "Mangan" D1, 1979, p. 108.

The structure and spectral properties of a pentacoordinated high-spin complex with N,N′-dimethylurea, $Mn(C_3H_8N_2O)_3Br_2$, have been investigated recently, see pp. 150/3.

24.4.1.1 Formation in Solution

Stability constants of 1:1 and 1:2 complexes in the system Mn^{2+}–urea–water at 25°C, K_1 = 4.1 and K_2 = 1.2 L/mol result from studies on the solubility of $MnCO_3$ and the electrical conductivity of $Mn(NO_3)_2$ [1, 2].

References:

[1] Stancheva, P. (Nauchn. Tr. Vissh. Pedagog. Inst. Plovdiv Mat. Fiz. Khim. Biol. **8** No. 1 [1970] 103/11, 105; C.A. **74** [1971] No. 60364).
[2] Perrin, D. D. (IUPAC Chem. Data Ser. B No. 22 [1979] 15).

24.4.1.2 Isolated Manganese(II) Compounds

24.4.1.2.1 Nitrates

$Mn(ur)_4(NO_3)_2$. The anhydrous complex was obtained by mixing equal volumes of aqueous solutions containing 1 M $Mn(NO_3)_2$ and 4 M urea at 40°C. The resulting crystals were washed and dried in a desiccator over H_2SO_4 [1]. The crystals are pale pink [2] and melt at 115°C. Their density D = 1.8537 g/cm³ was measured pycnometrically in CCl_4 at 25°C [1]. The complex starts to decompose at 160°C with release of ammonia as shown by thermogravimetric analysis. Its reaction with gaseous ammonia was studied at p_{NH_3} = 550 Torr and temperatures ranging from −13 to 100°C. It was shown that the reaction rate and number of ammonia molecules added to $Mn(ur)_4(NO_3)_2$ decrease at higher temperatures (from 7 at 0°C to 6 at 60°C and 3 at 100°C). The ammonia adducts obtained at 80 and 100°C are liquids. It is assumed that in the adducts the urea molecules in the inner complex sphere are replaced by ammonia [1].

$Mn(ur)_4(NO_3)_2 \cdot 2H_2O$ was detected as a stable crystal phase in the solubility isotherms of the ternary system $Mn(NO_3)_2$–urea–water at 20°C [3], 25°C [4, 5], 30°C [6], and of the quaternary system $Mn(NO_3)_2$–KNO_3–urea–H_2O at 20°C [7] together with the species $Mn(ur)_2(NO_3)_2 \cdot 4H_2O$ (see below), $Mn(NO_3)_2 \cdot 6H_2O$ and urea [3, 4, 7]. To prepare the compound, an aqueous solution containing $Mn(NO_3)_2$ and urea in a 1:4 mole ratio was evaporated slowly in a desiccator over H_2SO_4 [8, 9]. Crystallization occurred within 12 h when the components had been dissolved in a minimum volume of water [6]. The well-shaped light pink tabular crystals [5, 9] or rectangular prisms [6] are nonhygroscopic [6, 9] and rather stable to air [4].

X-ray studies (Laue, Weissenberg, oscillation methods with MoKα radiation) reveal a monoclinic lattice, space group $P2_1n$-C_{2h}^5 (No. 14) with the lattice constants $a = 7.70 \pm 0.01$, $b = 17.86 \pm 0.02$, $c = 6.53 \pm 0.01$ Å and $\beta = 92°38' \pm 30'$; $Z = 2$, $V = 897.07$ Å^3 which is isotypic with those of the corresponding complexes of Co, Ni, and Zn. Density D is 1.67 g/cm^3 measured at 20°C by flotation ($CHBr_3$ + benzene), 1.69 g/cm^3 calculated from X-ray data [9]; for other (strongly differing) values see [3, 4]. The complex melts at 111 to 116°C [6], 109°C [3]; for a far lower value (50°C) see [4]. Upon further heating up to 200°C it decomposes with release of water and NO_2 followed by ammonia above 200°C which attains a maximum concentration at 240°C. Heating up to 310°C causes explosion [3]. Loss of water occurs at 137°C and decomposition at 257°C, leaving MnO_2 at 500°C [4]. $Mn(ur)_4(NO_3)_2 \cdot 2H_2O$ is soluble in water, reaching 71.48 wt% at 23°C [6], 200 g/100 g H_2O at 25°C [4].

Cryoscopic measurements in aqueous solutions containing urea and in molten $Na_2SO_4 \cdot 10H_2O$ indicate considerable solvolysis as the apparent molecular weight is far below the expected value corresponding to a dissociation into the complex cation $[Mn(ur)_4]^{2+}$ and $2NO_3^-$. This indicates almost complete dissociation of the complex cation in both media. Upon addition of urea to these solutions the dissociation is reduced with increasing concentration of urea and attains almost the theoretical value in solutions saturated with urea [8]. In alcohol $Mn(ur)_4(NO_3)_2 \cdot 2H_2O$ is much less soluble than in water, and it is insoluble in ether [6].

$Mn(ur)_3(NO_3)_2 \cdot 3H_2O$. The complex appears as the only ternary crystal phase in the solubility isotherm of the system $Mn(NO_3)_2$–urea–water at 50°C. It may be isolated from its saturated solution at 50°C and purified by recrystallization from aqueous ethanol. It forms thin pale pink crystals which are stable in air for a longer time and melt at 135°C with loss of water. However, prolonged heating (48 h) at that temperature causes decomposition producing ammonia, CO_2, NO_2, MnO_2, and an unknown organic residue. Explosion will occur if the complex is heated gradually (dynamic conditions) to about 240°C [10].

$Mn(ur)_2(NO_3)_2 \cdot 4H_2O$. The complex appears as the second ternary crystal phase in the system $Mn(NO_3)_2$–urea–H_2O at 20 and 25°C together with $Mn(ur)_4(NO_3)_2 \cdot 2H_2O$ (see above) [3, 4]. It can be prepared by melting together manganese nitrate hexahydrate and urea in a 1:2 mole ratio at 75°C with constant stirring [11] (also see [5]). The pale pink crystals are rather stable to air [4]. Density D is 1.670 g/cm^3 measured at 20°C [3], 1.476 g/cm^3 at 25°C [4]. They melt at 42°C [3], 60°C [4]. The IR absorption spectrum of the complex suggests bonding of urea to manganese through the carbonyl oxygens [3, 4]. Upon heating it loses water at 117°C and decomposes at 272°C [4], 268°C [1]. The thermal decomposition proceeds similarly to $Mn(ur)_4(NO_3)_2 \cdot 2H_2O$; see above [3]. The solubility in water is 2500 g/kg H_2O at 25°C [4].

References:

[1] Lazerka, G. A., Girei, I. V., Zonau [Zonev], Yu. R. (Vestsi Akad. Navuk Belarusk. SSR Ser. Khim. Navuk **1968** No. 2, pp. 50/5, 52; C.A. **70** [1969] No. 41184).

[2] Stancheva, P. (Nauchn. Tr. Vissh. Pedagog. Inst. Plovdiv Mat. Fiz. Khim. Biol. **8** No. 1 [1970] 103/11, 106; C.A. **74** [1971] No. 60364).
[3] Runov, N. N. (Uch. Zap. Yaroslav. Gos. Pedagog. Inst. No. 79 [1970] 146/50; C.A. **76** [1972] No. 50784).
[4] Protsenko, P. I., Kallaeva, Kh. I. (Sb. Nauchn. Soobshch. Dagestan Univ. Kafedra Khim. **1971** No. 7, pp. 11/3; C.A. **77** [1972] No. 28351).
[5] Sarnovski, M., Ścieńska, I., Zygadło, J. (Roczniki Chem. **31** [1957] 949/57, 953/4; C.A. **1958** 19373), Sarnovski, M., Zygadło, J., Ścieńska, I. (Roczniki Chem. **29** [1955] 1139/40; C.A. **1956** 8361).
[6] Slashcheva, L. A., Petriichuk, D. I. (Tr. Inst. Khim. Akad. Nauk Kirg. SSR **1957** No. 8, pp. 191/4; C.A. **1960** 18150).
[7] Runov, N. N. (Uch. Zap. Yaroslav. Gos. Pedagog. Inst. No. 95 [1971] 69/71; C.A. **78** [1973] No. 48622).
[8] Sumarokova, T. N., Slashcheva, L. A., Petriichuk, D. I. (Izv. Akad. Nauk Kaz. SSR Ser. Khim. **16** No. 1 [1966] 10/3; C.A. **65** [1966] 14816).
[9] Durski, Z., Boniuk, H., Zych, J., Dąbrowska, J. (Roczniki Chem. **48** [1974] 1615/8; C.A. **82** [1975] No. 50082).
[10] Novikov, A. V., Runov, N. N. (Uch. Zap. Yaroslav. Gos. Pedagog. Inst. No. 120 [1973] 107/12, 108; C.A. **82** [1975] No. 48151).

[11] Boller, E. R., Graselli Chemical Co. (U.S. 1986495 [1933/35] 1/2; C.A. **1936** 1218).

24.4.1.2.2 Fluorides

No compound containing both MnF_2 and urea could be detected in the solubility isotherm of the ternary system MnF_2–urea–water at 25°C by solubility determinations, Moldobaeva, U. K., Rysmendeev, K. R., Moldobaev, S. M. (Sb. Tr. Aspirantov Soiskatelei Kirg. Univ. Estestven. Nauk **1975** No. 6, pp. 13/21, 14/6; C.A. **84** [1976] No. 127306).

24.4.1.2.3 Chlorides

$Mn(ur)_4Cl_2$ was detected as a stable solid phase in the solubility isotherms of the ternary system $MnCl_2$–urea–water at 20, 30 [1, 2], and 40°C [1], 25°C [3], 50°C [17], and of the quaternary system $CoCl_2$–$MnCl_2$–urea–H_2O at 25°C [4, 5] together with $MnCl_2 \cdot 4H_2O$ and urea [1 to 3], and the species $Mn(ur)_2Cl_2$ at 25 [3] up to 40°C [1] or $Mn(ur)_2Cl_2 \cdot H_2O$ at 20°C [1, 2]. Its existence was confirmed also by measurement of density and viscosity at 30°C [1, 2].

Preparation is carried out by stirring together ethanolic solutions of $MnCl_2$ and urea in the stoichiometric mole ratio at room temperature for 2 h and adding diethyl ether to induce crystallization [6], by isothermal evaporation of the corresponding solution of the components at 25°C [7] or grinding together the required amounts of $MnCl_2 \cdot 4H_2O$ and urea [8]. The crystals were washed with small portions of cold water [7], recrystallized and dried in vacuum [8] or in the air after pressing between filter paper [7]. $Mn(ur)_4Cl_2$ is pale pink [6], pink [9]. X-ray powder patterns (FeKα radiation) were recorded to assure the identity of the compound, and the interplanar spacings, which are similar in position to those of $Co(ur)_4Cl_2$, are given in [7]. The density, D = 1.730 g/cm^3, was measured pycnometrically [1].

$Mn(ur)_4Cl_2$ melts at 162°C [6], 160°C [17], 155°C [1], 150°C [10], 180°C with decomposition [3]. Significant absorption bands of the IR spectrum (KBr disks and Nujol mulls) in the 4000 to 400 cm^{-1} region were assigned as follows (wave numbers in cm^{-1}, wave numbers of ligand bands in parentheses):

ν_{as}(NH)	ν_s(NH)	ν(CO)	$\delta(NH_2)$	ν_{as}(C–N)	ν_s(C–N)	$\delta(NH_2)$	Ref.
3460(s)	3330(s)	1620(s, br)	—	1480(s)	1010(w)	—	[6]
3440	3340	1665	1630	1493	1020	780	[8]
3450	3350	1640	—	—	—	—	[9]
3460	3380 to 3280	1640	—	1470	—	—	[17]
(3454)	(3340)	(1680)	(1625)	(1464)	(1010)	(792)	[8]

In the far-IR spectrum (500 to 50 cm^{-1}) the bands observed at 236(sh) and 218(s, br) cm^{-1} (room temperature) or at 267(w), 249(s), and 227(vs) cm^{-1} (in liquid N_2) are assigned to ν(Mn–O) and those at 162 cm^{-1} (s, room temperature) or 161 cm^{-1} (vs, in liquid N_2) to ν(Mn–Cl) vibrations. For other bands (including a very strong broad band assignable to H bond vibrations at 400 cm^{-1}) see [11]. The observed shift of ν(CO) toward lower and of ν(C–N) to higher frequencies [4, 6, 8 to 10, 12] in comparison to the free ligand and the occurrence of 3 bands assignable to ν(Mn–O) [11] clearly indicate bonding of urea to manganese through the oxygen of the carbonyl group [4, 6, 8 to 13]. The occurrence of a single ν(Mn–Cl) band is consistent with a trans-octahedral structure where the oxygens of urea occupy the equatorial sites and the chlorine atoms are on the apical sites. The splitting of the ν(Mn–O) band in the 250 cm^{-1} region may indicate a small distortion of the assumed D_{4h} symmetry of $Mn(ur)_4Cl_2$ [11].

The electronic reflection spectrum shows a band at 24150 cm^{-1} assigned to the electron transition $^6A_{1g} \rightarrow {}^4E + {}^4A_1(^4G)$, nephelauxetic ratio $\beta = 0.90$. The observed strong red fluorescence is characteristic of an octahedral complex with a symmetry lower than O_h [6]. The X-ray spectrum shows the absorption band edge at $\lambda = 4383.9$ KX [14]. The ESR X-band spectrum gives the parameters g = 2.00(1), D (zero field splitting) = 0.11(1) cm^{-1} and $E/D \leqq 0.05$, which are consistent with the proposed trans-octahedral structure [11].

Susceptibility measurements at 293 K yield the magnetic moment $\mu_{eff} = 5.87\ \mu_B$ indicating a high-spin $Mn^{II}(d^5)$ complex [6]. The complex decomposes on heating at 210°C [10], 190°C [17], or 180°C [3]. The endothermic decomposition of the urea component starts at 237°C and proceeds very rapidly at about 300°C with formation of gaseous CO and NH_3 as shown by DTA and TG analysis [3, 17].

$Mn(ur)_4Cl_2$ is congruently soluble in water at 25°C [3]. The solubility is 1.7 wt% in absolute ethanol, 4.20 wt% in 96% ethanol, 0.050 wt% in acetone and traces in benzene. It behaves as a nonelectrolyte in dimethylformamide as shown by its low electrical conductivity [10]. It is insoluble in $CHCl_3$, CCl_4, and diethyl ether [1].

$Mn(ur)_2Cl_2$ appears as the second binary urea complex in the ternary system $MnCl_2$–urea–H_2O at 25°C [3], 50°C [17], or higher temperatures [1] together with $Mn(ur)_4Cl_2$; see above [1, 3]. Its existence was confirmed by measurement of density and viscosity [1, 2]. To prepare $Mn(ur)_2Cl_2$, urea was added to an ethanol solution of $MnCl_2$ in a 2:1 mole ratio under N_2 and the mixture was refluxed for 2 h, whereupon crystallization occurred [6]. It could be obtained also by isothermal evaporation of aqueous solutions of the components (stoichiometric ratio) at 25°C, washing with cold water and drying in the air [4, 7]. The complex is pink [6].

X-ray powder patterns (FeKα radiation) were recorded to assure its identity and the interplanar spacings were given in [7]. The density, 1.905 g/cm^3, was measured pycnometrically [1]. The complex melts at 190°C [10], 210°C [1], or 215°C with decomposition [6]; also see [3]. The IR spectrum (KBr disks and Nujol mulls) shows absorption bands assigned to ν(NH) at 3480(s), 3410(s), and 3365(s) cm^{-1}, to ν(CO) and $\delta(NH_2)$ at 1660(sh), 1620(s), and 1580(s) cm^{-1}, to ν_{as}(C–N) at 1475 cm^{-1}(s) and to ν_s(C–N) at 1015 cm^{-1}(w) [6]. In the far-IR, bands assigned to ν(Mn–O) occur at 270(s) and 215(vs) cm^{-1} (in liquid N_2 at 276 and 222 cm^{-1}), to ν(Mn–Cl) at 164(vs) cm^{-1} (at 160 cm^{-1} in liquid N_2). For other bands (including a very strong broad band assignable to H bond vibrations at 360 cm^{-1}) see [11]. The occurrence of a ν(Mn–O) band at 270 cm^{-1} (higher than in the octahedral monomers) and the presence of a single ν(Mn–Cl) band are consistent with the assumption of a linear chain structure of the complex [11]. The electron reflection spectrum shows three bands at 19450, 23800, and 26900 cm^{-1} assigned to the electron transitions $^6A_{1g} \rightarrow {}^4T_1(^4G)$, $\rightarrow {}^4E + {}^4A_1(^4G)$, and $\rightarrow {}^4T_2(^4D)$, respectively, nephelauxetic ratio β (=23800/26846)=0.89. The observed strong red fluorescence suggests an octahedral complex with symmetry lower than O_h [6]. The ESR X-band and Q-band spectra give signals at g=2.01(1) with line widths of 350(5) and 325(5) Gauss, respectively. Susceptibility measurements at temperatures ranging from 77 to 293 K yield $\chi_{mol}^{corr} = 14 \times 10^{-3}$ cm^3/mol at 293 K and the Weiss constant Θ=−42 K from which $\mu_{eff} = 6.12$ μ_B was calculated [6]. Measurements at 2 to 100 K reveal a maximum of χ (0.133(7) emu) at 8.2(2) K and Θ=−18(1) K, from which $\mu_{eff} = 5.92(5)$ μ_B was derived. The anomaly of χ is indicative of antiferromagnetic interaction, for which the exchange parameters J/k=−0.84(5) K and −1.00(5) K were calculated from T_{max} values, using different equations, whereas the ESR data yield J/K=−0.70(8) K [11]. For calculation procedures see [15]. The spectra and magnetic data of $Mn(ur)_2Cl_2$ are indicative of a linear chain structure consisting of MnO_2Cl_4 octahedra linked together by the bridging Cl^- ions in the equatorial plane with antiferromagnetic coupling of the Mn^{2+} ions; the urea ligands occupy the apical sites [6, 11].

$Mn(ur)_2Cl_2$ melts at 205°C and decomposes on heating at 260°C [10], 275°C [17], or 280°C with endothermic decomposition of the urea component as for $Mn(ur)_4Cl_2$ (see p. 136). It is congruently soluble in water at 25°C [3]. The solubility is 1.5 wt% in absolute ethanol, 3.60 wt% in 96% ethanol, 0.035 wt% in acetone, traces in benzene [1]. In deuteromethanol solution, partial solvolysis is suggested by the IR spectrum [16]. It is insoluble in $CHCl_3$, CCl_4, and diethyl ether [1].

$Mn(ur)_2Cl_2 \cdot H_2O$. The hydrate appears as the second urea complex in the ternary system $MnCl_2$–urea–H_2O at 20°C together with $Mn(ur)_4Cl_2$; see p. 136. Its existence was confirmed by measurement of density and viscosity. The density, D=1.842 g/cm^3, was measured pycnometrically. On heating it does not melt but decomposes with release of water and ammonia yielding a black residue. Like $Mn(ur)_2Cl_2$ (see p. 137) it is congruently soluble in water but somewhat more soluble in ethanol and acetone. In benzene, $CHCl_3$, CCl_4, and diethyl ether it behaves like $Mn(ur)_2Cl_2$ [1].

References:

[1] Druzhinin, I. G., Rysmendeev, K. (Izv. Akad. Nauk Kirg. SSR Ser. Estestven. Tekhn. Nauk **4** No. 9 [1962] 21/32; C.A. **59** [1963] 2216).

[2] Druzhinin, I. G., Rysmendeev, K. (Izv. Vysshikh Uchebn. Zavedenii Khim. Khim. Tekhnol. **5** No. 1 [1962] 8/11; C.A. **57** [1962] 2907).

[3] Pavlenko, A. I., Sulaimankulov, K. S., Shevchuk, V. G. (Ukr. Khim. Zh. **37** [1971] 415/8; Soviet Progr. Chem. **37** No. 5 [1971] 5/7).

[4] Pavlenko, A. I. (Ukr. Khim. Zh. **43** [1977] 94/5; Soviet Progr. Chem. **43** No. 1 [1977] 95/6).

[5] Pavlenko, A. I., Sulaimankulov, K. S., Shevchuk, V. G. (Zh. Neorgan. Khim. **15** [1970] 2531/4; Russ. J. Inorg. Chem. **15** [1970] 1309/11).

[6] Barbier, J. P., Hugel, R. (Inorg. Chim. Acta **10** [1974] 93/6).

[7] Pavlenko, A. I., Redchenko, G. A. (Zh. Neorgan. Khim. **20** [1975] 2493/5; Russ. J. Inorg. Chem. **20** [1975] 1381/3).

[8] Antonenko, N. S., Nuger, Ya. A. (Zh. Neorgan. Khim. **11** [1966] 1072/5; Russ. J. Inorg. Chem. **11** [1966] 578/80).

[9] Stancheva, P. (Nauchn. Tr. Vissh. Pedagog. Inst. Plovdiv Mat. Fiz. Khim. Biol. **8** No. 1 [1970] 103/11, 106/8; C.A. **74** [1971] No. 60364).
[10] Khodzhaev, O. F., Azizov, T. A., Ergeshbaev, D., et al. (Koord. Khim. **2** [1976] 304/7; Soviet. J. Coord. Chem. **2** [1976] 223/6, 225).

[11] Barbier, J. P., Hugel, R. P., van der Put, P. J., Reddijk, J. (Rec. Trav. Chim. **95** [1976] 213/6).
[12] Dzhorobekov, B., Rysmendeev, K. R. (Sb. Tr. Aspirantov Soiskatelei Kirg. Univ. Estestven. Nauk **1971** No. 5, pp. 19/24 from C.A. **77** [1972] No. 100300).
[13] Runov, N. N., Zakharova, V. P. (Uch. Zap. Yaroslav. Gos. Pedagog. Inst. No. 78 [1970] 103/6, 104; C.A. **76** [1972] No. 9915).
[14] Stelling, O. (Z. Physik. Chem. B **24** [1934] 282/92, 284).
[15] Witteveen, H. T., Nieuwenhuijse, B., Reedijk, J. (J. Inorg. Nucl. Chem. **36** [1974] 1535/41, 1536).
[16] Alieva, Z. F., Sulaimankulov, K. S., Agashkin, O. V., Matamarov, N. (Izv. Akad. Nauk Kaz. SSR Ser. Khim. **21** No. 2 [1971] 15/8; C.A. **75** [1971] No. 82061).
[17] Shchedrina, A. P., Khasova, P. D., Kirsanov, Yu. I., Ovechkii, V. T. (Mezhvuz. Sb. Nauchn. Tr. Yarosl. Gos. Pedagog. Inst. im. K. D. Ushinskogo No. 205 [1984] 60/3; C.A. **103** [1985] No. 166988).

24.4.1.2.4 Perchlorates

$Mn(ur)_6(ClO_4)_2$ was detected as a stable solid phase in the solubility isotherms of the ternary system $Mn(ClO_4)_2$–urea–water at 25°C [1, 7] and of the quaternary system $Mn(ClO_4)_2$–$TlClO_4$–urea–H_2O at 25°C [2] together with the species $Mn(ur)_4(ClO_4)_2 \cdot 2H_2O$, $Mn(ClO_4)_2 \cdot 6H_2O$, urea [1, 2] and $TlClO_4$ which does not form any ternary compound with $Mn(ClO_4)_2$ or urea [2]. The existence of $Mn(ur)_6(ClO_4)_2$ was confirmed also by measurement of density, viscosity, and electrical conductivity (minimum) of the solutions at the eutonic point [1]. To prepare the complex, stoichiometric amounts of $NaClO_4$ and urea were added to an ethanolic solution of $MnCl_2$ [3] or stoichiometric amounts of $Mn(ClO_4)_2 \cdot 6H_2O$ and urea were stirred (in ethanol) for 2 h at room temperature and cold diethyl ether was added to start crystallization [4]. Single crystals could be grown from aqueous solutions containing 60 g/L urea [1]. The colorless complex melts at 190°C [4], between 193 and 198°C. An endothermic effect observed at 116°C (by DTA) may be ascribed to a change in the inner or outer coordination sphere of the complex.

The density D = 2.0206 g/cm³ at 25°C was measured by pycnometry and the refractive indices $n_\alpha = 1.507$, $n_\beta = 1.561$, $n_\gamma = 1.574$ were found by the immersion method [6]. The IR spectrum of the complex in fluoroparaffin (LiF) or Vaseline (NaCl) mulls shows significant absorption bands assigned to ν(NH) at 3458 and 3350 cm^{-1}, to ν(CO) at 1666 cm^{-1}, to ν(CO) + $\delta(NH_2)$ at 1614 and 1589 cm^{-1} [6] or 1630 cm^{-1} [4], to ν_{as}(C–N) at 1470 and to ν_s(C–N) at 1010 cm^{-1} [4]. Bands at 785 and 734 cm^{-1}, were assigned to ν_s(C–N) by [6]. In the far-IR a very strong band occurs at 232 cm^{-1} (in liquid N_2 at 239 cm^{-1}). A very strong band assignable to H bond vibrations was observed at 375 cm^{-1} (in liquid N_2 doublet at 410 and 330 cm^{-1}) [5]. The broad band in the 1140 to 940 cm^{-1} region is due to the perchlorate ions [6].

The ${}^4T_{1g}$ ($t_{2g}^4 e_g^1$) → ${}^6A_{1g}$ ($t_{2g}^3 e_g^2$) phosphorescence spectrum of Mn^{II} in single crystals of $Mn(ur)_6(ClO_4)_2$ has been recorded in the temperature range of 298 to 4.2 K. With decreasing temperature the emission maximum is shifted from 17138 cm^{-1} (298 K) to 16809 cm^{-1} (4.2 K). This shift points to a linear thermal expansion coefficient of $\alpha = 24 \times 10^{-6}$ K^{-1}. The absorption spectrum of the complex in aqueous 10 M urea solution shows five bands at 19610, 23530, 24810, 27972, and 29586 cm^{-1} assigned to the electron transition: ${}^6A_{1g} \rightarrow {}^4T_{1g}$, $\rightarrow {}^4T_{2g}$, $\rightarrow {}^4A_{1g}$, ${}^4E_g$, $\rightarrow {}^4T_{2g}$, and $\rightarrow {}^4E_g$, respectively [3] which suggest a weak octahedral crystal field of

O_h symmetry containing 6-coordinated $Mn^{2+}(d^5)$ ions [3, 7]; nephelauxetic ratio $\beta_{35}=0.93$. Crystal field parameters of $D_q=748\ cm^{-1}$ and $\beta=827\ cm^{-1}$ have been calculated from the room temperature absorption and excitation spectra; D_q was calculated assuming a nephelauxetic shift of about $-2000\ cm^{-1}$ for all crystal-field terms with respect to the free ion states [3].

The ESR X-band spectrum of $[Mn(ur)_6]^{2+}$ ions diluted in solid $Cd(ur)_6(ClO_4)_2$ gives the parameters g=2.005(5), A=95.1(4) Gauss and D=0.0025(1) cm^{-1}, which are indicative of almost undistorted $[Mn(ur)_6]^{2+}$ octahedra [5]. Susceptibility measurements at 293 K yield the magnetic moment $\mu_{eff}=5.86\ \mu_B$ corresponding to high-spin $Mn^{II}(d^5)$ complexes [4].

Differential thermal analysis reveals three endothermic effects at 116, 193 to 198, and 273 cm^{-1} indicating polymorphic transition, melting and loss of urea, respectively, and a strong exothermic effect at 335°C due to complete decomposition of the complex. $Mn(ur)_6(ClO_4)_2$ is soluble in water. It is readily soluble in highly polar organic solvents such as ethyl acetate and ethylene glycol but insoluble in those of low polarity like benzene, toluene, xylene, $CHCl_3$, and ether. The electronic spectrum of a 1.5 M solution resembles that of $Mn(ClO_4)_2 \cdot 6H_2O$ since the complex is unstable in dilute aqueous solution (formation of $Mn(H_2O)_6^{2+}$ ions) [6].

$Mn(ur)_4(ClO_4)_2 \cdot 2H_2O$. The compound appears as the second urea complex in the ternary system $Mn(ClO_4)_2$–urea–H_2O at 25°C together with $Mn(ur)_6(ClO_4)_2$ (p. 139 [1, 7]), and its existence was confirmed by measurement of density, viscosity, and electrical conductivity [1]. The density D=1.8265 g/cm^3 was measured pycnometrically at 25°C. Refractive indices $n_\alpha=1.486$ and $n_\gamma=1.574$ were found by the immersion method. The IR spectrum (recorded as for $Mn(ur)_6(ClO_4)_2$) shows absorption bands assigned to ν(NH) at 3449, 3350, and 3224 cm^{-1}, to ν(CO) at 1665 cm^{-1} and to ν(CO)+δ(NH_2) at 1624 and 1604 cm^{-1}; the other bands are rather similar to those of $Mn(ur)_6(ClO_4)_2$. A broad band due to ionic perchlorate appears in the 1140 to 940 cm^{-1} region masking those of ν(C–N) and δ(NH_2) in that range [6]. An octahedral structure similar to $Mn(ClO_4)_2 \cdot 6H_2O$ was concluded from the electronic spectra of the compounds [7]. Differential thermoanalysis reveals two endothermic effects at 70°C when the complex melts and loses the water molecules and at 240°C due to loss uf urea and an exothermic effect at 347°C which indicates reaction of the organic residue with manganese perchlorate. The complex is readily soluble in polar and insoluble in nonpolar organic solvents [6].

$Mn^{III}(ur)_6(ClO_4)_3$. For preparation a hot solution of manganese(II) acetate obtained by dissolving powdered manganese (0.182 mol) in glacial acetic acid (400 mL) solid $KMnO_4$ (0.041 mol) was added with stirring followed (after 25 min) by perchloric acid (60%, 250 mL) saturated with urea. The resulting dark purple, microcrystalline solid was filtered off, washed with a mixture of equal volumes of absolute ethanol and ether and dried in a desiccator over $CaCl_2$. Single crystals were grown from a hot (75 to 80°C) solution of $HClO_4$ (60%) saturated with urea at room temperature. X-ray diffractometer studies (MoKα radiation) reveal a rhombohedral lattice, isomorphous with that of $Ti(ur)_6(ClO_4)_3$ and $Al(ur)_6(ClO_4)_3$. The space group is $R\bar{3}c$-D_{3d}^6 (No. 167) with the lattice constants a=18.124(4) and c=14.042(3) Å; Z=6. The structure was solved using the final positional parameters of the isomorphous Ti^{III} complex as the starting values and refined up to R=0.055. The positional parameters of Mn, Cl, O, N, and C are given in the paper. The geometry of the cation $[Mn(ur)_6]^{3+}$ is shown in **Fig. 16**. The crystal consists of these cations which have the site symmetry 3 with six equivalent Mn–O bonds (1.987(7) Å) and of the ClO_4^- anions (site symmetry 2) joined by weak hydrogen bonds which are characterized by long distances O···H and N···O (see paper). The analysis of the visible absorption spectrum (two prominent bands in the 5000 to 25000 cm^{-1} region assigned to the electron transitions $^5B_{1g} \rightarrow {}^5A_{1g}$ and $^5B_{1g} \rightarrow {}^5B_g$, respectively) and mean-square displacements of Mn and O in $[Mn(ur)_6]^{3+}$ are consistent with the existence of a dynamic Jahn-Teller effect in the cation. The Jahn-Teller radius is estimated to be about 0.33 Å with a stabilization energy of about 36 kJ. The density D=1.788 g/cm^3 was measured and D=1.80 g/cm^3 was calculated from the X-ray data. Suspectibility measurements yield the magnetic moment $\mu_{eff} = 5.02\ \mu_B$ of a high-spin $Mn^{III}(d^4)$ complex. The crystalline complex decomposes within a few days in air or in the X-ray beam; it is quickly decomposed by water [8].

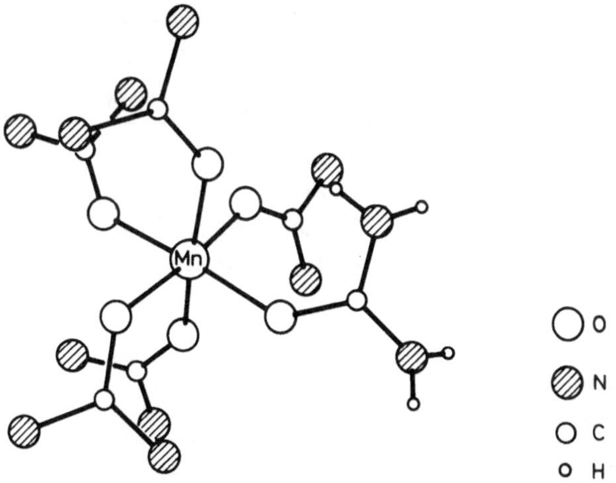

Fig. 16. Structure of the cation $[Mn(ur)_6]^{3+}$ [8]. (Most H atoms are missing for clarity.)

References:

[1] Runov, N. N., Goryunov, Yu. A. (Uch. Zap. Yaroslav. Gos. Pedagog. Inst. No. 95 [1971] 105/8; C.A. **77** [1972] No. 169377).
[2] Ovsyannikov, M. N., Ivanov, S. A. (Sb. Nauchn. Tr. Yaroslav. Pedagog. Inst. No. 164 [1977] 64/6 from Ref. Zh. Khim **1978** No. 10B727; C.A. **89** [1978] No. 66209).
[3] Koglin, E., Schenk, H. J. (Z. Naturforsch. **33a** [1978] 377/9).
[4] Barbier, J. P., Hugel, R. (Inorg. Chim. Acta **10** [1974] 93/6).
[5] Barbier, J. P., Hugel, R., van der Put, P. J., Reedijk, J. (Rec. Trav. Chim. **95** [1976] 213/6).
[6] Goryunov, Yu. A. (Sb. Vys. Sk. Chem. Technol. Praze Anal. Chem. H **16** [1981] 181/7, 182, 186; C.A. **97** [1982] No. 65413).
[7] Goryunov, Yu. A., Runov, N. N. (Sb. Vys. Sk. Chem. Technol. Praze Anal. Chem. H **16** [1981] 175/9; C.A. **97** [1982] No. 61337).
[8] Aghabozorg, H., Palenik, G. J., Stoufer, R. C., Summers, J. (Inorg. Chem. **21** [1982] 3903/7).

24.4.1.2.5 Bromides and Iodides

$Mn(ur)_nBr_2$. The complexes with n = 10, 6, 4, 2 were detected as the binary urea complexes in the solubility isotherms of the ternary system $MnBr_2$–urea–H_2O studied at 20 and 40°C [1]. $Mn(ur)_{10}Br_2$ was prepared by stirring together ethanolic solutions of $MnBr_2$ and urea in a 1 : >10 mole ratio at room temperature and addition of diethyl ether. To prepare $Mn(ur)_6Br_2$ and $Mn(ur)_4Br_2$, ethanol solutions of $MnBr_2$ and urea were combined in the stoichiometric ratios and stirred for 2 h as above. $Mn(ur)_2Br_2$ was obtained by adding the stoichiometric amount of urea to an ethanolic solution of $MnBr_2$. After 2 h of refluxing, cold diethyl ether was added to induce crystallization. The colorless complexes $Mn(ur)_{10}Br_2$ and $Mn(ur)_6Br_2$ melt at 117 and 130°C, respectively, the pale pink $Mn(ur)_4Br_2$ at 150°C and the pink $Mn(ur)_2Br_2$ at 207°C [2].

The IR spectra of all complexes show strong ν(NH) bands in the 3500 to 3200 cm^{-1} region, strong ν_{as}(C–N) bands around 1470 cm^{-1} and weak ν_s(C–N) bands around 1010 cm^{-1}. Other IR data obtained on samples in KBr disks or Nujol mulls are tabulated below together with magnetic moments resulting from susceptibility measurements at 293 K and absorption bands in the electronic reflection spectra (ν_{max} in cm^{-1}). For unassigned bands in the far-IR see [3].

complex	μ_{eff} in μ_B	IR data in cm^{-1} ν(CO) + δ(NH_2)	ν(Mn–O)	ν(Mn–Cl)	UV: $^6A_1 \rightarrow ^4E + ^4A_1(^4G)$ ν_{max}
$Mn(ur)_{10}Br_2$	5.93	1625 (s, b)	—	—	—
$Mn(ur)_6Br_2$	5.88	1610 (s, b)	235 (s)	—	24700
$Mn(ur)_4Br_2$	5.87	1625 (s, b)	250 (s), 229 (s)	144 (s), 132 (s)	24150
$Mn(ur)_2Br_2$	5.89*)	1625 (s), 1575 (s)	270 (s), 196 (vs)	153 (m), 116 (s)	23450**)
Ref.	[2]	[2]	[3]	[3]	[2]

*) Calculated with Θ = −22 K [2]. – **) Other bands were observed at 19550 ($\rightarrow ^4T_1(^4G)$), 26650 ($\rightarrow ^4T_2(^4D)$), and 28000 cm^{-1} ($\rightarrow ^4E(^4D)$) [2].

Susceptibility measurements on $Mn(ur)_2Br_2$ in the 2 to 100 K temperature region reveal a maximum of χ(0.18(1) emu) at 4.9(3) K and the Weiss constant Θ = −12(1) K, from which μ_{eff} = 5.92 μ_B was derived. The anomaly of χ is indicative of antiferromagnetic behavior as in $Mn(ur)_2Cl_2$ (p. 138). The exchange parameters J/k = −0.51(5) K and −0.60(5) K were calculated from T_{max} values, using different equations, whereas the ESR data yield J/k = −0.60(7) K [3]. The X-band in the ESR spectra gives the parameters g = 2.00(1), D = 0.18(2) cm^{-1}, and E/D = 0.10(3) for $Mn(ur)_4Br_2$ and g = 2.01(1) for $Mn(ur)_2Br_2$. An X-band line width of 1030(10) and a Q-band line width of 940(10) Gauss were reported for $Mn(ur)_2Br_2$. The exchange parameter J/K = −0.60(7) K was calculated from the line width differences between the X- and Q-band spectra [3].

The physical data suggest an octahedral structure with six-coordinate high-spin manganese(II) in all complexes. In $Mn(ur)_{10}Br_2$ and $Mn(ur)_6Br_2$ complexes manganese is coordinated only to the carbonyl oxygens of six urea ligands [2], whereas the structures of $Mn(ur)_4Br_2$ and $Mn(ur)_2Br_2$ closely resemble the trans-octahedral structure of $Mn(ur)_4Cl_2$ (see p. 137) and the linear chain structure of $Mn(ur)_2Cl_2$ (see p. 137), respectively [2, 3]. The complexes $Mn(ur)_{10}Br_2$ and $Mn(ur)_6Br_2$ are isostructural with the corresponding urea complexes of manganese(II) iodide, $Mn(ur)_{10}Br_2$ (which should be formulated as $Mn(ur)_6Br_2 \cdot 4(ur)$ also with $Co(ur)_{10}X_2$ (X = Br, I), and $Mn(ur)_2Br_2$ with $Mn(ur)_2Cl_2$ and $Co(ur)_2Cl_2$ [2].

$Mn(ur)_nI_2$ and $Mn(ur)_nI_2 \cdot xH_2O$. The complexes $Mn(ur)_{10}I_2$, $Mn(ur)_6I_2$, and the hydrates $Mn(ur)_4I_2 \cdot 2H_2O$, and $Mn(ur)I_2 \cdot 3H_2O$ were detected as the stable solid complexes in the solubility isotherms of the ternary system MnI_2–urea–water studied at 15 and 30°C together with the species $MnI_2 \cdot 4H_2O$ and urea. The urea complexes were identified by their X-ray powder diagrams. They are congruently soluble in water except for $Mn(ur)_4I_2 \cdot 2H_2O$ which decomposes [4]. $Mn(ur)_{10}I_2$ and $Mn(ur)_6I_2$ were prepared from MnI_2 and urea in the same way as the corresponding bromides; see p. 141. They are colorless and melt at 121 and 158°C, respectively. Significant absorption bands (in cm^{-1}), in the IR spectra (KBr disks, Nujol mulls) are shown below together with the magnetic moments μ_{eff} at 293 K:

complex	μ_{eff} in μ_B	ν(CO) + δ(NH_2)	ν_{as}(C–N)	ν_s(C–N)	ν(Mn–O)
$Mn(ur)_{10}I_2$	5.89	1625 (s, b)	1480, 1450 (s, b)	1010 (w)	—
$Mn(ur)_6I_2$	5.97	1610 (s, b)	1470 (s)	1015 (w)	234 (s)
Ref	[2]	[2]	[2]	[2]	[3]

The physical data of the complexes can be interpreted like those of $Mn(ur)_{10}Br_2$ or $Mn(ur)_6Br_2$ (see p. 142 [2, 3]).

References:

[1] Moldobaeva, U. K., Rysmendeev, K. R. (Tr. Kirg. Univ. Ser. Khim. Nauk **3** [1975] 18/24 from Ref. Zh. Khim. **1976** No. 12 B 1004; C.A. **85** [1976] No. 113199).
[2] Barbier, J. P., Hugel, R. (Inorg. Chim. Acta **10** [1974] 93/6).
[3] Barbier, J. P., Hugel, R., van der Put, P. J., Reedijk, J. (Rec. Trav. Chim. **95** [1976] 213/6).
[4] Adamkulov, K., Sulaimankulov, K. S., Chermashentseva, M. V. (Khim. Kompleksn. Soedin. Redk. Soputstv. Elem. **1970** 164/70; C.A. **76** [1972] No. 121000).

24.4.1.2.6 Sulfates

$Mn(ur)_6SO_4$ and $Mn(ur)_6SO_4 \cdot xH_2O$. The anhydrous complex was detected as a stable solid phase in the solubility isotherms of the ternary system $MnSO_4$–urea–water at 10°C [1, 3], 25°C [2, 3], and 50°C [3] together with the species $Mn(ur)_4SO_4$ at 50°C [3], $Mn(ur)SO_4 \cdot 4H_2O$ at 10°C [1], or $Mn(ur)SO_4 \cdot 2H_2O$ at 20 and 25°C [1, 4], and the hydrates $MnSO_4 \cdot 5H_2O$ at 10°C, $MnSO_4 \cdot 4H_2O$ at 25°C and $MnSO_4 \cdot 3H_2O$ at 50°C, and urea [3]; also see [1, 2]. $Mn(ur)_6SO_4$ appears also in the quaternary system $MnSO_4$–K_2SO_4–urea·H_2O at 20°C together with $Mn(ur)SO_4 \cdot 2H_2O$, $MnSO_4 \cdot 5H_2O$, $K_2Mn(SO_4)_2 \cdot 4H_2O$, K_2SO_4 and urea [5].

The anhydrous complex was prepared by combining aqueous 1 M $MnSO_4$ and 6 M urea solutions and evaporating the mixture over concentrated H_2SO_4 at 5 to 10°C. After 2 to 3 d fine crystals had formed which were washed and dried over H_2SO_4 [6]. Preparation from saturated solutions of the components at 50°C or a somewhat higher temperature was proposed by [3].

X-ray diffraction studies of the complex formulated as $Mn(ur)_6SO_4 \cdot H_2O$(?) yield the orthorhombic space group $Pbc2_1$-C_{2v}^5 (No. 29) or Pbcm-D_{2h}^{11} (No. 57) with the lattice constants a = 7.25 ± 0.03, b = 14.63 ± 0.02, and c = 20.41 ± 0.03 Å. The complex is assumed to be isotypic with the corresponding ones of Zn, Co, and Ni [7]. However, chemical analysis and X-ray powder patterns (spacings d in Å, see the paper) showed the complex to be anhydrous and crystalline [6], and isostructural with $Mn(ur)_4SO_4 \cdot 2H_2O$ (see below [8]). Pycnometric measurements yield the density D = 1.7158 g/cm³ for the anhydrous complex [6]. A value of D = 1.911 g/cm³ is given in [2]; D = 1.62 g/cm³ is reported for the monohydrate [7]. The IR spectrum of $Mn(ur)_6SO_4$ in KBr reported in [13] reveals bonding of the urea molecules through the carbonyl oxygen.

The refractive indices of $Mn(ur)_6SO_4$ obtained by the immersion method are $n_\alpha = 1.546$, $n_\beta = 1.558$, and $n_\gamma = 1.558$ [2].

$Mn(ur)_6SO_4$ melts at 187.5°C [2]. According to [6] it decomposes without melting around 160°C. A melting point of 80°C was given for the dihydrate [9].

Thermogravimetric analysis shows that the decomposition of $Mn(ur)_6SO_4$ starts around 160°C with release of ammonia; sublimation was observed at higher temperatures [6], and further heating affords blackening [2]. The reaction with gaseous ammonia was studied in vacuum under a constant pressure of NH_3 (500 mm Hg) at −13 to 100°C. The reaction rate (highest at the start but sharply decreasing after about 2 min) and the number of ammonia molecules added to $Mn(ur)_6SO_4$ decrease at higher temperature (from 7 at 0°C, to 6 at 60°C, and 3 at 100°C). X-ray analysis of the reaction products revealed the formation of the species

$Mn(ur)_6SO_4 \cdot 6NH_3$ which successively converts into the species $Mn(ur)_2(NH_3)_4SO_4$ and $Mn(NH_3)_6SO_4$ with replacement of urea by NH_3 [6]. The complex was proposed as a microelement additive for urea fertilizers and cattle food [1].

$Mn(ur)_4SO_4$ appears as the second anhydrous urea complex in the ternary system $MnSO_4$–urea–H_2O at 50°C together with $Mn(ur)_6SO_4$ (see above [3]) and was prepared by slow isothermal evaporation of an aqueous solution containing $MnSO_4$ and urea in a 1:4 mole ratio. The resulting solid was recrystallized from water [10]. It was obtained also from equal volumes of aqueous 1 M $MnSO_4$ and 4 M urea solutions at 40°C. The crystals which appeared after 2 to 3 d were washed and dried over H_2SO_4 [6]. The pink complex [11] is crystalline as shown by its X-ray powder pattern (spacings d in Å, see the paper). The density D = 3.1300 g/cm³ at 25°C was measured pycnometrically. On heating, the complex decomposes without melting [6]; according to [9] it melts at 110°C [9]. The IR absorption spectrum (KBr disks or Vaseline mulls) shows the following characteristic absorption bands (in cm^{-1}): ν(NH) at 3450, 3350, $\delta(NH_2)+\nu(CO)$ at 1665, $\delta(NH_2)$ at 1640, $\nu(CO)+\delta(NH_2)$ at 1590, ν(C–N) at 1490 [10].

Upon heating, the complex decomposes at 150°C [8], at 160°C with release of NH_3. The reaction with gaseous ammonia was studied as for $Mn(ur)_6SO_4$. It proceeds in about the same way. Thus 7 mol of NH_3 were added at −13°C, 6 mol up to 40°C, 3 mol at 60°C, and 1 mol at 80°C. X-ray analysis of the reaction products revealed the formation of the species $Mn(ur)_4SO_4 \cdot 6NH_3$ and $Mn(ur)_4SO_4 \cdot 3NH_3$ which successively convert into the species $Mn(NH_3)_6SO_4$. The reactivity with ammonia decreases in the series $Mn(ur)_6SO_4 > Mn(ur)_4SO_4 > MnSO_4$ and is lower for $Mn(ur)_4SO_4$ than for $Mn(ur)_4(NO_3)_2$ [6].

$Mn(ur)_4SO_4 \cdot 2H_2O$ was detected in the solubility isotherm of the ternary system $MnSO_4$–urea–water at 20°C [1] and 25°C [1, 2, 4]. Its identity was confirmed by X-ray powder diagrams [2, 8], which suggested that it is isostructural with $Mn(ur)_6SO_4$ [8]. The density D = 1.704 g/cm³ was measured by the flotation method. The compound melts at 213.5°C and decomposes on further heating, turning black (formation of higher manganese oxides). The refractive indices obtained by the immersion method are $n_\alpha = 1.536$, $n_\beta = 1.564$, and $n_\gamma = 1.593$ [2]. The compound was proposed as a microelement additive for urea fertilizers [4].

$Mn(ur)_2SO_4$. The pink compound was synthesized from manganese sulfate and urea, mole ratio 1:2, in aqueous solution. The IR spectrum shows an absorption band at 1640 cm^{-1} assigned to ν(CO) which on complexation had shifted to lower frequencies by 43 cm^{-1} [11].

$Mn(ur)SO_4 \cdot nH_2O$ (n = 2, 4). The tetrahydrate appears as a solid phase in the ternary system $MnSO_4$–urea–H_2O at 10°C [1, 3]; it is replaced by the dihydrate at 20°C [1, 5] and 25°C [1 to 4]; also see [12]. The densities measured by flotation and refractive indices obtained by the immersion method are D = 2.240 g/cm³, $n_\alpha = 1.534$, $n_\beta = 1.516$, $n_\gamma = 1.549$ for $Mn(ur)SO_4 \cdot 4H_2O$ and D = 2.053 g/cm³, $n_\alpha = 1.619$, $n_\beta = 1.565$, $n_\gamma = 1.643$ for $Mn(ur)SO_4 \cdot 2H_2O$.

The hydrated complexes have individual structures as suggested by their X-ray powder diagrams and on heating they do not melt but decompose with loss of water and ammonia followed by blackening [2].

References:

[1] Druzhinin, I. G., Duishenalieva, N. D. (Izv. Akad. Nauk Kirg.SSR Ser. Estestven. Tekhn. Nauk **2** No. 3 [1960] 85/92, 87, 89, 91; C.A. **57** [1962] 5347).

[2] Druzhinin, I. G., Duishenalieva, N. D. (Izv. Akad. Nauk Kirg.SSR Ser. Estestven. Tekhn. Nauk **4** No. 9 [1962] 123/7, 125; C.A. **59** [1963] 1264).

[3] Duishenalieva, N., Abdykerimova, G. (Geterogen. Ravnovesiya Sist. Neorgan. Org. Soedin. **1974** 48/51; C.A. **83** [1975] No. 153400).

[4] Duishenalieva, N., Druzhinin, I. G. (Izv. Akad. Nauk Kirg.SSR Ser. Estestven. Tekhn. Nauk **2** No. 3 [1961] 73/82 from C.A. **56** [1962] 9701).
[5] Runov, N. N. (Uch. Zap. Yaroslav. Gos. Pedagog. Inst. No. 95 [1971] 72/5; C.A. **77** [1972] No. 169335).
[6] Lazerka, G. A., Girei, I. V., Zonau [Zonov], Yu. R. (Vestsi Akad. Navuk Belarusk. SSR Ser. Khim. Navuk **1968** No. 2, pp. 50/5, 53; C.A. **70** [1969] No. 41184).
[7] Gałdecki, Z., Golinski, B. (Zeszyty Nauk. Politech. Lodz. Chem. **10** No. 36 [1961] 15/20, 17; C.A. **60** [1964] 1007).
[8] Alieva, Z. F., Shalamov, A. E., Lityakova, E. N., Sulaimankulov, K. S., Agashkin, O. V. (Izv. Akad. Nauk Kirg.SSR Ser. Khim. **22** No. 2 [1976] 76/8; C.A. **77** [1972] No. 10759).
[9] Khodzhaev, O. F., Azizov, T. A., Ergeshbaev, D., Parpiev, I. A., Sulaimankulov, K. (Koord. Khim. **2** [1976] 304/7; Soviet J. Coord. Chem. **2** [1976] 223/6).
[10] Sulaimankulov, K. S., Alieva, Z. F., Agashkin, O. V., Kushnikov, Yu. A. (Khim. Kompleksn. Soedin. Redk. Soputstv. Elem. **1970** 157/63, 161; C.A. **77** [1972] No. 26970).

[11] Stancheva, P. (Nauchn. Tr. Vissh. Pedagog. Inst. Plovdiv Mat. Fiz. Khim. Biol. **8** [1970] 103/11, 106; C.A. **74** [1971] No. 60364).
[12] Karnaukhov, A. S., Runov, N. N. (Issled. Khlornokislykh Khromovokislykh Solei Elem. Pervoi Gruppy Period. Sistem. D. I. Mendeleeva **1966** 46/52 from Ref. Zh. Khim. **1967** I No. 8 B 648; C.A. **67** [1967] No. 94441).
[13] Runov, N. N. (Uch. Zap. Yaroslav. Gos. Pedagog. Inst. No. 79 [1970] 142/5; C.A. **77** [1972] No. 40955).

24.4.1.2.7 Tetrafluoroborates

$Mn(ur)_6(BF_4)_2$. The colorless complex was prepared from $Mn(BF_4)_2 \cdot 6H_2O$ and urea dissolved in a minimum amount of ethanol with addition of triethyl orthoformate for dehydration. The compound melts at 178°C with decomposition. The IR spectrum (not reported in the paper) shows a shift of the ν(CO) absorption band toward lower frequencies and bands due to uncoordinated BF_4^- ions.

Reference:

De Bolster, M. W. G., Driesen, W. L., Groenefeld, W. L., Van Kerkwijk, C. J. (Inorg. Chim. Acta **7** [1973] 439/44, 441).

24.4.1.2.8 Isothiocyanates

$Mn(ur)_8(NCS)_2$ was prepared by refluxing a solution of $Mn(NCS)_2$ and urea (mole ratio 1:8 and higher) in anhydrous ethanol under N_2 for 10 min with addition of triethyl orthoformate followed by stirring at room temperature for several hours and addition of diethyl ether. The precipitated colorless solid was washed with diethyl ether and dried in vacuum (p = 0.2 Torr). $Mn(ur)_8(NCS)_2$ melts at 120°C. Its IR spectrum shows an absorption band assigned to the ν(CN) vibrations of the thiocyanate ion at 2095 cm^{-1} which indicates coordination of these ions to manganese through their nitrogen atoms and supports the formulation $[Mn(ur)_4(NCS)_2] \cdot 4ur$ with octahedrally coordinated manganese and four additional urea molecules hydrogen-bonded to the thiocyanate ions. The IR bands are broad and the ν(CN) frequency is increased by 30 cm^{-1}. The electronic spectrum shows an absorption band at 24100 cm^{-1}

assigned to the $^6A_1(^6S) \rightarrow {}^4E + {}^4A_1(G)$ electron transition from which the nephelauxetic ratio $\beta = 0.90$ was calculated. Susceptibility measurements at 293 K yield the magnetic moment $\mu = 5.87\ \mu_B$ of a high-spin $Mn^{II}(d^5)$ complex. The ESR spectrum (not reported in the paper) reveals a very distorted coordination geometry of the complex [1].

$Mn(ur)_4(NCS)_2$ was prepared by refluxing a solution of $Mn(NCS)_2$ and urea (mole ratio 1:7) in anhydrous ethanol and triethyl orthoformate under N_2 for 4 h; a small volume of diethyl ether was then added to induce crystallization [1]. According to [2] the mixture of $Mn(NCS)_2$ and urea (mole ratio 1:4) in ethanol solution was placed in a vacuum desiccator over $CaCl_2$. The colorless crystals, which appeared after 8 to 12 d, were washed with alcohol and ether and dried in vacuum over $CaCl_2$ [2]. Weissenberg X-ray goniometer studies show the complex to be monoclinic, space group $P2_1/b\text{-}C_{2h}^5$ (No. 14), with the lattice constants a = 7.89(2), b = 10.40(2), c = 9.74(2) Å and $\gamma = 114.0(5)°$; Z = 2. Structural analyses with refinement up to R = 17.7% reveal a structure consisting of discrete $Mn(ur)_4(NCS)_2$ octahedra. The center is occupied by Mn, the equatorial plane by the carbonyl oxygens of urea and the apical sites by the nitrogens of the linear NCS groups which enter into the inner coordination sphere. The bond lengths within the molecule are $Mn\text{–}O_1 = 2.03$, $Mn\text{–}O_2 = 2.04$, and Mn–N = 2.05 Å. The density D = 1.72 g/cm³ was measured by pycnometry whereas 1.87 g/cm³ was calculated from X-ray data [3]. The complex melts at 141°C [1]. Its IR spectrum (Vaseline mulls) shows a broad and intense absorption band around 1640 cm^{-1} due to a downward shift of the urea ν(CO) band (at 1687 cm^{-1}). Other intense bands at 1538 cm^{-1} were ascribed to $\nu(CO) + \delta(NH_2)$ and in the 1040 to 1015 cm^{-1} region to antisymmetrical and symmetrical δ(NCN) vibrations of coordinated urea. The bands of the coordinated thiocyanate anion assigned to ν(CN) and δ(NCS) vibrations occur at 2065 and 470 cm^{-1}, respectively, and indicate its coordination to Mn through nitrogen [1]; see also [2]. The observed shifts of the urea and thiocyanate bands are consistent with the trans-octahedral complex structure shown by the X-ray investigation above. A supplementary band in the far-IR at about 190 cm^{-1} can be assigned to (Mn–NCS) vibrations. The electronic spectrum shows an absorption band at 24300 cm^{-1} assigned to the $^6A_1(^6S) \rightarrow {}^4E + {}^4A_1(^4G)$ electron transition. The nephelauxetic ratio $\beta = 0.905$ calculated therefrom agrees with the results of the ESR spectra concerning the ionic character of the Mn-urea bond. The magnetic moment $\mu = 5.88\ \mu_B$ at 288 K indicates a high-spin $Mn^{II}(d^5)$ complex [1]. The compound gradually oxidizes and blackens in air; it is easily soluble in water, ethanol, methanol, ether, acetone, and dimethyl formamide but only slightly soluble in CCl_4 and benzene [2].

References:

[1] Barbier, J. P., Hugel, R. P. (J. Inorg. Nucl. Chem. **39** [1977] 2283/4).

[2] Tsintsadze, G. V., Matiashvili, M. G., Ugulava, M. M. (Soobshch. Akad. Nauk Gruz.SSR **70** [1973] 613/6; C.A. **79** [1973] No. 73060).

[3] Tsintsadze, G. V., Tsivtsivadze, T. I., Orbeladze, F. V. (Zh. Strukt. Khim. **15** [1974] 306/7; J. Struct. Chem. [USSR] **5** [1974] 282/3).

24.4.1.2.9 Acetates

$Mn(ur)_2(CH_3COO)_2$. The compound was detected as a stable solid phase in the solubility isotherm of the ternary system $Mn(CH_3COO)_2$–urea–water at 30°C together with the species $4Mn(CH_3COO)_2 \cdot 3ur \cdot 2H_2O$, $Mn(CH_3COO)_2 \cdot 4H_2O$ and urea. The pale pink complex has the density $D_4^{20} = 1.645$ g/cm³ (measured pycnometrically) [1]. The IR spectrum reveals a shift of the urea absorption bands due to ν(CO) and ν(C–N) occurring at 1685 and 1008 cm^{-1}, respectively, to lower frequencies by 17 cm^{-1} for ν(CO), and to higher frequencies by about 15 cm^{-1} for ν(C–N). Their ν(C–N) band, however, is overlapped by a broad band due to the symmetrical ν(COO) vibrations of the acetate ion. The magnetic moment $\mu_{eff}=5.60\ \mu_B$ resulting from suceptibility measurements at 278 to 279 K indicates a high-spin $Mn^{II}(d^5)$ complex. Conductivity measurements show that the compound is a nonelectrolyte in dimethyl formamide and thus it is assumed that the acetate groups located in the inner complex sphere are coordinated to Mn as bidentate ligands through both oxygens. Thermal analysis (DTA, TG, DTG) shows that the complex melts at 130°C, decomposes at 200 to 220°C; γ-Mn_2O_3 is formed above 440°C in an exothermic reaction as shown by X-ray analysis [2]. $Mn(ur)_2(CH_3COO)_2$ is congruently soluble in water, less soluble in alcohol and acetone (but more than the initial components) and almost insoluble in $CHCl_3$, CCl_4, benzene, and toluene [1].

4 $Mn(CH_3COO)_2 \cdot 3ur \cdot 2H_2O$ appears as the second urea complex in the system $Mn(CH_3COO)_2$–urea–H_2O at 30°C together with $Mn(ur)_2(CH_3COO)_2$. The density $D_4^{20}=1.6632$ g/cm³ was measured pycnometrically. The compound is incongruently soluble in water; in organic solvents it behaves like $Mn(ur)_2(CH_3COO)_2$ [1].

References:

[1] Ergeshbaev, D., Murzubraimov, B., Sulaimankulov, K., Rysmendeev, K. (Zh. Neorgan. Khim. **18** [1973] 1406/9; Russ. J. Inorg. Chem. **18** [1973] 744/7).

[2] Khodzhaev, O. F., Azizov, T. A., Ergeshbaev, D., Parpiev, I. A., Sulaimankulov, K. (Koord. Khim. **2** [1976] 304/7; Soviet J. Coord. Chem. **2** [1976] 223/6).

24.4.2 With Hydroxyurea $H_2NC(O)NHOH$ (= $CH_4N_2O_2$)

Complexes in Solution. The stability constants of the 1:1 and 1:2 complex in aqueous solutions of $Mn(ClO_4)_2$ and N-hydroxyurea (= HL) were determined potentiometrically (glass electrode) at 20°C and ionic strength I = 0.1 M ($NaClO_4$): log K_1 = 2.87, log K_2 = 2.17. A comparison with other metal complexes of the ligand yields the stability series of Irving and Williams (Mn < Co < Ni < Cu > Zn) [1].

$Mn(CH_3N_2O_2)_2$. An ethanol solution of 0.02 mol $Mn(OC_2H_5)_2$ was reacted with a methanol solution of the ligand (0.04 mol) at −30°C with stirring. The resulting light pink (or colorless) solid was washed with methanol and dried in a high vacuum for 10 h. The IR spectrum of the complex in KBr was investigated in the 4000 to 400 cm⁻¹ region and as Nujol mulls in the 600 to 150 cm⁻¹ region [1]. Characteristic absorption bands (in cm⁻¹) with their assignments are given below:

complex	3430	3330	1625 (vs)	1585 (vs)	1410 (w)
ligand	3430	3325	1648, 1595	1498	1413
assignment	$\nu_{as}(NH_2)$	$\nu_s(NH_2)$	amide I	amide II	$\nu_{as}(NCN)$

complex	985 (w)	1078 (m)	710 (m)	360 (w)
ligand	978	1115	760	—
assignment	$\nu_s(NCN)$	ν(NO)	γ(NH)	ν(Mn–O)

The ligand is bidentate and coordinates to Mn through both its oxygens without participation of the nitrogens [1]. The spectral and structural changes occurring upon coordination of the deprotonated ligand to Mn were discussed using the results of normal vibration analysis. The valency force constants of the CO, N–O, and Mn–O bonds calculated in this way are $K=10.82\times10^6$, 5.88×10^6, and 1.84×10^6 cm⁻², respectively. The spectral and structural changes on coordination of the deprotonated ligand by Mn^{II}, Fe^{III}, Co^{II}, and Ni^{II} have been discussed. The observations and calculations carried out indicate that coordination through the oxygen of the carbonyl group generally leads to an appreciable decrease in the force constant of the CO bond and to only a slight decrease in the ν(CO) wave number [2]. The magnetic moment $\mu_{eff}=5.56\ \mu_B$ indicates a high-spin $Mn^{II}(d^5)$ complex [1].

References:

[1] Berger, R., Fritz, H. P. (Z. Naturforsch. **27b** [1972] 608/16, 611, 613).
[2] Kharitonov, Yu. Ya., Sarukhanov, M. A., Slivko, S. A. (Zh. Neorgan. Khim. **29** [1984] 1505/8; Russ. J. Inorg. Chem. **29** [1984] 864/6).

24.4.3 With Methylurea $H_2NC(O)NHCH_3$ (= $C_2H_6N_2O$)

$Mn(C_2H_6N_2O)_6X_2$ (X = NO_3, Cl, I). The complexes were prepared by crystallization from an aqueous solution containing $Mn(NO_3)_2$, MnI_2 [1], or $MnCl_2$ [2] and N-methyl urea in a 1:10 mole ratio at room temperature [1, 2]. The almost colorless chloride forms large tabular crystals [2]. X-ray studies (Weissenberg and crystal rotation) show the complexes to be triclinic and isostructural with each other and with the corresponding complexes of Co and Ni [1, 2] which are of C_1 or C_2 symmetry [2]. The lattice constants are a = 7.27, b = 10.73, c = 11.73 Å and α = 110.4°, β = 101.3°, γ = 106.6° for the complex with X = NO_3 and a = 7.36, b = 10.82, c = 11.18 Å and α = 113.7°, β = 99.2°, γ = 107.4° for the complex with X = I [1]. Crystals of the chloride were unsuitable for measurements [2]. The densities D_4^{20} (in g/cm^3) were obtained by the flotation method ($CHBr_3$ + benzene): 1.38 for the nitrate [1], 1.40 for the chloride [2], 1.64 for the iodide [1]. Values calculated from X-ray data are 1.40 for the nitrate and 1.70 for the iodide [1].

$Mn(C_2H_6N_2O)_6SO_4$ and **$Mn(C_2H_6N_2O)_6S_2O_3$** complexes were obtained by slow evaporation of an aqueous solution containing $MnSO_4$ [3] or MnS_2O_3 [4] and N-methylurea in a 1:10 mole ratio [3, 4]. The sulfate forms rather large pseudocubic crystals which according to goniometer studies are hexagonal with the axial ratio c/a = 3.658 and the observed faces {0001}, {000$\bar{1}$}, {10$\bar{1}$2}, {10$\bar{1}\bar{4}$}, {01$\bar{1}$4}, and {01$\bar{1}\bar{2}$}. X-ray studies (Weissenberg and crystal rotation around c) reveal a rhombohedral lattice, space group R3c-C_{3v}^6 (No. 161) with lattice constants a = 10.99 ± 0.02 and c = 40.56 ± 0.08 Å (c/a = 3.691); Z = 6. The sulfate is isostructural with the corresponding complexes of Co, Ni, Cu, Zn, and Cd. The density D_4^{16} = 1.38 g/cm^3 was measured (by flotation) and 1.41 g/cm^3 was calculated from X-ray data. The crystals are optically uniaxial negative [3].

The thiosulfate crystals are of rhombohedral habitus and present the distinctly rhombohedral facies {0114} from which the axial ratio c/a = 3.542 was derived. X-ray studies confirm the rhombohedral symmetry with the lattice constants a = 11.24 (± 0.02) and c = 39.97 (± 0.08) Å (c/a = 3.556). The complex is isostructural with the corresponding ones of Co, Ni, Zn, and Cd. The density D = 1.39 g/cm^3 was calculated from X-ray data [4].

References:

[1] Nardelli, M., Coghi, L. (Ric. Sci. **29** [1959] 134/8; C.A. **1960** 1151).
[2] Cavalca, L., Nardelli, M., Coghi, L. (Gazz. Chim. Ital. **87** [1957] 903/6).
[3] Nardelli, M., Coghi, L. (Gazz. Chim. Ital. **88** [1958] 355/8).
[4] Nardelli, M., Coghi, L. (Ric. Sci. **28** [1958] 609/10; C.A. **1958** 15320).

24.4.4 With Butyl- and *tert*-Butylurea $H_2NC(O)NHR$ (= $C_5H_{12}N_2O$)

ligand 1 R = C_4H_9 ligand 2 R = $C(CH_3)_3$

$Mn(C_5H_{12}N_2O)_6X_2$. Complexes with ligand 1 and X = Cl, Br were prepared by heating the solution of MnX_2 and n-butylurea in hot methanol for 30 to 60 min. The mixture then was cooled and stirred to induce crystallization with addition of benzene or diethyl ether if required. The colorless crystals were recrystallized from methanol, thoroughly washed with solvents dissolving the unreacted starting material, and dried in vacuum. The chloride and bromide melt at 108 and 135°C, respectively. The IR spectrum of both complexes shows the ligand absorption band assigned to ν(CO) at 1650 cm^{-1} shifted to lower frequencies by 10 to 25 cm^{-1} whereas the ν(NH) band at 3400 cm^{-1} is unaffected. Susceptibility measurements at room temperature yield the magnetic moments μ_{eff} = 5.95 μ_B for the chloride and 5.97 μ_B for the bromide indicating octahedral spin-only $Mn^{II}(d^5)$ complexes which obey the Curie-Weiss law with the Curie constants Θ = −22 and −19 K, respectively. The X-ray powder pattern of $Mn(C_5H_{12}N_2O)_6Br_2$ (for spacings d see the paper) suggests isomorphism with the corresponding (octahedral) cobalt complex [1].

Complexes with ligand 2 and X = NO_3, ClO_4 were prepared by reaction of *tert*-butylurea in nitromethane with $Mn(NO_3)_2 \cdot 6H_2O$ in methanol or by reaction of the ligand with $Mn(ClO_4)_2 \cdot 6H_2O$ in 2-propanol solution. The magnetic moment μ_{eff} = 5.94 μ_B of the nitrate was calculated from susceptibility measurements at 295 K and μ_{eff} = 6.03 μ_B for the perchlorate at 293 K. The ν(CO) absorption band of the ligand at 1662 cm^{-1} is shifted on complexation to lower frequencies by 15 to 20 cm^{-1} [2].

References:

[1] Askalani, P., Bailey, R. A. (Can. J. Chem. **47** [1969] 2275/82, 2276/8).
[2] Kircheiss, A., Pretsch, U., Borth, U. (Z. Chem. [Leipzig] **20** [1980] 349/50).

24.4.5 With Phenylurea $H_2NC(O)NHC_6H_5$ (= $C_7H_8N_2O$)

$Mn(C_7H_8N_2O)_6X_2$ (X = NO_3, ClO_4). The colorless complexes were prepared by reacting phenylurea with the manganese(II) salts (previously dried in vacuum over P_4O_{10} for 48 h) in hot nitromethane solution. The only slight shift of the ν(CO) absorption band of the ligand in the IR spectrum (at 1665 cm^{-1}) to lower frequencies in the complexes (nitrate: 1 cm^{-1}, perchlorate: 5 cm^{-1}) is evidently due to the formation of hydrogen bridges in the solid ligand since solutions of the ligand in $CHCl_3$ show the ν(CO) band at 1697 cm^{-1}. Susceptibility measurements (Gouy method) at 295 yield μ_{eff} = 5.94 μ_B for X = NO_3 and 6.32 μ_B for X = ClO_4, Kircheiss, A., Gleichmann, I. (Z. Chem. [Leipzig] **16** [1976] 26/7).

24.4.6 With N,N-Dimethylurea $H_2NC(O)N(CH_3)_2$ (= $C_3H_8N_2O$)

$Mn(C_3H_8N_2O)_6(ClO_4)_2$ appears as a stable solid phase in the solubility isotherm of the ternary system $Mn(ClO_4)_2$–N,N-dimethylurea–water at 25°C together with $Mn(ClO_4)_2 \cdot 6H_2O$ and solid N,N-dimethylurea. The complex forms pale pink crystals which are isomorphous with those of $Ni(C_3H_8N_2O)_6(ClO_4)_2$ [1].

$Mn(C_3H_8N_2O)_nI_2$ and **$Mn(C_3H_8N_2O)I_2 \cdot 2H_2O$.** Anhydrous complexes with n = 6, 4, 3 and the dihydrate were detected as the solid complex species in the solubility isotherm of the ternary system MnI_2–$C_3H_8N_2O$–H_2O at 20°C together with $MnI_2 \cdot 4H_2O$ and solid N,N-dimethylurea. All complex species are incongruently soluble in water [2].

References:

[1] Volgina, N. V., Klyushina, L. I., Kocheshkova, N. G., Goryunov, Yu. A. (Sb. Nauchn. Tr. Yaroslav. Gos. Pedagog. Inst. No. 178 [1978] 43/5; Ref. Zh. Khim. **1979** No. 23 B 827; C. A. **92** [1980] No. 100235).

[2] Manaeva, R., Kochkorova, Z., Rysmendeev, K., Baichalova, S. (Kompleksn. Pererab. Prirodn. Nedefitsitnogo Syr'ya Kirg. Probl. Ekol. **1978** 25/8; C.A. **91** [1979] No. 146546).

24.4.7 With N, N'-Dimethyl- and N, N'-Diethylurea RNHC(O)NHR

$R=CH_3$ ($=C_3H_8N_2O$) $R=C_2H_5$ ($=C_5H_{12}N_2O$)

$Mn(C_3H_8N_2O)_6X_2$ ($X=ClO_4$, I). The perchlorate was prepared by refluxing $Mn(ClO_4)_2 \cdot 6H_2O$ (0.01 mol) with 2,2-dimethoxypropane (12 mL) for 20 min, and adding solid N, N'-dimethylurea (0.06 mol) with refluxing for another 10 min. Crystallization was induced by addition of diethyl ether and the crystals were recrystallized from anhydrous ethanol, washed with diethyl ether and dried in vacuum (0.2 Torr) at room temperature [1]. For preparation from the components in ethanol or 2-propanol, see [3]. The iodide was synthesized by adding a slight excess of N, N'-dimethylurea (0.064 mol) to a solution of MnI_2 (0.01 mol) in anhydrous ethanol obtained by treating $MnI_2 \cdot 4H_2O$ with 2,2-dimethoxypropane. After refluxing for 30 min the solid complex was precipitated, recrystallized from anhydrous ethanol, washed and dried in vacuum like the perchlorate [4]. Both complexes are colorless. The perchlorate melts at 151 to 153°C [1], 159°C [4], the iodide at 143°C [4].

The lowering of the ν(CO) and δ(NCO) frequencies and the raising of the ν(CN), δ(NH), and δ(NCN) frequencies in the IR spectrum of the complexes [4] as well as a band assignable to ν(Mn–O) in the far-IR [5] are indicative of coordination of the ligand to the hexacoordinated manganese ion through the oxygen of the carbonyl group [1, 4]. Most essential absorption bands (in cm^{-1}) of the complexes in KBr disks and Nujol mulls are shown below together with their assignments [4, 5]:

compound	ν(CO)	δ(NH)	ν(CN)	δ(NCN)	δ(NCO)	ν(Mn–O)
$Mn(C_3H_8N_2O)_6(ClO_4)_2$	1635 (vs)	1575 (vs)	1350 (s)	($\nu_3(ClO_4)$)	565 (s)	224 (s)
$Mn(C_3H_8N_2O)_6I_2$	1635 (vs)	1565 (vs)	1345 (s)	1175, 1145 (s)	555 (s)	204 (vs)

Other bands in the far-IR spectrum (400 to 50 cm^{-1} region) assignable to hydrogen bond or ligand vibrations occur around 370, 260, and 135 cm^{-1}; for unassigned bands in the region <100 cm^{-1} see the paper [5].

The absence of discrete bands in the electronic spectrum is consistent with a perfect octahedral O_h symmetry of the MnO_6 units (chromophores) in the complexes or due to the very low content of manganese [4]. An octahedral structure is suggested also by the high-spin magnetic moments $\mu_{eff}=6.10\ \mu_B$ at 295 K [3], 5.89 μ_B at 296 K [4], 5.79 μ_B at 298 K [1] resulting from the paramagnetic susceptibility of the perchlorate and $\mu_{eff}=5.9\ \mu_B$ at 298 K [4] reported for the iodide. A slight distortion of the MnO_6 octahedra is concluded from the ESR spectrum of Mn^{2+} ions doped into $Cd(C_3H_8N_2O)_6(ClO_4)_2$ which gives the parameters g = 2.004(5), A = 94.0(3) Gauss and the zero field splitting D = 0.0019(1) cm^{-1} [5]. Conductivity measurements on 0.001 M solutions of $Mn(C_3H_8N_2O)_6(ClO_4)_2$ in nitromethane at 25°C yield the molar electrical conductance $\Lambda=163\ cm^2 \cdot \Omega^{-1} \cdot mol^{-1}$ indicating a 1:2 electrolyte [1].

$Mn(C_3H_8N_2O)_3X_2$ (X = Cl, Br, I). The complexes were prepared, as described in [2] for the complexes with N, N'-diethylurea, by refluxing solutions of $MnX_2 \cdot 4H_2O$ salts (0.01 mol) in anhydrous ethanol with 2,2-dimethoxypropane (7 mL) for 20 min, adding N,N'-dimethylurea (0.03 mol) and refluxing for another 2 h. After cooling, diethyl ether was added to precipitate the complex which was recrystallized from anhydrous ethanol, washed with diethyl ether and dried in vacuum (0.2 mm Hg) at normal temperature. Single crystals of $Mn(C_3H_8N_2O)_3Br_2$ were obtained from a concentrated solution of the powdered complex in dry butanol and ethanol (50/50 v/v) which was very slowly evaporated at room temperature in a desiccator over silica gel; thin transparent plates appear after 1 month. The plates are very fragile and moisture-sensitive, and many of them were twinned [8]. The X-ray structure analysis (MoKα radiation) at 20°C shows the crystals to be monoclinic, space group C2/c-C_{2h}^6 (No. 15), with the lattice constants a = 13.211(3), b = 8.670(3), c = 16.593(4) Å and β = 106.23(3)°; Z = 4. The molecular structure was deduced from 856 independent reflections with refinement (except for H) up to R = 0.054 [6]. The final full matrix least-squares refinement, including 12 hydrogen atoms, reveals a weighted value of R_w = 0.0513 and R_F = 1.82. Some interatomic bond lengths and angles are tabulated below. For fractional atomic coordinates and isotropic temperature factors, see the paper [8]:

distance	Å	angle	(°)	angle	(°)
Mn–Br	2.563(2)	Br–Mn–Br	114.0(4)	O(1)–Mn–O'(1)	174.2(6)
Mn–O(1)	2.175(8)	Br–Mn–O(3)	123.0(2)	Br–Mn–O(1)	91.6(2)
Mn–O(3)	2.035(12)	O(1)–Mn–O(3)	87.1(3)	Mn–O(1)–C(1)	129.5(8)
O(3)–C(3)	1.249(21)	Mn–O(3)–C(3)	180		

The structure is shown in **Fig. 17**. The geometry of the coordination polyhedra is very close to trigonal bipyramidal. The molecules lie on a C_2 axis running through Mn, O(3) and C(3). The Br atoms occupy equatorial positions and the N,N'-dimethylurea ligands are oxygen-bonded in two types of coordinations. The axial ones are normally coordinated: the oxygen atom shares one lone (electron) pair with the metal ion forming an angle Mn–O(1)–C(1) of 129.5°. The equatorial ligand is, uncharacteristically, linearly bonded with the angle Mn–O(3)–C(3) = 180°. Such unusual bonding of carbonyl oxygen probably implies a delocalization of the electrons as the bond length Mn–O(3) is significantly shorter than the Mn–O(1) distance. Packing considerations or intermolecular hydrogen bonds apparently are not responsible for such difference in coordination. The density D = 1.739 g/cm³ was calculated from the X-ray data, D = 1.74(1) g/cm³ was measured by the flotation method [6, 8].

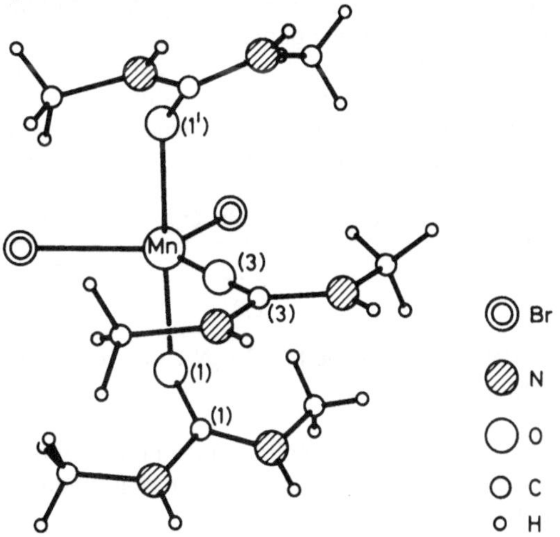

Fig. 17. Molecular structure of the complex with N,N'-dimethylurea, $Mn(C_3H_8N_2O)_3Br_2$ [6, 8].

The IR spectra of the complexes with X = Cl, Br in KBr and polyethylene pellets and the Raman spectra (using 250 to 800 mW power from the 488.0 nm line) were interpreted on the basis of the structural data reported for the bromide. Principal characteristic amide bands of the ligand and the complexes (in cm^{-1}) are shown below. Spectral bands of ligand solutions in $CHCl_3$ are given in parentheses:

assignment	ligand		$Mn(C_3H_8N_2O)_3Cl_2$		$Mn(C_3H_8N_2O)_3Br_2$	
	R	IR	R	IR	R	IR
ν(CO) + δ(NH) amide I band	—	1620 vs, br (1670 vs)	1630 vw, 1660 sh	1628 vs, br	1639 vw, 1660 sh	1629 vw, br
ν_{as}(CN) + δ(NH) amide II band	—	1528 sh	1525 vw, 1557 sh	1510 m	1522 vw, 1562 sh	1508 m
	1584 vw	1574 m (1560 vs)	1585 vw	1582 vw	1584 vw	1582 vw
$\delta(CO)_{out\text{-}of\text{-}plane}$	—	770 vw	778 sh, 798 vw	754 m, 774 m	787 vw	752 w, 774 w

The shifts of the amide I and amide II bands on complexation clearly indicate coordination of N,N′-dimethylurea to Mn through carbonyl oxygen. These shifts are consistent with the presence of resonant forms $^{+}N{=}C{-}O^{-}$, suggesting conformations of methylamino groups more favorable for π-electron delocalization. Appearance of weak multiple bands, found in the Raman spectra of the complexes and missing in the ligand spectrum, accounts for both effects, on ν_{as}(CN) and δ(NH), of dissymmetrical angular and of linear C–O → Mn bonds. These effects were shown up by the splitting of equal intensity (IR) of the band due to the out-of-plane (OCNH) bending of the CO group. One of the components is slightly shifted (+4 cm^{-1}) relative to that of the free ligand, whereas the other is shifted by about −17 cm^{-1}. The former, much less affected by coordination, is probably due to the equatorial ligand molecule, for which the linear bonding involves some rigidity, and the latter would be due to the axial N,N′-dimethylurea ligand. The spectra in the low-frequency range (700 to 0 cm^{-1}) have been assigned in terms of skeletal (metal–ligand) vibrations. Bands at 266 to 248 cm^{-1} (X = Cl) and 220 to 210 cm^{-1} (X = Br) were assigned to Mn–X stretching vibrations. In the Raman spectra the band at 613, 618 cm^{-1} (X = Cl, Br) refers to the stretching mode of the particular manganese to equatorial oxygen bond, $\nu_s(Mn\text{–}O_{eq})$. Bands of the antisymmetrical and symmetrical stretching vibrations of the manganese to axial oxygen bonds were found around 570 and 520 cm^{-1}. These values are close to those found for an octahedral MnO_6 framework and an identical angular bonding mode of N,N′-diethylurea in $[Mn(C_5H_{12}N_2O)_6][MnBr_4]$; see p. 153 [8]. Internal valence force constants f (in mdyn/Å) for the framework vibrations of $Mn(C_3H_8N_2O)_3X_2$ complexes have been deduced from the symmetry force constant calculations [8]:

X	$f(Mn\text{–}O_{eq})$	$f(Mn\text{–}O_{ax})$	f(Mn–X)
Cl	2.68	2.22	1.00
Br	2.72	2.22	1.11

IR data for the iodide $Mn(C_3H_8N_2O)_3I_2$ are reported in [4, 5] together with those of the chloride and bromide.

The reflectance spectra of the chloride and bromide have been interpreted on the basis of the structural data, see above. For assignments of the electronic spectral bands (in cm^{-1}), see below [8]:

assignment	$Mn(C_3H_8N_2O)_3Cl_2$	$Mn(C_3H_8N_2O)_3Br_2$
$^6A_1' \rightarrow$ $^4E'(^4G)$	—	19510
$\rightarrow$ $^4A_1''\equiv{}^4A_2''(^4G)$	23750	23530
$^4A_1'\equiv{}^4E'(^4G)$	24110sh	
$\rightarrow$ $^4A_2'(^4P)$	—	26110
$\rightarrow$ $^4E'(^4D)$	27990	27590
$\rightarrow$ $^4E''(^4P)$	—	28020
$\rightarrow$ $^4A_1'(^4D)$	—	29450
$\rightarrow$ $^4E''(^4D)$	—	30675
	37520*)	36760*)

*) This band may be a charge-transfer band.

Susceptibility measurements at temperatures in the 2 to 100 K range show that the complexes obey the Curie-Weiss law with the constant $\Theta = 0(1)$ K and their magnetic moments close to the spin-only value (5.90 μ_B) indicate mononuclear species with no significant magnetic exchange [5]. For the complexes with X = Cl, Br, I the following ESR data have been reported: g = 2.00(1), D = 0.24(2), 0.44(3), 0.68(4) cm^{-1}, E/D = 0.25(3), 0.27(3), 0.25(4), respectively [5].

The pale pink chloride melts at 192°C, the pink bromide at 202°C, the pink iodide at 164°C [4]. $Mn(C_3H_8N_2O)_3Br_2$ is very sensitive to moisture [8].

$Mn(C_5H_{12}N_2O)_6X_2$ (X = ClO_4, Br, I). The complexes were prepared from manganese perchlorate, bromide, or iodide, and N,N'-diethylurea by the methods recommended for the synthesis of $Mn(C_3H_8N_2O)_6X_2$ complexes, see p. 150. A slight excess of diethylurea was used for the preparation of the bromide and iodide. The mixtures were refluxed for 30 min. On prolonged refluxing only $Mn(C_5H_{12}N_2O)_3Br_2$ was obtained [2]. $Mn(C_5H_{12}N_2O)_6(ClO_4)_2$ was also prepared by reaction of the manganese(II) salt with the ligand in ethanol or 2-propanol [3]. The colorless complexes melt at 142°C (perchlorate), 111°C (bromide), and 139°C (iodide). The IR spectra of the complexes in KBr and Nujol reveal bands of ν(CO) at 1630 to 1635 cm^{-1}, of δ(NH) at 1565 to 1570 cm^{-1}, of ν(CN) at 1335, 1290 to 1295, and 1265 to 1275 cm^{-1}, of δ(NCN) at 1155 and 1140 cm^{-1}, of δ(NCO) at 540 to 550 cm^{-1}, and of ν(MnO) at 227 (X = ClO_4), 215 (X = Br), and 191 (X = I) cm^{-1}. The IR spectra and the magnetic data (magnetic moments of μ_{eff} = 5.92 to 6.10 μ_B have been calculated in the range 295 to 298 K) can be interpreted like those of $Mn(C_3H_8N_2O)_6X_2$ which suggests an octahedral complex structure with oxygen-bonded dialkylurea ligands [2, 5]. The ESR spectrum of Mn^{2+} ions doped into $Cd(C_5H_{12}N_2O)_6(ClO_4)_2$ (parameters g = 2.004(4), A = 94.1(3) Gauss and D = 0.0014 cm^{-1}), suggests a very slight distortion of the MnO_6 octahedra which is even less than in the N,N'-dimethylurea complex [5].

$[Mn(C_5H_{12}N_2O)_6][MnX_4]$ (X = Cl, Br). The yellowish complexes were prepared from the manganese halogenide and N,N'-diethylurea by the method recommended for the synthesis of $Mn(C_3H_8N_2O)_3X_2$ complexes, see p. 150. The tetrachloro complex melts at 147°C, the tetrabromo complex at 121°C [2]. The IR spectra of the complexes are very similar and resemble those of $Mn(C_5H_{12}N_2O)_6X_2$ which contain the oxygen-bonded ligands in an octahedral arrangement around the manganese ion [5]. IR absorption bands of ν(MnO) and ν(MnX) in cm^{-1} are shown below together with wave numbers of bands observed in the reflection spectra [2].

complex	ν(MnO)	ν(MnX)	electron transitions $^6A_{1g}$ $\rightarrow ^4T_2(^4G)$	$\rightarrow ^4E + ^4A_1(^4G)$	$\rightarrow ^4T_2 + ^4E(^4D)$	$\rightarrow ^4T_1(^4P)$
$[MnL_6][MnCl_4]$	196 (s, v, br)	283 (vs), 124 (vs, br)	22250	23150	—	—
$[MnL_6][MnBr_4]$	190 (sh, br)	223 (vs), 88 (m, br)	22000	22800	26550	27400
Ref.	[5]	[5]	[2]	[2]	[2]	[2]

The two ν(MnX) bands observed in the far-IR [2] and the electronic spectra are similar to those of other compounds containing tetrahedral $[MnX_4]^{2-}$ anions [2]. Since the far-IR spectra of the complexes appear to be simple additions of vibration modes due to $[Mn(C_5H_{12}N_2O)_6]^{2+}$ and those due to $[MnX_4]^{2-}$, the compounds can be represented by the formula $[Mn(C_5H_{12}N_2O)_6]$-$[MnX_4]$, see [2, 5].

$Mn(C_3H_8N_2O)_4(NCS)_2$ and **$Mn(C_5H_{12}N_2O)_4(NCS)_2$** complexes were prepared from $Mn(NCS)_2$ and N,N'-dimethyl- or N,N'-diethylurea in anhydrous ethanol in the same way as for $Mn(ur)_4(NCS)_2$; see p. 146. $Mn(C_3H_8N_2O)_4(NCS)_2$ melts at 67°C, $Mn(C_5H_{12}N_2O)_4(NCS)_2$ melts at 71°C. A magnetic moment of $\mu_{eff} = 5.86\ \mu_B$ was observed for both complexes. The band shifts in the IR spectra of the complexes suggest coordination of the N,N'-dialkylurea ligands to Mn through the carbonyl oxygen and of the thiocyanate ions through nitrogen. The sharp single ν(CN) band at 2070 and 2075 cm^{-1}, respectively, agrees with a trans arrangement of the two NCS^- ions, which is confirmed by the ESR spectrum of $Mn(C_5H_{12}N_2O)_4(NCS)_2$ indicating an axial complex symmetry. Bands observed in the electronic spectrum at 24200 cm^{-1} (dimethylurea complex) and 24100 cm^{-1} (diethylurea complex) were assigned to the $^6A_1(^6S) \rightarrow ^4E + ^4A_1(^4G)$ transition and the nephelauxetic ratio $\beta = 0.90$ was calculated [7].

References:

[1] Stonestreet, B. C., Bull, W. E, Williams, R. J. (J. Inorg. Nucl. Chem. **28** [1966] 1895/900).
[2] Barbier, J.-P., Hugel, R. P. (J. Inorg. Nucl. Chem. **35** [1973] 781/8).
[3] Kircheiss, A., Siodlaczek, J. (Z. Chem. [Leipzig] **19** [1979] 227).
[4] Barbier, J.-P., Hugel, R. P. (J. Inorg. Nucl. Chem. **35** [1973] 3026/30).
[5] Barbier, J.-P., Hugel, R. P., van der Put, P. J., Reedijk, J. (Rec. Trav. Chim. **95** [1976] 213/6).
[6] Delaunay, J., Kappenstein, C., Hugel, R. P. (J. Chem. Soc. Chem. Commun. **1980** 679/80).
[7] Barbier, J.-P., Hugel, R. P. (J. Inorg. Nucl. Chem. **39** [1977] 2283/4).
[8] Delaunay, J., Hugel, R. P. (Inorg. Chem. **25** [1986] 3957/61).

24.4.8 With N-Acetylurea $CH_3C(O)NHC(O)NH_2$ (= $C_3H_6N_2O_2$)

$[Mn(C_3H_6N_2O_2)_2Cl_2]_n$. Crystals of the complex (obtained from $MnCl_2$ and acetylurea in aqueous solutions at pH 3 and ~70°C) are monoclinic; space group C_c-C_s^4 (No. 9) with a = 14.542(6), b = 9.155(3), c = 7.334(1) Å, β = 120.25°, Z = 4, R = 0.055 for 993 reflections. The Cl atoms are bridging and the coordination polyhedron of Mn^{II} is tetragonal bipyramidal with the Cl atoms in the equatorial plane, Kharitonov, Yu. Ya., Gushchina, T. N., Chuklanova, E. B., Gusev, A. I. (Koord. Khim. **12** [1986] 1145/6 from C.A. **105** [1986] No. 125981).

24.4.9 With a Copolymer of Salicylic Acid, Urea, and Formaldehyde

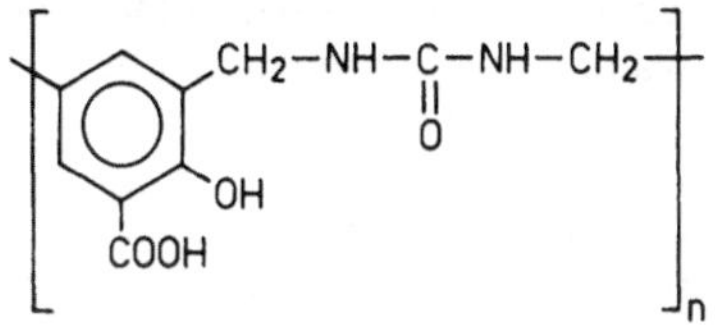

$(= (C_{10}H_{10}N_2O_4)_n)$

$[Mn(C_{10}H_9N_2O_4)_2(H_2O)_2]_n$. The gray copolymer complex was prepared by adding a solution of manganese(II) chloride in a mixture of dimethylformamide and water dropwise with stirring to that of the ligand in dimethylformamide (mole ratio 1:2). Then a saturated aqueous solution of CH_3COONa was added to precipitate the complex at pH 6. The product was digested on a water bath for some time, filtered off, washed successively with dimethylformamide, hot water and acetone, dried in air and finally in an oven at 60°C for 2 h. The IR spectrum of the complex shows a very broad absorption band in the 3500 to 2400 cm^{-1} region, ascribed to the presence of water and a medium band at 1642 cm^{-1} assigned to ν(CO) of the deprotonated and coordinated carboxylate group which in the parent ligand appears at 1670 cm^{-1}. The electronic spectrum shows two bands at 17090 and 20000 cm^{-1}, assigned to the electron transitions $^6A_{1g} \rightarrow {}^4T_{1g}$ and $^6A_{1g} \rightarrow {}^4T_{2g}$, respectively. Susceptibility measurements yield the magnetic moment $\mu_{eff} = 5.36\ \mu_B$ of a high-spin Mn^{II} complex. Basing on the spectral and magnetic data chelate structures with the metal coordinated to the oxygens of the carboxyl and phenol groups of the ligand and to those of two water molecules were proposed. The complex decomposes upon heating at 200°C (ligand at 210°C) and is insoluble in common organic solvents, Joshi, R. M., Patel, M. M. (Indian J. Chem. A **21** [1982] 637/9).

24.4.10 With Biuret $H_2NC(O)NHC(O)NH_2$ $(= C_2H_5N_3O_2)$

Complexes in Solution: The formation of a 1:1 complex with the stability constant log K_1 = 9.10 at 25°C [1, 2] and log K_1 = 9.02 at 35°C was determined potentiometrically (glass electrode) at ionic strength I = 0.1 ($NaClO_4$). The thermodynamic parameters were calculated from the temperature slope of the stability constant: ΔG = −12.40 kcal/mol at 25°C and −12.50 kcal/mol at 35°C, ΔH = −33.03 kcal/mol, ΔS = −67.75 cal·mol^{-1}·K^{-1} [1]. On comparison with the biuret complexes of other metals the stability series Cu > Hg > Ni > Co > Zn > Cd > Mn is derived [1, 2].

$Mn(C_2H_5N_3O_2)_2(NO_3)_2 \cdot 2H_2O$ crystallizes upon slow evaporation of an acidified aqueous solution of $Mn(NO_3)_2$ and biuret (mole ratio ~1:2) in vacuum over anhydrous $CaCl_2$ at room temperature. The pale pink crystals were washed with cold acidified water (pH ~3), then with acetone and ether and dried over $CaCl_2$. The IR spectrum recorded in the 4000 to 400 cm^{-1} region shows unaltered ν(NH) bands. Therefore no bonding to the amino groups of the ligand molecule is assumed. The bands of ν(NH) occurring at 3420, 3330, 3295, and 3210 cm^{-1} are almost unchanged, those of ν(CO) + δ(NH_2) at 1700, 1640, 1580, and 1500 cm^{-1} are shifted to lower wave numbers in comparison to the free ligand. The coordination of biuret seems to be bidentate via the oxygen atoms forming a six-membered ring with the manganese atom. A band of ν(C–N) was observed at 1475 cm^{-1}. The splitting of the band due to the nitrato group (ν(NO_3) at 1040 cm^{-1}, δ(NO_3) at 830 cm^{-1}) indicates that these groups are in the inner sphere of the complex [3].

As shown by thermal analysis $Mn(C_2H_5N_3O_2)_2(NO_3)_2 \cdot 2H_2O$ dehydrates in the range 90 to 130°C. $Mn(C_2H_5N_3O_2)_2(NO_3)_2$ loses one ligand molecule between 170 and 240°C. The species $Mn(C_2H_5N_3O_2)(NO_3)_2$ decomposes between 300 and 500°C, forming Mn_2O_3. The complex $Mn(C_2H_5N_3O_2)_2(NO_3)_2 \cdot 2H_2O$ is readily soluble in water, less so in methanol, but almost insoluble in other usual organic solvents. The electrical conductivity measurements in aqueous and methanolic solutions indicate that the nitrato groups are gradually displaced to the outer sphere by the solvent molecules as the conductivity increases with time. The molar conductance $\Lambda = 231\,cm^2 \cdot \Omega^{-1} \cdot mol^{-1}$ measured in water at 20°C and a dilution rate V = 1000 L/mol increases to 242 $cm^2 \cdot \Omega^{-1} \cdot mol^{-1}$ within 1 h after preparation [3].

$Mn(C_2H_5N_3O_2)_2X_2$ (X = Cl, Br). The chloride was detected as a stable but incongruently soluble solid phase in the solubility isotherm of the ternary system $MnCl_2$–biuret–water at 30°C together with the solid species $MnCl_2 \cdot 4H_2O$ and biuret [4]. It was prepared by combining hot ethanolic solutions of $MnCl_2 \cdot 4H_2O$ (0.01 mol) and biuret (0.02 mol) and cooling the mixture to room temperature [5] or by slow evaporation of an aqueous solution of the components [6]. The corresponding bromo complex (white) was prepared from $MnBr_2 \cdot 4H_2O$ and ligand [5, 10]. The resulting solid was washed with ethanol and dried in a vacuum over $Mg(ClO_4)_2$ [5]. The chloro complex forms small colorless cubes which are probably triclinic [6]. X-ray powder diagrams suggest isomorphism of the chloro and bromo complexes with the corresponding complexes of cobalt and nickel [10].

$Mn(C_2H_5N_3O_2)_2Cl_2$ melts at 260°C and $Mn(C_2H_5N_3O_2)_2Br_2$ melts at 270°C [10]. The IR spectrum (Nujol or hexachlorobutadiene mulls) shows a shift of the ν(CO) band to a lower wave number (1680 cm^{-1}) and of the ν(C–N) bands to a higher wave number in comparison to the free ligand. Coordination of biuret through the oxygens of both carbonyl groups [5] (also see [10]), was assumed with probably a bidentate cis conformation of the ligand [5, 7]. The far-IR spectrum of the complexes shows one band assigned to ν(Mn–Cl) at 198 cm^{-1} and of ν(Mn–Br) at 136 cm^{-1}. Two bands assigned to $\nu(Mn–O_{ligand})$ occur at 292 and 250 cm^{-1} for the chloride or at 305 and 250 cm^{-1} for the bromide. The occurrence of only one (Mn–X) band is indicative of a trans-octahedral structure [8, 10]. The electronic reflectance spectrum of the chloride shows a strong band at 36000 cm^{-1} ascribed to the ligand, and it is assumed that the complex has a formal D_{2h} symmetry. The magnetic moment $\mu_{eff} = 5.90\ \mu_B$ resulting from susceptibility measurements at 294 K indicates a spin-free octahedral $Mn^{II}(d^5)$ complex with a weak ligand field [5]. The X-ray photoelectron spectrum of the chloride recorded in vacuum ($p = 10^{-7}$ Torr) at room temperature yields the following binding energies (in eV, referring to the value of 285.0 eV for the C_{1s} level of carbon): 289.9 (C_{1s}), 400.4 (N_{1s}), 532.6 (O_{1s}), 198.5 ($Cl\,2p_{3/2}$) and 642.3 ($Mn\,2p_{3/2}$). The observed increase of the N_{1s}, C_{1s}, and O_{1s} binding energies by 0, 2, 0.4, and 0.4 eV, respectively, in the complex is consistent with the coordination of the two carbonyl oxygens of biuret to manganese. The $Cl(2p_{3/2})$ binding energy, increased by 0.3 eV relative to KCl, indicates a less ionic character of the Mn-Cl bond [9].

Susceptibility measurements (Curie-Cheveneau balance; correction for diamagnetism) at 20°C yield the magnetic moment $\mu_{eff} = 5.83\ \mu_B$ for X = Cl and 5.87 μ_B for X = Br, indicative of high-spin six-coordinate octahedral complex configuration [10].

$Mn(C_2H_5N_3O_2)_2X_2 \cdot H_2O$ (X = Cl, $SO_4/2$, CH_3COO). The complexes were prepared from manganese chloride, sulfate, or acetate and biuret in water in the same way as $Mn(C_2H_5N_3O_2)_2$-$(NO_3)_2 \cdot 2H_2O$ (p. 155) and their IR spectra are similar to those of that compound except for the anions. Thus the IR spectrum of the sulfato complex shows four strongly decoupled bands at 1150, 1130, 1108, and 1085 cm^{-1} (shoulder) due to $\nu(SO_4)$ and at 623 and 615 cm due to $\delta(SO_4)$ which indicate coordination of the sulfate ion to manganese. This is confirmed also by the rather slow reaction of the complex with $BaCl_2$ in aqueous solution. A similar slow reaction with $AgNO_3$ is shown by the chloro and acetato complex in the precipitation of AgCl or CH_3COOAg. Thus it must be assumed that the anions are situated in the inner coordination sphere [3].

The hydrated chloro complex shows bands ascribed to $\nu(H_2O)$ at 3600 cm^{-1} and to $\tau(H_2O)$ at 780 cm^{-1}, the acetato complex exhibits bands at 1580(?), 1420, and 900 cm^{-1} assigned to

$\nu_{as}(COO)$, $\nu_s(COO)$, and $\nu(CC)$ vibrations, respectively, of the acetato group. The thermal analysis of the complexes in a derivatograph from room temperature up to 1000°C in air suggests the following degradation steps ($L = C_2H_5N_3O_2$):

$$MnCl_2L_2\cdot H_2O \xrightarrow{180\ \text{to}\ 220°C} MnCl_2L_2 \xrightarrow{230\ \text{to}\ 280°C} MnCl_2L \xrightarrow{960\ \text{to}\ 1000°C} Mn_3O_4$$

$$MnSO_4L_2\cdot H_2O \xrightarrow{100\ \text{to}\ 140°C} MnSO_4L_2 \xrightarrow{180\ \text{to}\ 280°C} MnSO_4L \xrightarrow{280\ \text{to}\ 530°C} MnSO_4$$

$$Mn(CH_3COO)_2L_2\cdot H_2O \xrightarrow{80\ \text{to}\ 90°C} Mn(CH_3COO)_2L_2 \xrightarrow{100\ \text{to}\ 300°C} Mn(CH_3COO)_2 \xrightarrow{300\ \text{to}\ 470°C} Mn_2O_3$$

The complexes are air-stable and readily soluble in water but far less soluble or insoluble in common organic solvents such as methanol, ethanol, $CHCl_3$, CCl_4, ether, hexane, benzene, and toluene, except the acetato complex, which dissolves also in methanol and CCl_4, sparingly in $CHCl_3$. The electrical conductance Λ (in $cm^2\cdot\Omega^{-1}\cdot mol^{-1}$) = 213 for X = Cl, 93 for X = ½ SO_4 and 201 for X = CH_3COO measured in water as for $Mn(C_2H_5N_3O_2)_2(NO_3)_2\cdot 2H_2O$ (p. 155) increases to 223, 101, and 211 $cm^2\cdot\Omega^{-1}\cdot mol^{-1}$, respectively, within 1 h after preparation. A methanolic solution of the acetato complex (V = 1000 L/mol), yields Λ = 91 and 210 $cm^2\cdot\Omega^{-1}\cdot mol^{-1}$ within 1 h, respectively. Such behavior indicates gradual displacement of the acido groups from the inner sphere of the complexes by the solvent molecules [3].

References:

[1] Srivastava, P. C., Banerjee, B. K. (Indian J. Chem. A **17** [1979] 583/5).
[2] Sanyal, R. M., Srivastava, P. C., Banerjee, B. K. (J. Inorg. Nucl. Chem. **37** [1975] 343/5).
[3] Kharitonov, Yu. Ya., Ambroladze, L. N. (Zh. Neorgan. Khim. **31** [1986] 671/7; Russ. J. Inorg. Chem. **31** [1986] 382/6; Koord. Khim. **8** [1982] 1569/70; C.A. **98** [1983] No. 82764).
[4] Shyityeva, N., Sulaimankulov, K., Davranov, M. (Zh. Neorgan. Khim. **27** [1982] 1037/9; Russ. J. Inorg. Chem. **27** [1982] 583/4).
[5] McLellan, A. W., Nelson, G. A. (J. Chem. Soc. A **1967** 137/42, 140).
[6] Nardelli, M., Chierici, I. (J. Chem. Soc. **1960** 1952/3).
[7] Aida, K. (J. Inorg. Nucl. Chem. **23** [1961] 155/8).
[8] Nuttall, R. H., Nelson, G. H. (J. Inorg. Nucl. Chem. **31** [1969] 2979/81).
[9] Yoshida, T., Yamasaki, K., Sawada, Sh. (Bull. Chem. Soc. Japan **51** [1978] 1561/2).
[10] N' Dongui Mabiala, Barbier, J. P., Hugel, R. P. (Polyhedron **3** [1984] 99/106, 101, 105).

24.4.11 With Azodicarboxamide $H_2NC(O)N{=}NC(O)NH_2$ (= $C_2H_4N_4O_2$)

$Mn(C_2H_2N_4O_2)$. An aqueous solution of $MnCl_2\cdot 4H_2O$ was added to that of azodicarboxamide (mole ratio 1:2) in slightly alkaline (KOH, pH ~8 [1] or ammonia, pH 8 to 9 [2]) solution. The precipitated solid was digested on a water bath at 70 to 80°C for 1 h, then filtered off, washed with hot and cold water, alcohol, ether and dried in the air [1]. The yellowish brown fine crystalline solid is stable and insoluble in water and common organic solvents. Its IR spectrum, recorded in the 4000 to 250 cm^{-1} region, shows three absorption bands around 3375, 3285, and 3190 cm^{-1}, assignable to $\nu(NH)$ and three bands at ~1665, 1615, and 1595 cm^{-1}, assignable to $\nu(CN) + \delta(NH)$ vibrations of the coordinated ligand. A complex band at 1492 cm^{-1} may be referred to $\nu(OCN)$ vibrations suggesting an $HN{=}C(O^-){-}N{=}N{-}C(O^-){=}NH$ conformation of the ligand in the complex. In the region below 600 cm^{-1} some bands which are absent in the free ligand can be ascribed to metal-to-ligand vibrations. Thus it is assumed that anionic azodicarboxamide acts as a tetradentate, oxygen- and nitrogen-bonded ligand in a polymeric complex structure [1, 2]. Thermal analysis reveals an exothermic effect starting at 60°C and ending at 195°C with a weight loss of 36% [2].

References:

[1] Batyr, D. G., Baloyan, B. M., Popa, E. V., Kharitonov, Yu. Ya., Khitrova, A. V. (Koord. Khim. **7** [1981] 1454/9; C.A. **96** [1982] No. 78881).

[2] Batyr, D. G., Baloyan, B. M., Popa, E. V., Kharitonov, Yu. Ya (Koord. Khim. **7** [1981] 648/50; C.A. **95** [1981] No. 53947).

24.4.12 With 1,2-Hydrazinedicarboxamide $H_2NC(O)NHNHC(O)NH_2$ (= $C_2H_6N_4O_2$)

$Mn(C_2H_4N_4O_2)\cdot 2H_2O$. To an aqueous solution containing $MnCl_2\cdot 4H_2O$ and hydrazodicarboxamide (mole ratio 1:2) aqueous KOH solution was added up to pH ~9 and the mixture was heated on a water bath at 70 to 80°C for about 30 min. The resulting precipitate was filtered off while hot, washed with alcohol and ether, and dried in air. The anhydrous complex was obtained in the same way from aqueous alcohol solution. The IR spectrum, recorded in the 4000 to 250 cm^{-1} region, does not differ much from that of the free ligand, probably because of hydrolysis during irradiation which renders its interpretation difficult. However, the analogy of composition with that of other hydrazide complexes allows one to assume a bridging tetradentate ligand anion [1, 2] of the tautomeric conformation $H_2N{-}C(O^-){=}N{-}N{=}C(O^-)NH_2$ [1] in a polymeric complex structure since the compound is almost insoluble in water and common organic solvents [1, 2]. As shown by thermographic studies (up to 600°C), the hydrated complex loses its water between 50 and 90°C in an endothermic reaction. Upon further heating an exothermic effect is observed in the 100 to 200°C region followed by an endothermic one between 160 and 182°C and a further exothermic effect between 360 and 415°C which indicate a three-step decomposition leading to manganese oxide as the final product [2].

References:

[1] Batyr, D. G., Baloyan, B. M., Popa, E. V., Kharitonov, Yu. Ya. (Koord. Khim. **6** [1980] 1125/6; C.A. **93** [1980] No. 178638).

[2] Batyr, D. G., Baloyan, B. M., Popa, E. V., Kharitonov, Yu. Ya (Koord. Khim. **7** [1981] 905/7; C.A. **95** [1981] No. 90341).

24.5 Complexes with Derivatives of Guanidine and Related Compounds

With Dicyandiamide $H_2N{-}C(NH_2){=}N{-}C{\equiv}N \rightleftharpoons H_2N{-}C({=}NH){-}NH{-}C{\equiv}N$ (=1-Cyanoguanidine = $C_2H_4N_4$)

$Mn(C_2H_4N_4)Cl_2$ appeared in the shape of pale pink, flat prisms within one week, from a solution of the ligand in a concentrated aqueous solution of $MnCl_2$. The compound was washed and dried at 40°C. Its density is 1.661 g/cm^3 at 18°C. The molar conductance, $\Lambda = 51.2$ $cm^2\cdot\Omega^{-1}\cdot mol^{-1}$, of a 2 mM aqueous solution is practically that of a nonelectrolyte. Thermoanalytic studies reveal an endothermic effect at 100°C indicating melting and an exothermic effect at 240°C due to the complete decomposition of the complex. $Mn(C_2H_4N_4)Cl_2$ is soluble in water, sulfuric acid, and ethanol, but only slightly soluble in ether, benzene, and toluene [1].

$Mn(C_2H_4N_4)_4X_2$ (X = NO_3, Cl, ClO_4, Br, NCS). The complexes were prepared by combining solutions of the appropriate manganese salts and the ligand in absolute ethanol in a 1:2 mole ratio. On shaking (in some cases) or after refluxing for 30 min and standing overnight, the light pink complexes precipitated. Melting points are: 178°C for the nitrate, 185°C for the chloride, 195°C for the perchlorate, 206°C for the bromide, and 192°C for the isothiocyanate. The IR spectra of the complexes in Nujol show absorption bands (in cm^{-1}) which were assigned as follows: 3350 to 3000, ν(NH); 2260 to 2220, ν(C≡N) (free ligand 2170); 1660, δ(NH). The bands in the complexes are unchanged compared to the free ligand except for the ν(C≡N) band which has shifted to higher wave numbers on complexation. This suggests coordination of the ligand only through the nitrogen atom of the nitrile group. In the nitrate, bands assigned to $\nu_4(NO_3)$ and $\nu_1(NO_3)$ (at 1400 and 1280 cm^{-1}, respectively) indicate the presence of coordinated unidentate NO_3 groups. The perchlorate shows a broad $\nu(ClO_4)$ band around 1100 cm^{-1} indicative of ionic perchlorate, and the isothiocyanate exhibits a sharp band at 2090 cm^{-1} due to ν(C=N) which indicates the presence of terminal N-bonded isothiocyanate groups. The electronic spectra of the nitrate, bromide, and isothiocyanate show the well-known absorption bands in the 18400 to 18600, 22600 to 22800, 24800 to 25300 and 27800 cm^{-1} regions, assignable to d-d transitions of Mn^{II}, which presumably indicate a high-spin octahedral configuration. The perchlorate with bands at 20400, 22600, 23800, and 26600 cm^{-1} has possibly a tetrahedral geometry which would be in accord with the molar electric conductance value $\Lambda = 205\ cm^2 \cdot \Omega^{-1} \cdot mol^{-1}$ of a 0.001 M solution in acetone indicating a 1:2 electrolyte. The nitrate, bromide, and isothiocyanate are nonelectrolytes in acetone [2].

References:

[1] Avakyan, S. N., Eminyan, R. S. (Izv. Akad. Nauk Arm. SSR Khim. Nauki **16** No. 1 [1963] 13/7, 13/4; C.A. **59** [1963] 2372).
[2] Sahu, B. K., Mohapatra, B. K. (J. Indian Chem. Soc. **57** [1980] 936/7).

24.5.2 With *1H*-Benzimidazol-2-ylguanidine

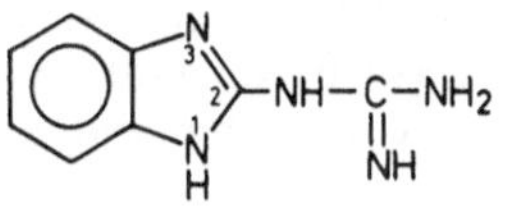

$(=C_8H_9N_5)$

$Mn(C_8H_9N_5)_2Cl_2$. To a solution of partially dehydrated $MnCl_2 \cdot n\,H_2O$ in a minimum volume of dry methanol, excess ligand in the same solvent was slowly added. The mixture was concentrated under vacuum until crystallization began. The green crystals were washed with methanol and dried in vacuum [1].

The electronic spectrum of the complex in butanol shows absorption bands at 16650, 18500, and 25000 cm^{-1} (all sh) with the molar extinction coefficients ε = 240, 320, and 1100 $L \cdot mol^{-1} \cdot cm^{-1}$. The absorptions were assigned to the transitions $^6A_{1g}(S) \rightarrow {}^4T_{1g}(G)$, $^6A_{1g}(S) \rightarrow {}^4T_{2g}(G)$, and $^6A_{1g}(S) \rightarrow {}^4E_g(G)$, respectively, in a typical octahedral Mn^{II} compound. Two other bands (at 28550 and 31250 cm^{-1}) were ascribed to Mn-to-ligand charge-transfer transitions. The magnetic moment $\mu_{eff} = 5.1\ \mu_B$ indicates a high-spin d^5 complex. It is assumed that the two ligands are bidentately coordinated to the Mn atom through the secondary imino N atoms of the benzimidazole moieties and the imide N atoms of the guanidine moieties forming a nearly square and perfectly planar arrangement around the Mn atom. The axial positions are occupied by the Cl^- ions [1].

On heating, the complex decomposes at 285°C [1]. Thermal analysis (DTA, TGA) at a heating rate of 5°C/min indicates the loss of 1 mol Cl_2 per mol of complex in the 150 to 250°C region to leave $Mn(C_8H_9N_5)_2$, which in turn loses the organic component between 320 and

650°C yielding a residue of MnO [2]. The compound is soluble in water (with a change of color due to solvolysis). It is also soluble in morpholine, moderately soluble in methanol, ethanol, and nitromethane and insoluble in chloroform, carbon tetrachloride, dioxane, benzene, and nitrobenzene. It is a 1:2 electrolyte in water or alcohols and a nonelectrolyte in morpholine [1].

References:

[1] Hussain, M. S., Ali, T., Ali, S. M. (Pakistan J. Sci. Ind. Res. **16** [1973] 96/9; C.A. **80** [1974] No. 115575).

[2] Ali, T., Wadud, A., Haroon-Al-Rashid (J. Chem. Soc. Pakistan **6** [1984] 1/9, 3; C.A. **102** [1985] No. 38643).

24.5.3 With N′,N″-Dialkyloxaldiamidines

$$\text{R-NH-C(=NH)-C(=NH)-NHR} \qquad R = C_2H_5;\ (=C_6H_{14}N_4) \qquad R = C_4H_9;\ (=C_{10}H_{22}N_4)$$

Complex in Solution. Spectrophotometric studies at 20°C on solutions of $MnCl_2 \cdot 4H_2O$ and dibutyloxaldiamidine (0.1M) in aqueous methanol or ethanol by the method of continuous variation reveal the formation of a 1:2 complex (Mn : ligand) which exhibits absorption maxima at 370, 380, and 400 nm [1].

$Mn(C_6H_{14}N_4)Cl_2 \cdot 4H_2O$. An aqueous solution of $MnCl_2 \cdot 4H_2O$ was added to a solution of diethyloxaldiamidine in ethanol. The sludge which formed immediately was filtered off, and the filtrate was allowed to stand overnight. Dark red crystal clusters separated out which were collected, washed with ethanol, and dried in vacuum. The molecular weight of 345 (340 theory) was determined. The complex decomposes at 212 to 214°C. It is soluble in water and insoluble in ethanol and ether [2].

References:

[1] Hoffman, W. E., Jacobs, M., Kennepohl, G., Parrot, D. W., Reed, P., Stout, T. R., Sundy, J. (Proc. Indiana Acad. Sci. **79** [1969/70] 129/33, 130; C.A. **73** [1970] No. 126513).

[2] Woodburn, H. M., Salvesen, R. H., Fisher, J. R., Hoffman, W. E., Graminski, E. L., Van Deusen, R. L. (J. Chem. Eng. Data **12** [1967] 615/7).

24.5.4 With N-Phenylbenzamidrazone

$$C_6H_5\text{-}C(=NH)\text{-}NH\text{-}NH\text{-}C_6H_5 \rightleftharpoons C_6H_5\text{-}C(\text{-}NH_2)=N\text{-}NH\text{-}C_6H_5\ (=C_{13}H_{13}N_3)$$

The formation of a 1:1 complex in aqueous ethanol (75 vol%) with the stability constant K_1 = 13600 L/mol was determined spectrophotometrically (ε=1830 $L \cdot mol^{-1} \cdot cm^{-1}$). The complex obeys the law of Lambert-Beer at concentrations of Mn^{2+} between 0.3 and 8.4 μg/mL, Pelova, R., Toleva, A. (Nauchni Tr. Polovdivski Univ. **12** [1974] 29/34, 33; C.A. **83** [1975] No. 85826).

24.5.5 With Biguanide and Derivatives of Biguanide

General

Highly stable Mn^{III} and Mn^{IV} complexes with biguanide and ethylenedibiguanide were first prepared by Rây and co-workers [1, 6, 7]. These authors assumed simple coordinative bonds between the Mn atom and the bidentate ligand units by one amino nitrogen and one nitrogen of a deprotonated imino group forming two six-membered chelate rings. However, this model could not explain the high stability of these complex compounds in unusual oxidation states of manganese. X-ray structural, ^1H NMR, and N(1s) photoelectron spectral studies of biguanide complexes of other transition metals performed later indicate planar arrangement of the chelate rings around the metal ion. They also reveal the existence of dπ(metal)–pπ(ligand) bonds in addition to the σ bonds and the existence of π-electron delocalization extending to all the nitrogen atoms including those outside the ring [2, 3]. These results are best represented by structure 1 for complexes with the deprotonated ligand and structure 2 for complexes with the nondeprotonated ligand reflecting the aromatic character of the chelate rings [4, 5]. Structure 2 takes into account that the positive charge is not localized on any nitrogen atom [4].

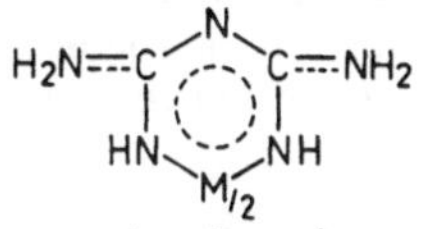

structure 1

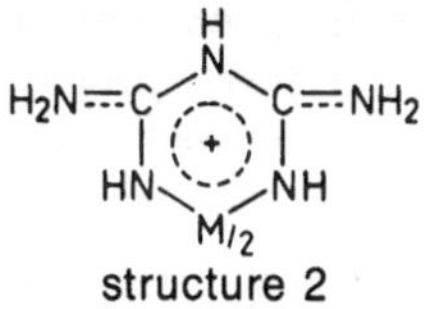

structure 2

In the octahedral Mn^{III} and Mn^{IV} complexes for which structural data are lacking up to now, oxygen atoms are coordinated to the Mn atom in addition to the biguanide molecules. Magnetic measurements indicate dimeric structures with O bridges as shown in Fig. 18, p. 163, for $[Mn^{IV}O(C_2H_6N_5)_2]\cdot 6H_2O$. The magnetic moments of two Mn^{III} complexes, one with ethylenedibiguanide and one with a derivative of piperazine (see pp. 164 and 166), indicate the unusual low-spin configuration of Mn^{III} establishing the strong-field character of these ligands [4]. The complexes are sparingly soluble or insoluble in water.

Previous authors, e.g. [7], differentiated between "complex bases", e.g., $[Mn^{IV}(OH)_2$-$(C_2H_7N_5)_2](OH)_2$ and complex salts, e.g., $[Mn^{IV}(OH)_2(C_2H_7N_5)_2]X_2$. However, spectral studies of the "complex bases" and of anhydrous compounds obtained from them by dehydration showed no difference in their spectra [8], so that the "complex bases" are better formulated as uncharged or neutral complexes with the deprotonated biguanide units [4].

References:

[1] Rây, P. (Chem. Rev. **61** [1961] 313/59, 346/7).
[2] Creitz, T. C., Gsell, R., Wampler, D. L. (J. Chem. Soc. Chem. Commun. **1969** 1371/2).
[3] Swartz Jr., W. E., Alfonso, R. A. (J. Electron Spectrosc. Relat. Phenom. **4** [1974] 351/4 from C.A. **81** [1974] No. 161831).
[4] Syamal, A. (J. Sci. Ind. Res. [India] **37** [1978] 661/85, 661/2).
[5] Schwarzenbach, G., Schwarzenbach, D. (J. Indian Chem. Soc. **54** [1977] 23/4).
[6] Ray, M. M., Rây, P. (J. Indian Chem. Soc. **35** [1958] 595/600).
[7] Ray, M. M., Rây, P. (J. Indian Chem. Soc. **35** [1958] 601/9; Sci. Cult. [Calcutta] **23** [1958] 158/60).
[8] Skabo, R. H., Smith, P. W. (Australian J. Chem. **22** [1969] 659/61).

24.5.5.1 With Biguanide $H_2N{-}C({=}NH){-}NH{-}C({=}NH){-}NH_2$ ($=C_2H_7N_5$)

Manganese(III) Complexes

$[Mn(OH)(C_2H_6N_5)_2]_2\cdot H_2O$. To a solution of $C_2H_7N_5\cdot H_2SO_4$ in aqueous NaOH, a solution of $MnSO_4\cdot 4H_2O$ was added and a rapid current of air was drawn through the mixture to convert the manganese(II) hydroxide first formed in suspension into dark chocolate-colored glistening crystals of the Mn^{III} complex. The crystals were collected, washed with ice cold water, and kept in vacuum over $CaCl_2$ and KOH. The complex could be obtained also from freshly prepared manganese(III) acetate and biguanide in alkaline solution [2, 9].

The IR spectrum (KBr disks and Nujol mulls) shows strong characteristic absorption bands (in cm^{-1}) which were assigned as follows: 3350 and 3200 to ν(NH); 1620 to ν(C=N) + δ(OH) + δ(NH); 1580 and 1480 to ν(NCN); 1120 to ν(C–N). The lowered ν(C=N) absorption, at 1620 cm^{-1} for the complex (1680, 1640 for the free ligand) suggests coordination of biguanide to manganese through the imino nitrogen atoms. A band at 1250 cm^{-1} (not present in the spectrum of the free ligand) was associated with the formation of a chelate ring. The electronic reflectance spectrum shows absorption bands at 425 and 725 nm, the first of which was assigned to the $^5E_g \rightarrow {}^5T_{2g}$ transition of Mn^{III} in an octahedral environment. The other band indicates distortion from octahedral symmetry [6]. Susceptibility measurements at 30°C yielded $\mu_{eff} = 4.70\ \mu_B$, indicating a high-spin Mn^{III} (d^4) complex [2]. The somewhat lower than theoretical value of μ_{eff} suggests a superexchange interaction between the two Mn atoms in a dimeric OH bridged structure $[(C_2H_6N_5)_2Mn\langle{}^{O(H)}_{O(H)}\rangle Mn(C_2H_6N_5)_2]\cdot H_2O$ [2, 6, 8]. Thermal analysis (DTA, DTG) shows endothermic peaks at 85 to 90°C and 315 to 330°C, indicating loss of water and the decomposition of the biguanide molecule [7]. The complex is insoluble in water, but it is hydrolyzed in contact with water, producing a strongly alkaline reaction. It dissolves in dilute acetic acid to give a brown solution which rapidly becomes turbid and separates hydrated manganese oxides. The solution in syrupy phosphoric acid is violet [2]. Polarographic studies on solutions of the compound in aqueous 0.1 M KOH reveal three distinct cathodic waves with $E_{1/2} = -0.237$, -0.570, and -1.720 V (versus SCE) corresponding to the one-electron reduction steps $Mn^{III} \rightarrow Mn^{II}$, $Mn^{II} \rightarrow Mn^{I}$, and $Mn^{I} \rightarrow Mn^{0}$, respectively [5].

Mixed ligand complexes $[Mn(OH)aca(C_2H_6N_5)(H_2O)]\cdot nH_2O$ and $[Mn(C_{10}H_9O_2)(C_2H_6N_5)_2]$ of biguanide (= $C_2H_7N_5$) and acetylacetone (= acaH) or benzoylacetone (= $C_{10}H_{10}O_2$) are described in "Mangan" D 1, 1979, pp. 106 and 119, respectively.

Manganese(IV) Complexes

The complex with the deprotonated ligand, $\mathbf{[Mn(OH)_2(C_2H_6N_5)_2]\cdot 2H_2O}$ or $\mathbf{[MnO(C_2H_6N_5)_2]_2 \cdot 6H_2O}$, was prepared by action of aqueous $KMnO_4$ on a solution of $C_2H_7N_5\cdot H_2SO_4$ in aqueous NaOH. Bright red crystals were obtained on cooling. The compound was also synthesized by oxidizing a solution of manganese(II) sulfate and $C_2H_7N_5\cdot H_2SO_4$ in aqueous NaOH at 10 to 15°C with H_2O_2 or $Na_2S_2O_8$. The resulting crystals were kept in a vacuum over $CaCl_2$ and KOH [1, 9]. The compound could be isolated also from the mother liquor of $[Mn(OH)(C_2H_6N_5)_2]_2\cdot H_2O$ (see p. 161) by keeping it in the air for 1 to 2 d [2].

By treating a solution of the $[Mn(OH)_2(C_2H_6N_5)_2]\cdot 2H_2O$ complex in dilute acetic or nitric acid with solutions of the appropriate alkali or ammonium salts, cinnamon to chocolate-brown salts of composition $\mathbf{[Mn(OH)_2(C_2H_7N_5)_2]X_2\cdot nH_2O}$ or $\mathbf{[MnO(C_2H_7N_5)_2]_2X_4\cdot(2n+2)H_2O}$ with n = 0 for X = NO_3, n = 2 for X = IO_3, n = 4.5 for $X_2 = SO_4$; n = 1.5 for $X_2 = HPO_4$, n = 0.5 for $X_2 = C_2O_4$, n = 3.5 for $X_2 = BeF_4$, and n = 3 for $X_2 = CrO_4$ have been prepared [1, 8, 9].

The IR spectra of $[MnO(C_2H_6N_5)_2]_2\cdot 6H_2O$ and $[MnO(C_2H_7N_5)_2]_2(NO_3)_2\cdot 2H_2O$ show absorption bands between 3350 and 3170 cm^{-1} (free ligand 3400 to 3200) which were assigned to ν(NH) vibrations. The bands in the 1680 to 1600, 1565 to 1510, and 1300 to 900 cm^{-1} regions were assigned to ν_{as}(CN), $\delta(NH_2)$, and ν_s(CN) vibrations, respectively. A band around 770 cm^{-1} may be assigned to a $\delta(NH)_{out\text{-}of\text{-}plane}$ vibration, and the broadening and increased intensity of the band in the 740 to 720 cm^{-1} region were ascribed to an overlap of $Mn\text{–}O_2\text{–}Mn$ and ligand bands. A band at 820 cm^{-1} for the nitrate was assigned to a vibration of the NO_3^- ion [4]. The reflectance spectrum of $[MnO(C_2H_6N_5)_2]_2\cdot 6H_2O$ shows an intense electron transfer band above 25000 cm^{-1}. Ligand field bands at 19230, 17856, and 15923 cm^{-1} were assigned to the electronic transitions $^4A_{2g} \rightarrow {^4T_{1g}}$, $^4A_{2g} \rightarrow {^2T_{1g}}$, and $^4A_{2g} \rightarrow {^4T_{2g}}$, respectively, in the cubic ligand field of Mn^{IV}. The ligand field parameter $\Delta = 10\ Dq = 15923\ cm^{-1}$ was calculated [3]. In the nitrate, the corresponding bands occur at 22200, 19040, and 17700 cm^{-1}, respectively, and indicate a complex symmetry not far from cubic. A band at 40000 cm^{-1} was ascribed to the biguanide ligand [4]. Measurements of the magnetic susceptibility at 25 to 30°C yielded effective magnetic moments of 2 to 2.6 μ_B which are considerably lower than the spin-only value of 3.87 μ_B expected for d^3 complexes [1, 4]. The susceptibility as a function of temperature in the 305 to 80 K range shows maxima at 290 K for the complex with the deprotonated ligand and at 250 K for the nitrate, then steadily decreases. The magnetic susceptibility at 90 K corresponds to values of $\mu_{eff} \approx 1\ \mu_B$ for $[MnO(C_2H_6N_5)_2]_2 \cdot 6H_2O$, $[MnO(C_2H_7N_5)_2]_2(NO_3)_2 \cdot 2H_2O$, and $[MnO(C_2H_7N_5)_2]_2SO_4 \cdot 11H_2O$. From a theoretical treatment, values for the exchange energy $-2\ J$ and the g-values of these three compounds were obtained: $-2\ J = 62.55\ cm^{-1}$, $g = 2.13$ for $[MnO(C_2H_6N_5)_2]_2 \cdot 6H_2O$; $-2\ J = 59.14\ cm^{-1}$, $g = 1.75$ for the nitrate; $-2\ J = 66.02\ cm^{-1}$, $g = 1.93$ for the sulfate. The high values of the exchange energy indicate strong magnetic exchange and probable metal-metal direct interaction in addition. A dimeric structure with bridging oxygen atoms as shown in **Fig. 18** for $[MnO(C_2H_6N_5)_2]_2 \cdot 6H_2O$ was therefore proposed. It is assumed that the electron density initially on t_{2g} orbitals of one Mn atom is transferred to the other Mn atom through 2p orbitals of the bridging oxygen atoms. The direct metal-metal interaction takes place through the normalized overlap of the d orbitals of the Mn atom. The maxima of the susceptibility at 290 and 250 K may be due to phase transitions at the Néel temperature [4].

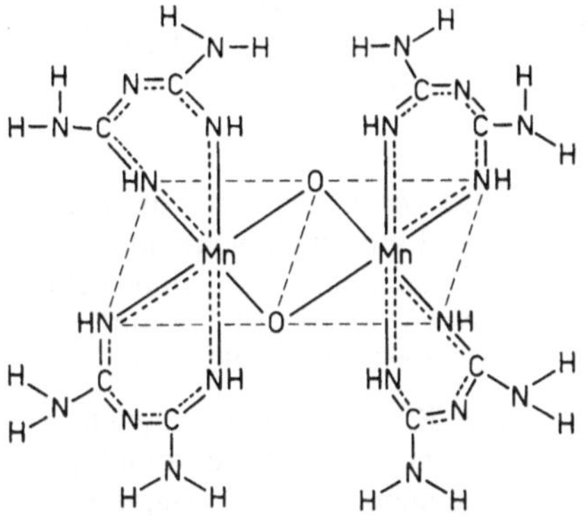

Fig. 18. Structure proposed for the dimeric complex unit in $[MnO(C_2H_6N_5)_2]_2 \cdot 6H_2O$ [4].

$[MnO(C_2H_6N_5)_2]_2 \cdot 6H_2O$ loses two molecules of water on being kept in vacuum or on heating at 90°C and decomposes slowly at higher temperature. The dehydrated complex is dark chocolate-red and regains its original color and weight on exposure to air [1]. Thermogravimetric analysis indicates endothermic decomposition at 180°C. The decomposition apparently occurs with loss of water (3 molecules per Mn atom), whereas $[MnO(C_2H_7N_5)_2]_2(NO_3)_2 \cdot 2H_2O$ decomposes at 150°C with release of water and nitric oxides [4]. $[MnO(C_2H_7N_5)_2]SO_4 \cdot 11H_2O$ is fairly stable at room temperature when dry [1].

The compounds are sparingly soluble in water except for the nitrate and fluoroberyllate which are fairly soluble. Aqueous solutions of the complex with the deprotonated ligand are slightly alkaline (pH ~ 8). They are stable for several hours [1]. Polarographic studies on solutions of $[MnO(C_2H_6N_5)_2]_2 \cdot 6H_2O$ in aqueous 0.1 KOH at 30°C reveal three distinct cathodic waves with the half-wave potentials $E_{1/2} = -0.232$, -0.543, and -1.740 V (versus SCE) corresponding to the reduction steps $Mn^{IV} \rightarrow Mn^{II}$, $Mn^{II} \rightarrow Mn^{I}$, and $Mn^{I} \rightarrow Mn^{0}$, respectively [5]. Solutions of the nitrate and sulfate are faintly acidic. Conductivity measurements on aqueous solutions of the nitrate indicate a 1:2 electrolyte [1].

References:

[1] Ray, M. M., Rây, P. (J. Indian Chem. Soc. **35** [1958] 601/9).
[2] Ray, M. M., Rây, P. (J. Indian Chem. Soc. **35** [1958] 595/600).
[3] Banerjee, R. S., Basu, S. (J. Inorg. Nucl. Chem. **27** [1965] 359/60).
[4] Bera, J., Sen, D. (Indian J. Chem. A **14** [1976] 880/3).
[5] Chakravarty, B. (J. Inorg. Nucl. Chem. **41** [1979] 1211/2).
[6] Babykutty, P. V., Prabhakaran, C. P., Anantaraman, R., Nair, C. G. R. (J. Inorg. Nucl. Chem. **36** [1974] 3685/8).
[7] Babykutty, P. V., Indrasenan, P., Anantaraman, R., Nair, C. G. R. (Thermochim. Acta **8** [1974] 271/82, 273/6, 280).
[8] Rây, P. (Chem. Rev. **61** [1961] 313/59, 346/7).
[9] Ray, M. M., Rây, P. (Sci. Cult. [Calcutta] **23** [1958] 158/60).

24.5.5.2 With Ethylenedibiguanide

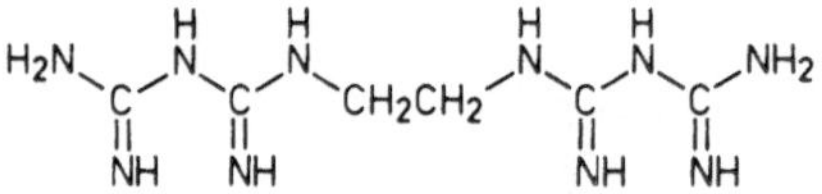

$(= C_6H_{16}N_{10})$

Manganese(III) Complexes

$[Mn(OH)(C_6H_{14}N_{10})(H_2O)] \cdot 0.5\,H_2O$ was prepared by oxidizing with a brisk current of air a mixture of aqueous Mn^{II} sulfate and ethylenedibiguanidium sulfate dissolved in 5% NaOH solution. The manganese(II) hydroxide initially formed in suspension was gradually oxidized and converted to dark chocolate-red crystals of the Mn^{III} complex which were collected, washed, and dried in vacuum over $CaCl_2$ and KOH [1, 3].

Susceptibility measurements at 30°C yielded the magnetic moment $\mu_{eff} = 3.26\ \mu_B$ which is far below the spin-only value of 4.90 μ_B expected for a high-spin Mn^{III} (d^4) complex. The low moment value suggests that the compound is a low-spin complex but with an incomplete quenching of its orbital momentum [1].

The complex is quite stable at room temperature for several months. It becomes anhydrous on being heated at 90°C. No significant further loss in weight occurred on heating up to 110°C. The compound is sparingly soluble in water. The aqueous solutions are quite stable and show alkaline reaction (pH ~ 8). Appreciable decomposition occurs only on prolonged boiling. In faintly acidic solutions it hydrolyzes forming the diaquo complex [1, 3].

$[Mn(C_6H_{16}N_{10})(H_2O)_2]_2X_6 \cdot n\,H_2O$ ($n = 6$ for $X_2 = SO_4$ or SeO_4 and $n = 5$ for $X_2 = CrO_4$). The compounds were prepared by reacting $[Mn(OH)(C_6H_{14}N_{10})(H_2O)] \cdot 0.5\,H_2O$ in dilute acetic acid with ammonium sulfate, sodium selenate, or potassium chromate. The silky yellow crystals were dried in vacuum over $CaCl_2$ [1, 3].

Measurements of the magnetic susceptibility yielded the magnetic moment $\mu_{eff} = 4.58\ \mu_B$ at ~30°C for both the sulfate and selenate indicating high-spin d^4 complexes [1, 3]. The sulfate gives off the water of crystallization on being heated to 70°C. It loses two molecules of coordinated water at 80°C without any decomposition to give a product of composition $[Mn^{III}SO_4(C_6H_{16}N_{10})(H_2O)]_2SO_4$. The compounds are sparingly soluble (the chromate) or almost insoluble in water (the sulfate and selenate). The selenate does not decompose even on boiling its suspension in water. It gives a violet solution in phosphoric acid [1].

Manganese(IV) Complex

$[Mn(OH)_2(C_6H_{14}N_{10})]\cdot1.5H_2O$ was prepared by reacting ethylenedibiguanidium sulfate dissolved in aqueous NaOH with a solution of Mn^{II} sulfate and sodium persulfate in water. On cooling of the mixture, reddish brown crystals deposited which were collected, washed, and dried in vacuum over $CaCl_2$ and KOH. Susceptibility measurements at 29°C yielded μ_{eff} = 2.50 μ_B which is considerably below the value expected for a high-spin Mn^{IV} (d^3) complex (3.87 μ_B) and may indicate metal-metal interactions. The compound is quite stable in the solid state. It is sparingly soluble in water. The reddish brown aqueous solution is alkaline to litmus (pH ~ 8). It is stable at room temperature. Addition of an alkali salt to a solution of the complex acidified by dilute acetic acid produces an opalescence [2, 3].

References:

[1] Ray, M. M., Rây, P. (J. Indian Chem. Soc. **35** [1958] 595/600).
[2] Ray, M. M., Rây, P. (J. Indian Chem. Soc. **35** [1958] 601/9, 603, 609).
[3] Ray, M. M., Rây, P. (Sci. Cult. [Calcutta] **23** [1958] 158/60).

24.5.5.3 With Hexamethylenedibiguanide

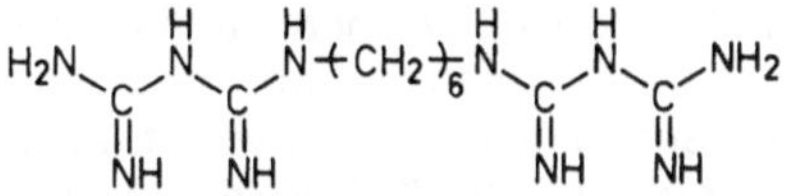

(= $C_{10}H_{24}N_{10}$)

$[Mn^{IV}(OH)_2(C_{10}H_{22}N_{10})]\cdot4H_2O$. A strong aqueous solution of $KMnO_4$ was added to a cooled solution obtained by dissolving hexamethylenedibiguanidium sulfate ($C_{10}H_{24}N_{10}\cdot2H_2SO_4\cdot H_2O$) in aqueous NaOH. The dark red crystals of the complex which separated within 20 min were filtered off after cooling for another 30 min, washed with cold water, and dried in vacuum over $CaCl_2$. Susceptibility measurements at 29°C yielded the magnetic moment μ_{eff} = 2.83 μ_B which is considerably below that expected for a high-spin Mn^{IV} (d^3) complex (3.87 μ_B). The compound decomposes on heating at 70°C. It is soluble in water. It oxidizes Fe^{II} to Fe^{III} and liberates iodine from KI in acid solution, Dutta, R. L. (J. Indian Chem. Soc. **37** [1960] 32/4).

24.5.5.4 With Other Biguanides and a Related Compound

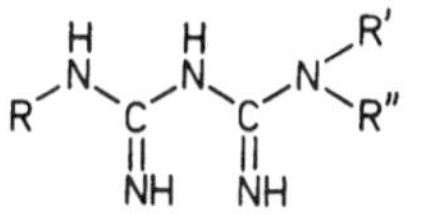

1) Phenylbiguanide = HL
(R = C_6H_5, R′ = R″ = H); (= $C_8H_{11}N_5$)

2) Glucosylbiguanides = H_2L
R = $OHCH_2(CHOH)_3C(OH)$=CH

R′ = R″ = H;	(= $C_8H_{17}N_5O_5$)	R′ = R″ = CH_3;	(= $C_{10}H_{21}N_5O_5$)
R′ = H, R″ = CH_3;	(= $C_9H_{19}N_5O_5$)	R′ = H, R″ = C_2H_5;	(= $C_{10}H_{21}N_5O_5$)
		R′ = R″ = C_2H_5;	(= $C_{12}H_{25}N_5O_5$)
		R′ = C_6H_5, R″ = H;	(= $C_{14}H_{21}N_5O_5$)

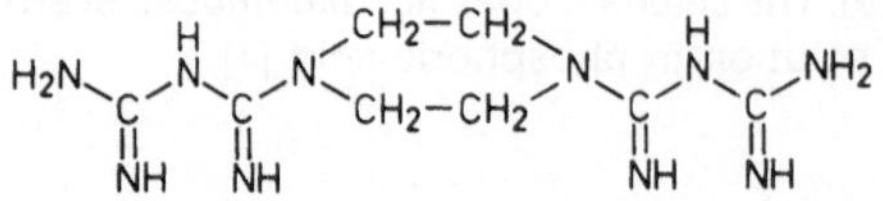

3) N, N″-Bis(aminoiminomethyl)-1,4-piperazinedicarboximidamide ($=C_8H_{18}N_{10}$)

A complex of composition $[Mn^{IV}(OH)_2(HL)_2](OH)_2$ ($=[MnOL_2]_2 \cdot 6H_2O$) was supposed to form on reacting a solution of phenylbiguanide hydrochloride with an aqueous Mn^{IV} compound in neutral or ammoniacal medium. The light maroon precipitate is of analytical interest [1].

Compounds of composition $[Mn^{II}L(H_2O)]$ were obtained by refluxing a manganese(II) salt and glucosylbiguanides in alkaline aqueous alcohol. The brown compounds have magnetic moments between 5.7 and 6.1 μ_B indicating high-spin Mn^{II} complexes. They are insoluble in water, but soluble in organic solvents. The ligands are assumed to be terdentate with two replaceable hydrogens [2].

$[Mn^{III}(OH)(C_8H_{16}N_{10})(H_2O)]$. A freshly prepared ethanol solution of manganese(III) acetate (1.5 g) was added to an ice-cold ethanol solution of ligand 3 obtained by reacting the corresponding sulfate with the stoichiometric amount of $NaOC_2H_5$ in ethanol. The precipitated brown crystalline solid was washed with ethanol and dried in vacuum over $CaCl_2$ and solid KOH. The electronic reflectance spectrum (Nujol mulls) shows three bands at around 43000, 40000, and 20000 cm^{-1} which were ascribed to charge-transfer transitions. Bands observed around 25500 and 24500 cm^{-1} may be due to a splitting of the spin-allowed electronic transition $^3T_{1g} \rightarrow {}^3E_g$ caused by the reduction of the complex symmetry from O_h to C_{4v}. Bands observed at 13000, 11800, and 11000 cm^{-1} can be assigned to the spin-forbidden transitions $^3T_{1g} \rightarrow {}^1E_g$, $^3T_{1g} \rightarrow {}^1T_{2g}$, and $^3T_{1g} \rightarrow {}^5E_g$, respectively. Susceptibility measurements at 27°C yielded the magnetic moment $\mu_{eff} = 2.88\ \mu_B$, close to the spin-only value (2.83 μ_B) of a low-spin Mn^{III} (d^4) system.

The complex is insoluble in water and common organic solvents. It is neutral to litmus. There is no loss in weight when the complex is heated to 110°C. It is stable in contact with water. The oxidation state of the metal ion was determined by oxidimetric titration [3].

References:

[1] Bosch Reig, F., Martinez Calatayud, J., Garcia Alvarez-Coque, M. C. (Afinidad **37** No. 366 [1980] 137/40, 139; C.A. **94** [1981] No. 24347).

[2] Poddar, S. N. (Sci. Cult. [Calcutta] **29** [1963] 212/3).

[3] Dey, K., Ray, K. C., Poddar, S. N., Podder, N. G. (Indian J. Chem. A **14** [1976] 205/6).

25 Complexes with Hydrazides

General. Complexes of composition MnL_nX_2 with hydrazides of monocarboxylic acids, $RC(O)NHNH_2$ (= L) are prepared in alcoholic and aqueous alcoholic solution. The IR and Raman spectra of the compounds were interpreted mostly in terms of bidentate ligation of the hydrazide ligands to Mn through the carbonyl oxygen atom and the nitrogen atom of the terminal NH_2 group, forming a five-membered metallocycle. This was confirmed by X-ray studies, as reported for the isothiocyanato complexes with aceto- (pp. 169/70) and benzohydrazide (pp. 175/6), and for the chloro complex with 2-pyridinecarbohydrazide (pp. 186/7). The magnetic moments of the MnL_nX_2 compounds are indicative of high-spin octahedral manganese(II) complexes, which are often polymeric.

A complex with isonicotinohydrazide, $[Mn(NO_3)_2C_6H_7N_3O(H_2O)]$ (pp. 190/1), contains seven-coordinate manganese(II) in a slightly distorted pentagonal bipyramidal environment. In this case the organic ligand is bonded to Mn through the carbonyl oxygen, the nitrogen of the NH_2 group, and the pyridine ring nitrogen. The H_2O molecule and one of the NO_3^- groups are bonded to Mn through one oxygen whereas the other NO_3^- group is coordinated as a bidentate ligand through two oxygens. Another complex with isonicotinohydrazide, the cation $[Mn_2(C_6H_7N_3O)_3Cl(H_2O)_3]^{3+}$, contains two independent manganese atoms each in octahedral environment. Two of the organic ligands are bridging tridentate; they are coordinated to one Mn atom through the terminal amine nitrogen and carbonyl oxygen and to a second metal atom through the pyridine nitrogen. The third ligand is bidentate and coordinated only through the carbohydrazide nitrogen and oxygen.

Hydrazides of the type RC(O)–NH–NH–C(O)R, e.g., diformyl-, diacetyl-, or dibenzoylhydrazine (= H_2L), form $Mn(H_2L)_2X_2 \cdot nH_2O$ complexes (type I) in neutral or weakly acid solutions. The ligand molecules are coordinated by the carbonyl oxygens and the nitrogens of the NH_2 groups. In weakly alkaline solutions (pH ~ 8 to 10) $MnL \cdot nH_2O$ complexes (type II), containing the doubly deprotonated tautomeric form of the ligands $RC(O^-)$=N–N=$C(O^-)R$, which are tetradentate bridging, are formed.

Coordination to Mn through both carbonyl oxygens and both nitrogens of the NH_2 groups was shown for the complex with malonodihydrazide, $Mn(C_3H_8N_4O_2)Cl_2 \cdot 2H_2O$; see p. 201. X-ray studies revealed the existence of two modifications. The chief difference between the α- and the β-forms of the chloro complex is caused by the way in which the Mn atoms are linked by the ligand molecules to give chains in the α-form and networks in the β-form.

The complexes with hydrazides are of interest because of their biological activity.

25.1 With Hydrazides of Monocarboxylic Acids

25.1.1 With Formohydrazide $HC(O)NHNH_2$ (= CH_4N_2O)

$Mn(CH_4N_2O)_nX_2$. Complexes with n = 3 for X = NO_3, Cl, or $SO_4/2$ and n = 2 for X = Cl, $SO_4/2$, and NCS were prepared by reaction of the manganese(II) salt with the appropriate amount of formohydrazide in ethanol or aqueous ethanol. Sometimes the mixtures were concentrated on a water bath and a small volume of absolute ethanol was added to induce crystallization. The crystals were washed with absolute ethanol and dried in air [1, 2]. In the case of $Mn(CH_4N_2O)_3SO_4$, the white mass left after complete evaporation of the solvent was treated several times with hot $CHCl_3$ and ether to remove excess ligand [3]. For the preparation of $Mn(CH_4N_2O)_2Cl_2$ and $Mn(CH_4N_2O)_2SO_4$, a ratio Mn : ligand = 1 : 4 was used. The crystals, which are highly soluble in water and insoluble in ethanol, were recrystallized from a water-ethanol (1:1) mixture and dried in vacuum [6] or in air [1].

The IR spectra of the complexes in KBr disks, fluorinated hydrocarbons or mineral oil mulls, recorded in the range 4000 to 400 cm^{-1}, were interpreted in terms of bidentate ligation of the ligand through the carbonyl oxygen atom and the nitrogen atom of the NH_2 group [1 to 7]. Some of the main vibrational wave numbers of the five-membered metallocycle were assigned from an analysis of the normal vibrations of the complex units, with C_s symmetry [4]. The NH stretching vibrations were assigned to two bands in the regions 3315 to 3345 and 3280 to 3300 cm^{-1}. Bands of $\nu_{as}(NH_2)$ were observed at 3220 to 3235 cm^{-1}, of $\nu_s(NH_2)$ at 3170 to 3190 cm^{-1} and ~3140 to 3145 cm^{-1}. The amide I band, consisting primarily of the CO stretching vibration, was observed in the region 1670 to 1680 cm^{-1} and found to be shifted to lower wave numbers (~30 cm^{-1}) in comparison to the free ligand. Bands at 1590 to 1620 cm^{-1} were assigned to $\delta(NH_2)$ vibrations and those at 1510 cm^{-1} to $\nu_{as}(CN)+\delta(NH)$ vibrations (amide II band) [1, 2]. The shape and position of the anion bands suggest noncoordination of NO_3^- and SO_4^{2-} in the complexes $Mn(CH_4N_2O)_3(NO_3)_2$ and $Mn(CH_4N_2O)_3SO_4$. Thus, an octahedral environment of the metal atom consisting of three bidentate hydrazide units can be assumed [4]. This is in agreement with the measured conductivities in aqueous solution at 25°C, Λ: 272 (nitrate), 338 (chloride), and 349 (isothiocyanate) $cm^2\cdot\Omega^{-1}\cdot mol^{-1}$ [1, 2]. The coordination of the thiocyanato group to Mn through the nitrogen atoms in $Mn(CH_4N_2O)_2(NCS)_2$ is demonstrated by the placement of the ν(CN) and ν(CS) modes [1, 7]. Absorption bands, observed in the range 520 to 575 cm^{-1} can be assigned to vibrations in which the deformation of the Mn–N and Mn–O bonds make the primary contribution. Absorptions at 415 to 435 cm^{-1} were assigned to vibrations with substantial participation of Mn–O bonds. Additional vibrational numbers and their assignments are reported in [4, 5, 7].

The complexes are stable in air. Melting points of several compounds are listed below:

complex	$Mn(CH_4N_2O)_3(NO_3)_2$	$Mn(CH_4N_2O)_3Cl_2$	$Mn(CH_4N_2O)_3SO_4$	$Mn(CH_4N_2O)_2(NCS)_2$
m.p. in °C ...	126	181	288	138
Ref.	[1]	[2]	[2]	[1]

The complex $Mn(CH_4N_2O)_3Cl_2$ is readily soluble in water and poorly soluble in ethanol at room temperature, the solubility increasing on heating, whereas the sulfate $Mn(CH_4N_2O)_3SO_4$ is moderately soluble in water and insoluble in ethanol. Both complexes are insoluble in $CHCl_3$, CCl_4, acetone, acetonitrile, dimethylformamide, and dimethyl sulfoxide [2].

Susceptibility measurements at room temperature yield the magnetic moment $\mu=6.00\ \mu_B$ for $Mn(CH_4N_2O)_2Cl_2$ and 5.92 μ_B for $Mn(CH_4N_2O)_2SO_4$, both indicative of high-spin octahedral Mn^{II} complexes with coordinated anions. The $Mn(CH_4N_2O)_2Cl_2$ is highly soluble in water and insoluble in ethanol but moderately soluble in a mixture of the two solvents. $Mn(CH_4N_2O)_2SO_4$ is moderately soluble in water and insoluble in ethanol. Conductivity measurements on ~0.001M aqueous solutions at 25°C yield $\Lambda=240$ and 245 $cm^2\cdot\Omega^{-1}\cdot mol^{-1}$ for the chloride and sulfate, respectively, indicating 1:2 electrolytes in each case with dissociation of the anions due to solvolysis [6].

$Mn(CH_4N_2O)_2(NCS)_2$ is soluble in water, dimethylformamide and dimethyl sulfoxide, insoluble in ethanol, $CHCl_3$, CCl_4, acetone, and acetonitrile. Conductivity measurements on aqueous solutions at 25°C as for $Mn(CH_4N_2O)_3Cl_2$ yield $\Lambda=282\ cm^2\cdot\Omega^{-1}\cdot mol^{-1}$, indicating a 1:2 electrolyte due to dissociation of the thiocyanate ions [1].

References:

[1] Kharitonov, Yu. Ya., Machkhoshvili, R. I., Metreveli, D. P., Pirtskhalava, N. I. (Koord. Khim. **3** [1977] 897/901; Soviet J. Coord. Chem. **3** [1977] 696/700, 698).

[2] Kharitonov, Yu. Ya., Machkhoshvili, R. I., Metreveli, D. P., Pirtskhalava, N. I. (Koord. Khim. **3** [1977] 1069/74; Soviet J. Coord. Chem. **3** [1977] 829/33, 831).

[3] Kharitonov, Yu. Ya., Machkhoshvili, R. I., Metreveli, D. P. (Koord. Khim. **2** [1976] 131/2; Soviet J. Coord. Chem. **2** [1976] 107).
[4] Kharitonov, Yu. Ya., Machkhoshvili, R. I., Pirtskhalava, N. I., Metreveli, D. P. (Zh. Neorgan. Khim. **22** [1977] 2768/73; Russ. J. Inorg. Chem. **22** [1977] 1502/6, 1505).
[5] Kharitonov, Yu. Ya., Machkhoshvili, R. I., Metreveli, D. P., Pirtskhalava, N. I. (Koord. Khim. **3** [1977] 1534/40; Soviet J. Coord. Chem. **3** [1977] 1196/202, 1198).
[6] Biswas, P. K., Dasgupta, M. K., Mitra, S., Ray Chaudhuri, N. (J. Coord. Chem. **11** [1982] 225/30, 227).
[7] Kharitonov, Yu. Ya., Metreveli, D. P., Machkhoshvili, R. I., Pirtskhalava, N. I. (Koord. Khim. **3** [1977] 1386/93; Soviet J. Coord. Chem. **3** [1977] 1082/9, 1083, 1087/8).

25.1.2 With Acetohydrazide $CH_3C(O)NHNH_2$ (= $C_2H_6N_2O$)

$Mn(C_2H_6N_2O)_nCl_2$ (n = 2, 3). To prepare the complex with n = 3 an ethanolic solution of 2 g $MnCl_2 \cdot 4H_2O$ and 2.96 g acetohydrazide (mole ratio 1:4) was heated to produce a white crystalline precipitate which was washed with ethanol and ether and recrystallized from water [1, 2]. The white complex with n = 2 was obtained from manganese chloride and acetohydrazide in a minimum volume of water as described in [4] for the malonodihydrazide complex $Mn(C_3H_8N_4O_2)_2Cl_2 \cdot 2H_2O$ (p. 200), or from methanolic or ethanolic solutions of the ligand and solid manganese chloride [3]. $Mn(C_2H_6N_2O)_3Cl_2$ is stable in air and on storage [2]; it melts at 204°C [1]. The IR spectrum of $Mn(C_2H_6N_2O)_2Cl_2$ (liquid paraffin and fluorinated hydrocarbon mulls) shows bands of $\nu_{as}(NH_2)$ at 3195 cm^{-1}, of $\nu_s(NH_2)$ at 3170 cm^{-1}, and a $\nu(CO)$ band at 1660 cm^{-1}. Coordination of acetohydrazide to Mn as a bidentate chelating ligand through carbonyl oxygen and the nitrogen of the NH_2 group [1, 2, 5] was suggested. The metal-to-ligand bond is less strong than in the corresponding complexes of Fe^{III} and Co^{II} as shown by the lower wave numbers of the ν(Mn–N) and ν(Mn–O) frequencies observed at ~526 and 485 cm^{-1} [5]. Coordination of Mn to the imino nitrogen instead of the amino nitrogen and no coordination of carbonyl oxygen was assumed by [3]. The visible spectrum of $Mn(C_2H_6N_2O)_2Cl_2$ in aqueous solution shows only weak absorption [3]. Susceptibility measurements on solid $Mn(C_2H_6N_2O)_2Cl_2$ at 24°C yield the magnetic moment μ_{eff} = 5.80 μ_B suggesting a high-spin octahedral configuration [3].

$Mn(C_2H_6N_2O)_2Cl_2$ is readily soluble in water to give an almost neutral solution, less readily soluble in ethanol, and insoluble in $CHCl_3$, CCl_4, acetone, benzene, and nitrobenzene [1]. Conductivity measurements on about 0.001 M aqueous solutions at 25°C yield the molar conductance Λ = 280 $cm^2 \cdot \Omega^{-1} \cdot mol^{-1}$ for $Mn(C_2H_6N_2O_3)_3Cl_2$ [1] and 203.2 $cm^2 \cdot \Omega^{-1} \cdot mol^{-1}$ for $Mn(C_2H_6N_2O)_2Cl_2$ [3] indicating 1:2 electrolytes in each case [1, 3]. Λ = 39.8 $cm^2 \cdot \Omega^{-1} \cdot mol^{-1}$ was observed for $Mn(C_2H_6N_2O)_2Cl_2$ in methanol [3].

$Mn(C_2H_6N_2O)_2(NCS)_2 \cdot 0.5(1)H_2O$ and **$Mn(C_2H_6N_2O)_2(NCSe) \cdot H_2O$**. The monohydrates were prepared from the manganese(II) salts and acetohydrazide in ethanol. The mixture was allowed to crystallize in a vacuum desiccator over $CaCl_2$ [6]. The isothiocyanate formulated as semihydrate was obtained also by mixing aqueous solutions of $Mn(NCS)_2$ and acetohydrazide in a 1:2 mole ratio at room temperature or with gentle heating. The precipitate was washed with cold water and ethanol and recrystallized from water [7]. It forms sky blue crystals [6] which are stable in the air and on storing [7]; the isoselenocyanate is pale green and finely crystalline [6].

X-ray diffractometer studies showed the complex $Mn(C_2H_6N_2O)_2(NCS)_2 \cdot 0.5H_2O$ to be orthorhombic, space group Pbcn-D_{2h}^{14} (No. 60), with lattice constants a = 19.409(4), b = 10.499(5), and c = 13.809(6) Å; Z = 8, V = 2826.16 $Å^3$. The structure was refined to R = 0.048. The Mn atom is octahedrally coordinated to the carbonyl oxygens and terminal NH_2 groups of the two bidentate ligand molecules and to the nitrogens of the two thiocyanate groups in *trans*-position with atomic distances of Mn–O = 2.130(4) and 2.178(4) Å, $Mn{-}N_{NCS}$ = 2.050(5) and 2.158(4) Å, $Mn{-}N_{ligand}$ = 2.249(5) and 2.313(5) Å. For other atomic distances and bonding angles see the paper. The octahedron around the metal atom is distorted because of steric hindrances due to the coordinated ligands which form five-membered chelate rings. The water molecules displaced on second-order axes form hydrogen bonds. The density of 1.57 g/cm^3 was calculated from X-ray data [11].

The IR and Raman spectra of the complexes reported in [7 to 10] also indicate, that the ligand is coordinated to Mn through the nitrogen of the NH_2 group and the carbonyl oxygen. The bands in the range 600 to 400 cm^{-1} were tentatively assigned to Mn–N and Mn–O vibration modes [8, 9]. The isothiocyanato complex loses its water of crystallization at about 110°C; it is readily soluble in water and ethanol [8], soluble in dimethylformamide and dimethyl sulfoxide [6], and insoluble in usual organic solvents [8] such as diethyl ether even when warmed [6]. The isoselenocyanate is soluble in water, dimethylformamide and dimethyl sulfoxide, less soluble in alcohol and acetone, and insoluble in ether [6].

References:

[1] Kharitonov, Yu. Ya., Machkhoshvili, R. I. (Zh. Neorgan. Khim. **16** [1971] 924/30; Russ. J. Inorg. Chem. **16** [1971] 492/5, 493).

[2] Kharitonov, Yu. Ya., Machkhoshvili, R. I. (Zh. Neorgan. Khim. **14** [1969] 3181; Russ. J. Inorg. Chem. **14** [1969] 1697/80).

[3] Ahmed, A. D., Ray Chaudhuri, N. (J. Inorg. Nucl. Chem. **33** [1971] 189/201, 192/4).

[4] Ahmed, A. D., Mandal, P. K., Ray Chaudhuri, N. (J. Inorg. Nucl. Chem. **28** [1966] 2951/9, 2952).

[5] Kharitonov, Yu. Ya., Machkhoshvili, R. I. (Zh. Neorgan. Khim. **16** [1971] 2697/704; Russ. J. Inorg. Chem. **16** [1971] 1438/42, 1440).

[6] Tsintsadze, G. V., Gogorishvili, P. V., Nagornaya, L. K., Tsivtsivadze, T. N., Machkhoshivili, R. I. (Issled. Obl. Khim. Kompleksn. Prostykh Soedin. Nek. Perekhodnykh Redk. Metal. No. 2 [1974] 194/6; C.A. **82** [1975] No. 25295).

[7] Kharitonov, Yu. Ya., Machkhoshvili, R. I., Generalova, N. B., Shchelokov, R. N. (Zh. Neorgan. Khim. **19** [1974] 1124/5; Russ. J. Inorg. Chem. **19** [1974] 613/4).

[8] Kharitonov, Yu. Ya., Machkhoshvili, R. I., Generalova, N. B., Shchelokov, R. N. (Zh. Neorgan. Khim. **20** [1975] 965/71; Russ. J. Inorg. Chem. **20** [1975] 540/4, 542).

[9] Kharitonov, Yu. Ya., Machkhoshvili, R. I., Tsintsadze, G. V., Gogorishvili, P. V., Nagornaya, L. K. (Zh. Neorgan. Khim. **19** [1974] 2769/73; Russ. J. Inorg. Chem. **19** [1974] 1512/5).

[10] Kharitonov, Yu. Ya., Machkhoshvili, R. I., Generalova, N. B. (Zh. Neorgan. Khim. **22** [1977] 3294/8; Russ. J. Inorg. Chem. **22** [1977] 1796/8).

[11] Tsintsadze, G. V., Tsivtsivadze, T. I., Turiashvili, T. N., Kvitashvili, A. I., Shkurpelo, A. I., Nagornaya, L. K. (Soobshch. Akad. Nauk Gruz. SSR **115** [1984] 537/40; C. A. **102** [1985] No. 123483).

25.1.3 With Derivatives of Acetohydrazide

1) $CH_3C(O)NHNHC_6H_5$ (= $C_8H_{10}N_2O$)
2) $[(CH_3)_3NCH_2C(O)NHNH_2]Cl$ (= $C_5H_{14}ClN_3O$)
3) $NCCH_2C(O)NHNH_2$ (= $C_3H_5N_3O$)
4) $C_6H_5OCH_2C(O)NHNH_2$ (= $C_8H_{10}N_2O_2$)

$Mn(C_8H_{10}N_2O)Cl_2 \cdot 2H_2O$ was prepared from manganese chloride and N′-phenylacetohydrazide as described in [1] for the malonodihydrazide complex $Mn(C_3H_8N_2O)_2Cl_2 \cdot 2H_2O$; see p. 200. The white compound melts at 104°C. The characteristic bands in the IR spectrum (Nujol mulls) were assigned as follows (wave numbers in cm^{-1}, values of ligand in parentheses): 1640 (1635) to amide I, 1608 (—) to $\beta(NH_2)$, 1582 (1587) to amide II, 1330 (1342), 1150 (1158), 1020 (1027) to $\delta(NH_2)$, 770 (772) to $\delta(NCO)$. The N-deuterated complex exhibits the bands assignable to ND vibration modes at lower frequencies. It was concluded that the ligand is monodentate and coordinated to Mn only through the terminal nitrogen atom. Susceptibility measurements (solid at 24°C) yield the magnetic moment $\mu_{eff} = 5.76\ \mu_B$ of a high-spin Mn^{II} complex. $Mn(C_8H_{10}N_2O)Cl_2 \cdot 2H_2O$ is soluble in water. The electronic spectrum of the complex in aqueous solution shows two absorption maxima at $\lambda = 211$ and 257 nm with molar extinction coefficients $\varepsilon = 6237$ and 208 $L \cdot mol^{-1} \cdot cm^{-1}$, respectively, which can be ascribed to the electron transitions $\pi \rightarrow \pi^*$ and n-π^*, respectively. Conductivity measurements on 0.001M aqueous solutions at 24°C yield $\Lambda = 211.7\ cm^2 \cdot \Omega^{-1} \cdot mol^{-1}$ indicating a 1:2 electrolyte [2].

$Mn(C_5H_{14}ClN_3O)Cl_3$ and **$Mn(C_5H_{14}ClN_3O)_2Cl_4$** were prepared by addition of ligand 2 to manganese chloride, both in ethanol, in a 1:1 or 2:1 mole ratio. The mixtures were refluxed on a water bath for about 30 min to induce crystallization and the crystals were washed with absolute ethanol and dried in vacuum over anhydrous $CaCl_2$. The IR spectrum of $Mn(C_5H_{14}N_3O)_2Cl_4$ in KBr disks shows absorption bands of the CO stretching vibration at 1630, of the NH_2 bending vibration at 1620 cm^{-1}, and the amide II vibration at 1480 cm^{-1}, which upon complexation had shifted to lower frequencies by 60, 20, and 80 cm^{-1}, respectively. Such shifts clearly indicate coordination of the ligand to Mn through the carbonyl oxygen and nitrogen of the NH_2 group. Similar coordination of the ligand to Mn is assumed also for $Mn(C_5H_{14}N_3O)Cl_3$. The complexes are strongly dissociated in aqueous solution as shown by their electrical conductance; they are insoluble in organic solvents such as $CHCl_3$, CCl_4, alcohol, and acetone [3].

$Mn(C_3H_5N_3O)_nCl_2 \cdot H_2O$ (n = 1, 2). The complexes were prepared by refluxing manganese chloride and cyanoacetohydrazide (ligand 3) in a mole ratio of 1:1 or 1:2 in absolute ethanol solution. The resulting yellow solids were collected, washed with ethanol, and dried in vacuum; they melt above 280°C. The IR spectrum of $Mn(C_3H_5N_3O)_2Cl_2 \cdot H_2O$ recorded on KBr disks shows characteristic absorption bands (in cm^{-1}) which were assigned as follows: 3280, 3200, 3170, 3050 to $\nu(NH)$, 2240 to $\nu(CN)$, 1670 to amide I, 1550 to amide II, and 1270 to amide III. Similar band positions at somewhat lower frequencies were observed for $Mn(C_3H_5N_3O)Cl_2 \cdot H_2O$; see the paper. The observed shift of the $\nu(NH)$ and amide III bands to higher, and of the $\nu(CN)$ and $\nu(CO)$ bands to lower, frequencies in comparison to the free ligand suggests bidentate chelate-like coordination of the ligand to Mn through the cyano group nitrogen and carbonyl oxygen to form a six-membered metallocycle without participation of the amine nitrogens. The coordination of the cyano nitrogen is supported by a band assignable to $\nu(Mn\text{–}N_{CN})$ in the 800 to 600 cm^{-1} region. Conductivity measurements in dimethylformamide solutions of $Mn(C_3H_5N_3O)_nCl_2 \cdot H_2O$ (0.001M) yield the molar electrical conductance $\Lambda = 23.6$ and 45.6 $cm^2 \cdot \Omega^{-1} \cdot mol^{-1}$ for n = 1 and 2, respectively, indicative of nonelectrolytes in that solvent [4].

$Mn(C_8H_{10}N_2O_2)_3X_2 \cdot H_2O$ (X = Cl, SO_4). Warm aqueous solutions of $MnCl_2$ or $MnSO_4$ were mixed with a hot ethanolic solution of ligand 4 in a 1:3 mole ratio. The mixtures were heated on a water bath for about 1 h and allowed to crystallize. The crystals, which had separated after several days, were collected, washed with warm ethanol, and dried in air. The beige chloride complex melts at 138°C, the white sulfate complex at 222°C.

The most characteristic IR absorption bands (in cm^{-1}) of the complexes in liquid paraffin or hexachlorobutadiene mulls are shown below:

assignment	ν(OH), ν(NH), $ν(NH_2)$	amide I	amide II	ν(C–O)	ν(Mn–N), (Mn–O)
chloride	3400, ~3260, ~3180	1660	~1540	1250	~510, 430
sulfate	3540, ~3300, 3215, 3120	1660	1550	~1250	510, 425
ligand	— 3320, ~3210, ~3100	1670, 1688*)	~1550	1250	—

*) Ligand in CCl_4: $ν_{as}(NH_2)$ at 3447, $ν_s(NH_2)$ at 3335 cm^{-1}.

The shift of the $ν(NH_2)$ and ν(CO) vibration modes toward lower wave numbers (28 cm^{-1} for ν(CO) = amide I) indicates bidentate coordination of the ligand to Mn through the carbonyl oxygen and terminal hydrazide nitrogen. The sulfate complex exhibits two bands assignable to the SO_4^{2-} ion at about 1090($ν_3$) and about 620($ν_4$) cm^{-1}, indicative of outer-sphere anions. This agrees with its molar conductance ($Λ = 296\ cm^2 \cdot Ω^{-1} \cdot mol^{-1}$) of a two-ion electrolyte in aqueous solution at 25°C, whereas the chloride behaves as a three-ion electrolyte ($Λ = 318\ cm^2 \cdot Ω^{-1} \cdot mol^{-1}$). The magnetic moment of the complexes $μ_{eff} = 6.16$ and 6.20 $μ_B$, respectively, is consistent with the octahedral structure concluded from the IR spectrum. The water molecules are water of crystallization. The complexes are readily soluble in water, less soluble in ethanol, and practically insoluble in other common organic solvents [5].

References:

[1] Ahmed, A. D., Mandal, P. K., Ray Chaudhuri, N. (J. Inorg. Nucl. Chem. **28** [1966] 2951/9, 2952).
[2] Ahmed, A. D., Ray Chaudhuri, N. (J. Indian Chem. Soc. **48** [1971] 747/52, 748, 751).
[3] Moussa, M. N. H., Taha, F. I. M., Mostafa, M. M. (Egypt. J. Chem. **16** [1973] 115/23, 116, 119/21; C.A. **80** [1974] No. 77730).
[4] Mostafa, M. M., Hassan, S. M., El-Asmy, A. F. A. (J. Indian Chem. Soc. **55** [1978] 529/31).
[5] Machkhoshvili, R. I., Metreveli, D. P., Mitaishvili, G. Sh., Shchelokov, R. N. (Zh. Neorgan. Khim. **29** [1984] 1020/6; Russ. J. Inorg. Chem. **29** [1984] 586/90).

25.1.4 With Other Hydrazides $RC(O)NHNH_2$ of Aliphatic Carboxylic Acids

1) $(CH_3)_2CHCH_2C(O)NHNH_2$ ($= C_5H_{12}N_2O$)
2) $C_6H_{13}C(O)NHNH_2$ ($= C_7H_{16}N_2O$)
3) $+CH_2C(CH_3)\{C(O)NHNH_2\}+_n$

The complexes $\mathbf{Mn(C_5H_{12}N_2O)_nCl_2 \cdot H_2O}$ (n = 1, 2) with the hydrazide of 3-methylbutanoic acid (ligand 1) were prepared by mixing ethanolic solutions of $MnCl_2 \cdot 4H_2O$ and ligand in a 1:1 or 1:2 mole ratio and refluxing. The pale pink crystals which appeared on cooling were washed with diethyl ether and dried in a vacuum over silica gel. The complex with n = 1 melts at 200°C, that with n = 2 at 230°C. The IR spectrum of both compounds shows a broad absorption band around 3400 cm^{-1} assigned to ν(OH) of H_2O or of the ligand. The coordinated ligand is assumed to be present in its hydroxy hydrazone form, since the band assignable to ν(CO) in the 1750 to 1650 cm^{-1} region is replaced by a new band at 1595 cm^{-1} assigned to ν(C=N). The shift of the band ascribed to ν(N–N) from 1087 to 1105 cm^{-1} in the complexes suggests coordination of the ligand to Mn via one of the nitrogens which evidently is that of the azomethine and not that of the NH_2 group, since the bands assignable to $\beta(NH_2)$ at 1321 cm^{-1} and to $\delta(NH_2)$ at 1159 cm^{-1} are largely unchanged. Bands at 1045 and 920 cm^{-1} assignable to ν_{as}(-C-O) and ν_s(-C-O), respectively, suggest the presence of a free OH group in the coordinated ligand which is supported also by the PMR spectrum of the corresponding diamagnetic complexes in d_6-dimethyl sulfoxide. Bands at 242 and 239 cm^{-1} were assigned to ν(Mn-Cl) vibrations of the 1:1 and 1:2 complexes, respectively. The manganese atom is assumed to be six-coordinate. Susceptibility measurements at room temperature yield the magnetic moments $\mu_{eff}=6.35$ and 6.09 μ_B for the complexes with n=1 and 2, respectively. The molar electrical conductance $\Lambda=46$ (n=1) or 57 (n=2) $cm^2 \cdot \Omega^{-1} \cdot mol^{-1}$ measured at 25°C in dimethylformamide solution is indicative of nonelectrolytes in that solvent. The complexes are insoluble in nonpolar solvents but dissolve in dimethylformamide and dimethyl sulfoxide [1].

The complexes **$Mn(C_7H_{16}N_2O)_3Cl_2 \cdot 0.5H_2O$** and **$Mn(C_7H_{16}N_2O)_3SO_4 \cdot H_2O$** were prepared by mixing aqueous ethanolic (50 vol%) solutions of $MnCl_2 \cdot 4H_2O$ or manganese sulfate and heptanohydrazide (ligand 2) in a 1:3 mole ratio and refluxing the mixture for 2 h. From the filtered solution fine crystalline, bright yellow (chloride) or white (sulfate) precipitates separated after standing for several days. They were washed with water and ethanol and dried in air. The chloride decomposes at 135°C, the sulfate at 201°C. The positions (shifts) of the ligand bands in the complexes suggest coordination of heptanohydrazide to Mn through the nitrogen of the NH_2 group and the oxygen of the carbonyl group. Bands of ν(NH) were observed at 3280, 3240, and 3100 cm^{-1} (chloride), 3280 and 3240 cm^{-1} (sulfate), those of ν(CO) at 1660 cm^{-1} (chloride) and 1645 cm^{-1} (sulfate). The SO_4^{2-} ion of the sulfato complex is in the outer sphere, since there is only one single band assignable to $\nu_3(SO_4^{2-})$. The IR spectra and high-spin magnetic moments of the complexes ($\mu_{eff}=6.10$ and 5.86 μ_B, respectively) indicate an apparently octahedral structure. The complexes decompose upon heating and are insoluble in water, ethanol, $CHCl_3$, CCl_4, acetone, benzene, and nitrobenzene at room temperature; upon heating they dissolve to some extent in water and ethanol [2].

A complex containing Mn and N in a 1:1 atom ratio was obtained by mixing about equal volumes of an aqueous 10% solution of $MnSO_4$ and a 1% solution of the ligand 3 (= polymethacrylohydrazide) at pH ~7. The mixture was heated on a water bath to improve coagulation of the precipitate, which was washed with water to remove unreacted $MnSO_4 \cdot nH_2O$ and ligand, and dried over H_2SO_4. The product has the density D=1.27 g/cm^3 measured by pycnometry with benzene. The magnetic moment $\mu_{eff}=4.8$ μ_B indicates a high-spin paramagnetic Mn^{II} complex which is insoluble in water and common organic solvents but soluble in concentrated aqueous ammonia and swells in hydrazine. It is assumed that the Mn-to-ligand bonds are ionic [3]. Manganese complexes of a copolymerizate of acrylohydrazide and divinylbenzene (varying ratios) are described with their physical properties (density, magnetic moment) by [4] and the IR spectra are given by [5]. It is assumed that they are high-spin complexes [3] and contain coordinative bonds between manganese and the terminal nitrogens of the hydrazide groups [5].

References:

[1] Mostafa, M. M., Nicholls, D. (Inorg. Chim. Acta **51** [1981] 35/8).

[2] Machkhoshvili, R. I., Mitaishvili, G. Sh., Pirtskhalava, N. I. (Zh. Neorgan. Khim. **27** [1982] 972/4; Russ. J. Inorg. Chem. **27** [1982] 545/7).

[3] Tolmachev, V. N., Lomako, L. A., Miroshnik, L. V. (Vysokomol. Soedin. B **10** [1968] 905/7; C.A. **70** [1969] No. 47973).

[4] Lomako, L. A., Tolmachev, V. N. (Ukr. Khim. Zh. **37** [1971] 1141/5; Soviet Progr. Chem. **37** No. 11 [1975] 55/8, 57).

[5] Lomako, L. A., Ishchenko, I. K., Tolmachov, V. M. (Vestn. Khar'kov. Univ. No. 84 [1972] 79/82; C.A. **78** [1973] No. 148416).

25.1.5 With Benzohydrazide $C_6H_5C(O)NHNH_2$ (= $C_7H_8N_2O$)

$Mn(C_7H_8N_2O)_3Cl_2 \cdot nH_2O$ (n = 2, 3). Yellow-brown crystals of the trihydrate were synthesized by mixing ethanol solutions of manganese(II) chloride and benzohydrazide in the appropriate mole ratio. The compound was recrystallized from hot ethanol [1]. Brown crystals containing two moles of water separated on prolonged standing (4 to 5 d) from an aqueous solution of $MnCl_2$ and the ligand [2]. The IR absorption spectrum was investigated in the range 4000 to 400 cm^{-1} using mulls in liquid paraffin and fluorinated liquid paraffin. The spectra reveal linkage between the ligand and manganese via the nitrogen atom of the NH_2 group and the carbonyl oxygen atom forming a five-membered chelate ring. Bands of ν(NH) observed at 3140 cm^{-1} and the amide I band at 1640 cm^{-1} are displaced to lower wave numbers in comparison to the free ligand. Bands in the region 600 to ~400 cm^{-1} were assigned tentatively to vibrations of the Mn–N and Mn–O bonds [3]. The observed shifts in ν(CO) of several $M(C_7H_8N_2O)_3Cl_2$ complexes with M = Ni^{II}, Co^{II}, Mn^{II} suggest the following order of M–O bond strengths in the complexes: $Ni^{II} > Co^{II} > Mn^{II}$. Water molecules are only present as lattice water, since the trihydrate does not change color upon dehydration and its IR spectrum shows no bands assignable to coordinated water which would disappear [1]. The magnetic moment $\mu_{eff} = 6.1\ \mu_B$ resulting from susceptibility measurements (Faraday method) at room temperature is indicative of a high-spin octahedral Mn^{II} complex [2].

The complex melts at 71°C [1]. Thermal analysis (TG) reveals two endothermic effects at 73 and 130°C due to the loss of the water molecules between 73 and 130°C and two exothermic effects at 345 and 474°C, indicating the removal of almost all benzohydrazide, which starts at 278°C [2]. $Mn(C_7H_8N_2O)_2Cl_2 \cdot 2(\text{or } 3)H_2O$ is readily soluble in water [1, 2], the solubility being 25.55 g/100 mL·H_2O and the molar electrical conductance of the solution measured at 25°C is $\Lambda = 265.5\ cm^2 \cdot \Omega^{-1} \cdot mol^{-1}$ at dilution rate V = 1000 L/mol indicating a 1:2 electrolyte [2]. The complex is sparingly soluble in ethanol [1, 2], insoluble in other organic solvents like $CHCl_3$, CCl_4, ether, benzene, nitrobenzene [1], but soluble in molten phenol to give a red solution. It is decomposed by mineral acids [2]. The electronic spectrum of the complex (in ethanolic solution) exhibits two absorption maxima at λ = 228 and 420 nm, the latter with the molar extinction coefficient $\varepsilon = 4.42\ L \cdot mol^{-1} \cdot cm^{-1}$ [1].

$Mn(C_7H_8N_2O)Cl_2$. The white complex was prepared from manganese chloride and benzohydrazide [4] by the method given in [5] for the synthesis of the malonodihydrazide complex $Mn(C_3H_8N_4O_2)_2Cl_2 \cdot 2H_2O$ (p. 200). The IR bands of the complex (Nujol mulls, ν in cm^{-1}) were assigned as follows: 1640 to ν(C=N), 1608 to β(NH_2), 1327 to ν(C=N), 1310 to τ(NH_2), 1085 to ω(NH_2), and 1040 to ν(C–O). The coordinated ligand was assumed to be present in its hydroxy hydrazone form and coordination to Mn only through the azomethine nitrogen suggested. The visible spectrum of the complex in aqueous solution shows only weak absorption. The magnetic moment of the solid obtained at 24°C, $\mu_{eff} = 5.59\ \mu_B$, suggests a high-spin configuration and the conductance of about 0.001M aqueous solutions measured at 25°C, $\Lambda = 208\ cm^2 \cdot \Omega^{-1} \cdot mol^{-1}$, indicates a 1:2 electrolyte [4].

$Mn(C_7H_8N_2O)_2SO_4 \cdot 2H_2O$. A hot aqueous solution of manganese(II) sulfate was mixed with a hot ethanolic solution of benzohydrazide in a 1:2 mole ratio [1, 6]. The mixture was refluxed on a water bath for 10 min. The pale rose-red [6] or shining white [1] fine crystals which appeared upon cooling were washed with ethanol [1] and dried in vacuum [1, 6]. The IR spectrum (KBr disks) shows bands of amide I at 1641 cm^{-1} and NH_2 deformation bands at 1605, 1333, and 990 cm^{-1}; the corresponding wave numbers of the ligand are 1660, 1618, 1355, and 996, respectively. The IR bands expected in the ν(NH_2) region are covered by those ascribed to the water molecules, and bands indicating the presence of sulfate occur in the 1200 to 1100

cm^{-1} region [6]. Since bands assignable to ionic sulfate (around 1230 cm^{-1}) are absent, a bidentate bridging function of the sulfato groups in an apparently dimeric structure of the complex ($L_2Mn(SO_4)_2MnL_2$) can be assumed; this is supported also by its high magnetic moment $\mu_{eff} = 7.2\ \mu_B$ possibly due to ferromagnetic impurities. The complex decomposes upon heating above 240°C; it is insoluble in water, ethanol, $CHCl_2$, CCl_4, ether, benzene, nitrobenzene, and other organic solvents [1].

$Mn(C_7H_8N_2O)_3S_2O_6 \cdot H_2O$ was prepared by reaction of MnS_2O_6, with benzohydrazide (mole ratio 1:3) in aqueous solution. The fine pink crystals which precipitated were collected after 10 d, washed with water and ethanol, and dried over P_2O_5 in a desiccator. They melt at 220°C. The IR spectrum of the complex in liquid paraffin and fluorinated hydrocarbons shows shifts of the ν(NH) and ν(CO) ligand bands, similar to the complexes above. The position of the $S_2O_6^{2-}$ stretching and deformation modes as compared with those of the alkalidithionates indicates the presence of $S_2O_6^{2-}$ ions in the outer sphere of the evidently octahedral complex structure. The water content is water of crystallization. $Mn(C_7H_8N_2O)_3S_2O_6 \cdot H_2O$ is readily soluble at room temperature in dimethylformamide and dimethyl sulfoxide, less readily soluble in water and ethanol, where the solublity increases on heating, and insoluble in $CHCl_3$, CCl_4, acetone, nitrobenzene, and tetrahydrofuran [7].

$Mn(C_7H_8N_2O)_2(NCO)_2 \cdot 1.5(CH_3)_2CO$. Solutions of manganese(II) nitrate in acetone and KOCN in aqueous acetone were mixed in a 1:2 mole ratio, the precipitated KNO_3 was removed and an acetone solution of benzohydrazide was added up to an Mn:ligand ratio = 1:3. The complex, which crystallized upon leaving the mixture in a vacuum over $CaCl_2$ for 2 to 3 d, was washed quickly with small portions of water (to remove any adherent KNO_3) and dried between filter paper. It forms a yellow crystalline powder which according to X-ray studies is isomorphous with $Co(C_7H_8N_2O)_2(NCO)_2 \cdot 1.5(CH_3)_2CO$ [8].

The IR spectrum of the complex (in liquid paraffin and fluorinated hydrocarbon mulls) reveals a displacement of the ν(NH) wave numbers towards lower values by about 150 cm^{-1} relative to the free ligand. The amide I band is located at 1618 cm^{-1}, while at 1645 cm^{-1} only an inflection was observed. (The amide I band is near 1665 cm^{-1} in the crystalline ligand and near 1672 cm^{-1} in the spectrum of its acetonitrile solution.) The position of the bands assigned to the OCN groups indicates coordination to Mn through the nitrogen atoms which complete the octahedral structure of the complex [9]. The complex is soluble in dimethylformamide, rather less soluble in dimethyl sulfoxide, and insoluble in water, alcohol, and acetone [8].

$Mn(C_7H_8N_2O)_2(NCS)_2$. Ethanolic solutions of manganese(II) nitrate and KSCN were mixed in a 1:2 mole ratio, the precipitated KNO_3 was removed, and an ethanolic solution of benzohydrazide was added in a 1:1 or higher mole ratio. The complex crystallized upon leaving the mixture in a vacuum over $CaCl_2$ for 2 to 3 d [8]. It was obtained also by reacting an aqueous solution of $Mn(C_7H_8N_2O)_3Cl_2 \cdot 2H_2O$ with the stoichiometric amount of NaSCN and allowing the mixture to stand for 4 to 5 d [10].

Transparent pink leaflets were obtained from alcohol [8] or coarse cinnamon-brown crystals from water, melting at 200°C [10]. X-ray diffractometer studies (MoKα radiation) showed the complex to be monoclinic, space group $P2_1/n\text{-}C_{2h}^5$ (No. 14) with the lattice constants a = 13.897(3), b = 10.704(4), c = 13.685(7) Å, and γ = 101.06(4)°; Z = 4, V = 1997.85 $Å^3$. The structure was solved from 1939 independent reflections and refined anisotropically up to R = 0.068. The complex is monomeric, and the Mn atom is coordinated to two oxygens and two terminal amide nitrogens of the two bidentate benzohydrazide ligands and the nitrogens of the two thiocyanate groups in a distorted *cis*-octahedral configuration, see **Fig. 19**, p. 176. The atomic distances within the coordination sphere of Mn are (in Å): Mn–O(1) = 2.188(4), Mn–O(2) = 2.205(4), $Mn\text{–}N(1)_{NCS}$ = 2.136(7), $Mn\text{–}N(2)_{NCS}$ = 2.125(6), Mn–N(3) = 2.355(5),

Mn–N(5) = 2.316(6). For other atomic distances and bond angles see the paper. The *cis*-arrangement of the thiocyanato groups is supported by a ν(CN) band doublet structure in the IR spectrum. The density of 1.574 g/cm³ was calculated from X-ray data [14].

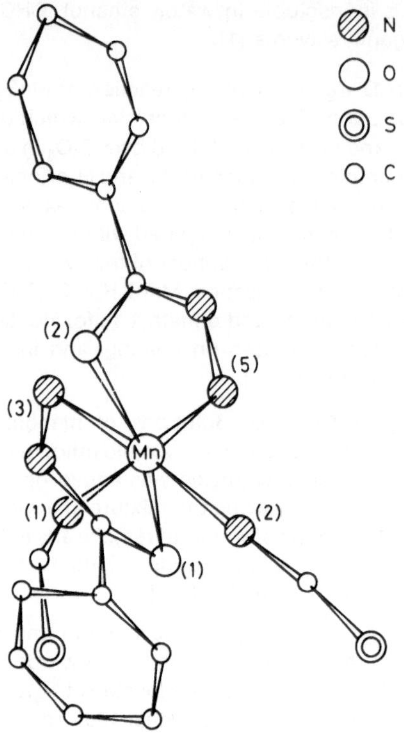

Fig. 19. Molecular structure of $Mn(C_7H_8N_2O)_2(NCS)_2$ reported in [15].

The refractive indices are $n_\alpha = 1.686$, $n_\beta = 1.737$, and $n_\gamma = 1.780$, and the magnetic moment $\mu_{eff} = 5.93\ \mu_B$ was measured at ~290 K [8]. The complex is soluble in water, alcohol, and dimethylformamide [8], and according to [10] only slightly soluble in water.

$Mn(C_7H_8N_2O)_2(NCSe)_2 \cdot 2C_2H_5OH$ and **$Mn(C_7H_5D_3N_2O)_3(NCSe)_2$**. The nondeuterated compound was prepared from manganese(II) nitrate, KSeCN and benzohydrazide in ethanol in the same way as $Mn(C_7H_8N_2O)_2(NCS)_2$ (see above). The N-deuterated, unsolvated complex was obtained by treating $Mn(C_7H_8N_2O)_3(NCSe)_2$ with D_2O in a sealed quartz ampule at 105 to 110°C for several hours and removing the excess of D_2O [11]. $Mn(C_7H_8N_2O)_3(NCSe)_2 \cdot 2C_2H_5OH$ forms transparent crystals which according to X-ray studies are isomorphous with the corresponding complexes of Co and Ni and effloresce in air within 1 d [8]. The IR spectrum is similar to those of the isocyanato and isothiocyanato complexes above. The position of the essentially unsplit bands assigned to the ν(CN) mode at 2095 and 2050 cm^{-1} and ν(CSe) mode at 618 cm^{-1} indicates coordination of the NCSe groups to Mn through terminal nitrogen and their probable arrangement on trans(apical) positions to each other in an assumed octahedral structure. The ethanol molecules are apparently solvent of crystallization [11]. The complex is almost insoluble in cold water but its solubility increases on heating; it is sparingly soluble in alcohol and acetone and readily soluble in dimethylformamide and dimethyl sulfoxide [8].

$Mn(C_7H_8N_2O)_3(C_6H_2N_3O_7)_2 \cdot 2H_2O$ and **$[Mn(C_7H_8N_2O)_2(H_2O)_2](C_6H_2N_3O_7)_2$** ($C_6H_2N_3O_7^-$ = anion of 2,4,6-trinitrophenol = picric acid). The tris complex was prepared by mixing aqueous solutions of manganese(II) picrate and benzohydrazide in an about 1:3 mole ratio. The precipitate was washed with water and dried in a vacuum. $[Mn(C_7H_8N_2O)_2(H_2O)_2](C_6H_2N_3O_7)_2$ forms when ethanolic solutions of the components were mixed in the same ratio, refluxed for about ½ h, and concentrated on a water bath. The crystals which separated on cooling were washed with cold ethanol, dried in vacuum, and recrystallized from ethanol or acetone. The complexes are yellow. $Mn(C_7H_8N_2O_3)_3(C_6H_2N_3O_7)_2 \cdot 2H_2O$ melts at 115°C, $[Mn(C_7H_8N_2O)_2(H_2O)_2]$-$(C_6H_2N_3O_7)_2$ at 120°C. The most significant absorption bands (in cm^{-1}) in the IR spectra (KBr disks or Nujol mull) were assigned as follows:

complex	ν(OH) of H_2O	ν(NH) + ν(NH_2)	ν(CO)	β(NH_2)	ν(NN)	ν(Mn–N)	ν(Mn–O)
$MnL_3X_2 \cdot 2H_2O$	3600 to 3325	3250, 3200	1642	1610	912	484	358
$[MnL_2(H_2O)_2]X_2$	3600 to 3320	3250, 3200	1640	1610	912	490	340
ligand in CH_3CN	—	3350, —	1672	1634	885	—	—

L = $C_7H_8N_2O$, X = $C_6H_2N_3O_7^-$

The shift of the ν(NH) and ν(CO) bands to lower and of the ν(NN) bands to higher frequencies in the complexes suggests coordination of benzohydrazide to Mn through carbonyl oxygen and the nitrogen of the NH_2 group. The observed magnetic moments μ_{eff} = 6.12 and 6.17 μ_B, respectively, indicate that the complexes are spin-free octahedral [12].

$Mn(C_7H_8N_2O)_3(C_6H_2N_3O_7)_2 \cdot 2H_2O$ loses its water at about 110°C and $[Mn(C_7H_8N_2O)_2(H_2O)_2]$-$(C_6H_2N_3O_7)_2$ does so around 150°C, which indicates lattice water in the former and coordinated water in the latter case. Upon heating to about 250°C the complexes will explode. They are insoluble in water but soluble in methanol, ethanol, and acetone and are 1:2 electrolytes in 0.001M methanol solution (Λ = 196 and 220 $cm^2 \cdot \Omega^{-1} \cdot mol^{-1}$, respectively) [12].

Complexes of composition **$Mn(C_7H_8N_2O)(C_6H_9N_3O_2)Cl_2$** (with $C_6H_9N_3O_2$ = histidine) were prepared by mixing aqueous solutions of $MnCl_2 \cdot 4H_2O$, benzohydrazide, and histidine and heating the mixture on a water bath. The precipitate was washed with water and benzene. The tetrahydrate, **$Mn(C_7H_8N_2O)(C_6H_9N_3O_2)Cl_2 \cdot 4H_2O$**, was obtained by adding histidine to a warm ethanolic solution of $MnCl_2 \cdot 4H_2O$ and benzohydrazide. The mixture was evaporated to leave a flesh-colored powder which was washed with ether. The ammonia-containing compound **$Mn(C_7H_8N_2O)(C_6H_9N_3O_2)Cl_2 \cdot NH_3 \cdot 2H_2O$** was obtained by dissolving the mixture of $MnCl_2 \cdot 4H_2O$, benzohydrazide, and histidine in aqueous ammonia with heating followed by evaporation. The syrupy residue was treated again with ammonia, evaporated and dried in a desiccator over H_2SO_4. The flesh-colored solid was washed with benzene. The ethanol solvate **$Mn(C_7H_8N_2O)(C_6H_9N_3O_2)_2Cl_2 \cdot C_2H_5OH$** was obtained from an ethanolic solution of the components which was evaporated and was washed like the tetrahydrate. In the following table are listed: the composition of the complexes, the melting points, solubility S in water and ethanol, and the pH values and molar electrical conductance Λ of aqueous solutions at 25°C, and a dilution rate V ~ 900 L/mol:

compound*)	m.p. in °C	S_{water} in g/100 mL	pH	Λ in $cm^2 \cdot \Omega^{-1} \cdot mol^{-1}$	$S_{ethanol}$ in g/100 mL
$Mn(C_7H_8N_2O)(C_6H_9N_3O_2)_4Cl_2$	80	18.87	6.90	218.6	11.43
$Mn(C_7H_8N_2O)(C_6H_9N_3O_2)_4Cl_2 \cdot 4H_2O$	45	9.16	6.84	243.2	3.81

compound*)	m.p. in °C	S_{water} in g/100 mL	pH	Λ in $cm^2 \cdot \Omega^{-1} \cdot mol^{-1}$	$S_{ethanol}$ in g/100 mL
$Mn(C_7H_8N_2O)(C_6H_9N_3O_2)_4Cl_2 \cdot NH_3 \cdot 2H_2O$	70*)	3.68	8.25	252	—
$Mn(C_7H_8N_2O)(C_6H_9N_3O_2)_4Cl_2 \cdot C_2H_5OH$	55	11.54	7.98	228	—

*) Color change at 60°C.

The complexes are 1:2 electrolytes in water. Their conductance increases on further dilution or warming. The ammonia and ethanol solvates dissolve in acetone upon heating [13].

References:

[1] Aggarwal, R. C., Narang, K. K. (Indian J. Chem. A **14** [1976] 64/7).

[2] Gogorishvili, P. V., Karkarashvili, M. V., Kalandarishvili, D. Z. (Zh. Neorgan. Khim. **14** [1969] 1516/20; Russ. J. Inorg. Chem. **14** [1969] 794/6).

[3] Kharitonov, Yu. Ya., Machkhoshvili, R. I., Gogorishivili, P. V., Karkarashvili, M. V. (Zh. Neorgan. Khim. **17** [1972] 1059/66; Russ. J. Inorg. Chem. **17** [1972] 550/3, 552).

[4] Dutta Ahmed, A., Ray Chaudhuri, N. (J. Inorg. Nucl. Chem. **33** [1974] 189/201, 192/4).

[5] Dutta Ahmed, A., Mandal, P. K., Ray Chaudhuri, N. (J. Inorg. Nucl. Chem. **28** [1966] 2951/9, 2952).

[6] Issa, R. M., El-Shazly, M. F., Iskander, M. F. (Z. Anorg. Allgem. Chem. **354** [1967] 90/7, 92, 95).

[7] Machkhoshvili, R. I., Pirtskhalava, N. I. Kharitonov, Yu. Ya., Kvernadze, M. S. (Zh. Neorgan. Khim. **23** [1978] 1018/23, 547/8; Russ. J. Inorg. Chem. **23** [1978] 563/6, 565, 304/5).

[8] Tsintsadze, G. V., Gogorishvili, P. V., Nagormaya, L. K., Tsivtsivadze, T. I. (Nauchn. Tr. Gruz. Politekhn. Inst. im. V. I. Lenina No. 153, Pt. 5 [1972] 34/9, 35/7; C.A. **80** [1974] No. 66253).

[9] Kharitonov, Yu. Ya., Machkhoshvili, R. I., Tsintsadze, G. V., Gogorishvili, P. V., Nagornaya, L. K. (Zh. Neorgan. Khim. **20** [1975] 1281/5; Russ. J. Inorg. Chem. **20** [1975] 720/2).

[10] Karkarashvili, M. V., Gogorishvili, P. V., Nagebashvili, Dzh. S. (Soobshch. Akad. Nauk Gruz.SSR **51** [1968] 101/6, 105; C.A. **69** [1968] No. 100085).

[11] Kharitonov, Yu. Ya., Machkhoshvili, R. I., Tsintsadze, G. V., Gogorishvili, P. V., Nagornaya, L. K. (Zh. Neorgan. Khim. **19** [1974] 1337/44; Russ. J. Inorg. Chem. **19** [1974] 727/32, 731).

[12] Aggarwal, R. C., Singh, N. K. (Def. Sci. J. **26** No. 4 [1976] 169/74, 170/1; C.A. **87** [1977] No. 15264).

[13] Karkarashvili, M. V., Beshkenadze, I. A. (Issled. Obl. Khim. Kompleksn. Prostykh. Soedin. Nek. Perekhodnykh Redk. Metal. No. 3 [1978] 44/51, 45/8; C.A. **90** [1979] No. 214471).

[14] Tsintsadze, G. V., Tsivtsivadze, T. I., Turiashvili, T. N., Kvitashvili, A. I., Shkurpelo, A. I., Nagornaya, L. K. (Soobshch. Akad. Nauk Gruz.SSR **111** [1983] 65/8; C.A. **99** [1983] No. 222804).

[15] Tsintsadze, G. V., Tsivtsivadze, T. I., Turiashvili, T. N., Kvitashvili, A. I., Shkurpelo, A. I., Nagonaya, L. K. (Soobshch. Akad. Nauk Gruz.SSR **115** [1984] 537/40; C.A. **102** [1985] No. 123483).

25.1.6 With 3-Nitrobenzohydrazide $3\text{-}NO_2C_6H_4C(O)NHNH_2$ ($=C_7H_7N_3O_3$)

$Mn(C_7H_7N_3O_3)_3Cl_2\cdot nH_2O$ (n = 1, 2). Solutions of $MnCl_2\cdot4H_2O$ (1 g) and the ligand in boiling alcohol (mole ratio 1:3) were allowed to crystallize. Pale pink crystals of the monohydrate, which are air-stable and have the refractive indices $n_\alpha=1.527$ and $n_\gamma>1.700$, formed slowly [1]. The IR spectrum of the complex (formulated as dihydrate) resembles those of the benzohydrazide complexes described before: coordination of the ligand to Mn through the terminal nitrogen and carbonyl oxygen is suggested [2]. Thermal analysis reveals an endothermic effect at 130°C indicating loss of water. Upon further heating the complex is oxidized and decomposes around 200°C with detonation. The solubility of the complex in water is 0.39 g/100 mL H_2O at 25°C, the solution has a pH of 6.1, and the molar electrical conductance, $\Lambda=299\ cm^2\cdot\Omega^{-1}\cdot mol^{-1}$, indicates a 1:2 electrolyte [1].

References:

[1] Gogorishvili, P. V., Karkarashvili, M. V., Shamilishvili, O. Kh. (Soobshch. Akad. Nauk Gruz.SSR **62** [1971] 61/4; C.A. **75** [1971] No. 54107).

[2] Kharitonov, Yu. Ya., Machkhoshvili, R. I., Gogorishvili, P. V., Shamilishvili, O. Kh. (Zh. Neorgan. Khim. **17** [1972] 1631/7; Russ. J. Inorg. Chem. **17** [1972] 843/7, 845).

25.1.7 With Hydroxybenzohydrazides $HOC_6H_4C(O)NHNH_2$ ($=C_7H_8N_2O_2$)

25.1.7.1 With 2-Hydroxybenzohydrazide (= HL)

Formation in Solution. The formation constant of the 1:1 chelate in 25:75 (v/v) dioxane-water medium was determined potentiometrically (glass electrode) at 25°C, I = 0.1 M $NaClO_4$: $\log K_1=2.76\pm0.05$. The pK_{a1} value of the ligand is 8.56. Corrections of the pH titrations in the nonaqueous solvent were applied. The free enthalpy of formation of the 1:1 complex was calculated as $\Delta G=-3.76$ kcal/mol [1].

The same complex was detected by spectrophotometric studies at $\lambda=220$, 240, and 500 nm (Job's method) on 0.001 M solutions of the components in water containing 10% dioxane and by conductometric titrations (varying ratios). The dissociation constant is reported as 9.7×10^{-4} mol/L at 26 ± 0.1°C. It is assumed that Mn is coordinated to the carbonyl oxygen and terminal nitrogen of the carbohydrazide group and to the deprotonated phenolic oxygen of the terdentate ligand and to at least one molecule of water [2]. Upon comparison with other metal complexes of the ligand the stability series: $Cu^{II}>Be^{II}>Hg^{II}>Ni^{II}>Co^{II}>Mn^{II}>Mg^{II}$ results [1].

$Mn(C_7H_7N_2O_2)_2$. An aqueous solution of a manganese(II) salt was added to that of 2-hydroxybenzohydrazide in aqueous ammonia (neutralized by dilute HCl) in an Mn:ligand = 1:4 mole ratio. The precipitate was filtered off, washed with water and hot dilute ethanol and dried in a desiccator. The IR spectrum of the light yellow complex (KBr disks) indicates involvement of the carbonyl group in coordination. The negative shifts of the ring-stretching modes (~10 to 15 cm^{-1}) can be attributed to the simultaneous involvement of the phenolate oxygen in bonding. The $\delta(NH_2)$ bands are unaffected or slightly shifted. Susceptibility measurements (Faraday method) yield $\mu_{eff}=6.0\ \mu_B$. The complex is believed to be polymeric since it decomposes upon heating above 200 to 250°C without melting and is insoluble in water and common organic solvents [3].

$Mn(C_7H_8N_2O_2)_2X_2\cdot2H_2O$ (X = Cl, $SO_4/2$). The chloride was prepared from $MnCl_2\cdot4H_2O$ and the ligand (mole ratio 1:2) in absolute ethanol. The mixture was partly evaporated on a water bath. Ether was added to precipitate a white fine-grained solid which was removed by suction, washed with ether and dried and stored in vacuum. The complex is strongly hygroscopic when moist but stable when dry [4]. The $Mn(C_7H_8N_2O_2)_2SO_4 \cdot 2H_2O$ was prepared by mixing the ligand in hot ethanol with an aqueous solution of manganese(II) sulfate. The pale brown product was dried in vacuum over silica gel [5]. The shifts observed in the IR spectra of the complexes in KBr disks (mainly those of bands assigned to ν(NH), ν(C=O), ν(C–O), and $\delta(NH_2)$ vibration modes) suggest coordination of the ligand to Mn only through the carbonyl oxygen and the terminal nitrogen of the carbohydrazide group without participation of the phenolic oxygen [4, 5]. Characteristic absorption bands of the chloride in cm^{-1} (ligand bands in parentheses) were assigned as follows: 3230 (3190) to ν(NH); 1638 (1640), 1628 (1625) to ν(C=O); 1610 (1600, 1585) to $\delta(NH_2)$; 1535 (1547, 1532), 1320 (1302) to CNH vibrations; 1352 (1367, 1350), 1198 (1250) to ν(C–O) + δ(OH); 1120 (1155, 1135) to $\omega(NH_2)$; 898 (952), 865 (880) to ν(N–N); 830 (843, 825), 795 (795) to $\tau(NH_2)$. It was assumed that the water molecules are linked to the terminal hydrazide nitrogen and (apparently coordinated) Cl^- ions through hydrogen bridges which exist also between the carbonyl and phenolic oxygens. An octahedral structure of the complex is proposed [4]. The observed shift of δ(OH) indicates a change in the H bond strength due to chelation of the carbonyl oxygen by Mn. Bands indicating the presence of sulfate ions occur in the 1200 to 1100 cm^{-1} region [5].

$Mn(C_7H_8N_2O_2)_2Cl_2 \cdot 2H_2O$ according to thermal analysis (DTA) loses its water between 40 and 115°C (endothermic peak at 85°C). Upon further heating two exothermic effects with maxima (peaks) at 280 and 455°C indicating decomposition and finally combustion are observed [4]. The sulfate decomposes without melting upon heating above 250°C [5].

$[Mn(C_7H_8N_2O_2)_2(H_2O)_2](C_6H_2N_3O_7)_2$. The complex was prepared from manganese(II) picrate and 2-hydroxybenzohydrazide in ethanol in the same way as the benzohydrazide complex $[Mn(C_7H_8N_2O)_2(H_2O)_2](C_6H_2N_3O_7)_2$ (p. 177). The yellow compound melts at 135 to 140°C [6].

The bands occurring in the IR spectrum (KBr disks and Nujol mulls) were assigned as for the benzohydrazide complex, p. 177. The ν(OH) vibration mode of 2-hydroxybenzohydrazide in acetonitrile, at 3620 to 3540 cm^{-1} is only shifted to 3600 or 3550 cm^{-1} indicating that this group is not involved in bonding. The small negative shift appears to be due to intermolecular hydrogen bonding between the phenolic group of the ligand and the nitro group of the picrate ion. The bands at 475 and 340 cm^{-1} may be assigned to Mn–O and Mn–N vibrations, respectively. Susceptibility measurements yield the magnetic moment $\mu_{eff} = 6.13\ \mu_B$ of a high-spin octahedral Mn^{II} complex. The water molecules are apparently coordinated since they are released only upon heating to 150°C. Upon further heating, the complex will explode at 250°C. It is insoluble in water but soluble in methanol, ethanol, and acetone; the molar electrical conductance $\Lambda = 216\ cm^2 \cdot \Omega^{-1} \cdot mol^{-1}$ of a 0.001M solution in methanol indicates a 1:2 electrolyte [6].

References:

[1] Grewal, S., Sekhon, B. S., Pannu, B. S., Chopra, S. L. (Indian J. Chem. **13** [1975] 623/4).
[2] Kachhawaha, M. S., Bhattacharya, A. K. (J. Inorg. Nucl. Chem. **25** [1963] 361/3).
[3] Gopal, R., Misra, V. N., Narang, K. K. (Indian J. Chem. A **14** [1976] 364/6).
[4] Mach, J. (Acta Univ. Palacki Olomuc. Fac. Rerum Nat. No. 41 [1973] 27/36, 29/32; C.A. **81** [1974] No. 44787).
[5] Issa, R. M., El-Shazly, M. F., Iskander, M. F. (Z. Anorg. Allgem. Chem. **354** [1967] 90/7, 91/2, 95/6).
[6] Aggarwal, R. C., Singh, N. K. (Def. Sci. J. **26** No. 4 [1976] 169/74; C.A. **87** [1977] No. 15264).

25.1.7.2 With 2-Hydroxybenzohydrazide and Benzohydrazide

The mixed ligand compound **Mn(HL)L'Cl$_2$·2H$_2$O** containing 2-hydroxybenzohydrazide (HL = $C_7H_8N_2O_2$) and benzohydrazide (L' = $C_7H_8N_2O$) was prepared by adding successively HL (1.54 g) and L' (1.30 g) to an aqueous (or ethanolic) solution of $MnCl_2 \cdot 4H_2O$ (4 g). The mixture was heated to afford complete dissolution of the reagents. The filtered solution was evaporated on a water bath to leave a flesh-colored powder. **Mn(HL)L'Cl$_2$·NH$_3$·H$_2$O** was obtained if the components were dissolved in aqueous ammonia (25%, 60 mL). To produce the acetone solvate, **Mn(HL)L'Cl$_2$·C$_3$H$_6$O**, a solution of benzohydrazide in acetone was added to the aqueous solution of $MnCl_2 \cdot 4H_2O$ and 2-hydroxybenzohydrazide. The mixture was evaporated and the residue was redissolved in acetone and evaporated again. The pink powder was washed with ethanol and ether. It changes color upon heating at 225°C.

The complex **MnLL'Cl·2H$_2$O**, containing 2-hydroxybenzohydrazide in the deprotonated form, was synthesized in the same way as the complex Mn(HL)L'Cl$_2$·2H$_2$O, see above, using a mixture of ethanol (60 mL) and aqueous ammonia (25%, 10 mL) instead of water. The straw-colored precipitate was washed with water and dried in a thermostat. It becomes bright yellow at 220°C and darkens at 270°C. In the following table are listed: the melting points, the solubility S in water and ethanol (in g/100 mL), and the pH values and molar electrical conductance Λ (in $cm^2 \cdot \Omega^{-1} \cdot mol^{-1}$) of aqueous solutions at 25°C and a dilution rate V ~ 900 L/mol:

compound	m.p. in °C	S_{H_2O}	$S_{C_2H_5OH}$	Λ	pH
Mn(LH)L'Cl$_2$·2H$_2$O	70	2.69	1.43	243.9	5.46
Mn(LH)L'Cl$_2$·NH$_3$·H$_2$O	60	1.54	—	189.9	7.75
Mn(LH)L'Cl$_2$·C$_3$H$_6$O	230	[a]	—	74.9	5.70
MnLL'Cl·2H$_2$O	>350	0.36	—	18.6	8.26

[a] Soluble, no data.

The dichloro complexes are 1:2 electrolytes in aqueous solution (the acetone solvate apparently only at higher temperature or dilution) whereas MnLL'Cl·2H$_2$O behaves as a nonelectrolyte. The conductance of the complexes increases considerably on dilution or warming (apparently due to hydrolysis). Thermal analysis (DTA, TG, heating rate 4.9°C/min) of Mn(HL)L'Cl$_2$·NH$_3$·H$_2$O and Mn(HL)L'Cl$_2$·C$_3$H$_6$O reveals three distinct degradation steps for each complex which in the case of the NH_3 adduct occur at 140 (indicating the loss of water), at 340 (loss of NH_3 and benzohydrazide), and at 510°C (loss of 2-hydroxybenzohydrazide and Cl_2) leaving Mn_3O_4 in the residue. The acetone solvate loses the acetone at 140°C.

Reference:

Beshkenadze, I. A., Karkarashvili, M. V., Gogorishvili, P. V. (Izv. Akad. Nauk Gruz.SSR Ser. Khim. **5** No. 2 [1979] 103/10, 104/8; C.A. **91** [1979] No. 221662).

25.1.7.3 With 2-Hydroxybenzohydrazide and Histidine

The mixed ligand compound **Mn(HL)$_2$(HA)Cl$_2$** containing 2-hydroxybenzohydrazide (HL = $C_7H_8N_2O_2$) and histidine (HA = $C_6H_9N_3O_2$) was prepared by dissolving $MnCl_2 \cdot 4H_2O$, the hydrazide and histidine (mole ratio 1:2:1) in water and evaporating the mixture on a water bath. The bright yellow residue was washed first with a mixture of benzene and ethanol, then with ethanol. To prepare **Mn(HL)(HA)Cl$_2$·4H$_2$O** a solution of $MnCl_2 \cdot 4H_2O$ and the hydrazide in ethanol was mixed with an aqueous solution of histidine (mole ratio 1:1:1). The mixture was filtered and evaporated to leave a dark flesh-colored solid. **$Mn(HL)(HA)Cl_2 \cdot 2NH_3$** was obtained from the components in aqueous ammonia. The mixture was evaporated and treated again with ammonia to give a cinnamon-colored waxy solid which was dried in a desiccator over H_2SO_4 (conc.), then washed with benzene and ether. The complex Mn(HL)(A)Cl (with deprotonated histidine) was prepared by adding concentrated aqueous ammonia to the mixture used for the synthesis of $Mn(HL)(HA)Cl_2 \cdot 4H_2O$ (see above). The dark cinnamon-brown powder which formed during 2 d of standing at room temperature was collected, dried at room temperature and washed with water.

The flesh-colored dihydrate $Mn(L)(A) \cdot 2H_2O$, where both ligands are deprotonated, was prepared by mixing boiling aqueous solutions of the components (mole ratio 1:1:1) and leaving the mixture at room temperature for several days. $Mn(L)(A) \cdot C_3H_6O \cdot 3H_2O$ was obtained by evaporating the mixture used above to a small volume, adding acetone (20 mL) and allowing the solution to deposit white crystals at room temperature.

The IR spectrum of $Mn(HL)(HA)Cl_2$ suggests coordination to Mn through the terminal nitrogen and carbonyl oxygen of the hydrazide and through the imidazole-3 nitrogen and one carbonyl oxygen of histidine. Melting point or decomposition point (dec.) of the different complexes, their solubility S in water or ethanol (in g/100 mL), the pH value and the molar electrical conductivity of aqueous solutions at 25°C, and a dilution rate V ~ 900 L/mol are given below:

compound	m.p. in °C	S_{H_2O}	$S_{C_2H_5OH}$	pH	Λ in $cm^2 \cdot \Omega^{-1} \cdot mol^{-1}$
$Mn(HL)_2(HA)Cl_2$	70	a)	—	6.61	241.0
$Mn(HL)(HA)Cl_2 \cdot 4H_2O$	55	7.78	4.48	6.71	226.2
$Mn(HL)(HA)Cl_2 \cdot 2NH_3$	80 b)	3.06	—	8.15	277.9
Mn(HL)(A)Cl	120 (dec.)	0.04	—	7.29	—
$Mn(L)(A) \cdot 2H_2O$	>350	—	0.08	7.36	4.0
$Mn(L)(A) \cdot C_3H_6O \cdot 3H_2O$	240	0.04	—	6.45	9.3

a) Soluble, no data. – b) Boiling point at 80°C.

Thermal analysis (DTA, TG, heating rate 4.9°C/min) of $Mn(HL)_2(HA)Cl_2$ reveals that the complex loses one molecule of 2-hydroxybenzohydrazide at 300°C and the other at 520°C. Histidine is released at 580°C to leave $MnCl_2$ which at 600°C converts into manganese oxide.

Mn(HL)(A)Cl does not melt but decomposes upon heating with strong blowup in volume. $Mn(L)(A) \cdot 2H_2O$ changes its color at 340°C; at 350°C it becomes dark cinnamon-brown and melts upon further heating. The dichloro complexes are readily soluble in water whereas the chelates containing the deprotonated ligands are insoluble or sparingly soluble in water and insoluble in organic solvents. $Mn(HL)(HA)Cl_2 \cdot 4H_2O$ dissolves in acetone upon warming. The dichloro complexes (with neutral ligands) are 1:2 electrolytes in aqueous solution; those containing deprotonated ligands are rather stable nonelectrolytes at room temperature but their conductance increases on dilution or warming, especially in the case of $Mn(L)(A) \cdot C_3H_6O \cdot 3H_2O$ which becomes a 1:2 electrolyte at 40°C and V ~ 1850 L/mol ($\Lambda = 250.8\ cm^2 \cdot \Omega^{-1} \cdot mol^{-1}$).

Reference:

Karkarishvili, M. V., Beshkanadze, I. A. (Issled. Obl. Khim. Kompleksn. Prostykh. Soedin. Nek. Perekhodnykh Redk. Metal. No. 3 [1978] 32/43, 33, 35/40; C.A. **90** [1979] No. 214470).

25.1.7.4 With 3-Hydroxybenzohydrazide

$Mn(C_7H_8N_2O_2)_2(NCS)_2 \cdot 2H_2O$. A warm aqueous solution of the hydrazide was added to an aqueous solution of $Mn(NCS)_2$ (mole ratio 2:1) and the mixture was evaporated to half its original volume and allowed to crystallize. The finely crystalline beige precipitate was washed with water, ethanol, and ether and dried in air; it is stable in air and on storing. The IR spectrum (liquid paraffin and fluorocarbon oil mulls) shows shifts of ν(NH) and ν(CO) bands in comparison to the free ligand. Manganese is bonded only through the terminal nitrogen and carbonyl oxygen of the carbohydrazide group and not through the phenolic oxygen. The SCN^- ions are bonded to Mn through the nitrogen atoms and occupy the *trans* (apical) positions in the octahedrally configurated complex. $Mn(C_7H_8N_2O_2)_2(NCS)_2 \cdot 2H_2O$ loses its water of crystallization completely upon heating at 110°C. It is sparingly soluble in water and ethanol at room temperature and almost insoluble in the usual organic solvents.

Reference:

Kharitonov, Yu. Ya., Machkhoshvili, R. I., Generalova, N. B., Shchelokov, R. N. (Zh. Neorgan. Khim. **20** [1975] 965/71, **19** [1974] 1124/5; Russ. J. Inorg. Chem. **20** [1975] 540/4, 542, **19** [1974] 613/4).

25.1.8 With 2-, 3-, or 4-Methoxybenzohydrazide $CH_3OC_6H_4C(O)NHNH_2$ (= 2-, 3-, or 4-$C_8H_{10}N_2O_2$)

$Mn(2\text{-}C_8H_{10}N_2O_2)_3Cl_2 \cdot 2H_2O$ and **$Mn(2\text{-}C_8H_{10}N_2O_2)_2Cl_2$**. The dihydrate was prepared by heating gently an aqueous solution of $MnCl_2 \cdot 4H_2O$ with 2-methoxybenzohydrazide in a 1:3 mole ratio. After 1 d brown crystals had separated which were washed with cold water and ethanol and dried in air. For preparation of $Mn(2\text{-}C_8H_{10}N_2O_2)_2Cl_2$, an ethanolic solution of the components was treated as above and the white fine crystalline precipitate which deposited soon was filtered off, washed with ethanol and dried. The IR spectra are recorded. The thermal analysis of $Mn(2\text{-}C_8H_{10}N_2O_2)_3Cl_2 \cdot 2H_2O$ reveals an endothermic effect between 60 and 140°C indicating the loss of its water of crystallization. Decomposition of both complexes with loss of ligand occurs between ~300 and 600°C. The complexes are soluble in water and ethanol (less) and insoluble in most other usual organic solvents [1].

$Mn(2\text{- or }3\text{-}C_8H_{10}N_2O_2)_2(NCS)_2$ and **$Mn(4\text{-}C_8H_{10}N_2O_2)_2(NCS)_2 \cdot 2.5H_2O$**. The complexes were obtained by mixing aqueous solutions of $Mn(NCS)_2$ and the hydrazides in a 1:2 mole ratio. In the case of 2-methoxybenzohydrazide, a white fine crystalline precipitate resulted. In the case of 3-methoxybenzohydrazide, a sticky mass separated which became a pale brown crystalline powder within 15 to 20 min. To isolate the beige complex of 4-methoxybenzohydrazide the solution of the components was evaporated to about a third of its initial volume and allowed to crystallize. The complexes were washed with water, ethanol, and ether and dried in air. They are stable in air and during storage. The IR spectra were investigated. The SCN^- ions are coordinated to Mn through the nitrogen atoms. Since the $\nu(C{=}N)_{NCS}$ band at 2090 cm^{-1} appears as a singlet, a *trans*-octahedral configuration of the complexes with the thiocyanate groups in apical positions is assumed. The complexes are sparingly soluble in water and ethanol at room temperature but far more soluble upon heating, and insoluble in other usual organic solvents [2].

References:

[1] Shamilishvili, O. Kh., Machkhoshvili, R. I., Kharitonov, Yu. Ya., Dzhibladze, T. G. (Zh. Neorgan. Khim. **20** [1975] 3003/6; Russ. J. Inorg. Chem. **20** [1975] 1661/3).

[2] Kharitonov, Yu. Ya., Machkhoshvili, R. I., Generalova, N. B., Shchelokov, R. N. (Zh. Neorgan. Khim. **20** [1975] 693/700, **19** [1974] 1124/5; Russ. J. Inorg. Chem. **20** [1975] 387/92, 391, **19** [1974] 613/4).

25.1.9 With Cinnamohydrazide $C_6H_5CH{=}CHC(O)NHNH_2$ (= $C_9H_{10}N_2O$)

$Mn(C_9H_{10}N_2O)_2Cl_2$ and **$Mn(C_9H_{10}N_2O)Cl_2 \cdot 2H_2O$** complexes were prepared by refluxing manganese chloride with stoichiometric amounts of cinnamohydrazide in ethanol solution. The mixture was digested on a water bath for about 1 h and the resulting white precipitate was washed with hot ethanol and dried in a desiccator over anhydrous $CaCl_2$. Melting points (m.p.), the most characteristic absorption bands (in cm^{-1}) of the IR spectrum with their assignment and the molar electrical conductance Λ of 0.001M solutions in dimethylformamide at 26°C are shown below:

complex	m.p. in °C	ν(CO), amide I	$\beta(NH_2)$	amide II	amide III	Λ in $cm^2 \cdot \Omega^{-1} \cdot mol^{-1}$
$Mn(C_9H_{10}N_2O)_2Cl_2$	235	1660	1620	1555	1250	12.50
$Mn(C_9H_{10}N_2O)Cl_2 \cdot 2H_2O$	210	1660	1610	1550	1250	13.50
free ligand	—	1680	1640	1555	1260	—

Reference:

Mostafa, M. M., Hassan, S. M., El-Asmy, A. A. (Egypt. J. Chem. **22** [1979/80] 383/8, 384/6; J. Indian Chem. Soc. **57** [1980] 127/9).

25.1.10 With the Hydrazides of N-Benzoylglycine and 1-Tyrosine

ligand 1 $C_6H_5C(O)NHCH_2C(O)NHNH_2$ (= $C_9H_{11}N_3O_2$)

ligand 2 $HOC_6H_4CH_2CH(NH_2)C(O)NHNH_2$ (= $C_9H_{13}N_3O_2$)

$Mn(C_9H_{11}N_3O_2)_2Cl_2$. A pasty product was obtained after refluxing an ethanol solution of $MnCl_2 \cdot 4H_2O$ and ligand 1 for 30 min and decanting the supernatant liquid. When macerated with acetonitrile a white microcrystalline solid results which was repeatedly washed with acetone and ether and dried at 70 to 80°C for about 30 min and stored in a desiccator. The complex melts at 198°C [1]. The IR spectrum indicates that the carbonyl groups of the N-benzoyl moiety are not involved in complexation. Coordination through the carbonyl oxygens and the NH_2 group of the hydrazide moiety is assumed. Susceptibility measurements by the Cahn-Faraday electrobalance method at room temperature yield the magnetic moment $\mu_{eff} = 5.62\ \mu_B$ of a spin-free tetrahedral or octahedral Mn^{II} complex. The molar electrical conductance $\Lambda = 6.49\ cm^2 \cdot \Omega^{-1} \cdot mol^{-1}$ of a 0.001M solution in dimethyl sulfoxide indicates its nonionic nature. The complex is insoluble in water and common organic solvents but soluble in dimethyl sulfoxide and dimethylformamide [1].

$Mn(C_9H_{13}N_3O_2)_2X_2$ (X = OH, Cl). The chloro complex was obtained by mixing hot methanolic solutions of $MnCl_2$ and ligand 2 with vigorous stirring. Addition of diethyl ether precipitated the compound which was washed successively with ethanol and ether. The hydroxo complex was prepared by adding an aqueous solution of $MnCl_2$ to that of the ligand and KOH in a 1:2:2 mole ratio. The precipitate was washed successively with water, ethanol, and ether. Both compounds were dried in a vacuum. The brownish yellow chloro complex decomposes at 202°C; the flesh-colored hydroxo complex melts above 300°C. The IR spectrum shows characteristic absorption bands assigned to ν(C–C) + NH_2 scissoring, NH_2 wagging and NH_2 rocking modes which upon complexation shifted toward lower wave numbers. The negative shifts indicate the involvement of both the nitrogen atoms of the amino group and the terminal nitrogen of the hydrazide group in coordination. The bands assigned to vibrations of the phenolic OH group (around 1240 and 660 cm^{-1}) are almost unshifted and indicate noncoordination of that OH group. The hydroxo complex exhibits a new band in the 1140 to 1160 cm^{-1} region assignable to δ(Mn–OH) vibrations. The nonligand bands in the 490 to 330 and 390 to 320 cm^{-1} regions may be (tentatively) ascribed to ν(Mn–O) and ν(Mn–N) vibrations, respectively. Susceptibility measurements (Cahn-Faraday electrobalance) at room temperature yield the magnetic moment μ_{eff} = 5.76 μ_B for the complex with X = Cl and 5.74 μ_B for X = OH, indicative of spin-free octahedral Mn^{II} complexes. The molar conductance Λ = 0.76 to 5.9 $cm^2 \cdot \Omega^{-1} \cdot mol^{-1}$ of 0.001M solutions in dimethyl sulfoxide indicates a nonionic nature of the complexes which are insoluble in water and common organic solvents but readily soluble in dimethyl sulfoxide [2].

$Mn(C_9H_{12}N_3O_2)(OH) \cdot 2H_2O$. The dark brown complex was prepared in the same way as $Mn(C_9H_{13}N_3O)_2(OH)_2$ (see above), except that the $MnCl_2$:ligand:KOH ratio was 1:1:2. After vigorous stirring the complex was collected, washed and dried as above. It melts above 300°C. The IR spectrum shows the same characteristic absorption bands as $Mn(C_9H_{13}N_3O_2)_2(OH)_2$ (see above), except that the bands due to the phenolic OH group have disappeared because of deprotonation and subsequent coordination of the phenolic oxygen. The magnetic moment μ_{eff} = 5.86 μ_B (measured as above) is indicative of a spin-free octahedral (or tetrahedral) Mn^{II} complex. It is insoluble in water and common organic solvents but readily soluble in dimethyl sulfoxide, in which it is nonconducting [2].

References:

[1] Rao, T. R., Sahay, M., Aggarwal, R. C. (Syn. React. Inorg. Metal-Org. Chem. **15** [1982] 209/22, 212/3, 217).

[2] Rao, T. R., Sahay, M., Aggarwal, R. C. (Indian J. Chem. A **23** [1984] 214/8).

25.1.11 With 1-Naphthohydrazide

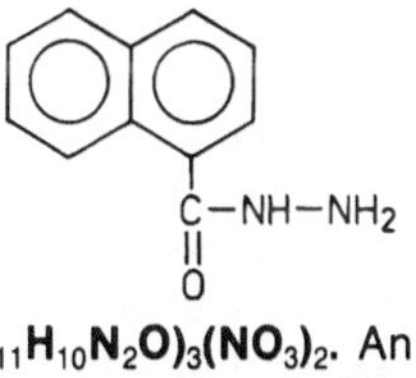

(= $C_{11}H_{10}N_2O$)

$Mn(C_{11}H_{10}N_2O)_3(NO_3)_2$. An ethanolic solution of $Mn(NO_3)_2 \cdot 6H_2O$ was added to that of the threefold molar amount of 1-naphthohydrazide. The resulting fine white crystals were washed with ethanol and ether. The shifts of the ν(NH) and ν(CO) ligand bands of the IR spectrum to lower frequencies in the complex suggest bidentate coordination of 1-naphthohydrazide to Mn through the terminal nitrogen and the carbonyl oxygen. The magnetic moment μ_{eff} = 5.78 μ_B resulting from susceptibility measurements at 299 K indicates a high-spin octahedral configuration of the complex where all six coordination sites of the metal atom are occupied by the three bidentate ligand molecules. $Mn(C_{11}H_{10}N_2O)_3(NO_3)_2$ is soluble in dimethylformamide and dimethyl sulfoxide, Chundak, S. Ya., Manole, S. F., Butsko, S. S., Gérbéléu, N. V. (Koord. Khim. **1** [1975] 840/4; Soviet J. Coord. Chem. **1** [1975] 711/4, 712).

25.1.12 With 2-Furohydrazide

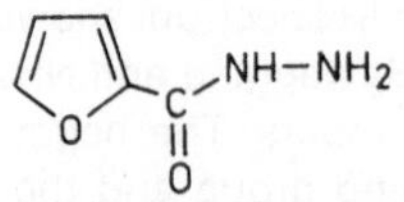

$(=C_5H_6N_2O_2)$

$Mn(C_5H_6N_2O_2)_2Cl_2$. Ethanolic solutions of $MnCl_2$ and ligand were mixed in a 1:2 mole ratio and ether was added to precipitate the complex. The white solid, which was washed with ethanol and ether and dried in a vacuum, melts at 200°C. The IR spectrum of the ligand shows characteristic absorption bands of the amide I, $\beta(NH_2)$, amide II, and ν(N–N) vibration modes at 1690, 1640, 1530, and 950 cm^{-1}, respectively. The amide I, $\beta(NH_2)$ and amide II bands shift on complexation toward lower frequencies, the band of ν(N–N) toward higher frequencies. Therefore bidentate bonding of the ligand to Mn through the carbonyl oxygen and the terminal nitrogen was suggested. The molar electrical conductance $\Lambda = 201$ to 278 $cm^2 \cdot \Omega^{-1} \cdot mol^{-1}$ of an aqueous 0.001M solution indicates a 1:2 electrolyte nature of the complex which is in agreement with the results of vapor pressure osmometric determination of the molecular weight in water. Susceptibility measurements (Cahn-Faraday electrobalance) at room temperature yield the magnetic moment $\mu_{eff} = 5.67\ \mu_B$ of a spin-free octahedral Mn^{II} complex. $Mn(C_5H_6N_2O_2)_2Cl_2$ is soluble in methanol, dimethylformamide, and dimethyl sulfoxide but insoluble in other common organic solvents, Singh, B., Singh, R. N., Aggarwal, R. C. (Indian J. Chem. A **23** [1984] 1016/20).

25.1.13 With 2-Pyridinecarbohydrazide (= Picolinohydrazide)

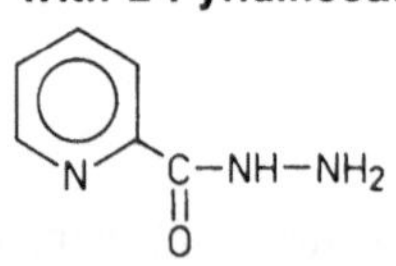

$(=C_6H_7N_3O)$

$Mn(C_6H_7N_3O)_2Cl_2$. Ethanolic solutions of manganese chloride and picolinohydrazide were mixed in a 1:2 mole ratio. The white precipitate was washed with ethanol and dried in air. The N-deuterated complex was obtained by heating $Mn(C_6H_7N_3O)_2Cl_2$ with D_2O in a sealed quartz ampule and recrystallization from D_2O [1]. X-ray diffractometer analyses (MoKα radiation) show the yellow crystals of $Mn(C_6H_7N_3O)_2Cl_2$ to be monoclinic, space group C2/c(B2/b)-C_{2h}^6 (No. 15) with the lattice constants a = 16.960(2), b = 14.124(3), c = 6.664(1) Å and $\beta(\gamma) = 98.9°$; Z = 4, V = 1577.01 $Å^3$. The structure was refined to R = 0.054. The atomic coordinates are shown in the paper and bond lengths and bonding angles around Mn in **Fig. 20**. The crystal structure consists of discrete $[Mn(C_6H_7N_3O)_2Cl_2]$ molecules. The central Mn atom lies on the twofold axis and is octahedrally surrounded by the two NH_2 nitrogens and the two carbonyl oxygens of the bidentate organic ligands and two Cl^- ions which are *cis* placed with an angle Cl–Mn–Cl = 95°. The oxygen of the nonplanar ligand lies in the plane of the planar pyridine ring and the carbohydrazide nitrogens are displaced significantly from the pyridine ring plane. The measured density is 1.681 g/cm^3; 1.685 g/cm^3 was calculated from X-ray data [2].

$Mn(C_6H_7N_3O)_2Cl_2$ melts at 250°C. The IR spectrum of the complex in liquid paraffin, fluorinated hydrocarbon, or hexachlorobutadiene and the Raman spectrum excited with an ionized Ar laser ($\lambda = 4880$ and 5145 Å) have been investigated in the 4000 to 40 cm^{-1} region and compared to the isothiocyanato complex, p. 187, and similar complexes with nicotinohydrazide; see p. 188. In the IR spectrum of $Mn(C_6H_7N_3O)_2Cl_2$ intense ν(NH) bands were found at ~3224 and 3174 cm^{-1}. The high-frequency ν(NH) component at ~3345 cm^{-1} appears in the form of a low-intensity band. No high-frequency ν(NH) component was found above 3300 cm^{-1}

in the Raman spectrum. Coordination by means of the nitrogen atom of the NH_2 group can be postulated if the intense ν(NH) components in the IR spectrum and the data from the Raman spectrum are taken into account. An intense amide I band appears in the IR spectrum near 1680 cm^{-1}. This almost unshifted band in comparison to the free ligand has suggested the absence of Mn–O bonds contrary to the results of X-ray investigation. When $Mn(C_6H_7N_3O)_2Cl_2$ is deuterated, the intensity of this band decreases appreciably, and a band appears at ~1676 cm^{-1}. The IR spectrum of the deuterated sample has an intense band at ~1635 cm^{-1}, assigned to ν(CO). The unshifted strong band of ν(ring) at 1593 cm^{-1} indicates no participation of the pyridine nitrogen in complex bonding. Bands of the ν(Mn–Cl) vibration mode were found in the Raman spectrum at 233 and 215 cm^{-1}, of δ(Cl–Mn–Cl) at 173 cm^{-1}. The molar electrical conductance $\Lambda = 249\ cm^2 \cdot \Omega^{-1} \cdot mol^{-1}$ was measured in aqueous solution at V = 1000 L/mol [1].

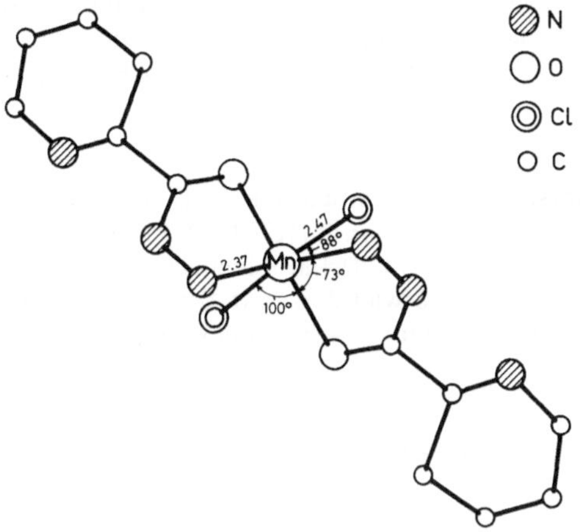

Fig. 20. Molecular structure of $Mn(C_6H_7N_3O)_2Cl_2$ with bond lengths (in Å) and bond angles around the Mn atom [2].

$Mn(C_6H_7N_3O)Cl_2 \cdot 2C_2H_5OH$. The light yellow complex obtained from cold ethanolic solutions of $MnCl_2$ and picolinohydrazide (mole ratio ~1:2) was washed with ethanol and dried at room temperature. It melts at 234°C. The IR spectrum was discussed in connections with the spectra of similar complexes of Cu^{II}, Ni^{II}, Co^{II}, and Fe^{II}, and coordination of the ligand to Mn through the carbonyl oxygen and the pyridine ring nitrogen was suggested. Susceptibility measurements (Cahn-Faraday method) at room temperature yield the magnetic moment $\mu_{eff} = 5.87\ \mu_B$ indicating a high-spin octahedral Mn^{II} complex as concluded also from other physical data. The molar conductance $\Lambda = 5.60\ cm^2 \cdot \Omega^{-1} \cdot mol^{-1}$ measured in dimethyl sulfoxide solution (0.001M) is that of a nonelectrolyte. The complex is soluble in water and dimethyl sulfoxide but insoluble in other common organic solvents [3].

$Mn(C_6H_7N_3O)_n(NCS)_2$ (n = 2, 3). The complex with n = 2 was obtained by adding the hydrazide to an aqueous solution of $Mn(NCS)_2$ in a 1:1 up to 1:6 mole ratio with constant stirring. The fine crystalline, lemon-colored precipitate was washed with water and dried in air. The complex with n = 3 was formed if the aqueous solution of $Mn(NCS)_2$ was added carefully to that of the ligand up to a ratio Mn:ligand = 1:6; a white crystalline solid separated within 24 h. $Mn(C_6H_7N_3O)_3(NCS)_2$ melts at 170°C, $Mn(C_6H_7N_3O)_2(NCS)_2$ at 220°C.

The IR and Raman spectra were investigated and compared to the spectra of $Mn(C_6H_7N_3O)_2Cl_2$ and to the similar complexes with nicotinohydrazide, p. 188. The ν(NH) component with the highest frequency both in the IR and Raman spectra is found at ~3298 cm^{-1} and is characterized by its high intensity. The shift of the ν(NH) band and of the amide I band (IR and Raman) to lower frequencies indicates coordination of the ligand to Mn through the NH_2 group nitrogen and carbonyl oxygen for $Mn(C_6H_7N_3O)_2(NCS)_2$. The strong ν(ring) vibration mode (at ~1595 cm^{-1}) suggests the absence of bonds between Mn and the pyridine ring nitrogen. The unsplit band of $\nu(CN)_{NCS}$ at 2092 cm^{-1} and of $\nu(CS)_{NCS}$ at ~792 cm^{-1} indicate bonding of the thiocyanate groups to Mn through the nitrogen atom in a *trans* arrangement. A band observed in the Raman spectrum at 280 cm^{-1} was assigned to ν(Mn–NCS), and bands at 140 and 102 cm^{-1} to δ(SCN–Mn–NCS) vibrations. The IR and Raman spectra of $Mn(C_6H_7N_3O)_3$-$(NCS)_2$ are different from those of $Mn(C_6H_7N_3O)_2Cl_2$. It was suggested that some of the ligand molecules are not coordinated to Mn through the terminal nitrogen and the carbonyl oxygen but through the pyridine ring nitrogens since the ν(ring) vibration mode (at 1593 cm^{-1}) is fairly intense and the ν(NH) and ν(CO) bands occur at somewhat higher frequencies than in the bis complex. Conductivity measurements in 0.001M aqueous solution yield Λ = 212 and 210 $cm^2 \cdot \Omega^{-1} \cdot mol^{-1}$ for $Mn(C_6H_7N_3O)_2(NCS)_2$ and $Mn(C_6H_7N_3O)_3(NCS)_2$, respectively [1].

References:

[1] Tsivadze, A. Yu., Tsintsadze, G. V., Petriashvili, Zh. D., Kharitonov, Yu. Ya. (Koord. Khim. **3** [1977] 1052/9; Soviet J. Coord. Chem. **3** [1977] 815/21, 816).
[2] Tsintsadze, G. V., Tsivtsivadze, T. I., Dzhavakhishvili, Z. O., Ilinskii, A. I., Orbeladze, F. V. (Koord. Khim. **5** [1979] 909/11; Soviet J. Coord. Chem. **5** [1979] 718/20).
[3] Aggarwal, R. C., Rao, T. R. (J. Inorg. Nucl. Chem. **40** [1978] 1177/8).

25.1.14 With 3-Pyridinecarbohydrazide (= Nicotinohydrazide)

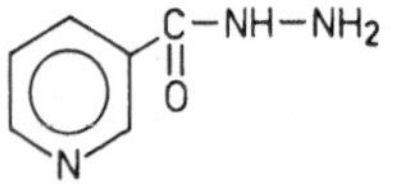

(= $C_6H_7N_3O$)

$Mn(C_6H_7N_3O)_nX_2$ (X = Cl, $SO_4/2$, n = 1, 2) and **$Mn(C_6H_7N_3O)(NCS)_2 \cdot 2H_2O$**. The chloro complexes with n = 1, 2 were prepared from manganese(II) chloride and nicotinohydrazide in methanol or acetone solution in a 1:2 up to 1:4 mole ratio for $Mn(C_6H_7N_3O)_2Cl_2$ and in a 1:1 mole ratio for $Mn(C_6H_7N_3O)Cl_2$. The white precipitates were washed with ether and dried in a vacuum. $Mn(C_6H_7N_3O)Cl_2$ was obtained also from hot ethanolic solutions of the components in a 1:2 mole ratio [2] after refluxing the mixture for about 30 min [6]. The precipitate was washed with aqueous ethanol and dried at room temperature [2]. The N-deuterated complex **$Mn(C_6H_4D_3N_3O)Cl_2$** was obtained by heating $Mn(C_6H_7N_3O)Cl_2$ with D_2O in a sealed quartz ampule [3].

$Mn(C_6H_7N_3O)SO_4$ precipitates upon mixing solutions of manganese sulfate and nicotinohydrazide in aqueous ethanol. The white complex is soluble in water [5].

$Mn(C_6H_7N_3O)(NCS)_2 \cdot 2H_2O$ was obtained by reacting $Mn(NO_3)_2 \cdot 6H_2O$ and KNCS in ethanol, removing the precipitated KNO_3, and adding the equimolar amount of the ligand in ethanol. The mixture was allowed to evaporate in vacuum over $CaCl_2$ for 3 d and water was added to the syrupy residue to salt out the complex. The resulting white powder was washed with water and dried in vacuum over $CaCl_2$ [1]. Heating with D_2O in a sealed quartz ampule yields the N-deuterated complex **$Mn(C_6H_4D_3N_3O)(NCS)_2 \cdot 2D_2O$**.

The IR spectra of the chloro and isothiocyanato complexes and their deuterated species have been investigated in the range 4000 to 400 cm^{-1} [3] and compared with the Raman spectrum of the isothiocyanato complex, reported in [4]. The shift of the $\nu_{as}(NH)$, $\nu_s(NH)$, and ν(CO) ligand bands to lower frequencies indicates coordination of nicotinohydrazide to Mn through the NH_2 nitrogen and carbonyl oxygen with formation of five-membered chelate rings. Since the ν(ring) band at 1598 cm^{-1} is not shifted in $Mn(C_6H_7N_3O)_2Cl_2$ the pyridine ring is assumed to be not coordinated although a weaker band at 1610 cm^{-1} would support such coordination or may be due to a splitting of the ν(ring) band as a consequence of an asymmetry of the complex framework. Coordination of the pyridine ring nitrogen is supported also by a displacement of the band assigned to γ(CNC) at 643 (ligand 632) cm^{-1}. Thus, it is assumed that in $Mn(C_6H_7N_3O)Cl_2$ the ligand can form both metal-containing chelate rings and bridges simultaneously [3]. Bands at 400, 358, and 292 cm^{-1} were assigned to ν(Mn–O), ν(Mn–N), and ν(Mn–Cl) modes, respectively. However, no coordination of the pyridine ring (no shift of the pyridine ring vibration modes at 670 and 390 cm^{-1}) and chelation by the hydrazide group only with formation of a tetrahedral structure involving the Cl^- ions is concluded by [2].

The Raman spectrum of $Mn(C_6H_7N_3O)_2(NCS)_2 \cdot 2H_2O$ (excited with an ionized argon laser of $\lambda = 4880$ and 5145 Å) shows bands assignable to $\nu(CS)_{NCS}$ at 803 and 794, to ν(Mn–ligand) at 518, to $\nu(Mn–N_{ring})$ at 250, 201, and to $\delta(N_{NCS}–Mn–N_{NCS})$ at 190 cm^{-1} [4]. The splitting of the $\nu(CN)_{NCS}$ and $\nu(CS)_{NCS}$ Raman bands suggests a *cis* arrangement of the N-bonded isothiocyanate groups in the complex [4].

The magnetic moment, $\mu_{eff} = 5.97\ \mu_B$, of $Mn(C_6H_7N_3O)Cl_2$ (from susceptibility measurements at room temperature) indicates either a tetrahedral or octahedral geometry. $Mn(C_6H_7N_3O)Cl_2$ is assumed to be polymeric because of its high thermal stability and low solubility [2]. Thus thermal analysis (DTA) in the 20 to 800°C region reveals an endothermic peak at 355°C indicating dechlorination. The exothermic peak at 400°C suggests a lattice rearrangement and the peak at 430°C indicates melting of the complex. The sharp exothermic peak at 520°C indicates the decomposition of the complex along the chelate bonds and loss of the organic components leaving a residue of MnO_2 [6].

The chloro and isothiocyanato complexes are readily soluble in water and upon warming also in methanol, ethanol, dimethylformamide [1], and dimethyl sulfoxide but insoluble in other organic solvents [1, 2]. The low molar electrical conductance $\Lambda = 5.49\ cm^2 \cdot \Omega^{-1} \cdot mol^{-1}$ of a 0.001M solution of $Mn(C_6H_7N_3O)Cl_2$ in dimethyl sulfoxide indicates a nonelectrolyte. The compound decomposes in alkaline medium [2].

References:

[1] Tsintsadze, G. V., Manvelidze, G. V., Mdivani, M. A., Petriashvili, Zh. D., Chigogidze, N. Sh., Kiguradze, R. A., Dzhavakhishvili, Z. O. (Tr. Gruz. Politekhn. Inst. **1975** No. 4, pp. 5/18, 11/2; C.A. **86** [1977] No. 47263).

[2] Aggarwal, R. C., Rao, T. R. (Transition Metal Chem. [Weinheim] **2** [1977] 201/4).

[3] Tsivadze, A. Yu., Tsintsadze, G. V., Petriashvili, Zh. D., Kharitonov, Yu. Ya. (Koord. Khim. **1** [1975] 1472/7; Soviet J. Coord. Chem. **1** [1975] 1223/7, 1224).

[4] Tsivadze, A. Yu., Tsintsadze, G. V., Petriashvili, Zh. D., Kharitonov, Yu. Ya. (Koord. Khim. **3** [1977] 1052/9; Soviet J. Coord. Chem. **3** [1977] 815/21, 816, 821).

[5] Pilicec, S., Vasiliev, R., Maurer, A., Mracec, M., Havlik, J., Costişor, O., Jitaru, M., Topciu, V., Csaki, N. (Timisoara Med. **26** No. 3 [1981] 29/34, 30; C.A. **97** [1982] No. 66073).

[6] Abou Sekkina, M. M., Abou El-Azm, M. G. (Thermochim. Acta **77** [1984] 211/8, 215).

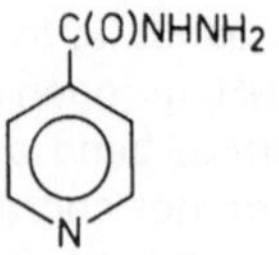

25.1.15 With 4-Pyridinecarbohydrazide (= Isonicotinohydrazide) (= $C_6H_7N_3O$)

25.1.15.1 Complexes in Solution

The formation of 1:1 complexes was shown by pH-potentiometric titrations of aqueous solutions containing $Mn(NO_3)_2$ and the ligand (mole ratios 20:1, 1:1, or 1:20) with 1M NaOH at 25°C and I = 1.0M (KNO_3). The logarithmic values of the equilibrium constants K = $[Mn(C_6H_7N_3O)^{2+}]/[Mn^{2+}][C_6H_7N_3O]$ and K' = $[Mn(C_6H_6N_3O)^+][H^+]/[Mn(C_6H_7N_3O)^{2+}]$ are log K = 1.04 and log K' = 4.14, respectively [1, 2].

References:

[1] Nagano, K., Kinoshita, H., Tamura, Z. (Chem. Pharm. Bull. [Tokyo] **11** [1963] 999/1013, 1006, 1013).

[2] Ishidate, M. (Anal. Chem. Proc. Intern. Symp., Birmingham, Engl., 1962 [1963], pp. 178/82, 180; C.A. **59** [1963] 10967).

25.1.15.2 Isolated Compounds

$Mn(C_6H_7N_3O)_2(NO_3)_2$ precipitates from ethanol solutions of manganese(II) nitrate and isonicotinohydrazide (mole ratio 1:2). The cream-colored crystalline powder was collected after 3 d, washed with ethanol and dried in vacuum over $CaCl_2$. The ν(NH) bands in the IR spectrum were found at ~3306 and ~3250 cm^{-1}, which indicates coordination of the ligand through the nitrogen atom of the NH_2 group. The inflection on the high-frequency side is apparently due to hydrogen bonds. The amide I band was observed at ~1676 cm^{-1} while the corresponding line in the Raman spectrum was found at 1660 cm^{-1}. Compared to the solution of the ligand in a nonpolar solvent, in which hydrogen bonds are absent, the ν(CO) frequency of the complex is decreased by ~40 cm^{-1}. This fact indicates coordination through the oxygen atom. The intense band at ~1612 cm^{-1} and a low-intensity band at ~1616 cm^{-1} are related to ν(ring) and δ(NH_2) vibration modes. An intense ν(ring) vibration observed in the Raman spectrum at ~1626 cm^{-1} indicates coordination of the ligand through the nitrogen hetero atom. The intense ν(NO_3) band at 1340 cm^{-1} suggests coordination also of the nitrato group. Bands assignable to free NO_3^- ions are absent. $Mn(C_6H_7N_3O)_2(NO_3)_2$ is quite soluble in water and dimethylformamide but insoluble in benzene [1].

$[Mn(NO_3)_2C_6H_7N_3O(H_2O)]$. Solutions of manganese(II) nitrate and the ligand in 95% ethanol were mixed in a 1:1 mole ratio and allowed to stand 7 or 8 d for crystallization. The cream-colored crystals which deposited were washed and dried as the compound above. X-ray diffractometer studies (MoKα radiation) of the thin plate-shaped crystals yield an orthorhombic lattice, space group Pcca-D_{2h}^8 (No. 54), with lattice constants a = 8.214(2), b = 13.162(9), and c = 22.270(9) Å; Z = 8. The structure was anisotropically refined to R = 0.053. Bond lengths (in Å) and some bonding angles are shown in **Fig. 21**. The structural unit contains the isonicotinohydrazide molecule as a terdentate ligand bonded to Mn through the carbonyl oxygen, the terminal hydrazidic NH_2 and pyridine ring nitrogens. The H_2O molecule and one of the NO_3^- groups are bonded to Mn through one oxygen whereas the other NO_3^- group is coordinated as a bidentate ligand through two oxygens. Thus the complex contains seven-coordinate manganese(II) in a slightly distorted pentagonal bipyramidal environment. It is assumed that the pyridine ring of the ligand and the NO_3^- groups are essentially planar. The complex units in the structure are interlinked by NH···O (2.69 to 3.43 Å) and OH···O (2.72 to 3.33 Å) hydrogen bonds with participation of almost all hydrogens of the (NH) and (OH) bonds. The measured density is 1.840 g/cm³, 1.849 g/cm³ was calculated from X-ray data [1].

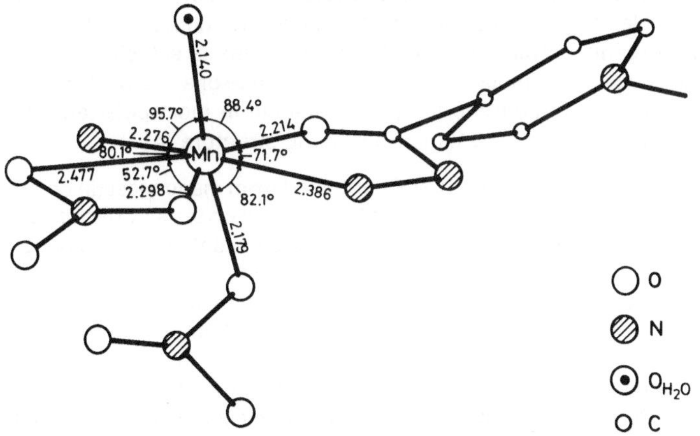

Fig. 21. Molecular structure of $[Mn(NO_3)_2C_6H_7N_3O(H_2O)]$ with bond lengths in Å and bond angles [1].

The IR spectrum contains a doublet at ~3332 and 3308 cm^{-1} with an inflection on the high-frequency side and a band at ~3254 cm^{-1}. Since only a line at ~3308 cm^{-1} is observed in the Raman spectrum it may be assumed that the band at 3332 cm^{-1} and the inflection at ~3400 cm^{-1} are related to the ν(OH) frequency of water molecules. The reduced ν(OH) value indicates the presence of coordinated water molecules. The assignment of the bands at 3308 and 3254 cm^{-1} to ν(NH) indicates the coordination of isonicotinohydrazide through the nitrogen atom of the NH_2 group. An intense amide I band is found at 1670 cm^{-1}; an intense band at ~1620 cm^{-1} in the IR spectrum (1622cm^{-1} in the Raman spectrum) can be assigned to ν(ring). The increased value of ν(ring) indicates coordination of the nitrogen hetero atom [1].

The compound is quite soluble in water and dimethylformamide [1].

$Mn(C_6H_7N_3O)_2X_2$ (X = Cl, Br). The complexes were prepared by addition of an ethanolic solution of the metal halide to an ethanolic solution of the ligand in a 2:1 (ligand to metal) mole ratio [2, 3]. The mixture of the chloro compound was heated to boiling for 1 h, the yellow precipitate was washed with ethanol and dried over $CaCl_2$ in vacuum [2]. The white bromo complex, which was washed with ethanol and ether and dried in a similar way [3], was also described in [4].

The IR spectra of chloro complexes containing isonicotinohydrazide are discussed in [2] and coordination of the ligand to the metal atom through the terminal (NH_2) nitrogen and carbonyl oxygen is suggested. The IR spectrum of the bromo complex in KBr disks and in a polyethylene matrix (600 to 200 cm^{-1} region) shows characteristic absorption bands (in cm^{-1}, ligand in parentheses) which are assigned as follows: 3280 (3275) to $\nu_{as}(NH_2)$, 3168 (3095) to $\nu_s(NH_2)$, 1645 (1660) to ν(CO), 1600 (1599), 1546 (1560) to ν(ring), 1540 (1535) to ν_s(C–N) and at 234 to ν(Mn–Br). Since the ν(NH) and ν(CO) bands are almost unshifted, coordination of the ligands to Mn apparently occurs only through the pyridine nitrogen. Thus it is assumed that $Mn(C_6H_7N_3O)_2Br_2$ has a polymeric structure consisting of bromine-bridged chains of metal atoms with isonicotinohydrazide molecules above and below the plane of the Mn–Br chain and hydrogen bonding between the carbohydrazide groups of adjacent ligand molecules [3].

The visible spectrum of $Mn(C_6H_7N_3O)_2Cl_2$ shows a broad absorption band at $\lambda = 400$ nm which is assigned to a charge transfer [2]. Susceptibility measurements at 27°C yield the magnetic moment $\mu_{eff} = 5.94$ μ_B for the chloro complex [2] and 6.20 μ_B for the bromo complex [3] suggesting high-spin octahedral structures [2, 3]. The chloro complex is stable up to 250°C, it is slightly soluble in dimethylformamide and dimethyl sulfoxide but insoluble in water or other common organic solvents [3]. The bromo complex is soluble in water with apparent dissociation as shown by its molar electrical conductance of a 1:2 electrolyte. It is insoluble in ethanol, ether, $CHCl_3$, and benzene [5], poorly soluble in most polar and nonpolar organic solvents in agreement with the assumed polymeric structure [4].

$[Mn_2(C_6H_7N_3O)_3Cl(H_2O)_3]Cl_3 \cdot 0.5\,C_6H_7N_3O$, obtained from manganese(II) chloride and isonicotinohydrazide in aqueous solution, forms yellow triclinic crystals. X-ray diffractometer studies yield the space group $P\bar{1}$-C_i^1 (No. 2) with the lattice constants a = 10.572(4), b = 12.125(7), c = 14.411(6) Å and $\alpha = 90.22(4)°$, $\beta = 109.41(3)°$, $\gamma = 104.75(4)°$; Z = 2, V = 1676.91 Å^3. The structure was refined anisotropically to R = 0.097. Atomic coordinates are given in the paper. The structure of the complex cation $[Mn_2Cl(C_6H_7N_3O)_3(H_2O)_3]^{3+}$ shown in **Fig. 22** with several bond lengths (in Å) and bond angles (around Mn) contains two independent manganese atoms, each in an octahedral environment. One of the metal atoms is coordinated to three organic ligands and one Cl^- ion and the other one to two organic ligands, two H_2O molecules and one Cl^- ion. Electron density analysis reveals that the oxygens of H_2O can replace the coordinated Cl^- ions statistically and that vice versa the sum of Cl^- ions and H_2O molecules is constant. Two of the organic ligands are bridging tridentate; they are coordinated to one Mn atom through the terminal amine nitrogen and carbonyl oxygen forming five-membered chelate rings and to a second metal atom through the pyridine ring nitrogen. The third ligand is bidentate and coordinated only through the carbohydrazide nitrogen and oxygen. A fourth uncoordinated molecule of nicotinohydrazide occupies the symmetry center of the complex in a disordered arrangement. The structure contains also three uncoordinated Cl^- ions which participate in the formation of hydrogen bonds. The complex cation $[Mn_2Cl(C_6H_7N_3O)_3(H_2O)_3]^{3+}$ is essentially planar with exception of the bidentate ligand, two Cl^- ions and two water oxygens which occupy the apices of the two octahedra. The five-membered metallocycles formed by each coordinated ligand are all almost planar and form dihedral angles of 3.55(0.89)°, 5.9(2.0)°, and 25.09° with the corresponding pyridine ring planes. The measured density is 1.61 g/cm^3, 1.69 g/cm^3 was calculated from X-ray data [5].

$Mn(C_6H_7N_3O)Cl_2$. Ethanolic solutions of manganese chloride and isonicotinohydrazide were mixed in a 1:2 up to 3 mole ratio [6] and refluxed for 36 min [7]. The precipitated complex was washed with ethanol [3, 6] and ether [3] and dried over $CaCl_2$ [3, 6]. The air-stable white powdery complex melts at 256°C; its identity was confirmed by X-ray powder diagrams which suggest isomorphism with the related copper complex [6]. The IR spectrum of the complex reveals shifts of the ν(NH) and ν(CO) ligand bands toward lower and of some ν(ring) bands to higher frequencies. Coordination of isonicotinohydrazide as a terdentate bridging ligand to one manganese atom through the terminal nitrogen and the carbonyl oxygen of the ligand forming a five-membered chelate ring and to a second Mn atom through the nitrogen of the pyridine ring was assumed thus producing infinite chains in a polymeric structure with five-coordinate manganese. A band at 290 cm^{-1} was assigned to ν(Mn–Cl) vibrations [6]. A tetrahedral structure with four-coordinate manganese and a bidentate, N and O bonded ligand was proposed by [3]. Susceptibility measurements yield the magnetic moment μ_{eff} = 5.49 μ_B due to five unpaired electrons in Mn which is consistent with such structure [3]. $Mn(C_6H_7N_3O)Cl_2$ is soluble in water [4]; the molar electrical conductance $\Lambda = 259$ cm^2 $\cdot \Omega^{-1} \cdot mol^{-1}$ of a 0.001M solution measured at 25°C [6] indicates a 1:2 electrolyte due to dissociation. It is insoluble in usual organic solvents such as alcohol, ether, $CHCl_3$, and benzene [4].

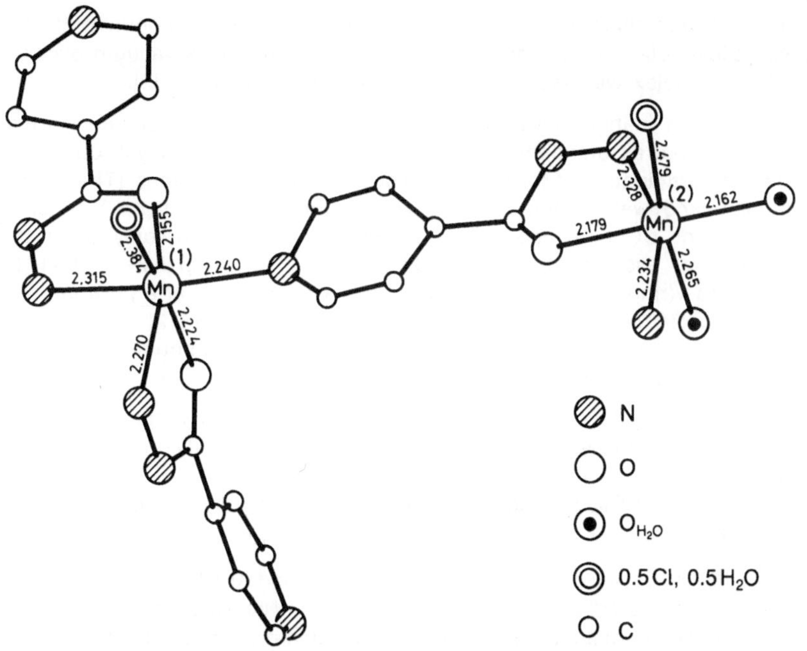

Fig. 22. Structure of the complex cation $[Mn_2(C_6H_7N_3O)_3Cl(H_2O)_3]^{3+}$ with bond lengths in Å and bond angles around the Mn^{2+} ions [5].

$Mn(C_6H_7N_3O)SO_4 \cdot 3H_2O$ precipitates from a concentrated aqueous solution of $MnSO_4$ and isonicotinohydrazide. It forms very fine needle-shaped microcrystals which are far more soluble in water than many other metal complexes [8].

$Mn(C_6H_7N_3O)(NCO)_2$. Solutions of manganese cyanate (obtained by mixing ethanolic solutions of manganese nitrate and KNCO in a 1:2 mole ratio and removing the precipitated KNO_3) and isonicotinohydrazide in ethanol were mixed and the lemon-yellow precipitate which had formed after about 1 h was washed rapidly with water to remove KNO_3 [10], ethanol, and ether and dried in vacuum over (anhydrous) $CaCl_2$ [9, 10]. The air-stable complex is apparently isostructural with the corresponding ones of nickel, cobalt, and zinc as shown by the X-ray powder diagrams [9]. The IR spectrum and the Raman spectrum were recorded in the 4000 to 30 cm^{-1} region. Shifts of the ν(NH) and the ν(ring) vibration modes were observed in comparison to the free ligand, but no shift of the amide I band. Therefore coordination of isonicotinohydrazide to Mn only through the amine (NH_2) and pyridine ring nitrogens with apparently no participation of the carbonyl oxygen (?) was suggested. It was assumed that the ligand is bidentate bridging between two Mn atoms and that the cyanato groups are bonded to Mn through the nitrogen atoms as suggested by the placement and intensity of the bands assignable to them [10].

The complex is soluble in water and insoluble in ethanol, ether, acetone, dioxane, dimethylformamide, and dimethyl sulfoxide [9].

$Mn(C_6H_7N_3O)_2(NCS)_2$. To an aqueous solution of $MnCl_2 \cdot 4H_2O$, aqueous solutions of NH_4NCS and isonicotinohydrazide were added. The precipitate which appeared after 10 to 15 min was redissolved by warming. The far more crystalline cream-white solid which separated on cooling was filtered off, washed twice with water and ethanol and dried on porous clay [11]. The complex could be obtained also from $Mn(NO_3)_2 \cdot 6H_2O$ and KNCS in ethanolic solution by removing the precipitate of KNO_3 and adding the two-fold molar amount of the ligand. The precipated yellow solid was washed with ethanol and dried in a vacuum over $CaCl_2$. The identity of the complex was ascertained by X-ray powder diagrams [9].

The IR spectrum (mineral oil and fluorinated oil mulls) was investigated in the 4000 to 400 cm^{-1} range. It was assumed that the ligand is coordinated to Mn only through the carbonyl oxygen with no participation of the NH_2 group or pyridine ring nitrogens. (The ν(CO) vibration mode of the ligand was shifted to lower wave numbers ($\Delta\nu \approx 40\ cm^{-1}$) on coordination whereas the bands of ν(NH) and ν(ring) remained unchanged or only slight shifts were observed.) However, assignment of the band at 1618 cm^{-1} to ν(ring) instead of $\delta(NH_2)$ and the band ascribed to ν(ring) at 1025 cm^{-1} support the assumption that a portion of the ligand molecules is coordinated also through the pyridine ring nitrogen [12]. The intense $\nu(CN)_{NCS}$ band (at 2090 cm^{-1}) and weak ν(CS) band (at 803 cm^{-1}) indicate bonding of the thiocyanate groups to Mn through the nitrogen atoms [12].

$Mn(C_6H_7N_3O)_2(NCS)_2$ decomposes upon heating at 230°C [13]. It is sparingly soluble in cold water [9, 11], the solubility being 0.0030 mol/L at 20 to 22°C (solubility product 1.1×10^{-7}) [13]. It is far more soluble in hot water and readily soluble in pyridine [11], insoluble in ethanol, ether [9, 11], acetone [9], $CHCl_3$, and benzene [11], sparingly soluble in dimethylformamide and dimethyl sulfoxide [9].

$Mn(C_6H_7N_3O)_2(NCSe)_2$. Solutions of $Mn(NO_3)_2 \cdot 6H_2O$ and KNCSe in a minimum volume of ethanol were mixed in a 1:2 mole ratio, the precipitated KNO_3 was removed and an ethanolic solution of the ligand (stoichiometric amount) was added dropwise to the filtrate at pH 6. The precipitate was filtered off and the filtrate was allowed to crystallize for 2 to 3 d in a vacuum desiccator. The crystals were washed with small portions of ethanol and dried in vacuum over $CaCl_2$. The complex is rather air-stable. Its IR spectrum (Vaseline mulls) shows characteristic absorption bands of $\nu(NH_2)$ at 3148, ν(CN) at 2078, ν(CO) at 1660, ν(ring) at 1605, 1495, and 1017 cm^{-1}; the latter bands had shifted to higher frequencies upon complexation therefore coordination of the pyridine ring nitrogen to Mn was suggested. Coordination of the SeCN groups to Mn through the nitrogen is supported by a band assigned to ν(CSe) at 600 cm^{-1}. For other unassigned bands see the paper. The complex exhibits antituberculosis properties [17].

$Mn(C_6H_7N_3O)(C_6H_5COO)_2$. Aqueous solutions of $MnCl_2 \cdot 4H_2O$, sodium benzoate and isonicotinohydrazide were mixed and evaporated on a water bath for about 30 min. The crystals which separated on cooling were removed, washed and dried as far as possible; they are yellowish white [14], or maroon [13], air-stable up to 100°C [14] and melt at 210°C [15]. The IR spectrum (KBr pellets) was investigated in the 4000 to 700 cm^{-1} region, and coordination of isonicotinohydrazide to Mn through the nitrogen of the NH_2 group and carbonyl oxygen was suggested [15].

The complex decomposes upon heating at 150°C [13]. It is soluble in water [14], the solubility being 0.0022 mol/L at 20 to 22°C (solubility product 4.1×10^{-8}) [15]. It dissolves also in ammonia and pyridine but is sparingly soluble in alcohol, ether, $CHCl_3$, acetone, and butanol [14]. The complex is being studied in view of its efficiency in antituberculosis therapy [15].

$Mn(C_6H_7N_3O)_2HgI_4$. A warm aqueous solution of K_2HgI_4 (obtained from $HgCl_2$ and KI) was treated first with an aqueous solution of $MnSO_4 \cdot 4H_2O$, then with a solution of isonicotinohydrazide. The mixture was concentrated on a water bath and allowed to crystallize. The microcrystalline yellow-green air-stable precipitate which had formed after several hours was washed with a little water, alcohol, and ether and dried in vacuum or in air. The complex is sparingly soluble in water, alcohol, ether, $CHCl_3$, and acetone, but soluble in ammonia, aqueous HCl, and (somewhat more) in pyridine. A comparison with other related metal complexes shows that the solubility in water increases in the series Cu<Cd<Fe<Ni<Co<Mn [16].

References:

[1] Narimanidze, A. P., Tsivadze, A. Yu., Dzhavakhishvili, Z. O., Il'inskii, A. L., Tsivtsivadze, T. I., Kharitonov, Yu. Ya. (Koord. Khim. **4** [1978] 233/8; Soviet J. Coord. Chem. **4** [1978] 175/9).
[2] Sanke Gowda, H., Janardhan, R. (Proc. Indian Acad. Sci. Chem. Sci. **91** [1982] 339/41; C.A. **98** [1983] No. 10673).
[3] Allan, J. R., Baillie, G. M., Baird, N. D. (J. Coord. Chem. **13** [1984] 83/8, 84).
[4] Khakimov, Kh. Kh., Azizov, M. A., Khamraev, A. D. (Mater. Yubileinoi Resp. Nauchn. Konf. Farm. Posvyashch. 50-Letiyu Obrazov. SSSR, Tashkent 1972, pp. 151/2; C.A. **82** [1975] No. 164327).
[5] Tsintsadze, G. V., Dzhavakhishvili, Z. O., Aleksandrov, G. G., Struchkov, Yu. T., Narimanidze, A. P. (Koord. Khim. **6** [1980] 785/92; C.A. **93** [1980] No. 35435).
[6] Narimanidze, A. P., Dzhashiashvili, T. K., Kharitonov, Yu. Ya. (Tr. Gruz. Politekhn. Inst. **1972** No. 2, pp. 9/15, 9/11, 14; C.A. **92** [1980] No. 68844).
[7] Abou Sekkina, M. M., Abou El-Azm, M. G. (Thermochim. Acta **79** [1984] 47/53, 49).
[8] Neuzil, E., Segonne, J. (Bull. Soc. Pharm. Bordeaux **93** [1955] 49/66, 58; C.A. **1956** 9200).
[9] Petriashvili, Zh. D., Narimanidze, A. P., Mamulashvili, A. M. (Sakartvelos Politeknikuri Institutis Shromebi [Tr. Gruz. Politekhn. Inst.] **3** No. 167 [1974] 13/4; C.A. **83** [1975] No. 141223).
[10] Tsivadze, A. Yu., Kharitonov, Yu. Ya., Tsintsadze, G. V., Petriashvili, Zh. D. (Koord. Khim. **1** [1975] 525/30; Soviet J. Coord. Chem. **1** [1975] 425/30, 427).

[11] Hlevca, M. (Acad. Rep. Populare Romine Studii Cercetari Chim. **9** [1961] 547/56, 550; C.A. **57** [1962] 12090).
[12] Narimanidze, A. P., Tsivadze, A. Yu., Kharitonov, Yu. Ya., Dzhashiashvili, T. K. (Koord. Khim. **1** [1975] 936/41; Soviet J. Coord. Chem. **1** [1975] 793/8, 794).
[13] Grecu, I. (Omagiu Raluca Ripan **1966** 236/7; C.A. **67** [1967] No. 49969).
[14] Grecu, I., Curea, E., Pitiş, M. (Acad. Rep. Populare Romine Filiala Cluj Studii Cercetari Chim. **13** [1962] 213/8, 216; C.A. **61** [1964] 10027).
[15] Doadrio, A., Crăciunescu, D., Suarez, M., Shohet, J., Castro, F. (Acta Pharm. Fennica **85** [1976] 19/27, 22/3; C.A. **85** [1976] No. 117301).
[16] Grecu, I., Pitiş Neamţu, M. (Farmacia [Bucharest] **11** [1963] 655/61, 659; C.A. **60** [1964] 14108).
[17] Narimanidze, A. P., Dzhashiashvili, T. K., Kharitonov, Yu. Ya. (Nauchn. Tr. Gruz. Politekhn. Inst. Khim. Koord. Soedin. No. 12 [1982] 14/8; C.A. **100** [1984] No. 44421).

25.2 Complexes with Ligands of the Type RC(O)NHNHC(O)R

25.2.1 With N, N'-Diformylhydrazine

HC(O)–NH–NH–C(O)H $\rightleftharpoons$ HC(OH)=N–N=C(OH)H (= $C_2H_4N_2O_2$)

$Mn(C_2H_2N_2O_2)\cdot nH_2O$ (n = 2, 0). To an aqueous solution containing $MnCl_2\cdot 4H_2O$ and diformylhydrazine in a 1:2 mole ratio, an aqueous solution of KOH was added up to pH ~ 9. The fine crystalline beige dihydrate was heated in its mother liquor on a water bath for about 30 min, filtered while hot, washed thoroughly first with hot then with cold water and alcohol, and dried in air. Upon heating above 180°C all water was lost to leave the anhydrous complex. The IR spectra of both complexes (mineral oil and fluorinated oil mulls) were recorded in the 4000 to 400 cm^{-1} region. The dihydrate shows bands assignable to ν(OH) of coordinated water at 3395 and 3300 to 2900 cm^{-1}. The $\delta(H_2O)$ bands are superimposed on the high-frequency side. The bands due to water disappear on dehydration. Considering the occurrence of IR bands assignable to ν(C=N) vibrations at ~1550 cm^{-1} and ν(C–O) vibrations at 1280 cm^{-1}, it is assumed that diformylhydrazine is present in its deprotonated hydroxy hydrazone form which is coordinated to Mn as a tetradentate bridging ligand through two imino nitrogens and two deprotonated oxygens. The coordination sphere around the Mn atom is completed up to an octahedron by coordinated water, with the organic ligand adopting a *cis* or *trans* configuration. Coordination of water in the dihydrate is supported also by thermal analysis which shows an endothermic effect at 180°C indicating the loss of water. An exothermic effect around 240°C indicates decomposition of the anhydrous complex with formation of manganese oxide as the final product in air. The (apparently polymeric) compounds are practically insoluble in water, ethanol, CH_2Cl_2, $CHCl_3$, acetone, acetonitrile, benzene, nitrobenzene, and dimethylformamide, Kharitonov, Yu. Ya., Machkhoshvili, R. I., Goeva, L. V., Shchelokov, R. N. (Koord. Khim. **1** [1975] 333/41; Soviet J. Coord. Chem. **1** [1975] 263/9, 264, 267).

25.2.2 With N,N′-Diacetylhydrazine

$H_3CC(O){-}NH{-}NH{-}C(O)CH_3 \rightleftharpoons H_3CC(OH){=}N{-}N{=}C(OH)CH_3 \quad (=C_4H_8N_2O_2)$

$Mn(C_4H_8N_2O_2)Cl_2 \cdot H_2O$. A stream of gaseous HCl was passed slowly for about 30 min over a thin layer of $Mn(C_4H_6N_2O_2) \cdot 2H_2O$ (see below), which becomes light pink. The product was left in air for 10 to 12 h to remove excess absorbed HCl. The IR spectrum of the compound in liquid petrolatum or fluorinated oil mulls was recorded in the 4000 to 400 cm^{-1} region. The shifts of the ν(NH) vibration modes observed at 3170 cm^{-1} and of the ν(CO) band at 1680 and 1638 cm^{-1} to lower frequencies by about 50 cm^{-1} in comparison to the free ligand indicate coordination of the tetradentate chelating ligand in its oxo hydrazine tautomeric form to Mn through the carbonyl oxygen and nitrogen atoms, which is supported also by bands assignable to ν(Mn–N) and ν(Mn–O) vibration modes below 500 cm^{-1}. Since $Mn(C_4H_8N_2O_2)Cl_2 \cdot H_2O$ retains its water on heating up to 320°C it is assumed that the water molecule is strongly bonded in the inner complex sphere. The conversion into $Mn(C_4H_6N_2O_2)$ at higher temperature (with release of HCl) is accompanied by the simultaneous formation of manganese oxides [1].

$Mn(C_4H_6N_2O_2) \cdot nH_2O$ (n = 1, 2). The dihydrate was prepared by addition of aqueous ammonia to an aqueous solution of $MnCl_2 \cdot 4H_2O$ and the ligand (mole ratio 1:2) until pH of 9 was reached. The light beige precipitate was decanted, washed several times with hot and cold water and dried in air, where it becomes brown. The anhydrous complex is obtained by heating the dihydrate at about 150°C or keeping it in vacuum over P_2O_5 for 10 to 12 h. The IR spectrum of the dihydrate, recorded as for $Mn(C_4H_8N_2O_2)Cl_2 \cdot H_2O$ (see above), shows characteristic absorption bands (in cm^{-1}) which were assigned as follows: 3400 to 3200 to $\nu(H_2O)$, 1547 to ν(C=N), 1332 to ν(C–O), 941 to $\nu(C{-}CH_3)$, 535, 515(sh) to ν(Mn–ligand) [1]. Except for the band due to H_2O the IR spectrum of the anhydrous complex is almost identical. The absence of bands assignable to NH vibrations indicates that diacetylhydrazine behaves as a negatively charged deprotonated ligand bonded to Mn through the oxygens of its tautomeric hydroxy hydrazone form. Upon heating, the dihydrate starts to lose water at about 110°C and is completely dehydrated at about 140 to 150°C, apparently with some decomposition leading to the formation of manganese oxides especially on further heating. Since the complex is practically insoluble in water and usual organic solvents a dimeric or polymeric structure involving Mn–O and Mn–N bonds must be assumed [2]. Treatment with gaseous HCl affords the formation of $Mn(C_4H_8N_2O_2)Cl_2$, see above [1].

References:

[1] Kharitonov, Yu.Ya., Machkhoshvili, R. I., Goeva, L. V. (Koord. Khim. **5** [1979] 1150/5; Soviet J. Coord. Chem. **5** [1979] 904/8, 907).

[2] Kharitonov, Yu.Ya., Machkhoshvili, R. I., Goeva, L. V. (Koord. Khim. **1** [1975] 1449/57; Soviet J. Coord. Chem. **1** [1975] 1205/11, 1206).

25.2.3 With N,N'-Dibenzoylhydrazine

$C_6H_5C(O)$–NH–NH–$C(O)C_6H_5 \rightleftharpoons C_6H_5C(OH)$=N–N=$C(OH)C_6H_5$ (= $C_{14}H_{12}N_2O_2$)

$Mn(C_{14}H_{12}N_2O_2)Cl_2 \cdot 2H_2O$. A stream of gaseous HCl was passed slowly over $Mn(C_{14}H_{10}N_2O_2) \cdot 1.5H_2O$ (see below) for about 30 min. The resulting pale pink product was left in air for 10 to 12 h to remove any adsorbed HCl. The IR spectrum shows bands of ν(NH) at 1675 cm^{-1} and of ν(C=O) at 1642 cm^{-1}. The shifts of these bands to lower frequencies in comparison to the free ligand by 150 to 100 and ~40 cm^{-1}, respectively, suggests coordination of the tetradendate ligand to Mn through the nitrogen and the carbonyl oxygen atoms. A polymeric structure in which the ligands are bridging was assumed. The water molecules are obviously coordinated to Mn since the complex does not lose water upon heating up to 380°C, whereupon it decomposes completely without previous reversion to the anhydrous compound. $Mn(C_{14}H_{12}N_2O_2)Cl_2 \cdot 2H_2O$ dissolves in water, alcohols, and acetone with decomposition; it is insoluble in CH_2Cl_2, $CHCl_3$, benzene, and nitrobenzene [1].

$Mn(C_{14}H_{10}N_2O_2) \cdot 1.5H_2O$. To a hot ethanolic solution of benzoylhydrazine which had been made slightly alkaline (pH ~ 8.5) by addition of aqueous ammonia, a hot ethanolic solution of $MnCl_2 \cdot 4H_2O$ was added. The white precipitate was filtered off while hot, washed with hot ethanol followed by acetone and dried in air. The IR spectrum of the complex shows bands assignable to ν(C=N) and ν(C–O) modes at ~1510 and 1415 cm^{-1}, respectively, but no bands due to ν(NH) vibrations. It was concluded that the deprotonated ligand has four coordination sites (N, N, O, O) and is possibly bridging between the metal atoms in a dimeric or polymeric structure. The hydrated complex becomes anhydrous upon heating at about 100°C and starts to decompose gradually around 240°C. Decomposition is complete at about 400°C with formation of manganese oxides as shown by IR studies. The complex is practically insoluble in water or common organic solvents [2].

References:

[1] Kharitonov, Yu. Ya., Machkhoshvili, R. I., Goeva, L. V. (Koord. Khim. **5** [1979] 1352/8; Soviet J. Coord. Chem. **5** [1979] 1055/60, 1058).

[2] Kharitonov, Yu. Ya., Machkhoshvili, R. I., Goeva, L. V. (Koord. Khim. **2** [1976] 1481/9; Soviet J. Coord. Chem. **2** [1976] 1137/43, 1142).

25.2.4 With N-Acyl-N'-Salicyloyl- or N,N'-Disalicyloylhydrazines

RC(O)–NH–NH–C(O)R' $\rightleftharpoons$ RC(OH)=N–N=C(OH)R'

ligand 1 R = 2-HO–C_6H_4, R' = CH_3; (= $C_9H_{10}N_2O_3$)

ligand 2 R = 2-HO–C_6H_4, R' = C_6H_5; (= $C_{14}H_{12}N_2O_3$)

ligand 3 R = R' = 2-HO–C_6H_4; (= $C_{14}H_{12}N_2O_4$)

$Mn(C_9H_8N_2O_3)$ and **$Mn(C_{14}H_{10}N_2O_3) \cdot 2H_2O$.** For preparation, solutions of the ligands in aqueous ammonia were neutralized with dilute HCl. To the clear filtrate a manganese(II) salt in aqueous solution was added, using a mole ratio Mn:ligand up to 1:4. The dirty white complexes, which precipitated immediately, were washed with dilute ethanol and dried in a desiccator. The IR spectra of the complex (KBr disks) show a broad absorption band in the 3300 cm^{-1} region assignable to mixed ν(OH) and ν(NH) vibrations which upon complexation had shifted to lower frequencies. The band around 1600 cm^{-1} may be ascribed to coordinated carbonyl and C=N groups, the latter apparently being due to ligand enolization. A band occurring in the 1520 to 1500 cm^{-1} region can be assigned to vibrations of coordinated $N{=}CO^-$ groups. Considering the IR data, the decomposition points and the insolubility of the complexes, a polymeric structure appears to be likely. Susceptibility measurements (Faraday method) yield the magnetic moments $\mu_{eff} = 5.7$ μ_B for $Mn(C_9H_8N_2O_3)$ and 6.2 μ_B for $Mn(C_{14}H_{10}N_2O_3) \cdot 2H_2O$. Upon heating, the complexes decompose without melting at 200 or 300°C, respectively. They are insoluble in water and common organic solvents, except for pyridine, in which they yield intensely colored solutions [1].

$Mn_3(C_{14}H_{10}N_2O_4)_2(OH)_2$ prepared from a manganese(II) salt und N,N'-disalicyloylhydrazine (mole ratio 1:2) as described above was washed with water and hot dilute ethanol. The dark green paramagnetic complex has the magnetic moment $\mu_{eff} = 5.8$ μ_B. Since it decomposes above 200 to 250°C without melting and is insoluble in water and common organic solvents, a polymeric structure was suggested [2].

References:

[1] Gopal, R., Misra, V. N., Narang, K. K. (Indian J. Chem. **13** [1975] 186/8).
[2] Gopal, R., Misra, V. N., Narang, K. K. (Indian J. Chem. A **14** [1976] 364/6).

25.2.5 With N-Acyl-N'-Isonicotinoylylhydrazines

$$3\text{-}C_5H_4NC(O){-}NH{-}NH{-}C(O)R \rightleftharpoons 3\text{-}C_5H_4NC(OH){=}N{-}N{=}C(OH)R$$

ligand 1 $R = CH_3$; $(= C_8H_9N_3O_2)$

ligand 2 $R = C_6H_5$; $(= C_{13}H_{11}N_3O_2)$

$[Mn(C_8H_7N_3O_2)(H_2O)_2]$. Aqueous solutions of manganese nitrate or sulfate and ligand 1 were mixed in about 1:1 mole ratio and an NH_4Cl-ammonia buffer solution was added until a slight turbidity appeared. The mixture was then refluxed on a water bath for 0.5 h to complete the reaction. The dirty yellow precipitate was collected, washed with water and ethanol and dried at room temperature. The IR spectrum, recorded as Nujol mulls, shows an absorption band in the 3470 to 3300 cm^{-1} region which is assigned to ν(OH) of coordinated water since it disappears upon dehydration above 160°C as well as a further band in the 755 to 720 cm^{-1} region. The ligand bands assignable to ν(NH) and to ν(C=O) disappear and are replaced by bands assignable to ν(C=N), ν($N{=}C{-}O^-$), and to ν(C–O) in the complex. These changes indicate enolization of both oxo groups with formation of C–O–Mn bonds which are supported by the absence of any anion band. A band in the 1035 to 1015 cm^{-1} region which is assigned to ν(N–N) bridging had shifted to higher frequencies, whereas the ν(ring) bands are practically unaffected and exclude coordination of the pyridine ring nitrogen. The observed vibration modes show that N-acetyl-N'-isonicotinoylhydrazine behaves as a tetradentate ligand, bonded to Mn through two enolic oxygens and two azomethine nitrogens. Six-coordination of manganese in an octahedral environment is completed by the two oxygens of the water molecules. The electronic reflectance spectrum shows two bands in the 26700 to 29400 and 34480 to 35000 cm^{-1} regions due to intra-ligand transitions. The magnetic moment $\mu_{eff} = 5.85$ μ_B (from susceptibility data obtained by the Cahn-Faraday method) is consistent with the proposed octahedral structure. The high temperature of decomposition (>250°C) and the insolubility of the complex in common organic solvents suggest an either ionic or polymeric structure [1].

[Mn($C_{13}H_{10}N_3O_2$)$_2$(H_2O)$_2$] was obtained from hot aqueous solutions of manganese sulfate or nitrate and N-benzoyl-N′-isonicotinoylhydrazine (mole ratio 1:2). Analytical data and the IR spectrum shows that N-benzoyl-N′-nicotinoylhydrazine acts as a monodeprotonated bidentate ligand. The electronic reflectance spectrum shows bands at 28570, 35100, and 37700 cm^{-1} of intra-ligand transitions. The magnetic moment $\mu_{eff}=6.20\ \mu_B$ (from susceptibility data) suggests an octahedral complex geometry. [Mn($C_{13}H_{10}N_3O_2$)$_2$(H_2O)$_2$] loses its water upon heating at about 160°C with change in color and decomposes above 250°C. It is insoluble in common organic solvents such as ethanol, $CHCl_3$, CCl_4, acetone, benzene, tetrahydrofuran, dimethylformamide, and dimethyl sulfoxide [2].

References:

[1] Aggarwal, R. C., Singh, N. K., Prasad, L. (Indian J. Chem. A **14** [1976] 325/7).
[2] Aggarwal, R. C., Singh, N. K., Prasad, L. (Indian J. Chem. A **14** [1976] 181/3).

25.3 With Hydrazides of Dicarboxylic Acids

25.3.1 With Oxalodihydrazide $H_2NNHC(O)C(O)NHNH_2$ (=$C_2H_6N_4O_2$)

Mn($C_2H_6N_4O_2$)$Cl_2\cdot2H_2O$. Aqueous solutions of manganese chloride and the ligand were mixed in a 1:1 mole ratio and a few drops of concentrated aqueous HCl were added. The mixture was heated on a water bath for about 2 h and ethanol was added to precipitate the complex which was washed with ethanol and ether and dried at room temperature. The white solid melts at 215°C. The IR absorption spectrum reveals shifts of the ligand absorption bands assigned to amide I, ν(CO) at 1680, to β or δ(NH_2) at 1608, and to amide II at 1540, to lower frequencies (for amide I ~25 cm^{-1}, for β(NH_2) ~20 cm^{-1}, and for amide II ~40 cm^{-1}) and a shift to higher frequencies (~40 cm^{-1}) for the ν(N–N) vibration mode of the ligand at 940 cm^{-1}. Bonding of both symmetrical carbonyl and both symmetrical NH_2 groups of the tetradentate ligand to manganese was suggested. The bands observed in the 385 to 300, 310 to 240 and 300 cm^{-1} regions were tentatively assigned to ν(Mn–O), ν(Mn–N), and ν(Mn–Cl) modes, respectively. Susceptibility measurements at room temperature yielded the magnetic moment $\mu_{eff}=6.10\ \mu_B$ indicative of a high-spin Mn^{II} (d^5) complex. The molar electrical conductivity $\Lambda=1.65\ cm^2\cdot\Omega^{-1}\cdot mol^{-1}$ measured in dimethyl sulfoxide solution is that of a nonelectrolyte. Upon heating, Mn($C_2H_6N_4O_2$)$Cl_2\cdot2H_2O$ loses its water at 100 to 110°C. It is soluble in cold dimethyl sulfoxide but insoluble in water and common organic solvents [1].

[Mn($C_2H_4N_4O_2$)(H_2O)$_2$]. Hot aqueous solutions of manganese chloride and oxalodihydrazide were mixed in about 1:1 mole ratio and the pH of the mixture was raised to about 8 by adding dilute NaOH [1] or NH_3 [2] solution. The precipitate was digested on a water bath for about 2 h, washed first with dilute acetic acid, then with water and alcohol and dried at room temperature [1]; also see [2]. The light yellow [1] or mustard-colored complex has the refractive indices $n_\alpha=1.629$ and $n_\gamma=1.682$; X-ray studies (Debyeograms) show it to be amorphous [2]. The IR spectrum (liquid paraffin and fluorinated hydrocarbon mulls) recorded in the 4000 to 400 cm^{-1} region shows characteristic absorption bands (in cm^{-1}) assigned as follows: ~3300 to ν(NH_2) or ν(H_2O), 1622 to ν(C=N) or δ(NH_2) and 1340 to ν(C–O) [3]. The disappearing of the strong ligand amide I and amide II bands and appearing of new bands assignable to ν(N=C–O) in the 1575 to 1550 and 1320 to 1290 cm^{-1} regions indicate the formation of Mn–O–C bonds presumably through the enolization of the ligand oxo groups. The shift of β(NH_2) to lower frequencies by about 20 cm^{-1} and of ν(N–N) to higher ones by about 50 cm^{-1} indicates coordination also of the two symmetrical NH_2 groups. The occurrence of a band indicative of coordinated water in the 760 to 665 cm^{-1} region clearly demonstrates coordination of the water molecules in $[Mn(C_2H_4N_4O_2)(H_2O)_2]$ [1]. Thus it is concluded from the IR data that the complex contains five-membered metallocyclic rings where oxalodihydrazide acts as a bridging tetradentate ligand forming a dimeric or, more probably, a polymeric complex [3] with the manganese(II) ion in a spin-free octahedral environment. This is supported also by the magnetic moment $\mu_{eff} = 5.98$ μ_B [1]. The dihydrate loses its water completely at 130 to 150°C to leave the anhydrous complex which in turn decomposes at 230°C [1], at 366°C with oxidation [2]. It is insoluble in water and common organic solvents and only slightly soluble in hot dimethyl sulfoxide [1].

References:

[1] Aggarwal, R. C., Singh, B. (Z. Anorg. Allgem. Chem. **445** [1978] 227/32, 229).
[2] Gogorishvili, P. V., Karkarashvili, M. V., Shamilishvili, O. Kh. (Soobshch. Akad. Nauk Gruz.SSR **60** No. 1 [1970] 89/92; C.A. **74** [1971] No. 60333).
[3] Kharitonov, Yu. Ya., Machkhoshvili, R. I., Gogorishvili, P. V., Shamilishvili, O. Kh. (Zh. Neorgan. Khim. **17** [1972] 2992/5; Russ. J. Inorg. Chem. **17** [1972] 1574/6).

25.3.2 With Malono- and Methylmalonodihydrazide

$H_2NNHC(O)CHRC(O)NHNH_2$ R = H; $(=C_3H_8N_4O_2)$
R = CH_3; $(=C_4H_{10}N_4O_2)$

Complexes in Solution. Spectrophotometric studies on aqueous solutions of $Mn(NCS)_2$ and malonodihydrazide at 288 nm by the method of continuous variation reveal the formation of 1:1 and 1:2 complexes for which values of $pK_1 = 2.00$ and $pK_2 = 1.52$ have been calculated [1].

$Mn(C_3H_8N_4O_2)_3(NO_3)_2 \cdot C_2H_5OH$ crystallizes from alcoholic solutions of manganese(II) nitrate and malonodihydrazide when the mixture is placed in a desiccator over anhydrous $CaCl_2$ for 2 d. The colorless crystals were washed with ethanol and acetone and dried in air; they melt at 110 to 120°C [2].

$Mn(C_3H_8N_4O_2)_nCl_2$ (n = 1, 2). The complexes were obtained from ethanolic solutions of manganese chloride and malonodihydrazide in a 1:1 or 1:2 mole ratio; they were washed with ethanol and ether and dried in air. They melt or decompose in the 131 to 241°C temperature range. The spectra of the complexes in Nujol were compared to the spectrum of the ligand in acetonitrile solution. It was assumed that in $Mn(C_3H_8N_4O_2)Cl_2$ the ligand is tetradentate and coordinated to Mn through both carbonyl oxygens and both terminal nitrogens whereas in $Mn(C_3H_8N_4O_2)_2Cl_2$ it is bidentate and coordinated only through the carbonyl oxygens. The magnetic moments of both complexes are between $\mu_{eff} = 5.90$ and 6.05 μ_B. The complexes are slightly soluble in cold or hot dimethyl sulfoxide but insoluble in other common organic solvents. The molar conductance ($\Lambda < 7$ $cm^2 \cdot \Omega^{-1} \cdot mol^{-1}$) of solutions in dimethyl sulfoxide is that of nonelectrolytes [3].

$Mn(C_3H_8N_4O_2)_2X_2 \cdot nH_2O$ (X = Cl, $SO_4/2$). The complexes with n = 2 precipitate from saturated aqueous solutions of the manganese(II) salts (mole ratio 1:2) after standing for some time or upon adding methanol to the aqueous mixture; they were also obtained from methanolic sulutions of the components and could be purified by precipitating saturated aqueous solutions with methanol. The chloro complex may also be purified by heating it with dimethylformamide, filtering and washing with hot methanol [4]. The complex $Mn(C_3H_2D_6N_4O_2)_2SO_4$ of the N-deuterated ligand was obtained by cooling the solution of $Mn(C_3H_8N_4O_2)_2SO_4$ in D_2O (>99.4%) in a freezing mixture for some time followed by slow evaporation to dryness under anhydrous conditions [5].

The light yellow crystals of $Mn(C_3H_8N_4O_2)_2Cl_2 \cdot 2H_2O$ are orthorhombic and dimorphic. X-ray studies show that the α-form crystallizes in space group F222-D_2^7 (No. 22) with the lattice constants a = 7.322(3), b = 14.563(3), and c = 15.070(3) Å; Z = 4. The β-form crystallizes in space group Pbam-D_{2h}^9 (No. 55) with a = 7.865(2), b = 13.130(2), and c = 16.104(2) Å; Z = 4. The structure of the α-form was solved by the heavy-atom technique from 1599 reflections and that of the β-form by direct methods from 1377 reflections. Least squares refinement of both structures was done by anisotropic patterns. The H atoms were located from difference Fourier syntheses and two last refinement cycles (with isotropic temperature factors for H and anisotropic temperature factors for the other atoms) led to a final reliability R = 0.039 for the α-form and R = 0.024 for the β-form. The atomic coordinates of both forms (with Mn in 0, 0, 0 for α and 0, 0, 0.2503(3) for β) are given in the paper. The atomic distances (in Å) and bond angles derived therefrom are shown in **Fig. 23**a, p. 202, for the cation of the α-form and in **Fig. 23**b, p. 202, for the β-form. In the structure of both forms malonodihydrazide acts as a bis-bidentate chelating ligand. It is bound to Mn through the carbonyl oxygens and the nitrogen atoms of the NH_2 groups. The C–C and C=O bond lengths are normal. Partial delocalization of the lone-pair electrons in the NH group caused shortening of the N–N and C–N bonds. The chelate rings have an envelope conformation with the Mn atom at the midpoint of the envelope. The angle of the fold in the envelope along the $O \cdots NH_2$ line is 8° in the α-form but 10° and 23° for the two independent chelate rings in the β-form. The increase of the dihedral angles between the O=C–(NH)–CH_2 planes from 93° (free ligand) to 101° in the α-form and to 118°C in the β-form arises from packing effects. In both forms the Mn atom is eight-coordinated and forms an eight-vertex coordination polyhedron. The structure is intermediate between a dodecahedron and an Archimedean antiprism, but in the β-form the structure is closer to a dodecahedron than in the α-form. A projection of both structures on the xz or xy plane is given in the paper. The structure of the α-form can be described as a chain polymer with the polymeric $[Mn(C_3H_8N_4O_3)_2^{2+}]_n$ chains extended along the x axis and linked together in the y direction by the water molecules which form H bonds with the hydrazide nitrogens and in the z direction by the outer-sphere Cl^- ions which also are involved in H bonds with the NH_2 and NH groups of ligand molecules in neighboring chains. In the β-form the Mn atoms are linked crosswise by ligand bridges to form nets parallel to the xy plane. The mesh points are displaced by ½ along the x and y axes. The Cl^- ions and the water molecules lying between the layers at z = 0 are linked in infinite chains extending along the x axis by $Cl \cdots H–O–H \cdots Cl$ hydrogen bonds. The other Cl^- ions and water molecules form similar chains along the y axis and there are also H bonds between these Cl^-–water chains and the $[Mn(C_3H_8N_4O_2)_2^{2+}]_n$ nets. The chief difference between the α- and β-forms of the complex thus is caused by the way in which the metal atoms are linked by the ligand molecules to give chains in the α-form and networks in the β-form. This difference, rather than the coordination of the metal or structural function of the ligand, leads to a variation in the form of the polyhedron from a dodecahedron in the β-form towards an antiprism in the α-form [6].

The light yellow crystals of the sulfato complex (formulated as a 4-hydrate in the paper) are also orthorhombic in the space group F222-D_2^7 (No. 22) as shown by X-ray studies. The lattice constants are a = 11.582(2), b = 19.306(2), and c = 8.142(1) Å; Z = 4, V = 1820.7(6) $Å^3$. The structure was solved from 681 reflections and refined in the same way as for α-$Mn(C_3H_8N_4O_2)_2Cl_2 \cdot 2H_2O$, (see above) up to a final reliability R = 0.034. The atomic coordinates (with Mn in 0, 0, 0) are given in the paper. The atomic distances (in Å) and bond angles derived therefrom are shown in **Fig. 24**, p. 202, for the metallocyclic unit. The main structural components of the complex are polymeric networks of $\{[Mn(C_3H_8N_4O_2)_2]^{2+}\}_\infty$ units with eight-coordinated manganese sites, tetrahedral SO_4^{2-} ions and water of crystallization. The coordination polyhedron has eight vertices and can be described as a dodecahedron distorted towards an Archimedean antiprism. The ligand molecule is bis-bidentate bridging and coordinates to Mn through the carbonyl oxygens and hydrazidic NH_2 groups to form five-membered rings. The OCNN grouping in the rings is approximately planar with maximum deviations of ± 0.002 Å from the mean-square plane; the Mn atom lies 0.286 Å beyond this plane. The metallocyclic ring can be described as an envelope with a 7.1° fold along the O–NH_2 line. The OCNN planes adjacent to different Mn atoms form an angle of 85.3° which is lower than in the free ligand (93°) and apparently due to steric factors, in particular the necessity of providing a dodecahedral coordination of the metal atoms. Coordination of the ligand does not affect its other bond lengths except for the (C–NH) bond which is somewhat shortened and indicates increase of π-interaction. The OSO bond angles in the SO_4^{2-} groups have the usual tetrahedral values. A projection of the structure of $[Mn(C_3H_8N_4O_2)_2SO_4 \cdot 4H_2O]_\infty$ onto the x0z plane is given in the paper. The polymeric manganese dihydrazide networks form layers parallel to (010) at y = 0 and ½. The holes formed within the layers by the meshes of the network are occupied by water molecules which are linked to the polymer by the hydrogen bond O_{water}–H···O and N–H···O_{water}. Between the networks at y = ¼ and ¾ there are layers of SO_4^{2-} ions and water molecules linked to each other by the H bonds O_{water}–H···O_{SO_4} to form a network with junctions at the S atoms. The manganese dihydrazide and water layers are linked to each other by the H bonds N–H···O_{water}. The structure thus contains a three-dimensional system of hydrogen bonds [7].

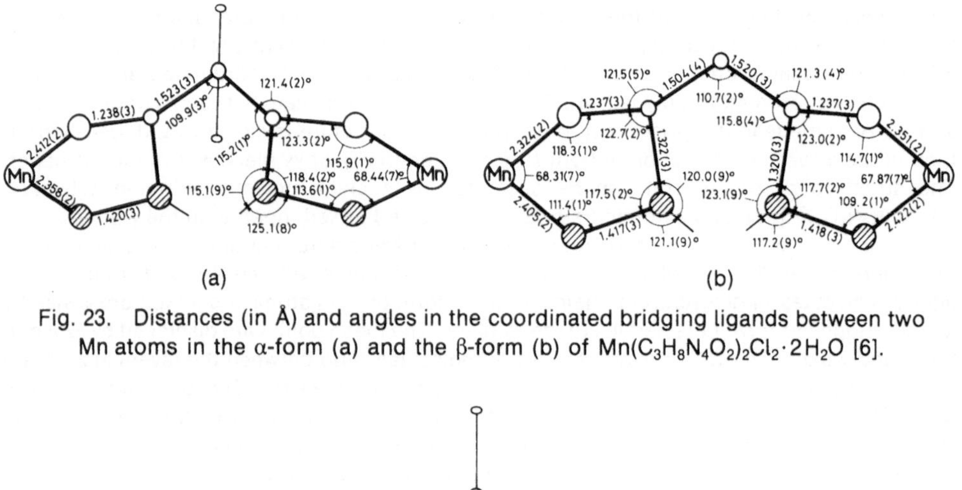

Fig. 23. Distances (in Å) and angles in the coordinated bridging ligands between two Mn atoms in the α-form (a) and the β-form (b) of $Mn(C_3H_8N_4O_2)_2Cl_2 \cdot 2H_2O$ [6].

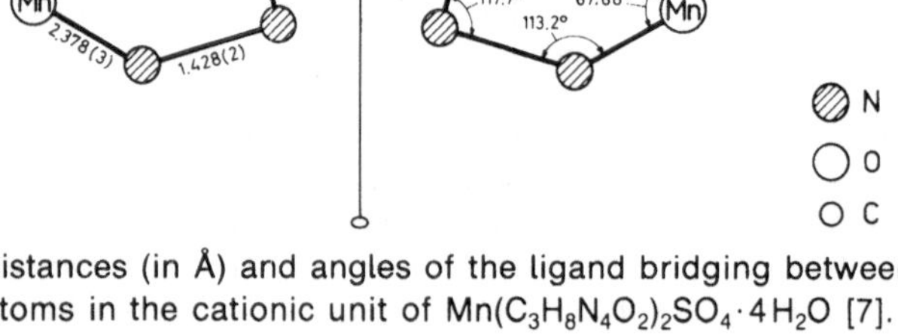

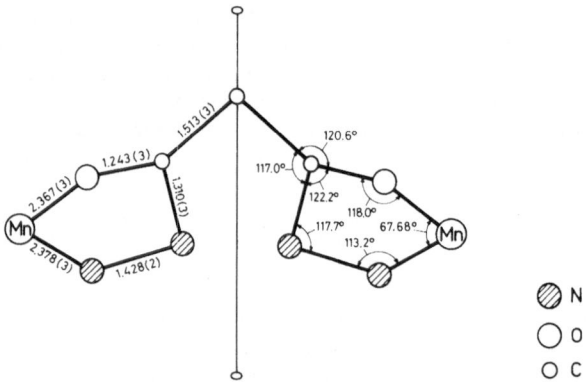

Fig. 24. Distances (in Å) and angles of the ligand bridging between two Mn atoms in the cationic unit of $Mn(C_3H_8N_4O_2)_2SO_4 \cdot 4H_2O$ [7].

$Mn(C_3H_8N_4O_2)_2Cl_2 \cdot 2H_2O$ melts at 128°C and $Mn(C_3H_8N_4O_2)_2SO_4 \cdot 2H_2O$ melts at 191°C with decomposition [4]. The IR spectra of the complexes in Nujol mulls [3] or KBr disks [4] have been recorded. The amide I and amide II bands of the ligand were found to be almost unchanged in the complexes, whereas the amide III band at ~1250 cm^{-1} shifts to higher frequency. The observed magnetic moments resulting from susceptibility measurements at 24°C (μ_{eff} = 5.64 and 6.13 μ_B, respectively) are those of high-spin octahedral Mn^{II} complexes. The compounds are poorly soluble in solvents other than water or methanol (for the chloro complex). The molar electrical conductivities of both complexes in aqueous solution (Λ = 198.7 and 182.1 $cm^2 \cdot \Omega^{-1} \cdot mol^{-1}$, respectively) or of the chloride in methanol solution (Λ = 106.47 $cm^2 \cdot \Omega^{-1} \cdot mol^{-1}$) show the complexes to be electrolytes in these solvents. The electronic spectrum of aqueous solutions shows absorption bands at 47400 nm with ε = 6428 $L \cdot mol^{-1} \cdot cm^{-1}$ (chloro complex) and at 47620 nm with ε = 6437 $L \cdot mol^{-1} \cdot cm^{-1}$ (sulfato complex) [4].

$Mn(C_3H_8N_4O_2)_2(NCS)_2 \cdot H_2O$. The large pale lilac-colored crystals which separated from aqueous solutions of $Mn(NCS)_2$ and malonodihydrazide (mole ratio 1:1) after standing for 2 to 3 d were washed with ethanol and ether and dried in air. The IR spectrum, recorded as liquid paraffin and fluorinated hydrocarbon mulls in the 4000 to 400 cm^{-1} range, shows characteristic absorption bands (in cm^{-1}) which were assigned as follows: 3295, 3180, 3130 to $\nu(OH)$, $\nu(NH)$, or $\nu(NH_2)$; 2090 to $\nu(CN)_{NCS}$; 1670 to amide I, (mainly $\nu(CO)$); ~1618 to $\delta(NH_2)$; 1535 to amide II; 1335 to $\omega(NH_2)$; 1160 to $\delta(NH)$; 1065 to $\tau(NH_2)$; 750 to $\nu(CS)_{NCS}$; 537 to ν(Mn–N); 475 to ν(Mn–O), and ~468 to δ(NCS). The ligands are bonded to Mn through the carbonyl oxygens and the nitrogen atoms of the NH_2 groups. It is concluded from the bands assignable to the thiocyanate groups that these are bonded to Mn through the nitrogen atoms. The complex, which probably has a dimeric or polymeric structure, decomposes upon heating to 210°C. It is soluble in water, the solubility being 5.54 g/100 mL H_2O and the pH of the solution 5.95. Conductivity measurements at 25°C and a dilution rate V = 1000 $L \cdot mol^{-1}$ yield the molar electrical conductance Λ = 224.1 $cm^2 \cdot \Omega^{-1} \cdot mol^{-1}$ indicating a 1:2 electrolyte in water [8].

$Mn(C_3H_8N_4O_2)SO_4 \cdot 3H_2O$ was obtained by mixing saturated aqueous solutions of manganese sulfate and the ligand as described for $Zn(C_3H_8N_4O_2)SO_4 \cdot 3H_2O$ in [11]. X-ray structure analyses showed it to be isostructural with the related complexes of Zn^{II}, Fe^{II}, Ni^{II}, and Co^{II}. The structural data of $Co(C_3H_8N_4O_2)SO_4 \cdot 3H_2O$ are given in the paper [10].

$Mn(C_4H_{10}N_4O_2)SO_4$ was prepared from manganese(II) sulfate and methylmalonodihydrazide in the same way as for $Mn(C_3H_8N_4O_2)_2SO_4 \cdot 2H_2O$ (see p. 200). The IR spectrum (recorded as for that compound) shows absorption bands assigned to amide I around 1650 cm^{-1}, to amide II around 1535 cm^{-1}, and to amide III around 1235 cm^{-1} of which only the latter one shifted to slightly higher frequency upon coordination. The electronic spectrum of the complex in aqueous solution shows two absorption maxima at λ = 213 and 290 nm with ε = 3268 and 50 $L \cdot mol^{-1} \cdot cm^{-1}$, respectively. The magnetic moment μ_{eff} = 5.88 μ_B indicates a high-spin Mn^{II} complex and the molar conductance Λ = 195.3 $cm^2 \cdot \Omega^{-1} \cdot mol^{-1}$ of a 0.001 M aqueous solution at 25°C corresponds to a 1:2 electrolyte due to aquation [9].

References:

[1] Shvelashvili, A. E., Zhorzholiani, N. B., Svanidze, O. P., Zedelashvili, E. N. (Soobshch. Akad. Nauk Gruz. SSR **119** No. 1 [1985] 105/8; C.A. **104** [1986] No. 25039).

[2] Gobedzhishvili, K. M., Nagebashvili, S. Sh., Dolidze, Ts. G. (Nauchn. Tr. Gruz. Politekhn. Inst. im. V. I. Lenina No. 11 [1980] 12/6, 14; C.A. **96** [1982] No. 134784).

[3] Aggarwal, R. C., Singh, B. (Current Sci. [India] **47** [1978] 679/80; C.A. **89** [1978] No. 208305).

[4] Dutta Ahmed, A., Mandal, P. K., Ray Chaudhuri, N. (J. Inorg. Nucl. Chem. **28** [1966] 2951/9, 2952/6).
[5] Dutta Ahmed, A., Ray Chaudhuri, N. (J. Inorg. Nucl. Chem. **31** [1969] 2545/56, 2550).
[6] Vardosanidze, T. O., Sokol, V. I., Sobolev, A. N., Shvelasvili, A. E., Porai-Koshits, M. A. (Zh. Neorgan. Khim. **30** [1985] 1745/51; Russ. J. Inorg. Chem. **30** [1985] 992/5).
[7] Vardosanidze, T. O., Shvelashvili, A. E., Sobolev, A. N., Porai-Koshits, M. A. (Zh. Neorgan. Khim. **30** [1985] 364/7; Russ. J. Inorg. Chem. **30** [1985] 203/6).
[8] Gogorishvili, P. V., Machkhoshvili, R. I., Tsitsishvili, L. D., Kharitonov, Yu. Ya., Shvelashvili, A. E., Tsutsunava, T. I. (Zh. Neorgan. Khim. **22** [1977] 3275/9; Russ. J. Inorg. Chem. **22** [1977] 1784/7).
[9] Dutta Ahmed, A., Ray Chaudhuri, N., Saha, U. (Indian J. Chem. **11** [1973] 1036/7).
[10] Vardosanidze, T. O., Sobolev, A. N., Shvelashvili, A. E., Charelishvili, L. Sh. (Soobshch. Akad. Nauk Gruz. SSR **119** [1985] 305/8; C. A. **104** [1986] No. 140948).

[11] Machkhoshvili, R. I., Vekua, N. N., Gogorishvili, P. V., et al. (Koord. Khim. **3** [1977] 332/41; Soviet J. Coord. Chem. **3** [1977] 244/52, 246).

25.3.3 With Succinodihydrazide $H_2NNHC(O)CH_2CH_2C(O)NHNH_2$ (= $C_4H_{10}N_4O_2$)

Complexes in Solution. Spectrophotometric studies on aqueous solutions of $Mn(NCS)_2$ and the ligand at λ = 280 nm (method of continuous variation) reveal the formation of 1:1 and 1:2 complexes, for which values of $pK_1 = 2.08$ and $pK_2 = 1.52$ have been calculated [1].

$Mn(C_4H_{10}N_4O_2)_nX_2 \cdot mH_2O$. Compounds with X = Cl, $SO_4/2$, NCS, n = 1 or 2, and different water contents (see the table below) have been prepared by reaction of manganese(II) salts with the appropriate amounts of the ligand in aqueous solution. The mixtures were concentrated or evaporated on a water bath, the precipitates were washed with ethanol and ether and dried in air. Some of the complexes were also precipitated by addition of alcohol to the aqueous solution, or obtained from methanolic solutions of the components. The compositions of the different complexes are shown below together with their melting or decomposition points (t_{dec}) the solubility S in water and the molar conductance of aqueous solutions at 25°C:

compound	t_{dec} in °C	S in g/100 mL	Λ in $cm^2 \cdot \Omega^{-1} \cdot mol^{-1}$	Ref.
$Mn(C_4H_{10}N_4O_2)_2Cl_2$	—	1.48	250.1	[3]
$Mn(C_4H_{10}N_4O_2)Cl_2$	176	1.48	250, 151*)	[1, 2]
	200	1.30	230.4	[4]
$Mn_2(C_4H_{10}N_4O_2)Cl_4 \cdot 9H_2O$	—	0.16	420.3	[3]
$Mn(C_4H_{10}N_4O_2)_2SO_4 \cdot 1(2)H_2O$	—	—	321.6	[3, 4]
$Mn(C_4H_{10}N_4O_2)SO_4 \cdot 1(3)H_2O$	—	0.75	254.9, 247.9*)	[2, 3]
$Mn(C_4H_{10}N_4O_2)(NCS)_2 \cdot 3H_2O$	280	2.75	217.5	[5]

*) At 30°C.

The IR spectra were recorded in the 4000 to 400 cm^{-1} region. Characteristic absorption bands (in cm^{-1}) of $Mn(C_4H_{10}N_4O_2)Cl_2 \cdot 3H_2O$ were assigned as follows: 7 bands between 3620 and 3050 to ν(OH, NH, NH_2); 1650 to amide I (predominantly ν(CO)); 1595 to δ(NH_2); 1550 to amide II (combination of ν(CN) + δ(NH_2)); 1325 to ω(NH_2); 1185, 1155 to δ(NH); 1065 to τ(NH_2); 500 to ν(Mn–N), and 430 to ν(Mn–O). Similar bands were observed in the spectra of

$Mn(C_4H_{10}N_4O_2)_2Cl_2 \cdot 4H_2O$ and $Mn(C_4H_{10}N_4O_2)_2SO_4 \cdot 2H_2O$. The shift of the bands ascribed to $\nu(NH_2)$ and $\nu(CO)$ toward lower frequencies in the complex by about 100 and 30 cm^{-1}, respectively, indicate that both the nitrogen atoms of the NH_2 groups and the oxygen atoms of the ligand are coordinated to Mn, forming five-membered chelate rings. The two absorption bands of the SO_4^{2-} ion are singlets and demonstrate its location in the outer complex sphere [3]. The IR data of $Mn(C_4H_{10}N_4O_2)(NCS)_2 \cdot 3H_2O$ (pale lilac crystals) show the isothiocyanate groups to be bonded through the nitrogen atom [4]. IR data of the chlorides and sulfates were reported also in [1, 2]. A dimeric or polymeric structure was suggested for the compounds, which are soluble in water and insoluble in inert solvents [1]. $Mn(C_4H_{10}N_4O_2)Cl_2 \cdot 3H_2O$ is reported to be insoluble in acetone, benzene, toluene, and dimethylformamide [3]. The molar conductivities in aqueous solution show the complexes to be 1:2 electrolytes [1 to 5].

References:

[1] Shvelashvili, A. E., Zhorzholiani, N. B., Svanidze, O. P., Zedelashvili, E. N. (Soobshch. Akad. Nauk Gruz.SSR **119** No. 1 [1985] 105/8; C.A. **104** [1986] No. 25039).

[2] Ahmed, A. D., Ray Chaudhuri, N. (J. Inorg. Nucl. Chem. **31** [1969] 2545/56, 2548, 2551).

[3] Tsitsishvili, L. D., Gogorishvili, P. V., Tsutsunava, T. I. (Issled. Obl. Khim. Kompleksn. Prostykh. Soedin. Nek. Perekhodnykh Metal. **2** [1974] 145/51, 148; C.A. **82** [1975] No. 38076).

[4] Kharitonov, Yu. Ya., Tsitsishvili, L. D., Tsutsunava, T. I., Chumakova, T. M., Stoppe, M. G. (Koord. Khim. **8** [1982] 666/73; Soviet J. Coord. Chem. **8** [1982] 357/63, 358, 362).

[5] Gogorishvili, P. V., Machkhoshvili, R. I., Tsitsishvili, L. D., Kharitonov, Yu. Ya., Shvelashvili, A. E., Tsutsunava, T. I. (Zh. Neorgan. Khim. **22** [1977] 3275/9; Russ. J. Inorg. Chem. **22** [1977] 1784/7).

25.3.4 With Dihydrazides of Other Aliphatic Dicarboxylic Acids

$H_2NNHC(O)(CH_2)_nC(O)NHNH_2$

Complexes in Solution. Spectrophotometric studies on aqueous solutions of $Mn(NCS)_2$, and ligand 1 (see the table below) at $\lambda = 280$ nm (method of continuous variation) and pH-potentiometric titrations of solutions, containing $Mn(NCS)_2$ and the ligands 2 to 4, with HCl at 55°C revealed the existence of 1:1 and 1:2 complexes, for which the stability constants pK_1 and pK_2 have been calculated [1]:

ligand No.	n	formula	acid	pK_1 of complex	pK_2 of complex
1	3	$C_5H_{12}N_4O_2$	glutaric acid	2.01	1.52
2	4	$C_6H_{14}N_4O_2$	adipic acid	2.02	1.52
3	7	$C_9H_{20}N_4O_2$	azelaic acid	2.12	1.58
4	8	$C_{10}H_{22}N_4O_2$	sebacic acid	2.16	1.59

$\mathbf{Mn(C_6H_{14}N_4O_2)_2Cl_2 \cdot 4H_2O}$. To an aqueous solution of $MnCl_2 \cdot 4H_2O$ the solid ligand 2 was added up to a 1:2 mole ratio. The mixture was heated to precipitate a white solid, which was washed with alcohol and ether and dried in air. The solubility of the complex in water is 2.66 g/100 mL H_2O to give a solution of pH 6.3. Conductivity measurements at 25°C and a dilution rate V = 1000 L/mol yield the molar electrical conductance $\Lambda = 338\ cm^2 \cdot \Omega^{-1} \cdot mol^{-1}$ of a 1:2 electrolyte, the enhanced value of Λ being due to hydration [2].

$Mn(C_6H_{14}N_4O_2)Cl_2 \cdot nH_2O$ (n = 1, 2). To a heated aqueous solution of $MnCl_2 \cdot 4H_2O$, ligand 2 was added up to a 1:1 mole ratio and the mixture was evaporated to a third of its original volume. After some time of standing a white solid separated [2]. The IR spectrum is similar to that of $Mn(C_4H_{10}N_4O_2)Cl_2 \cdot 3H_2O$, see p. 2. The octahedral environment of the metal atoms is apparently completed by the water molecules and not by the Cl^- ions, which are assumed to be in the outer sphere [3]. The complex is readily soluble in water up to 4.12 g/100 mL H_2O. The solution has a pH of 6.2. The molar conductance $\Lambda = 253\ cm^2 \cdot \Omega^{-1} \cdot mol^{-1}$ (measured as for $Mn(C_6H_{14}N_4O_2)_2Cl_2 \cdot 4H_2O$, see above), indicates a 1:2 electrolyte [2].

$Mn(C_6H_{14}N_4O_2)SO_4 \cdot 3H_2O$. An aqueous solution containing $MnSO_4 \cdot 5H_2O$ and ligand 2 was evaporated to half its original volume and than cooled. Alcohol was added to precipitate gradually a white microcrystalline solid which upon heating at 120°C (starting at 60°C) loses all water in an endothermic reaction to give the anhydrous complex as shown by thermal analysis. Upon further heating, the ligand is split off to leave $MnSO_4$ at 565°C (see diagram in the paper). The solubility of the complex in water is 1.3 g/100 mL H_2O, the solution has a pH of 6.1. The molar conductance $\Lambda = 253\ cm^2 \cdot \Omega^{-1} \cdot mol^{-1}$ (measured as for $Mn(C_6H_{14}N_4O_2)_2Cl_2 \cdot 4H_2O$) indicates a 1:1 electrolyte [4].

$Mn(C_{10}H_{22}N_4O_2)Cl_2$ and **$Mn(C_{10}H_{22}N_4O_2)_{1.5}SO_4 \cdot H_2O$.** The chloro complex was prepared by evaporating a filtered aqueous solution of $MnCl_2 \cdot 4H_2O$ and ligand 4 on a water bath until a sticky pink precipitate had formed which was dried at 60°C, then treated with methanol, filtered off, washed several times with ether and dried in air at room temperature. The sulfato complex was obtained by evaporating an aqueous solution of $MnSO_4 \cdot 4H_2O$ and ligand 4 (mole ratio 1:2). The resulting precipitate was treated with a mixture of water and ethanol, washed by decantation and dried in air. The white complexes are finely crystalline; the chloro complex melts at 195 to 200°C, the sulfato complex at 180°C [11].

The IR spectra show similar shifts of ν(NH) and ν(CO) as reported for the complexes with malono- or succinodihydrazide. The sulfato compound also exhibits three bands at 1065, 1050, and 1020 cm^{-1} assigned to $\nu(SO_4)$ and one band at 610 cm^{-1} due to $\delta(SO_4^{2-})$. Upon heating, $Mn(C_{10}H_{22}N_4O_2)Cl_2$ loses the ligand at about 340°C with subsequent decomposition of the residue. The complexes are poorly soluble in water, ethanol, and dimethylformamide and insoluble in acetone, benzene, and toluene [5]. The solubility in water at 25°C is 0.48 g/100 mL for the chloride and 0.24 g/100 mL for the sulfate. The chloride behaves as 1:2 electrolyte, the sulfate as 1:1 electrolyte in aqueous solution ($\Lambda = 203$ and $185\ cm^2 \cdot \Omega^{-1} \cdot mol^{-1}$, respectively) [11].

$MnL(NCS)_2 \cdot nH_2O$ (L = ligand 1 to 4). To prepare $Mn(C_5H_{12}N_4O_2)(NCS)_2$ hot aqueous solutions of $Mn(NCS)_2$ and ligand 1 were mixed in a 1:1 mole ratio and heated on a water bath for 15 min. The solution was filtered and a small quantity of ethanol was added to induce crystallization. After 24 h fine white crystals had formed which were collected, washed with ethanol and ether, and dried in air. $Mn(C_6H_{14}N_4O_2)(NCS)_2 \cdot H_2O$ was obtained in the same way from $Mn(NCS)_2$ and ligand 2 [6] or by reacting ethanolic solutions of $Mn(NCS)_2$ and ligand 2 in a 1:1 mole ratio. The fine crystalline solid which had separated after 10 d was washed with alcohol, and dried in air [4].

To prepare $Mn(C_9H_{20}N_4O_2)(NCS)_2 \cdot 2H_2O$ and $Mn(C_{10}H_{22}N_4O_2)(NCS)_2 \cdot 3H_2O$, an aqueous solution of $Mn(NCS)_2$ was treated with a hot solution of ligand 3 or 4, respectively, in a 1:1 mole ratio and a small volume of acetone was added to induce crystallization. After some time a sticky mass separated which gradually became crystalline, it was washed and dried as above [6]. The complexes form fine white crystals [4, 6]. The most characteristic absorption bands in the IR spectrum of the complexes were assigned similarly to $Mn(C_3H_8N_4O_2)_2(NCS)_2 \cdot H_2O$ (see p. 203). The temperature of decomposition (t_{dec}), the solubility in water, the pH, and the molar electrical conductance Λ of aqueous solutions at 25°C and a dilution rate V = 1000 L/mol, reported in [5], are shown below:

complex	t_{dec} in °C	S in g/100 mL	pH	Λ in $cm^2 \cdot \Omega^{-1} \cdot mol^{-1}$
$Mn(C_5H_{12}N_4O_2)(NCS)_2$	260	1.12	5.94	231.1
$Mn(C_6H_{14}N_4O_2)(NCS)_2 \cdot H_2O$	245	0.34	5.90	217.5
$Mn(C_9H_{20}N_4O_2)(NCS)_2 \cdot 2H_2O$	245	1.31	5.95	217.5
$Mn(C_{10}H_{22}N_4O_2)(NCS)_2 \cdot 3H_2O$	330	0.55	6.29	195.0

References:

[1] Shvelashvili, A. E., Zhorzholiani, N. B., Svanidze, O. P., Zedelashvili, E. N. (Soobshch. Akad. Nauk Gruz. SSR **119** No. 1 [1985] 105/8; C.A. **104** [1986] No. 25039).
[2] Gogorishvili, P. V., Tsitsishvili, L. D., Chrelashvili, M. V. (Issled. Obl. Khim. Kompleksn. Soedin. Nek. Perekhodnykh Redk. Metal No. 1 [1970] 5/13, 6/7, 10; C.A. **74** [1971] No. 82616).
[3] Machkhoshvili, R. I., Kharitonov, Yu. Ya., Gogorishvili, P. V., Tsitsishvili, L. D. (Koord. Khim. **3** [1977] 402/7; Soviet J. Coord. Chem. **3** [1977] 307/11, 310).
[4] Gogorishvili, P. V., Tsitsishvili, L. D., Chrelashvili, M. V. (Issled. Obl. Khim. Kompleksn. Prostykh. Soedin. Nek. Perekhodnykh Redk. Metal No. 1 [1970] 27/36, 28, 32, 35; C.A. **74** [1971] No. 60339).
[5] Tsutsunava, T. I., Kharitonov, Yu. Ya., Tsitsishvili, L. D., Stoppe, M. G., Chumakova, T. M. (Koord. Khim. **8** [1982] 786/93; Soviet J. Coord. Chem. **8** [1982] 420/6, 424).
[6] Kharitonov, Yu. Ya., Tsitsishvili, L. D., Tsutsunava, T. I., Chumakova, T. M., Stoppe, M. G. (Koord. Khim. **8** [1982] 666/73; Soviet J. Coord. Chem. **8** [1982] 357/63).
[7] Gogorishvili, P. V., Machkhoshvili, R. I., Tsitsishvili, L. D., Kharitonov, Yu. Ya., Shvelashvili, A. E., Tsutsunava, T. I. (Zh. Neorgan. Khim. **22** [1977] 3275/9; Russ. J. Inorg. Chem. **22** [1977] 1784/7).

25.3.5 With 2,6-Pyridinebis(carbohydrazide) (= 2,6-Dipicolinodihydrazide)

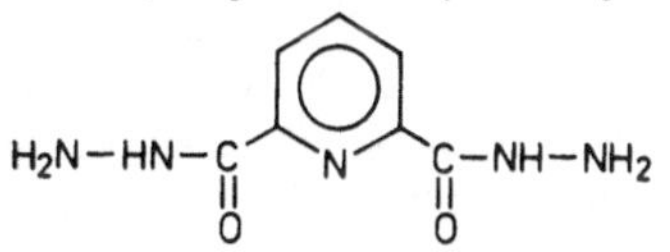

(= $C_7H_9N_5O_2$)

$Mn^{III}(C_7H_9N_5O_2)(CH_3COO)_3 \cdot 2H_2O$. Stoichiometric amounts of $Mn(CH_3COO)_3 \cdot 2H_2O$ and ligand in methanolic solution were mixed and refluxed on a water bath for 2.5 h and the solution was then concentrated to about half of the original volume. The yellow precipitate was washed thoroughly with methanol and dried at 85°C in an oven.

The IR spectrum (KBr disks and Nujol mulls) recorded in the 4000 to 200 cm^{-1} region shows a broad absorption band between 3200 and 3030 cm^{-1} assigned to ν(NH) vibrations and an amide I band mainly assignable to ν(CO) at 1650 to 1640 cm^{-1} which upon complexation had shifted to lower frequencies. The band ascribed to amide II (ν(CN) + δ(NH) vibrations) at 1540 to 1535 cm^{-1} shifted to higher frequencies and the amide III band split into two components occurring at 1388 to 1380 and 1210 to 1200 cm^{-1} (ligand at 1260 cm^{-1}). The observed band shifts and splits including those ascribed to the pyridine ring vibration modes indicate coordination of 2,6-dipicolinodihydrazide to Mn as a pentadentate chelating ligand through the terminal nitrogen atoms and carbonyl oxygens of the two symmetrical carbohydrazide side chains and through the pyridine ring nitrogen. This appears to be confirmed by far-IR bands assignable to $\nu(Mn–N_{amide})$ at 470 cm^{-1}, $\nu(Mn–O)$ at 375 cm^{-1}, and $\nu(Mn–N_{py})$ at 250 cm^{-1}. The electronic reflectance spectrum shows four bands at 12200, 15210 (=10 Dq), 20400 and 24000 cm^{-1} assigned to the electron transitions $^5B_1 \rightarrow {}^5A_1$, $^5B_1 \rightarrow {}^5B_2$, $^5B \rightarrow {}^5E$ and a charge transfer, respectively, which support the assumption of a five-coordinate square pyramidal complex geometry with a severe tetragonal distortion due to a Jahn-Teller effect in addition to an inequality of the donor atoms. Susceptibility measurements at room temperature (300 K) yield the magnetic moment $\mu_{eff} = 4.92\ \mu_B$ of a high-spin Mn^{III} (d^4) complex. The air-stable complex is soluble in ethanol and acetone to give stable solutions. The molar electrical conductance indicates a 1:3 electrolyte with ionic acetate as suggested also by the IR spectrum, Sahni, S. K., Gupta, S. P., Sangal, S. K., Rana, V. B. (J. Indian Chem. Soc. **54** [1977] 200/5, 204).

25.4 Complexes with Semicarbazides

ligand 1 $H_2NC(O)NHNH_2$ (= CH_5N_3O)

ligand 2 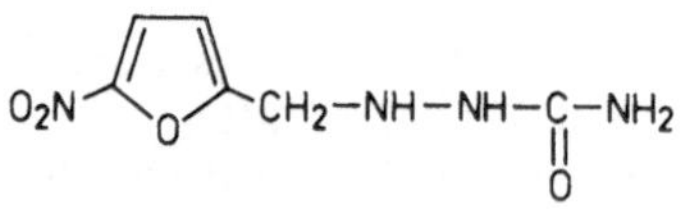 (= $C_6H_8N_4O_4$)

The stability constants of 1:1 and 1:2 complexes, with ligand 2, $\log K_1 = 5.90$ and $\log K_2 = 4.81$, in aqueous ethanol (50 vol%) were determined in N_2 atmosphere at 35°C. (Ionization constant of the ligand $pK_a = 9.28 \pm 0.02$.) The comparison with other divalent metals shows a decrease of both, the log K_1 and log K_2 values, in the order Cu > Zn > Mn > Mg > Ca [5].

$Mn(CH_5N_3O)_2(NO_3)_2$ precipitates on mixing solutions of $Mn(NO_3)_2 \cdot 6H_2O$ and semicarbazide (mole ratio 1:2) in boiling ethanol. The slightly pinkish precipitate was washed with hot ethanol. Susceptibility measurements (Faraday method) at room temperature yield the magnetic moment $\mu_{eff} = 5.8\ \mu_B$ indicating an octahedral high-spin Mn^{II} (d^5) complex [1].

$Mn(CH_6N_3O)_nCl_{2+n}$ (n = 1, 2) and **$Mn(CH_6N_3O)_2Cl_4 \cdot 2H_2O$**. The compounds were detected as stable solid phases in the solubility isotherms of the ternary system $MnCl_2$–$CH_5N_3O \cdot HCl$–water recorded at 20 to 60°C. $Mn(CH_6N_3O)_2Cl_4$ was detected at 60°C while the dihydrate $Mn(CH_6N_3O)_2Cl_4 \cdot 2H_2O$ appeared at 40 and 50°C and $Mn(CH_6N_3O)Cl_3$ was observed in the whole range from 20 to 60°C together with the species $MnCl_2 \cdot 4H_2O$ at 20, 40, and 50°C, $MnCl_2 \cdot 2H_2O$ at 60°C and solid semicarbazide hydrochloride. $Mn(CH_6N_3O)Cl_3$ does not form below 14°C [2].

$Mn(CH_6N_3O)_2Cl_4$ was obtained from $Mn(CH_6N_3O)_2Cl_4 \cdot 2H_2O$ which lost its water upon heating at 60°C [3]; it can be isolated also from solutions of the system $MnCl_2$–semicarbazide·HCl–water containing about 30 wt% $MnCl_2$ and 30 wt% $CH_5N_3O \cdot HCl$ at 60°C [2]. The IR spectrum resembles that of $CH_5N_3O \cdot HCl$ in the 3600 to 400 cm^{-1} region with almost no shift of the $\nu(CO)$ and $\nu(NH_2)$ absorption bands and may be formulated as a chloromanganate(II) complex, $(CH_6N_3O)_2[MnCl_4]$ [4]; also see [6]. Thermal analysis reveals three endothermic effects at 110, 135, and 170°C corresponding to the melting of $(CH_6N_3O)_2MnCl_4$, the decomposition into $MnCl_2$ and $CH_5N_3O \cdot HCl$ and the fusion of $CH_5N_3O \cdot HCl$, respectively. Further heating causes decomposition of that compound, leaving a residue of $MnCl_2$ at 550°C [3].

$Mn(CH_6N_3O)Cl_3$ can be isolated from solutions of the system $MnCl_2$–$CH_5N_3O\cdot HCl$–water containing ~40 wt% $MnCl_2$ and ~15 wt% $CH_5N_3O\cdot HCl$ at temperatures between 20 and 60°C [2]. It melts with release of HCl (decomposition) at 170°C [3]. The deuterated compound was prepared by heating $Mn(CH_6N_3O)Cl_3$ with D_2O in a sealed ampule and slow evaporation of the solvent in a desiccator [6]. The IR absorption spectra of both complexes are reported in [6]. The observed shift of the bands due to ν(CO) and ν(NH_2) to lower frequencies in the complex suggests bidentate coordination of semicarbazide hydrochloride to manganese through carbonyl oxygen and the nitrogen of the NH_2 group forming a five-membered chelate ring [4, 6]. A dimeric tetrahedral structure with bridging Cl atoms and a polymeric chlorine-bridged octahedral network structure of the complex were discussed in [6]. Thermal analysis reveals, in addition to the effect at 170°C due to melting, three overlapping endothermic effects at 300, 390, and 460°C accompanied by the liberation of N_2H_4, CO_2, and NH_3, leaving a residue of $MnCl_2$ at 560°C [3].

References:

[1] Nikolaev, A. V., Savel'eva, Z. A., Larionov, S. V., Leonova, T. G., Shklyarev, A. A. (Koord. Khim. **2** [1976] 1234/7; Soviet J. Coord. Chem. **2** [1976] 944/6).

[2] Murzubraimov, B., Shtrempler, G. I., Rysmendeev, K. (Zh. Neorgan. Khim. **23** [1978] 779/83; Russ. J. Inorg. Chem. **23** [1978] 430/2).

[3] Shtrempler, G. I., Murzubraimov, B. (Zh. Neorgan. Khim **26** [1981] 3297/9; Russ. J. Inorg. Chem. **26** [1981] 1768/70).

[4] Murzubraimov, B., Shtrempler, G. I. (Zh. Neorgan. Khim. **27** [1982] 1473/5; Russ. J. Inorg. Chem. **27** [1982] 829/30).

[5] Agrawal, Y. K., Patel, D. R. (J. Pharm. Sci. **75** [1986] 190/2).

[6] Murzubraimov, B., Toktomatov, A. (Koord. Khim. **10** [1984] 1531/5; C.A. **102** [1985] No. 35645).

25.5 With Carbonohydrazide $CO(NHNH_2)_2$ (= CH_6N_4O)

$Mn(CH_6N_4O)_3(NO_3)_2$. The complex precipitates upon adding ethanol to a hot aqueous solution of $Mn(NO_3)_2\cdot 6H_2O$ and carbonohydrazide (mole ratio 1:3). It was washed with ethanol and dried in air. The IR spectrum (mulls in liquid paraffin or fluorinated oil) shows four absorption bands at 3380, 3340, 3280, and 3200 cm^{-1} ascribed to ν(NH) vibrations. Further bands at 1657, 1600, 1545, and 445 cm^{-1} are assigned to amide I, δ(NH_2), and amide II vibrations of the ligand and to ν(Mn–N), respectively. The only slight shift of the ν(CO) band indicates nonbonding of the carbonyl oxygen. The more distinct (upward) shift of the ν(NH) band suggests the existence of ligand-to-anion (NO_3^-) hydrogen bonds. The two bands due to the nitrate ion at ~1360 (ν_1) and 830 (ν_3) cm^{-1} show the presence of uncoordinated NO_3^- ions and the absence of coordinated nitrato groups [1].

$Mn(CH_6N_4O)_3Cl_2$. A solution of $MnCl_2$ in dehydrated ethanol was added to that of the ligand (mole ratio 1:3). The mixture was heated for 15 min and cooled to room temperature. The resulting colorless crystals were washed with anhydrous methanol, and dried in air. Bands observed in the IR spectrum (Nujol mulls; in cm^{-1}, ligand bands in parentheses) were assigned as follows: 3340 (3360) to ν_{as}(NH); 3180 (3200) to ν_s(NH) of $-NH_2$; 3300 (3300) to ν(NH); 1635 (1625) to ν(CO), and 945 (915) to ν(CN). The observed shift of the ν(NH_2) bands to lower and of the ν(C–N) band to higher frequencies in the complex suggests coordination of the bidentate ligand only through the nitrogens of the NH_2 groups to manganese, thus forming six-membered chelate rings. Susceptibility measurements yield the magnetic moment $\mu_{eff} = 6.0\ \mu_B$

of a high-spin octahedrally configurated Mn^{II} complex and the molar electrical conductivity $\Lambda = 290\ cm^2 \cdot \Omega^{-1} \cdot mol^{-1}$ measured at 25°C in aqueous solution indicates a 1:2 electrolyte. $Mn(CH_6N_4O)_3Cl_2$ is highly soluble in water but insoluble in organic solvents like methanol, nitromethane, or acetone [2].

$Mn(CH_6N_4O)_2Cl_2$ crystallizes upon cooling a hot aqueous solution of $MnCl_2 \cdot 4H_2O$ and carbonohydrazide, mole ratio 1:2. The crystals were washed with ethanol and dried in air. The IR spectrum (recorded as for $Mn(CH_6N_4O)_3(NO_3)_2$, see above) shows four absorption bands at 3360, 3317, 3270, and 3195 cm^{-1} ascribed to ν(NH) vibrations of the NH_2 and NH groups of the ligand. The amide I band is observed at 1650 cm^{-1} in the IR and at 1640 cm^{-1} in the Raman spectrum. Its shift toward higher frequencies indicates the absence of Mn–O bonds and coordination of the hydrazide to manganese through the nitrogen of an NH_2 group. The band observed at 250 cm^{-1} (in the Raman spectrum) can be assigned to ν(Mn–Cl) vibrations; its reduced frequency does not rule out the possibility of chloride bridges [3].

$Mn(CH_6N_4O)_2(NCS)_2$ precipitates from an ethanolic solution of $Mn(NCS)_2$ and the ligand in stoichiometric amounts. The white complex was washed with ethanol and dried in air; it melts at 181 to 182°C. X-ray diffractometer (CuKα radiation) studies (for spacings d see the paper) showed that the structure of $Mn(CH_6N_4O)_2(NCS)_2$ differs from $Cd(CH_6N_4O)_2(NCS)_2$. The IR spectrum (KBr disks, Vaseline and fluoroparaffin mulls) exhibits several characteristic absorption bands in cm^{-1} assigned as follows: 2065, 2053 to ν(C≡N), 1672, 1650 to amide I, 1610 to $\delta(NH_2)$, 1530 to amide II, 1230, 1215 to $\tau(NH_2)$, 1150 to amide III, 780(sh), 760 to ν(C=S), and 483 to δ(NCS). The splitting of the amide I band and the small shift to higher frequencies compared to the free ligand (1648 cm^{-1}) indicates no coordination to the carbonyl oxygen but coordination of the ligand to Mn through the terminal nitrogens of the two hydrazide moieties. The splitting of the ν(CN) and ν(CS) bands suggests N-coordination of the thiocyanate groups in *cis*-positions [4]. $Mn(CH_6N_4O)_2(NCS)_2$ is readily soluble in water but insoluble in alcohol and other common organic solvents [4].

References:

[1] Ivanov, M. G., Kalinichenko, I. I. (Zh. Neorgan. Khim. **26** [1981] 2134/7; Russ. J. Inorg. Chem. **26** [1981] 1150/1).
[2] Dutta, R. L., Sarkar, A. K. (J. Inorg. Nucl. Chem. **43** [1981] 2557/9).
[3] Ivanov, M. G., Kalinichenko, I. I. (Zh. Neorgan. Khim. **26** [1981] 2159/62; Russ. J. Inorg. Chem. **26** [1981] 1162/4).
[4] Ivanov, M. G., Kalinichenko, I. I., Savitskii, A. M. (Koord. Khim. **11** [1985] 45/8; C.A. **102** [1985] No. 196726).

26 Complexes with Derivatives of Hydroxylamine

26.1 Complexes with Hydroxamic Acids

Survey

The hydroxamic acids RC(O)N(OH)R′, also known as N-acylhydroxylamines, are bidentate chelating ligands containing the bifunctional moiety –C(O)N(OH)– with various substituents R and R′. In neutral and weakly alkaline media, complexation occurs by replacement of the acid proton and by coordination of the carbonyl oxygen with formation of a five-membered chelate ring. Mn^{II}: ligand mole ratios of 1:1 and 1:2 are reported for these complexes. In alkaline media, complexes are formed with the enolic tautomer RC(OH)=NOH (R′ = H). The reaction is accompanied by oxidation of Mn^{II} to Mn^{III} in the presence of oxygen or H_2O_2. To prevent the oxidation reaction, the Mn^{II} complexes were prepared under nitrogen. Intensely red-colored Mn^{III} solution complexes were obtained with aromatic and heterocyclic hydroxamic acids. For these complexes, a 1:3 Mn:ligand ratio was also found. A complex $Mn(HL)X_2$ with a neutral ligand was obtained at a very low pH value.

For closely related ligands, a linear relation between logarithmic complex stability constants and ligand pK_a values in the same solvent was observed (see for example p. 215). The ratio of successive stability constants, log K_1/K_2, is generally positive. However, a logarithmic ratio <1 indicates simultaneous formation of 1:1 and 1:2 complexes with low hindrance for the second ligand in most cases. The insolubility in water of hydroxamates with aryl groups for R and R′ requires mixed aqueous organic solvents for the determination of the stability constants; aqueous dioxane is generally used. The pH corrections for solvents and the conversion to zero ionic strength to obtain thermodynamic stability constants (0 corr) were generally made by use of an empirical formula published in [1]. The stability constants increase linearly with growing mole fractions of the dioxane in the mixed solvent.

The magnetic moment ~5.9 μ_B prevailing for the solid Mn^{II} complexes is typical for high-spin d^5 complexes with octahedral coordination. In the IR spectra, the ν(OH) bands (3300 to 3150 cm^{-1}) and the δ(OH) bands of the free ligands disappear. The ν(NO) bands appear at ~920 to 940 cm^{-1} and the ν(CO) bands at ~1600 cm^{-1}, the latter shifted to lower wave numbers compared to the free ligands.

Manganese hydroxamates have mostly been studied together with those of the other 3d transition metals. Stabilities were found to follow the Irving-Williams sequence $Mn^{II} < Co^{II} < Ni^{II} > Zn^{II}$. This sequence is consistent with the order of ionic radii and ionization potentials.

Several studies have been made on the analytical application of Mn hydroxamates. Use of the colored Mn^{III} complexes with aromatic or heterocyclic hydroxamic acids was proposed for the spectrophotometric determination of manganese; a procedure is given in [2]. Mn^{II} complexes with N-aryl substituted arylhydroxamic acids were studied in view to gravimetric applications because of their poor solubility in aqueous solution. All the complexes can be extracted from aqueous solution or suspension into solvents such as chloroform. Analytical separation processes were based on this feature.

A review for complexes with hydroxamic acids is given in [3].

References:

[1] van Uitert, C. G., Haas, C. G. (J. Am. Chem. Soc. **75** [1953] 451/5).
[2] Umland, F. (Methoden der Analyse in der Chemie: Theorie und praktische Anwendung von Komplexbildnern, Vol. 9, Frankfurt a. M. 1971, p. 460).
[3] Agrawal, Y. K. (Usp. Khim. **48** [1979] 1773/803; Russ. Chem. Rev. **48** [1979] 948/63).

26.1.1 With Aliphatic and Alicyclic Hydroxamic Acids RC(O)NHOH

26.1.1.1 Formation in Solution

Binary Manganese(II) Complexes. Stability constants of the complexes MnL^+ and MnL_2 with the listed ligands HL in aqueous or aqueous-organic solution were determined potentiometrically (glass electrode):

ligand No.	R	formula	t in °C	I in mol/L	log K_1	log K_2	Ref.
1	CH_3	$C_2H_5NO_2$	20	0.1 ($NaNO_3$)	4.0	2.9	[1]
			25	0.1 ($NaClO_4$)	3.80	3.05	[2]
2	$C_6H_5CH_2$	$C_8H_9NO_2$	25	0.1 ($NaClO_4$)	3.86	3.60	[3]
			30	0.1 ($NaClO_4$)	3.45	—	[4]
			30	0.05 ($NaClO_4$)	3.59	—	[4]
			30	→0 ($NaClO_4$)	4.01	—	[4]
			40	0.1 ($NaClO_4$)	3.38	—	[4]
			50	0.1 ($NaClO_4$)	3.31	—	[4]
3*)	$(C_6H_5)_2CH$	$C_{14}H_{13}NO_2$	20	0.1 (KNO_3)	4.65	3.87	[5]
			30	0.1 (KNO_3)	4.60	3.80	[5]
			30	0.05 (KNO_3)	4.62	3.97	[5]
			30	→0 (KNO_3)	4.73	4.40	[5]
			40	0.1 (KNO_3)	4.54	3.75	[5]
			50	0.1 (KNO_3)	4.46	3.66	[5]
4	C_3H_7	$C_4H_9NO_2$	25	0.1 ($NaClO_4$)	4.23	3.35	[2]
5	(*E*, *E*)-$CH_3CH{=}CHCH{=}CH$	$C_6H_9NO_2$	25	0.1 ($NaClO_4$)	4.50	3.51	[2]

*) In 50% (v/v) ethanol-water.

Stability constants determined at other ionic strengths or temperatures for complexes with ligands 2 and 3 are presented in [4, 5].

Thermodynamic parameters, ΔG and ΔH in kcal/mol, ΔS in $cal \cdot mol^{-1} \cdot K^{-1}$, were calculated from the stability constants at 30°C for ionic strength I = 0.1 M and the temperature dependences of the log K values:

ligand No.	ΔG_1	ΔH_1	ΔS_1	ΔG_2	ΔH_2	ΔS_2	Ref.
2	−4.78	−3.05	5.71	−3.97	−3.67	0.98	[4,6]
3	−6.37	−2.74	11.98	−5.21	−2.54	8.97	[5]

Separation of ΔG and ΔH values for the complexes with ligand 2 into electrostatic and nonelectrostatic components (values given) shows that nonelectrostatic bonding forces are stronger for both complexes and that these forces prevail to a larger extent for the 1:2 complexes [6].

Ternary Manganese(II) Complexes. Equilibrium constants K_a, for the reaction $MnA^+ + L^- \rightleftharpoons MnAL$, and K_{ab}, for the reaction $Mn^{2+} + A^- + L^- \rightleftharpoons MnAL$, with HA = an amino acid of the following table and HL = phenylacetohydroxamic acid (= $C_8H_9NO_2$ = ligand 2), in aqueous or aqueous-organic media were determined with a glass electrode at 30°C:

ligand HA	formula	I in mol/L	log K_a	log K_{ab}	Ref.
glycine	$C_2H_5NO_2$	0.1 ($NaClO_4$)	—	7.00	[7]
α-alanine	$C_3H_7NO_2$	0.1 ($NaClO_4$)	—	6.90	[7]
β-alanine	$C_3H_7NO_2$	0.1 ($NaClO_4$)	—	5.85	[7]
phenylalanine	$C_9H_{11}NO_2$	0.1 ($NaClO_4$)	—	6.70	[7]
histidine[a)]	$C_6H_9N_3O_2$	0.1 (KNO_3)	3.30	—	[8]

[a)] At pH 6 to 8.

Stabilization parameters (equilibrium constants) which were calculated from the stability constants show that the ternary complexes are energetically favored over their binary complexes MnA_2 except for that with β-alanine [7, 10]. (Stability constants for binary Mn^{II} complexes with amino acids HA are reported in "Manganese" D 4, 1985, pp. 248/76.)

Equilibrium constants for the reaction $Mn\,bpy^{2+} + L^- \rightleftharpoons Mn\,bpy\,L^+$ with HL = phenylacetohydroxamic acid (= $C_8H_9NO_2$ = ligand 3) were determined in aqueous solution of pH 6 to 8 or with HL = diphenylacetohydroxamic acid (= $C_{14}H_{13}NO_2$ = ligand 4) in 50% (v/v) ethanol-water.

ligand HL	t in °C	I in mol/L	log K	Ref.
$C_8H_9NO_2$	30	0.1 ($NaClO_4$)	3.25 ± 0.03	[4]
	30	→0 ($NaClO_4$)	3.75 ± 0.03	[4]
	50	0.1 ($NaClO_4$)	3.12 ± 0.03	[4]
$C_{14}H_{13}NO_2$	30	0.1 ($NaNO_3$)	4.40	[9]
	30	→0 ($NaNO_3$)	5.14	[9]
	50	0.1 ($NaNO_3$)	4.25	[9]

Stability constants for other ionic strengths and temperatures are presented in [4, 9]. (Equilibrium constants for binary Mn^{II} complexes with 2,2'-bipyridine are tabulated in "Manganese" D 3, 1982, p. 203.) Thermodynamic parameters for the addition of the ligand anion L^- at 30°C and I = 0.1 mol/L were calculated from the log K values; ΔG and ΔH in kcal/mol; ΔS in cal·mol^{-1}·K^{-1} are: −4.50, −2.88, 5.35 for $Mn\,bpy(C_8H_8NO_2)^+$ [4]; −6.09, −2.96, 10.33 for $Mn\,bpy(C_{14}H_{12}NO_2)^+$, respectively [9]. Separation of the ΔG and ΔH values into their electrostatic and nonelectrostatic (cratic) components for the complex with phenylacetohydroxamic acid ($\Delta G_e = -2.92$, $\Delta G_c = -4.02$, $\Delta H_e = 1.12$, $\Delta H_c = -4.00$ kcal/mol) indicates a somewhat more ionic character of the manganese hydroxamate bonds, compared with the binary 1:1 complex [10].

References:

[1] Anderegg, G., L'Eplattenier, F., Schwarzenbach, G. (Helv. Chim. Acta **46** [1963] 1400/8, 1407, 1409/22).

[2] Liu, C.-Y., Chang, H.-J., Uang, S.-S., Sun, P.-J. (J. Chinese Chem. Soc. [Taipei] **22** [1975] 225/35; C.A. **84** [1976] No. 53322).

[3] Reddy, D., Sethuram, B., Navaneeth Rao, T. (Indian J. Chem. A **14** [1976] 67/8).

[4] Reddy, D., Sethuram, B., Navaneeth Rao, T. (Indian J. Chem. A **15** [1977] 333/6).

[5] Subba Reddy, V. V., Sethuram, B., Navaneeth Rao, T. (Indian J. Chem. A **17** [1979] 199/201).
[6] Reddy, D., Sethuram, B., Navaneeth Rao, T. (Indian J. Chem. A **19** [1980] 246/8).
[7] Reddy, D., Sethuram, B., Navaneeth Rao, T. (Indian J. Chem. A **20** [1981] 150/3).
[8] Janardhan Rao, M., Sethuram, B., Navaneeth Rao, T. (Indian J. Chem. A **19** [1980] 379/80).
[9] Subba Reddy, V. V., Sethuram, B., Navaneeth Rao, T. (Indian J. Chem. A **20** [1981] 1138/40).
[10] Reddy, D., Sethuram, B., Navaneeth Rao, T. (Indian J. Chem. A **20** [1981] 533/5).

26.1.1.2 Isolated Manganese(II) Compounds

No.	R	formula	name of RC(O)NHOH
1	C_7H_{15}	$C_8H_{17}NO_2$	octanohydroxamic acid
2	$(C_6H_5)_2CH$	$C_{14}H_{13}NO_2$	diphenylacetohydroxamic acid
3	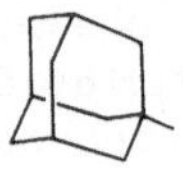	$C_{11}H_{17}NO_2$	adamantohydroxamic acid

Manganese(II) octanohydroxamate is reported to precipitate at pH 8.7 from a solution containing an Mn^{II} salt and ligand 1 (mole ratio 1:1) on addition of KOH [1]. The IR spectrum of the complex in KBr shows absorption bands at 1600, 1520, 1460, 1360, 1290, 1110, and 1050 cm^{-1} [1, 2]. The formation of a basic manganese(II) complex with the ligand on the surface of manganese dioxide [1] or manganese silicate ore [2] was indicated by the IR spectra during studies of the adsorption and flotation behavior of these ores in the presence of octanohydroxamate.

$Mn(C_{14}H_{12}NO_2)_2$ was obtained by refluxing methanolic solutions of manganese(II) nitrate and ligand 2 in a mole ratio of 1:20 for 2 to 3 h. Methanolic ammonia was then added to precipitate the ash-colored complex at pH 5 to 8. The compound was washed with 20% aqueous methanol and dried in vacuum; the decomposition temperature is 117°C [3].

The effective magnetic moment is 5.91 μ_B at 30°C and indicates a high-spin Mn^{II} complex. The IR spectrum shows two sharp peaks near 3300 and 1600 cm^{-1} assigned to ν(NH) and ν(CO) vibrations, the latter shifted by ~60 cm^{-1} to lower wave numbers compared to the free ligand. A square planar complex geometry is assumed [3].

$[Mn(C_{11}H_{17}NO_2)_2(H_2O)_2]Cl_2 \cdot H_2O$ was prepared by reacting 0.01 mol of $MnCl_2 \cdot 4H_2O$ in 30 mL hot ethanol with 0.02 mol of ligand 3 in 10 mL hot ethanol at 50 to 60°C for 15 min. The solution was concentrated until crystals appeared, then cooled to 0°C. After separation, the white crystals were washed with alcohol and ether and dried in vacuum. The yield was 74.2 % [4].

IR absorption bands are indicative of coordinated water and coordinated neutral ligand. The bands were assigned as follows (wave numbers in cm^{-1}): 3383, $\nu(OH_{H_2O})$; 1700 $\delta(OH_{H_2O})$; 3212, ν(NH); 1604, 1542, 1292, amide group vibrations; 942, ν(NO). (Respective free ligand wave numbers: 3320, ν(NH); 1645, 1610, 1320, amide group vibrations; 982 ν(NO) [4].)

References:

[1] Natarajan, R., Fuerstenau, D. W. (Intern. J. Mineral. Process. **11** [1983] 139/53; C.A. **99** [1983] No. 161964).

[2] Palmer, B. R., Gutierrez B., G., Fuerstenau, M. C. (Trans. Soc. Mining Eng. AIME **258** [1975] 257/63).
[3] Subba Reddy, V. V., Sethuram, B., Navaneeth Rao, T. (Indian J. Chem. A **22** [1983] 622/3).
[4] Demeter, E. S., Buzash, V. M., Butsko, S. S., et al. (Fiziol. Akt. Veshchestva **13** [1981] 20/4; C.A. **96** [1982] No. 134765).

26.1.2 With Aromatic Hydroxamic Acids RC(O)NHOH or Their Tautomers

26.1.2.1 Manganese(II) Complexes

Formation and Properties in Solution. The following table lists stability constants of 1:1 and 1:2 complexes with the ligands HL in 50% (v/v) aqueous dioxane determined potentiometrically (glass electrode) together with the pK_a values of the ligands.

No.	R		formula of HL	t in °C	I in mol/L	pK_a	log K_1	log K_2	Ref.
1	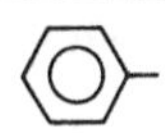		$C_7H_7NO_2$	25	0.1 (KCl)	10.35	4.90[a]	3.96[a]	[2]
				35	0 corr	10.57	5.97	4.52	[1]
2	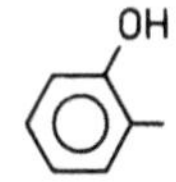		$C_7H_7NO_3$	30	0.1 ($NaClO_4$)	8.83	3.98	—	[3]
3	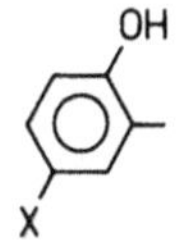	X = NO_2	$C_7H_6N_2O_5$	30	0.1 ($NaClO_4$)	5.56	2.83	—	[3]
4		X = Cl	$C_7H_6ClNO_3$	30	0.1 ($NaClO_4$)	8.13	3.83	—	[3]
5		X = Br	$C_7H_6BrNO_3$	30	0.1 ($NaClO_4$)	7.74	3.37	—	[3]
6		X = CH_3	$C_8H_9NO_3$	30	0.1 ($NaClO_4$)	9.09	4.54	—	[3]
7	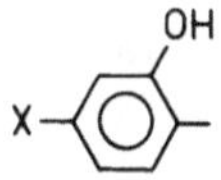	X = Cl	$C_7H_6ClNO_3$	30	0.1 ($NaClO_4$)	7.77	3.32	—	[3]
8		X = Br	$C_7H_6BrNO_3$	30	0.1 ($NaClO_4$)	7.73	3.28	—	[3]
9	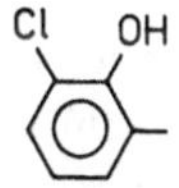		$C_7H_6ClNO_3$	30	0.1 ($NaClO_4$)	7.38	2.98	—	[3]
10	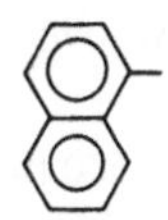		$C_{11}H_9NO_2$	25	0 corr	7.77	3.50[b]	2.30[b]	[4]
				35	0 corr	7.50	3.30[b]	2.09[b]	[4]

[a] In 70% (v/v) aqueous dioxane. – [b] In water.

Complexation by displacement of the amide hydrogen atom by the Mn^{2+} ion is assumed for the 1:1 complexes only formed with ligands 2 to 9. A linear relationship between the log K_1 values of the complexes and the acidity constants pK_a of these ligands is discussed [3].

Thermodynamic parameters of formation (ΔG and ΔH in kcal/mol, ΔS in $cal \cdot mol^{-1} \cdot K^{-1}$) were calculated for the 1-naphthohydroxamates (complexes with ligand 10): $\Delta G_1 = -4.78$ (25°C), −4.50 (35°C), $\Delta H_1 = -8.40$, $\Delta S_1 = -12.14$ (25°C), −12.65 (35°C), and $\Delta G_2 = -3.14$ (25°C), −2.85 (35°C), $\Delta H_2 = -8.80$, $\Delta S_2 = -18.99$ (25°C), −19.30 (35°C) [4].

For the complex with benzohydroxamic acid, the magnetic moment 5.79 μ_B was measured in the absence of air. Initial increase of this value to 6.24 μ_B in the presence of air suggests the formation of a transitional Mn^{II} complex with the O_2 molecule [6]. The Mn^{II} complexes are oxidized in basic solution by oxidizing agents such as O_2 or H_2O_2 to give the Mn^{III} complexes [5, 6], (see below).

$Mn(C_7H_6NO_2)_2 \cdot 0.5H_2O$. The solid complex with benzohydroxamic acid (= $C_7H_7NO_2$) was prepared by reacting an aqueous solution of an Mn^{II} salt with an aqueous solution of 25 to 50% excess ligand in 0.5 mM $HClO_4$. A 1% solution of $H_2NOH \cdot HCl$ (1%) was added to prevent oxidation of Mn^{II} to Mn^{III}. The resulting solution was treated with 1 M NaOH by dropwise addition until precipitation occurred. TGA and DTA investigations of the compound in N_2 or O_2 revealed two exothermic peaks starting at ~150 and ~300°C. The first peak corresponds to the melting point (with decomposition), the second one to the complete decomposition with formation of MnO_2. The heat of decomposition, ~15 kcal/mol, was estimated from the DTA curve in N_2 [7].

References:

[1] Agrawal, Y. K., Tandon, S. G. (J. Inorg. Nucl. Chem. **34** [1972] 1291/5).
[2] Jaimni, J. P. C., Sogani, N. C. (Bull. Acad. Polon. Sci. Ser. Sci. Chim. **17** [1969] 157/62; C. A. **71** [1969] No. 42897).
[3] Deshpande Ratnakar, G., Jahagirdar, D. V. (J. Inorg. Nucl. Chem. **39** [1977] 1385/9).
[4] Agrawal, Y. K., Mudaliar, A. (J. Indian Chem. Soc. **54** [1977] 757/8).
[5] Miller, D. O., Yoe, J. H. (Talanta **7** [1960] 107/16).
[6] Ksandr, Z. (Collection Czech. Chem. Commun. **27** [1962] 31/40).
[7] Lapatnick, L. N., Hazel, J. F., McNabb, W. M. (Anal. Chim. Acta **36** [1966] 366/71).

26.1.2.2 Manganese(III) Complexes in Solution

1)	$C_6H_5C(O)NHOH$	benzohydroxamic acid	(= $C_7H_7NO_2$)
2)	$2\text{-}HOC_6H_4C(O)NHOH$	salicylohydroxamic acid	(= $C_7H_7NO_3$)
3)	$2\text{-}NH_2C_6H_4C(O)NHOH$	anthranilohydroxamic acid	(= $C_7H_8N_2O_2$)
4)	$2\text{-}ClC_6H_4C(O)NHOH$	2-chlorobenzohydroxamic acid	(= $C_7H_6ClNO_2$)

Deeply red-colored solutions containing a manganese(III) complex with an Mn:ligand ratio of 1:3 were obtained by reaction of Mn^{2+} ions with ligand 1 in a large excess (Mn:ligand ratio = 1:20 to 1:25) in aqueous solution or organic solvents (DMF, methanol, ethanol) at pH 8 to 11 in the presence of O_2 or H_2O_2 [1, 2, 4, 6]. Red solutions were also obtained with ligands 2 [3, 5, 7] and 3 [3] under similar conditions.

Using atmospheric oxygen as an oxidizing agent, maximum absorbance at 500 nm was attained almost instantaneously for aqueous solutions containing ppm amounts of Mn, excess benzohydroxamic acid (>25-fold) and >0.5 mol/L ammonia. In DMF containing equivalent quantities of ammonia, complex formation is faster than in aqueous solution [1]. Solutions of pH 9.5 to 9.8 containing 2.5×10^{-3} mol/L Mn and 5×10^{-2} mol/L ligand 1 were oxidized with air and showed maximum absorbance at 490 nm only after 32 h. Magnetic measurements yielded the magnetic moment 4.97 μ_B after 3.5 h and 4.02 μ_B after 24 h. Due to the latter value and the number of the released protons, determined by alkali titration, the formation of an Mn^{IV} complex (?) (spin-only magnetic moment 3.88 μ_B) was suggested. An Mn:ligand ratio of 1:3 was evaluated by the method of continuous variation [2].

Complex formation with the doubly deprotonated enol form of benzohydroxamic acid $C_6H_5C(OH)$=NOH to give the $Mn(C_7H_5NO_2)_3^{3-}$ complex [1, 11] is indicated by the following facts: the compound only forms in basic solution; analogous complexes do not form from N-substituted hydroxamic acids; and the compound is strongly retained on an anion exchange resin. An octahedral structure containing three five-membered chelate rings was proposed [1].

All the complexes show a broad absorption band at 490 to 500 nm [1, 2, 3, 7, 8] and an additional absorption in the 320 to 360 nm region [1, 2]. For the complex with ligand 1, the absorbance at 500 nm increased with growing ligand concentration up to a mole ratio of 1:12 (Mn:ligand). For solutions with an Mn:ligand ratio of 1:25 and an ammonia content exceeding 0.35 M, Beer's law was obeyed over an Mn concentration range from 0.4 to 10 ppm in aqueous solution and in DMF. Such solutions remained unchanged for more than 24 h [1].

The complexes with ligand 1 [9 to 11], ligand 2 [7, 10], and ligand 4 [10] can be extracted from basic aqueous solution into toluene [7, 9, 10] or methyl isobutyl ketone solutions [11] of trioctylmethylammonium chloride (Adogen 464). Spectrophotometric measurements revealed the formation of ion associates Mn:ligand:Adogen 464 = 1:3:2 for ligand 1 [9] and 1:2:2 for ligand 2 [7]. The extraction of the complexes with ligand 1 and 2 was almost quantitative at pH $\geqq$8 [7, 9] with a molar excess of 60 for the ligand and 16 for Adogen 464 [9]. Procedures for the spectrophotometric determination of manganese by use of complex formation with ligand 3 [8] and by ion associate formation of the complexes of ligands 1 and 2 with Adogen 464 [7, 9, 11] are described.

References:

[1] Miller, D. O., Yoe, J. H. (Talanta **7** [1960] 107/16).
[2] Ksandr, Z. (Collection Czech. Chem. Commun. **27** [1962] 31/40).
[3] Dutta, R. L. (J. Indian Chem. Soc. **37** [1960] 167/70).
[4] Das Gupta, A. K., Singh, M. M. (J. Sci. Ind. Res. [India] B **11** [1952] 268/73).
[5] Sankar Bhaduri, A. (Z. Anal. Chem. **151** [1956] 109/18).
[6] Miller, D. O., Yoe, J. H. (Anal. Chim. Acta **26** [1962] 224/9).
[7] Salinas, F., March, J. G. (Quim. Anal. [Barcelona] **3** [1984] 11/8).
[8] Salinas, F., Llinas, M., Estela, J. M. (Quim. Anal. [Barcelona] **2** [1983] 38/44 from C.A. **102** [1985] No. 124740).
[9] Salinas, F., March, J. G. (Ann. Chim. [Rome] **75** [1985] 271/7 from C.A. **103** [1985] No. 152900).
[10] Salinas, F., Cantallops, J., Estela, J. M. (Quim. Anal. [Barcelona] **2** [1983] 96/111 from C.A. **102** [1985] No. 142414).

[11] March, J. G. (Microchem. J. **32** [1985] 338/41).

26.1.3 With Heterocyclic Hydroxamic Acids RC(O)NHOH or Their Tautomers

ligand No.	1	2	3	4
R	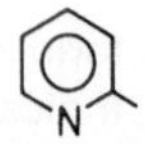	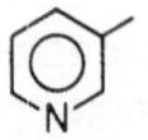		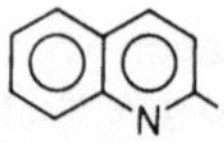
ligand formula	$C_6H_6N_2O_2$	$C_6H_6N_2O_2$	$C_6H_6N_2O_2$	$C_{10}H_8N_2O_2$

Red-violet aqueous solutions were obtained on reaction of Mn^{2+} ions with 2-, 3-, or 4-pyridinecarbohydroxamic acids (ligands 1 to 3) in at least 1:15 mole ratio [1, 2]. Air oxidation to Mn^{III} was inferred from the accelerating effect of traces of H_2O_2 and by the position of λ_{max} observed at 480 to 490 and at 470 to 480 nm in reaction solutions with ligands 2 and 3, respectively [1]. These ligands form stable intensely colored complexes [1, 2]; see also [3]. The Mn:ligand ratio is 1:3, as shown by Job's method of continuous variation [1]. With ligand 1, an unstable complex was observed only on prolonged agitation of the reaction solution in air. The ease of formation and solubility of the complexes in aqueous solution increase on going from ligand 1 to 3 whereas the extractability into organic solvents decreases [2]. A pale yellow solid, presumably an Mn^{II} complex, is obtained with ligand 4 in aqueous ammonia (pH ≧ 9) in the presence of air [2].

References:

[1] Dutta, R. L. (J. Indian Chem. Soc. **34** [1957] 311/6).
[2] Dutta, R. L. (J. Indian Chem. Soc. **36** [1959] 339/45).
[3] Bass, V. C., Yoe, J. H. (Talanta **13** [1966] 735/44, 738).

26.1.4 With Hydroxamic Acids of the Type HO(O)CRC(O)NHOH or HOHN(O)CRC(O)NHOH

No.	ligand	formula	oxidation state of Mn	Mn:ligand ratio	λ_{max} in nm (log ε)	Ref.
1	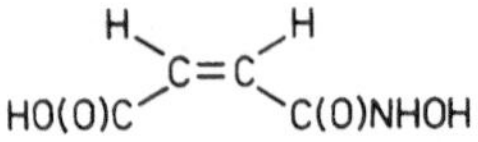	$C_4H_5NO_4$	Mn^{II}	1:1	—	[1]
2	HO(O)C−CH₂CH₂−C(O)NHOH	$C_4H_7NO_4$	Mn^{III}	1:3	450	[2, 7]
3		$C_8H_7NO_4$	Mn^{III}	1:2	465	[3, 4]
4	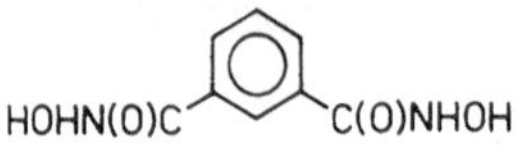	$C_8H_8N_2O_4$	Mn^{III}	1:2	480*) (3.93)	[5]
			Mn^{III}	1:2	490 (3.58)	[8]
5	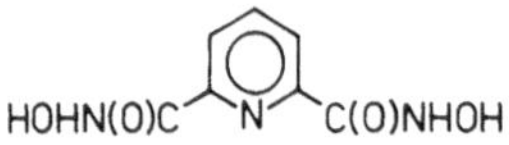	$C_7H_7N_3O_4$	Mn^{III}	—	500*) (3.61)	[6, 9]

*) Absorptions were measured on extracts into solutions of trioctylmethylammonium chloride (Adogen 464) in an organic solvent.

The formation constant log K = 7.14, of the manganese(II) complex $Mn(C_4H_2NO_4)^-$ with ligand 1 (see the table above) was determined by potentiometric pH measurement at 25°C for I = 0.25 mol/L (KNO_3) [1].

Deeply red-colored Mn^{III} complexes with ligands 2 to 5 are formed in aqueous ammoniacal solution in the presence of atmospheric oxygen [2 to 9]. A large excess of the ligands was applied to obtain the color reactions. No color was observed in the presence of hydrazine or hydroxylamine [7]. The absorbance of the complex with ligand 3 is greatly increased by addition of DMF. This is accounted for by the shift of the equilibrium from the keto form to the enol form of the ligand in this medium [4]. The stability constant log K = 7.79 for the complex $Mn^{III}(C_8H_5N_2O_4)_2^{3-}$ in ammoniacal solution of Mn^{II} and ligand 4 was evaluated spectrophotometrically after air oxidation of the Mn^{2+} ions. The ionic nature of the complex was indicated by extraction studies into benzene or toluene solutions of trioctylmethylammonium chloride (Adogen 464) wherein ion associates with the mole ratio 1:2:3 (Mn:ligand:Adogen 464) were formed [8].

References:

[1] Ivanov, V. A., Syrbu, V. A. (Koord. Soedin. Perekhodnykh Elem. Vopr. Khim. Khim. Tekhnol. **1983** 30/5; C.A. **99** [1983] No. 164761).
[2] Bhargava, S. P., Sogani, N. C. (Z. Anal. Chem. **255** [1971] 210/1).
[3] Salinas, F., Estela, J. M. (Anales Quim. B **79** [1983] 101/8).
[4] Salinas, F., Estela, J. M., Forteza, R. (Afinidad **40** [1983] 359/62).
[5] Salinas, F., Jiménez-Arrabal, M., Mahedero, M. C. (Anal. Letters A **16** [1983] 1449/55).
[6] Salinas, F., Forteza, R. (Anales Quim. B **79** [1983] 424/8).
[7] Bhargava, S. P., Sogani, N. C. (J. Inst. Chem. [Calcutta] **43** [1971] 172/4; C.A. **76** [1972] No. 80556).
[8] Salinas, F., Jiménez-Arrabal, M. (Microchem. J. **31** [1985] 113/7).
[9] Salinas, F., Forteza, R., March, J. G. (Quim. Anal. [Barcelona] **2** [1983] 283/8 from C.A. **102** [1985] No. 142416).

26.1.5 With N-Aryl Substituted Aliphatic or Alicyclic Hydroxamic Acids RC(O)N(OH)R′

26.1.5.1 Manganese(II) Complexes in Aqueous-Organic Solution

The stability constants of manganese(II) complexes with the hydroxamic acids listed in Table 3 were determined potentiometrically (glass electrode).

Table 3
Stability Constants of Complexes MnL^+ and MnL_2 with Hydroxamic Acids $RC(O)N(OH)C_6H_4X$ (=HL) in Aqueous Dioxane or Ethanol.

No.	ligand HL R	X	formula	°C	I in mol/L	% (v/v) dioxane	pK_a	log K_1	log K_2	Ref.
1	CH_3	H	$C_8H_9NO_2$	25	0.1 ($NaClO_4$)	50	9.76	5.07	4.42	[1]
				25	0.1 (KCl)	70	11.01	5.15	3.96	[2]
2	CH_3	4-Cl	$C_8H_8ClNO_2$	25	0.1 (KCl)	70	10.91	4.89	3.97	[3]
3	$C_6H_5CH_2$	H	$C_{14}H_{13}NO_2$	30	0.1 (KNO_3)	50*)	9.80	4.46	3.63	[4]
4	C_3H_7	H	$C_{10}H_{13}NO_2$	25	0.1 ($KClO_4$)	50	11.15	5.25	4.09	[7]
				25	0 corr	50	11.45	6.23	4.53	[5]
				35	0 corr	50	11.33	6.22	4.46	[6]

Table 3 (continued)

No.	ligand HL R	X	formula	°C	I in mol/L	% (v/v) dioxane	pK_a	log K_1	log K_2	Ref.
5	$CH_3C(O)CH_2$	H	$C_{10}H_{11}NO_3$	20	? ($NaClO_4$)	70	10.93	6.19	3.62	[8]
6	$CH_3C(O)CH_2$	4-Cl	$C_{10}H_{10}ClNO_3$	20	? ($NaClO_4$)	70	10.89	4.96	4.48	[8]
7	$CH_3C(O)CH_2$	4-Br	$C_{10}H_{10}BrNO_3$	20	? ($NaClO_4$)	70	10.90	5.57	3.22	[8]
8	$CH_3C(O)CH_2$	4-CH_3	$C_{11}H_{13}NO_3$	20	? ($NaClO_4$)	70	11.00	5.19	3.65	[8]
9	(*E*)-$CH_3CH{=}CH$	H	$C_{10}H_{11}NO_2$	35	0 corr	50	10.94	6.41	4.59	[9]
10	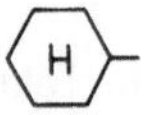	H	$C_{13}H_{17}NO_2$	25	0.1 ($NaClO_4$)	50	10.17	5.33	4.70	[10]

*) Aqueous ethanol.

From the stability constants as well as from spectrophotometric studies, it was inferred that the β-carbonyl group is not enolized in the complexes with ligands 5 to 8. The ligands thus behave as bidentate in contrast to the terdentate nature that could be expected [8].

Thermodynamic parameters of stepwise ligand addition are given for the complexes with ligand 3: $\Delta G_1 = -6.18 \pm 0.11$, $\Delta H_1 = -5.22 \pm 0.19$ kcal/mol, $\Delta S_1 = 2.35 \pm 0.04$ cal·mol^{-1}·K^{-1} and $\Delta G_2 = -5.50 \pm 0.12$, $\Delta H_2 = -5.36 \pm 0.21$ kcal/mol, $\Delta S_2 = 0.578 \pm 0.003$ cal·mol^{-1}·K^{-1}. Separation of ΔG and ΔH values into electrostatic and nonelectrostatic components (values given) shows that nonelectrostatic bonding forces are stronger for both complexes and that these forces prevail to a larger extent for the 1:2 complexes [11].

References:

[1] Bag, S. P., Lahiri, S. (Current Sci. [India] **43** [1974] 648/50).
[2] Sharma, K. K., Jaimini, J. P. C., Sogani, N. C. (J. Inst. Chem. [Calcutta] **43** [1971] 191/5).
[3] Jaimni [Jaimini], J. P. C., Sogani, N. C. (J. Indian Chem. Soc. **45** [1968] 59/62).
[4] Janardhan Rao, M., Sethuram, B., Navaneeth Rao, T. (Indian J. Chem. A **20** [1981] 1136/7).
[5] Shukla, J. P., Tandon, S. G. (Talanta **19** [1972] 711/3).
[6] Shukla, J. P., Tandon, S. G. (J. Electroanal. Chem. Interfacial Electrochem. **33** [1971] 195/200).
[7] Shukla, J. P., Tandon, S. G. (Bull. Chem. Soc. Japan **45** [1972] 3073/5).
[8] Kettrup, A., Seshadri, T., Cramer, M. (Talanta **26** [1979] 303/7).
[9] Bhura, D. C., Tandon, S. G. (J. Inorg. Nucl. Chem. **32** [1970] 2993/7).
[10] Bag, S. P., Lahiri, S. (Sci. Cult. [Calcutta] **41** [1975] 312/3).

[11] Janardhan Rao, M., Sethuram, B., Navaneeth Rao, T. (Indian J. Chem. A **24** [1985] 327/9).

26.1.5.2 Isolated Manganese(II) Compounds with Ligands of the Type $RC(O)N(OH)C_6H_4X$

No.	R	X	formula	No.	R	X	formula
1	$C_6H_5CH_2$	H	$C_{14}H_{13}NO_2$	8	$H_2C{=}C(CH_3)$	H	$C_{10}H_{11}NO_2$
2	$H_2C{=}CH$	H	$C_9H_9NO_2$	9	$H_2C{=}C(CH_3)$	4-Cl	$C_{10}H_{10}ClNO_2$
3	$H_2C{=}CH$	3-Cl	$C_9H_8ClNO_2$	10	$H_2C{=}C(CH_3)$	4-Br	$C_{10}H_{10}BrNO_2$
4	$H_2C{=}CH$	4-Cl	$C_9H_8ClNO_2$	11	$H_2C{=}C(CH_3)$	$4-CH_3$	$C_{11}H_{13}NO_2$
5	$H_2C{=}CH$	4-Br	$C_9H_8BrNO_2$	12	$H_2C{=}C(CH_3)$	$4-CH_3CO$	$C_{12}H_{13}NO_3$
6	$H_2C{=}CH$	$4-CH_3$	$C_{10}H_{11}NO_2$	13	$C_6H_5HC{=}CH$	H	$C_{15}H_{13}NO_2$
7	$H_2C{=}CH$	$4-CH_3CO$	$C_{11}H_{11}NO_3$				

No. 14

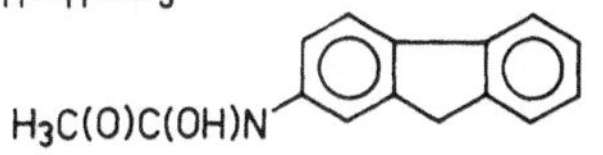

($=C_{15}H_{13}NO_2$)

Complex formation on reaction of Mn^{2+} ions in aqueous solution of pH ~ 6 with ligand 1 in ethanol was indicated by precipitation of a yellow solid which was extractable into chloroform forming yellow flakes [1]. When ligand 13 in alcohol was added to the neutral aqueous solution of an Mn^{II} salt, an orange precipitate was obtained extractable into chloroform [4]. Microcrystalline yellow compounds of compositions $\mathbf{Mn(C_9H_7NO_2X)_2}$ with ligands 2 to 7 and $\mathbf{Mn(C_{10}H_9NO_2X)_2}$ with ligands 8 to 12 (X = H, Cl, Br, CH_3, CH_3CO) were deposited on mixing methanol solutions of $MnCl_2 \cdot 4H_2O$, one of the ligands, and sodium acetate in the mole ratios 1:2:2. They were washed with methanol and dried in vacuum. Their magnetic moment is ~5.4 μ_B at room temperature. From ESR spectra at 77 and 300 K a g-factor of 2.01 was obtained. In the IR spectra of the complexes in Vaseline mulls, the ν(OH) bands (3250 to 3140 cm^{-1}) and the δ(OH) bands (1440 to 1404, 1284 to 1252 cm^{-1}) observed for the free ligands have disappeared. This indicates a substitution of the NOH proton by manganese. On coordination, the ν(CO) bands of the free ligands are shifted by 28 to 58 cm^{-1} to lower wave numbers; the bands range from 1582 to 1562 cm^{-1} for the complexes [3].

The compounds are stable when heated up to 150°C [3], but are decomposed by mineral acids. They are insoluble in water or organic solvents [3].

The complex $\mathbf{Mn(C_{15}H_{12}NO_2)_2}$ separated on mixing ethanolic solutions of $Mn(CH_3COO)_2 \cdot 4H_2O$ and ligand 14. It was washed with ethanol and ether and dried over $CaCl_2$. At temperatures above 150°C, the compound starts decomposing. On rapid heating, it melts with decomposition at 247 to 248°C. Its carcinogenic activity was investigated [2].

References:

[1] Zharovskii, F. G., Ostrovskaya, M. S. (Ukr. Khim. Zh. **32** [1966] 893/8; Soviet Progr. Chem. **32** [1966] 674/6).
[2] Poirier, M. M., Miller, J. A., Miller, E. C. (Cancer Res. **25** [1965] 527/33).
[3] Manole, S. F., Stratulat, A. A., Starysh, M. P. (Izv. Akad. Nauk Mold. SSR Ser. Biol. Khim. Nauk **1983** No. 6, pp. 66/7; C.A. **100** [1984] No. 192031).
[4] Zharovskii, F. G., Sukhomlin, R. I. (Ukr. Khim. Zh. **30** [1964] 750/3).

26.1.6 With N-Substituted Aromatic Hydroxamic Acids $XC_6H_4C(O)N(OH)C_6H_4Y$

Manganese compounds with N-phenylbenzohydroxamic acid have been studied most thoroughly and were the first known of all hydroxamate complexes. The influence of substituents on either of its phenyl groups on the stability of manganese complexes was a special point of investigation. The effect of substituents at the hydroxylamine part of the ligand can be seen from the stability constants in Table 4, No. 1 to 5. These substituents directly affect the ability for deprotonation of the ligand and the formation of an Mn–ON bond. Substitution on the benzoyl group, however, primarily influences the chelating donor property of the carbonyl group, as shown in the remainder of Table 4.

26.1.6.1 Manganese(II) Complexes in Solution

Stability constants of MnL^+ and MnL_2 complexes in aqueous dioxane were determined potentiometrically (glass electrode) at 25 and 35°C. Values of log K_1 and log K_2 at 25°C and solvent-dependent pK_a values of the ligands are collected in Table 4. Complex stability constants at 35°C are reported in references marked by asterisks in the table or in further references for the following ligands: 1, 3, 4, and 5 [24], 6, 14, 18, 22, 26, 27, and 33 [25]. Values for 60% (v/v) aqueous dioxane are reported in [11, 14, 15]. Sufficient solubility was achieved by use of solvent mixtures containing 50 to 70% (v/v) dioxane. Additional precautions to keep the complexes in solution such as use of low Mn^{II} concentrations and addition of some mineral acid were provided especially for ligands with *ortho*-F and *para*-NO_2, -Cl, or -Br substituents on the benzoyl side of the ligand molecule (ligands 12 to 14, 20, 21, 24, and 25).

Table 4

Stability Constants of Mn^{II} Complexes with N-Substituted Aromatic Hydroxamic Acids $XC_6H_4C(O)N(OH)C_6H_4Y$ (=HL) in Aqueous Dioxane at 25°C.

No.	X	Y	formula	I in mol/L	% (v/v) dioxane	pK_a	log K_1	log K_2	Ref.
1	H	H	$C_{13}H_{11}NO_2$	0.1 ($NaClO_4$)	50	10.05	5.08	4.52	[2]
				?	50	10.45	5.9	4.9	[3]
				0.1 (KCl)	70	10.96	5.52	4.32	[1]
				0 corr	50	11.04	6.02	5.15	[4]
2	H	4-Cl	$C_{13}H_{10}ClNO_2$	0.1 (KCl)	70	10.78	5.23	4.34	[5]
3	H	2-CH_3	$C_{14}H_{13}NO_2$	0.1 ($NaClO_4$)	50	10.09	5.51	4.74	[6]
				0.1 (KCl)	70	11.01	5.58	4.37	[7]
				0 corr	50	11.12	6.25	5.20	[8]
4	H	3-CH_3	$C_{14}H_{13}NO_2$	0 corr	50	11.12	insol.	insol.	[4]
				0 corr	70	13.41	7.37	6.60	[9]
5	H	4-CH_3	$C_{14}H_{13}NO_2$	0.1 ($NaClO_4$)	50	10.17	5.57	4.75	[6]
				0 corr	50	11.13	insol.	insol.	[4]
				0.1 (KCl)	70	11.08	5.78	4.54	[10]
				0 corr	70	13.15	8.90	8.13	[11]*)
6	2-NO_2	H	$C_{13}H_{10}N_2O_4$	0 corr	50	10.79	5.90	4.54	[12]
				0 corr	50	10.80	5.89	4.74	[13]*)

*) Values at 35°C are also given in the paper.

Table 4 (continued)

No.	X	Y	formula	I in mol/L	% (v/v) dioxane	pK_a	log K_1	log K_2	Ref.
7	2-NO_2	4-CH_3	$C_{14}H_{12}N_2O_4$	0 corr	50	10.71	5.82	4.70	[13]*)
8	3-NO_2	H	$C_{13}H_{10}N_2O_4$	0 corr	50	10.45	5.72	4.60	[13]*)
9	3-NO_2	3-CH_3	$C_{14}H_{10}N_2O_4$	0 corr	50	10.25	5.90	5.78(?)	[13]*)
10	3-NO_2	4-CH_3	$C_{14}H_{10}N_2O_4$	0 corr	50	10.45	5.74	4.61	[13]*)
11	4-NO_2	2-CH_3	$C_{14}H_{12}N_2O_4$	0 corr	50	10.67	5.80	4.90	[8]*)
12	4-NO_2	3-CH_3	$C_{14}H_{12}N_2O_4$	0 corr	70	—	5.60	4.40	[14]*)
13	4-NO_2	4-CH_3	$C_{14}H_{12}N_2O_4$	0 corr	70	—	5.65	4.40	[15]*)
14	2-F	H	$C_{13}H_{10}FNO_2$	0 corr	50	10.74	6.30	5.08	[12]
15	4-F	2-CH_3	$C_{14}H_{12}FNO_2$	0 corr	50	11.08	6.00	5.01	[8]*)
16	4-F	3-CH_3	$C_{14}H_{12}FNO_2$	0 corr	50	—	5.00	4.02	[14]*)
				0 corr	70	—	8.00	6.63	[14]*)
17	4-F	4-CH_3	$C_{14}H_{12}FNO_2$	0 corr	50	10.99	5.50	4.42	[15]*)
				0 corr	70	12.96	8.44	7.40	[15]*)
18	2-Cl	H	$C_{13}H_{10}ClNO_2$	0 corr	50	10.67	6.22	5.11	[12]
19	4-Cl	2-CH_3	$C_{14}H_{12}ClNO_2$	0 corr	50	11.01	5.98	4.99	[8]*)
20	4-Cl	3-CH_3	$C_{14}H_{12}ClNO_2$	0 corr	60	—	6.00	5.05	[14]*)
				0 corr	70	—	7.15	5.67	[14]*)
21	4-Cl	4-CH_3	$C_{14}H_{12}ClNO_2$	0 corr	60	12.13	6.50	5.55	[15]*)
				0 corr	70	12.75	7.53	6.05	[15]*)
22	2-Br	H	$C_{13}H_{10}BrNO_2$	0 corr	50	10.66	6.25	4.94	[12]
23	4-Br	2-CH_3	$C_{14}H_{12}BrNO_2$	0 corr	50	10.98	5.92	4.96	[8]*)
24	4-Br	3-CH_3	$C_{14}H_{12}BrNO_2$	0 corr	70	—	7.09	5.61	[14]*)
25	4-Br	4-CH_3	$C_{14}H_{12}BrNO_2$	0 corr	60	11.58	6.48	5.45	[15]*)
				0 corr	70	12.78	7.48	5.95	[15]*)
26	2-I	H	$C_{13}H_{10}INO_2$	0 corr	50	10.75	6.19	4.84	[12]
27	2-OCH_3	H	$C_{14}H_{13}NO_3$	0 corr	50	11.09	6.98	5.85	[12]
28	4-OCH_3	H	$C_{14}H_{13}NO_3$	0.1 (KCl)	70	11.18	5.93	4.62	[16]
29	4-OCH_3	2-CH_3	$C_{15}H_{15}NO_3$	0 corr	50	11.51	6.35	5.30	[8]*)
30	4-OCH_3	3-CH_3	$C_{15}H_{15}NO_3$	0 corr	50	11.41	6.34	5.28	[14, 17]*)
				0 corr	70	13.79	9.04	7.98	[14, 17]

*) Values at 35°C are also given in the paper.

Table 4 (continued)

No.	X	Y	formula	I in mol/L	% (v/v) dioxane	pK_a	log K_1	log K_2	Ref.
31	4-OCH_3	4-CH_3	$C_{15}H_{15}NO_3$	0.1 ($NaClO_4$)	50	10.34	5.38	4.55	[18]
				0 corr	50	11.30	7.00	6.05	[15*), 17]
				0 corr	70	13.28	9.64	8.50	[15]*)
32	2-OC_2H_5	H	$C_{15}H_{15}NO_3$	0.1 ($NaClO_4$)	50	10.35	5.50	4.82	[19]
33	2-CH_3	H	$C_{14}H_{13}NO_2$	0 corr	50	10.90	6.61	4.92	[12, 20]
				0 corr	70	13.18	8.45	6.76	[20]*)
34	3-CH_3	H	$C_{14}H_{13}NO_2$	0 corr	50	—	—	—	[21]*)
35	4-CH_3	2-CH_3	$C_{15}H_{15}NO_2$	0 corr	50	11.23	6.30	5.25	[8]*)
36	4-CH_3	3-CH_3	$C_{15}H_{15}NO_2$	0 corr	50	11.33	6.50	5.31	[14, 22]*)
				0 corr	70	13.66	8.65	6.88	[14, 22]
37	4-CH_3	4-CH_3	$C_{15}H_{15}NO_2$	0 corr	50	11.24	6.16	5.21	[23]*)
				0 corr	70	13.21	8.92	8.15	[23]*)

*) Values at 35°C are also given in the paper.

Stability constants in 50% (v/v) ethanol at 25°C, with I = 0.1 M ($NaClO_4$) were determined for the complex with ligand 37: log K_1 = 5.16, log K_2 = 4.38, pK_a (ligand) = 10.08 [18].

An approximately linear relationship between log K_1 (or log $K_1 \cdot K_2$) values of the complexes and pK_a values of the ligands was observed for N-arylbenzohydroxamic acids [6] and its derivatives substituted at the *para*-position of the benzoyl group [8, 14, 15], whereas significant deviation from linearity was observed for *ortho* substituted mesomers [25]. Slopes <1 of the plots log K_1 vs. pK_a indicate that the manganese complexes are less affected by substitution than the ligands themselves [14, 15]. The complex stability order is $OCH_3 > CH_3 > F > Cl \approx Br < NO_2$ for *para*-substituents X at the benzoyl group [8, 14, 15]. Increased complex stabilities on substitution with the OCH_3 or OC_2H_5 group (ligands 28, 31, and 32) are attributed to mesomeric effects [16, 18, 19]; increased stabilities on substitution with the CH_3 group are accounted for by inductive electron pushing [6, 18].

Thermodynamic parameters of formation calculated from the stability constants at 25 and 35°C are presented in Table 5. Other values are given for the complex with ligand 5 in 60% (v/v) dioxane in [11], for the complexes with ligands 36 or 37 in 70% dioxane in [22] and [23], respectively.

Table 5

Thermodynamic Parameters[a)] of Formation for Mn^{II} Complexes with Several N-Substituted Aromatic Hydroxamic Acids $XC_6H_4C(O)N(OH)C_6H_4Y$ in Aqueous Dioxane at 25°C.

No.[b)]	X	Y	% (v/v) dioxane	$-\Delta G_1$	$-\Delta H_1$	ΔS_1	$-\Delta G_2$	$-\Delta H_2$	ΔS_2	Ref.
4	H	3-CH_3	70	10.08	3.79	21.09	9.03	3.79	17.57	[9]
5	H	4-CH_3	70	12.2	4.6	25.3	11.1	4.2	23.2	[11]
6	2-NO_2	H	50	8.05	6.73	4.43	6.41	6.73	−1.07	[12]
14	2-F	H	50	8.60	8.41	0.64	7.17	8.41	−4.16	[12]

Table 5 (continued)

No.[b]	X	Y	% (v/v) dioxane	$-\Delta G_1$	$-\Delta H_1$	ΔS_1	$-\Delta G_2$	$-\Delta H_2$	ΔS_2	Ref.
18	2-Cl	H	50	8.49	7.57	3.08	7.21	7.57	−1.21	[12]
22	2-Br	H	50	8.53	7.57	3.22	6.97	7.57	−2.03	[12]
26	2-I	H	50	8.45	7.15	4.36	6.83	6.73	0.34	[12]
27	$2\text{-}OCH_3$	H	50	9.53	7.57	6.57	8.24	7.57	5.60	[12]
33	$2\text{-}CH_3$	H	50	9.02	6.31	5.73	6.94	6.31[c]	2.15	[12, 20]
36	$4\text{-}CH_3$	$3\text{-}CH_3$	50	8.87	4.20	15.66	7.25	4.20	10.16	[22]
37	$4\text{-}CH_3$	$4\text{-}CH_3$	50	8.41	5.01	11.14	7.11	2.10	16.80	[23]

[a] ΔG and ΔH in kcal/mol, ΔS in $cal \cdot mol^{-1} \cdot K^{-1}$. (Signs for ΔG and ΔH were corrected by the editor except for those of [20].) – [b] Numbers refer to those in Table 4, pp. 222/4. – [c] In [20]: −5.73 kcal/mol (misprint?).

Manganese(II) can be quantitatively extracted at pH 9.8 from aqueous solution with a 0.01 M solution of ligand 1 in chloroform [26 to 28], isopropyl ether, or nitrobenzene. With other organic solvents only partial extraction of the complex was achieved at optimal pH values of 7 to 10 [28]. The extraction may be used for the separation of Mn^{II} from other elements [26 to 29].

References:

[1] Jaimni [Jaimini], J. P. C., Sogani, N. C. (Z. Naturforsch. **22b** [1967] 922/4).
[2] Bag, S. P., Lahiri, S. (J. Inorg. Nucl. Chem. **38** [1976] 1611/3).
[3] Brydon, G. A., Ryan, D. E. (Anal. Chim. Acta **35** [1966] 190/4).
[4] Shukla, J. P., Tandon, S. G. (Talanta **19** [1972] 711/3).
[5] Jaimni [Jaimini], J. P. C., Sogani, N. C. (Z. Anorg. Allgem. Chem. **355** [1967] 332/6).
[6] Bag, S. P., Lahiri, S. (J. Indian Chem. Soc. **52** [1975] 36/8).
[7] Jaimni [Jaimini], J. P. C., Sogani, N. C. (J. Inst. Chem. [India] **40** II [1968] 52/6; C.A. **69** [1968] No. 70672).
[8] Agrawal, Y. K., Mudaliar, A. (Transition Metal Chem. [Weinheim] **4** [1979] 252/4).
[9] Sharma, T. P., Agrawal, Y. K. (J. Inorg. Nucl. Chem. **37** [1975] 1880/1).
[10] Jaimni [Jaimini], J. P. C., Sogani, N. C. (Bull. Acad. Polon. Sci. Ser. Sci. Chim. **17** [1969] 157/62; C.A. **71** [1969] No. 42897).

[11] Agrawal, Y. K., Khare, V. P. (Z. Physik. Chem. [Leipzig] **258** [1977] 337/43).
[12] Agrawal, Y. K. (Thermochim. Acta **18** [1977] 245/9).
[13] Verma, P. C., Khadikar, P. V., Agrawal, Y. K. (J. Inorg. Nucl. Chem. **39** [1977] 1847/8).
[14] Agrawal, Y. K. (Bull. Soc. Chim. Belges **86** [1977] 565/79).
[15] Agrawal, Y. K., Khare, V. P. (Bull. Soc. Chim. Belges **86** [1977] 429/43).
[16] Jaimini, J. P. C., Sogani, N. C. (J. Indian Chem. Soc. **48** [1971] 1107/11).
[17] Agrawal, Y. K. (Transition Metal Chem. [Weinheim] **4** [1979] 109/11).
[18] Lahiri, S. (Acta Ciencia Indica **1** [1975] 171/4; C.A. **83** [1975] No. 105396).
[19] Lahiri, S. (Current Sci. [India] **43** [1974] 717/9).
[20] Agrawal, Y. K. (J. Inorg. Nucl. Chem. [1977] 2011/3).

[21] Agrawal, Y. K., Tandon, S. G. (Z. Physik. Chem. [Leipzig] **255** [1974] 644/50).
[22] Agrawal, Y. K. (Monatsh. Chem. **108** [1977] 713/23).

[23] Agrawal, Y. K., Khare, V. P. (J. Inorg. Nucl. Chem. **38** [1976] 1663/7).
[24] Shukla, J. P., Tandon, S. G. (J. Electroanal. Chem. Interfacial Electrochem. **33** [1971] 195/200).
[25] Agrawal, Y. K., Tandon, S. G. (J. Inorg. Nucl. Chem. **36** [1974] 869/73).
[26] Chwastowska, J. (Proc. Conf. Appl. Phys. Chem. Methods Chem. Anal., Budapest 1966, Vol. 1, pp. 45/53; C.A. **68** [1968] No. 56251).
[27] Chwastowska, J. (Chem. Anal. [Warsaw] **12** [1967] 469/78).
[28] Chwastowska, J., Kosiarska, E., Maciejko, G. (Chem. Anal. [Warsaw] **22** [1977] 927/34).
[29] Förster, H. (J. Radioanal. Chem. **6** [1970] 11/26, 23).
[30] Agrawal, Y. K. (Bull. Soc. Chim. Belges **87** [1978] 89/91).

26.1.6.2 Isolated Manganese(II) Compounds

Mn^{2+} ions form a light yellow precipitate with N-phenylbenzohydroxamic acid (ligand 1, Table 4) in aqueous solutions of pH 5.5 to 9.5 [1 to 4]. Precipitates were also obtained with a series of derivatives of this ligand [2]. The compounds are only poorly soluble in aqueous solution. Therefore, studies with a view to gravimetric application were performed [1 to 4].

$Mn(C_{13}H_{10}NO_2)_2$ was prepared by adding a hot alcohol solution of N-phenylbenzohydroxamic acid (= $C_{13}H_{11}NO_2$) to the hot (40°C) aqueous solution of an Mn^{II} salt (3:1 mole ratio) and raising the pH value by dropwise addition of 2 M aqueous ammonia. The precipitate was filtered, washed with hot water, and dried at 110 to 120°C [5]. Similar procedures are reported by [6 to 8]. An Mn : ligand ratio of 1:2 and a boiling Mn^{II} salt solution were used by [6]; only a 5 to 10% ligand excess was used by [8]. After precipitation by adjusting the pH to between 5 and 6, the reaction mixture was digested for 1 to 2 h in the heat [6, 8, 9]. In a procedure for the gravimetric determination of manganese, aqueous $NH_2OH \cdot HCl$ and tartrate were added to prevent oxidation and hydrolysis of Mn^{II} [9].

The X-ray powder diagram (d-values and intensities reported) and magnetic susceptibility measurements at 302 K, resulting in $\mu_{eff} = 6.10\ \mu_B$, indicate a high-spin octahedral complex. On account of the bidentate ligand nature and the poor solubility of the complexes, a polymeric structure was proposed with triply bonded bridging oxygen atoms of the NO^- groups, as shown below [5].

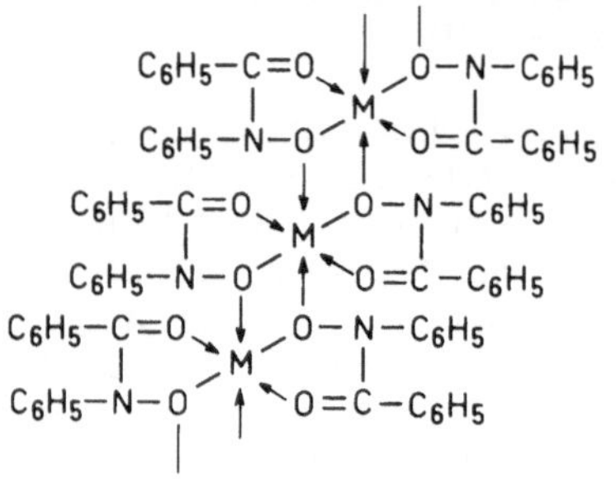

In the IR spectrum of the complex in Nujol, the ν(OH) band at 3106 cm^{-1} of the free ligand has disappeared, while the ν(NO) band at 917 cm^{-1} was intensified and has shifted to 927 cm^{-1} on complexation. The ν(CO) band (at 1631 cm^{-1} for the free ligand) is displaced to 1582 cm^{-1} and is superimposed to bands due to C=C vibrations which are unaffected by chelation [6]. An anomalously large shift of the ν(CO) band from 1632 to 1550 cm^{-1} was observed for KBr pellets [5]. Other bands for the complex and the free ligand are published in [10].

The complex melts with decomposition at 238°C [6, 7, 9]. DTA and TG studies in N_2 or O_2 with a gas flow rate of 10 mL/min and a heating rate of 5°C/min show an exothermic peak at ~230°C and an abrupt weight loss at this temperature. Benzanilide, Mn^{II} benzoate, Mn^{II} oxide, and tars were identified as the major products. In O_2, a second peak in the DTA curve at ~325°C represents the exothermic oxidation of Mn^{II} benzoate and the degradation of tars with a final weight loss at ~400°C. In N_2 [8] or air [9], weight loss continues to 500°C [8, 9]. $Mn(C_{13}H_{10}NO_2)_2$ is insoluble in hot water (80°C) and only sparingly soluble in common organic solvents [5, 9]. It is extractable with chloroform [1, 4, 14], forming yellow flakes [1, 14].

$Mn(C_{14}H_{12}NO_2)_2$. The complex with N-*p*-tolylbenzohydroxamic acid (ligand 5 in Table 4, p. 222) was prepared as described for $Mn(C_{13}H_{10}NO_2)_2$. The yellow solid melts at 253°C [6]. (No preparations are reported for the Mn^{II} complexes with the *ortho*- and *meta*-isomers of N-tolylbenzohydroxamic acids (ligands 3 and 4 in Table 4.)

In the IR spectrum, the ν(CO) band of the complex with ligand 3 in KBr was observed at 1540 cm^{-1} (free ligand 1628 cm^{-1}) [12]. The ν(CO) and ν(NO) bands of the other complexes appear at 1570 and 918 cm^{-1} for ligand 4 (sample technique and free ligand wave numbers not given) [11] and at 1592 and 938 cm^{-1} for ligand 5 in Nujol (free ligand 1603 and 920 cm^{-1}) [6]. The magnetic moment of the complex with the *m*-tolyl compound is $\mu_{eff} = 6.00$ μ_B at 298 K [11].

$Mn(C_{15}H_{14}NO_3)_2$ and **$Mn(C_{15}H_{14}NO_2)_2$.** The complexes with N-*m*-tolyl-*p*-methoxy(or -*p*-methyl)-benzohydroxamic acids (= ligands 30 and 36, Table 4, pp. 223/4) have the magnetic moment 5.97 μ_B (either complex). The ν(CO) and ν(NO) bands are observed at 1575 and 915 cm^{-1} for the complex with ligand 30 and at 1565 and 915 cm^{-1} for that with ligand 36 [11].

$Mn(C_{21}H_{26}NO_3)_2$. The compound with N-phenyl-3,5-di-*tert*-butyl-4-hydroxybenzohydroxamic acid (= $C_{21}H_{27}NO_3$) was prepared at 40°C by slow addition of ethanolic $MnCl_2$ or $MnSO_4$ to an alkali salt of the ligand in ethanol in a 1:2 mole ratio. The precipitate formed was washed with water and ethanol and, after drying, boiled with petroleum ether. The compound is useful as a stabilizer for organic compounds against heat and air [13].

References:

[1] Zharovskii, F. G. (Ukr. Khim. Zh. **25** [1959] 245/8).
[2] Lutwick, G. D., Ryan, D. E. (Can. J. Chem. **32** [1954] 949/55).
[3] Shome, S. C. (Analyst **75** [1950] 27/32).
[4] Förster, H. (J. Radioanal. Chem. **6** [1970] 11/26, 23).
[5] Bag, S. P., Lahiri, S. (J. Inorg. Nucl. Chem. **38** [1976] 1611/3).
[6] Agrawal, D. R., Tandon, S. G. (J. Indian Chem. Soc. **48** [1971] 571/3).
[7] Chan, F. L., Moshier, R. W. (PB-161496 [1959] 1/25; C.A. **1961** 26819).
[8] Meyer, R. A., Hazel, J. F., McNabb, W. M. (Anal. Chim. Acta **31** [1964] 419/25).
[9] Bag, S. P. Lahiri, S. (J. Indian Chem. Soc. **52** [1975] 593/5).
[10] Lahiri, S. (Diss. Calcutta 1972 from [5], [12]).

[11] Agrawal, Y. K. (Transition Metal Chem. [Weinheim] **4** [1979] 109/11).
[12] Bag, S. P., Lahiri, S. (J. Indian Chem. Soc. **52** [1975] 36/8).
[13] Avar, L., Hofer, K., Preiswerk, M., Sandoz Ltd. (Ger. Offen. 2318118 [1973] 1/25, 7/11, 15/6; C.A. **80** [1974] No. 27001).
[14] Tsê Yün-hsiang (Diss. Univ. Moscow 1960 from Alimarin, I. P., Sudakov, F. P., Golovkin, B. G., Usp. Khim. **31** [1962] 989/1003; Russ. Chem. Rev. **31** [1962] 466/74, 467).

26.1.7 With N-Substituted 2-Furohydroxamic Acids

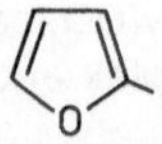
C(O)N(OH)R

Stepwise stability constants and the corresponding changes in enthalpy on complexation for binary Mn^{II} complexes with N-substituted 2-furohydroxamic acids are presented below. The stability constants were determined by pH-potentiometric titration (glass electrode) in 50% (v/v) aqueous dioxane at 25°C under nitrogen. Enthalpy values (in kcal/mol) were calculated from the stability constants for 25 and 35°C:

No.	R	ligand HL	I in mol/L	pK_a	log K_1	log K_2	$-\Delta H_1$*)	$-\Delta H_2$*)	Ref.
1	C_6H_5	$C_{11}H_9NO_3$	0.1 ($NaClO_4$)	9.60	4.84	4.02	—	—	[1]
			0 corr	10.73	5.13	4.39	—	—	[2, 7]
			0 corr	10.59	5.10	4.37	2.5	1.3	[3, 4]
			—	10.48	5.02	4.34	—	—	[10]
2	$C_6H_4CH_3$-2	$C_{12}H_{11}NO_3$	0 corr	10.73	4.90	4.01	3.4	3.8	[3]
3	$C_6H_4CH_3$-3	$C_{12}H_{11}NO_3$	0 corr	10.64	5.12	4.39	2.9	1.7	[3]
4	$C_6H_4CH_3$-4	$C_{12}H_{11}NO_3$	0 corr	10.68	5.14	4.39	3.7	2.9	[4]

*) Signs corrected by the editor of this handbook, except for those of [4].

An undefined soluble manganese(II) complex with N-methyl-2-furohydroxamic acid (= $C_6H_7NO_3$) was formed in aqueous solution of pH 9. It was adsorbed on a special resin column and eluted by HCl in acetone. The complex can be used for separation and concentration of Mn^{2+} ions from very dilute solutions [5].

Ternary complexes are known with bipyridine or phenanthroline and ligand 1. The equilibrium constant of the reaction $[MnL']^{2+} + L^- \rightleftharpoons [MnLL']^+$ was determined with a glass electrode: log K = 5.15 for L' = bipyridine and log K = 5.27 for L' = phenanthroline. The equilibrium constant for $Mn^{2+} + L^- + L' \rightleftharpoons [MnLL']^+$ is log β = 10.86 for L' = bipyridine and log β = 12.22 for L' = phenanthroline. The ternary complexes are more stable than the binary complex MnL^+ [10].

$Mn(C_{11}H_8NO_3)_2$. The yellow-brown precipitate with ligand 1 is formed in aqueous solution of pH 5 to 8.5 [6 to 9]. For preparation, the ligand in ethanol was added to aqueous $MnSO_4$ and the pH was adjusted to between 5 and 6 by dilute aqueous ammonia. The granular residue was digested over a steam bath for two hours, filtered, washed with hot water, 50% aqueous ethanol, and dried at 100 to 110°C [6, 7]. The complex melts at 223°C. Constant weight up to 170°C in TG analysis indicated the absence of water [7].

In the IR spectrum of the complex in Nujol, the ν(OH) band of the strongly hydrogen bonded ligand at 3145 cm^{-1} has disappeared [6, 7]. The ligand ν(CO) band at 1603 cm^{-1} is displaced to 1592 cm^{-1} on complex formation. It is superimposed on one of the bands due to C=C vibrations [6]. A much larger shift, from 1626 to 1566 cm^{-1}, was observed by other authors for KBr disks [1]. A weak ν(NO) band appears unshifted at 920 cm^{-1} [6]. The complex is extractable into chloroform [9].

References:

[1] Bag, S. P., Lahiri, S. (Indian J. Chem. **13** [1975] 1214/6).
[2] Shukla, J. P., Agrawal, Y. K. (J. Electroanal Chem. Interfacial Electrochem. **45** [1973] 492/5).
[3] Abbasi, S. A. (Thermochim. Acta **38** [1980] 335/9).
[4] Abbasi, S. A. (J. Electroanal. Chem. Interfacial Electrochem. **73** [1976] 115/8).
[5] Al-Biaty, I. A., Fritz, J. S. (Anal. Chim. Acta **146** [1983] 191/200).
[6] Agrawal, D. R., Tandon, S. G. (J. Indian Chem. Soc. **48** [1971] 571/3).

[7] Shukla, J. P., Agrawal, Y. K., Agrawal, D. R. (Indian J. Chem. **12** [1974] 534/6).
[8] Lutwick, G. D., Ryan, D. E. (Can. J. Chem. **32** [1954] 949/55, 950).
[9] Pilipenko, A. T., Shpak, E. A., Ruban, P. P. (Ukr. Khim. Zh. **29** [1963] 1209/14, 1212/4; C.A. **60** [1964] 6193).
[10] Abbasi, S. A. (Polish J. Chem. **58** [1984] 61/4).

26.1.8 With N-Substituted Dihydroxamic Acids $C_6H_5N(OH)C(O)RC(O)N(OH)C_6H_5$ (= H_2L)

ligand No.	1	2	3
R	–HC=CH–	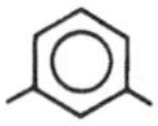	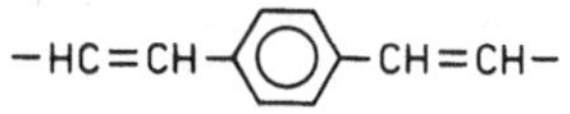
formula of H_2L	$C_{16}H_{14}N_2O_4$	$C_{20}H_{16}N_2O_4$	$C_{24}H_{20}N_2O_4$

Green and dark brown complexes of composition $Mn_5L_4(OH)_2 \cdot 2H_2O$ (?) were prepared by reacting manganese(II) acetate with ligand 1 or 3, respectively, in DMF. A greenish black compound $Mn_4L_3(OH)_2 \cdot 2H_2O$ (?) was reported to form with ligand 2.

Moderate intensity of the bands in the electronic reflectance spectra is indicative of tetrahedral complex geometry. The absorption maxima were assigned as listed in the table below:

complex with ligand No.	electronic reflectance spectrum $\bar{\nu}_{max}$ in cm^{-1}	transition	IR absorption spectrum $\bar{\nu}_{max}$ in cm^{-1}, assignments
1	27770	$^6A_1(S) \rightarrow {}^4E(D)$	1555, 1550, 1530, ν(CO); 960, ν(NO); 580, 470, ν(Mn–O)
2	25640	$^6A_1(S) \rightarrow {}^4E(G)$	1580, 1540, ν(CO); 970, ν(NO); 460, ν(Mn–O)
3	28750	$^6A_1(S) \rightarrow {}^4E(D)$	1550, ν(CO); 950, ν(NO); 615, 555, 450, ν(Mn–O)?
	23800	$^6A_1(S) \rightarrow {}^4E(G)$	

In the IR spectrum of the complexes, the presence of water is indicated by a new ν(OH) band in the 3600 to 3200 cm^{-1} region. The free ligand ν(OH) absorption (in the 3180 to 3095 cm^{-1} region) disappears on complexation, and the ν(CO) bands shift to lower wave numbers. Magnetic moments of the complexes are 5.16, 5.28, and 5.42 μ_B, respectively.

The complexes are insoluble in a wide variety of solvents. A linear chain structure with bridging tetradentate chelating ligands is proposed with possible formation of polymers by a two- or three-dimensional network via hydroxyl bridging groups, Gandhi, N. R., Munshi, K. N. (J. Indian Chem. Soc. **59** [1982] 1290/5).

26.1.9 With a Cyclic Hydroxamic Acid Radical $C_7H_{13}N_2O_3^{\bullet}$ (= HL)

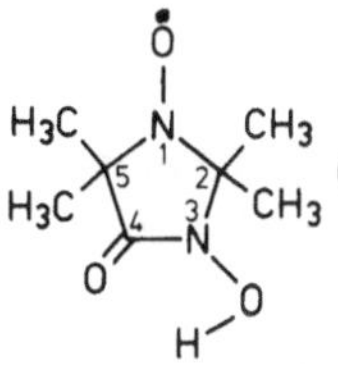

$Mn(C_7H_{12}N_2O_3^{\bullet})_2$ precipitated on evaporation of an aqueous solution of $MnSO_4$, the ligand, and KOH in the mole ratio 1:2:2. The crystalline complex was washed with water and dried over $MgClO_4$ (anhydrone) at 50 to 70°C. Its magnetic moment, $\mu_{eff} = 6.2\ \mu_B$, compares well with the theoretical spin-only value 6.40 μ_B for a high-spin state with independent orientation in the magnetic field of unpaired electron spins of the Mn^{2+} ion and the two radical centers. This indicates that the radical sites of the ligands are retained during complexation and are not involved in coordination. Deprotonation of the ligand NOH group is indicated by disappearance of the ν(OH) band in the IR spectrum of the complex. A new absorption in the 500 to 435 cm^{-1} region (in KBr pellets) was assigned to ν(Mn–O) vibrations. The strong ν(CO) band is shifted from 1715 cm^{-1} for the free ligand to 1644 cm^{-1} on chelation. The complex is soluble in water. It is stable both in solution and in solid form. Thermogravimetric and differential thermoanalytic studies show decomposition above 190°C. The thermal stability of the complex is higher than that of the free ligand, Larionov, S. V., Mironova, G. N., Ovcharenko, V. I., Volodarskii, L. B. (Izv. Akad. Nauk SSSR Ser. Khim. **1980** 977/82; Bull. Acad. Sci. USSR Div. Chem. Sci. **1980** 686/90).

26.2 Complexes with Isomeric Methylhydroxylamines

1) N-Methylhydroxylamine $CH_3NHOH \rightleftharpoons CH_3NH_2^{(+)}–O^{(-)}$ (= CH_5NO)

2) O-Methylhydroxylamine CH_3ONH_2 (= CH_5NO)

Mn^{II} complexes with hydroxylamine are described in "Manganese" D 3, 1982, pp. 75/6.

$Mn(CH_3NHOH)_2Cl_2$ was prepared by addition of ligand 1 to a hot alcoholic $MnCl_2$ solution containing some $CH_3NHOH \cdot HCl$. The transparent crystals were washed with alcohol and ether and dried over $CaCl_2$. They are soluble in water, insoluble in alcohol and other organic solvents; the melting point is 170°C [1]. The X-ray electron spectrum yielded the bond energies 402.8 eV for the N 1s electrons, 198.7 eV for the Cl $2p_{3/2}$ electrons, and 641.8 eV for the Mn $2p_{3/2}$ electrons (against the C 1s line (285 eV) as the standard) [2]. The IR spectrum of the compound in KBr or mulls shows absorption bands at 3115, 3032, or 1582 cm^{-1} which were assigned to $\nu_{as}(NH_2^+)$, $\nu_s(NH_2^+)$, and $\delta(NH_2^+)$ vibrations, respectively. IR data and bond energies indicate that the ligand is coordinated through the O atom of its N-oxide form and that the Cl atoms are also coordinated [1 to 3].

$Mn(CH_3ONH_2)_2Cl_2$ was prepared by dropwise adding ligand 2 to an alcohol solution of $MnCl_2$; the yield was 34%. The pale rose crystals are soluble in water and alcohol, but are insoluble in other organic solvents. Characteristic bands (in cm^{-1}) in the IR spectrum of the complex in KBr or vaseline mulls were assigned as follows: 3253, $\nu_{as}(NH_2)$; 3130, $\nu_s(NH_2)$; 1613, 1578, $\delta(NH_2)$; 1030, $\nu_{as}(CON)$, 872, $\nu_s(CON)$. Coordination of the ligand through the N atom is inferred from a comparison of the IR spectrum with those of the Ni, Co, and Pt complexes, for which an analysis of the normal vibrations had indicated N-coordination [4].

References:

[1] Sarukhanov, M. A., Val'dman, S. S., Parpiev, N. A. (Zh. Neorgan. Khim. **18** [1973] 838/9; Russ. J. Inorg. Chem. **18** [1973] 439/40).
[2] Salyn', Ya. V., Nefedov, V. I., Sarukhanov, M. A., Kharitonov, Yu. Ya. (Koord. Khim. **1** [1975] 945/9; Soviet J. Coord. Chem. **1** [1975] 802/5).
[3] Sarukhanov, M. A., Val'dman, S. S., Parpiev, N. A. (Zh. Neorgan. Khim. **17** [1972] 2584/5; Russ. J. Inorg. Chem. **17** [1972] 1355).
[4] Sarukhanov, M. A., Val'dman, S. S., Parpiev, N. A. (Uzb. Khim. Zh. **19** No. 6 [1975] 11/4).

27 Complexes with Oximes and Nitroso Compounds

Survey

This section deals mainly with manganese(II) complexes of chelating oximes containing complex-forming groups in addition to the hydroxyimino group. Only very few complexes are known with nonchelating aldehyde or ketone oximes. In solution, 1:1 and 1:2 complexes are observed. Because of the low solubility of the chelates in aqueous medium, studies were performed in aqueous-organic media, mainly in aqueous dioxane. The necessary pH corrections (not always mentioned) and in some cases corrections to zero ionic strength (0 corr) were performed by use of an empirical formula [1]. In many cases, the intensely colored solutions are suitable for the spectrophotometric determination of manganese in low concentrations.

Solid complexes with the deprotonated ligands are most frequently of the type $MnL_2 \cdot nH_2O$ or $Mn(HL)_2 \cdot nH_2O$. Complexes of the type $MnL \cdot nY$ with additional neutral ligands are also known, as well as $Mn(HL)_2X_2$ compounds with the nondeprotonated ligands. In general, solid compounds were prepared in aqueous alcohol.

In the IR spectrum, shifts of the free ligand ν(C=N) and ν(NO) bands indicate coordination of the hydroxyimino nitrogen atom for most of the complexes with oximes. Additional coordination of the hydroxyimino oxygen atom is suggested, e.g., for polymeric complexes with 2-hydroxybenzaldehyde oxime. Bidentate coordination only through the oxygen atoms is observed for the complexes with N-nitroso-N-phenylhydroxylamine (NH_4 salt cupferron). On the whole, the IR spectra are complicated, so that different assignments for the pertinent bands are given in the various references.

Anhydrous complexes $Mn(HL)_2$, particularly those with hydroxyaldehyde or hydroxyketone oximes, are in most cases *trans*-square planar. Their structure (see p. 235) comprises four chelate rings, two of them formed by H bonds between donor O atoms of the two coordinated ligands. For these complexes, magnetic moments suggest partial spin-pairing, i.e. $S = 3/2$ for the Mn atom. The complexes $Mn(HL)_2 \cdot 2H_2O$ and $MnL_2 \cdot 2H_2O$ have the usual monomeric octahedral structures known for Mn^{II} complexes. A low-spin configuration was revealed by the magnetic moment for the tetrabutylammonium salt of the anionic complex with ethylnitrosolic acid (see p. 276).

Most of the manganese(II) complexes are insoluble or sparingly soluble in water, but more soluble in organic solvents. They are nonelectrolytes in polar organic solvents.

Complex formation of manganese(II) with nonchelating oximes and carboxylic acids presumably occurs in the solvent extraction using mixtures of these agents. Nonchelating oximes utilized were, e.g.: octanal, 2-ethylhexanal, cyclohexanone, or acetophenone oxime. Carboxylic acids utilized were, e.g.: naphthenic, 3,5-diisopropylsalicyclic, or decanoic acid. A significant synergistic effect was observed in the extraction of transition metals and the depression of alkali earth metals using the above mixtures [2].

Mn^{III} and Mn^{IV} complexes are of less importance than the Mn^{II} complexes. However, a very stable brown-red Mn^{IV} complex with deprotonated formaldehyde oxime, $Mn(CH_2NO)_6^{2-}$, exists. It is formed by oxidation and depolymerization of an unstable 1:2 Mn^{II} complex with the cyclic trimer of formaldehyde oxime. The Mn^{IV} complex was extensively studied because of its applicability to the spectrophotometric determination of manganese in low concentrations in the presence of other metals.

References:

[1] van Uitert, C. G., Haas, C. G. (J. Am. Chem. Soc. **75** [1953] 451/5).
[2] Preston, J. S. (MINTEK M 234 [1985] 1/16 from C.A. **105** [1986] 27605).

27.1 Complexes with Monoximes

27.1.1 With Formaldehyde Oxime $H_2C{=}NOH$ (= CH_3NO) or Its Cyclic Trimer, 1,3,5-Trihydroxyhexahydro-1,3,5-triazine (= $C_3H_9N_3O_3$)

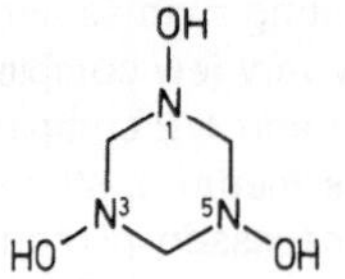

Complexes in Aqueous Solution. Mn^{2+} ions form colorless complexes with the trimeric ligand (but not with the monomer) in neutral or alkaline aqueous solutions under exclusion of air, as shown by polarographic [5], conductometric [10], electrophoretic [3, 4], and ^1H NMR studies [14]. While the formation of a species $Mn(C_3H_6N_3O_3)^-$ in the pH range 5.5 to 8.75 is assumed on the basis of an electrophoretic study (air exclusion not mentioned) [3, 4], polarographic [5] and conductometric studies [10] indicate the species $Mn(C_3H_9N_3O_3)_2^{2+}$, $Mn(C_3H_8N_3O_3)_2$, and $Mn(C_3H_7N_3O_3)_2^{2-}$. The latter two are formed by deprotonation of the trimeric ligand with increasing pH value [5, 10]. The course of the polarographic curves indicates additional complexes of lower stability and a complicated system in the solution [5], see also [3, 4, 18]. The stability constant log K_1 = 20.7 was calculated for the $Mn(C_3H_6N_3O_3)^-$ ion from the electrophoretic results [4]. Magnetic measurements on complex solutions yielded the effective magnetic moment 5.49 μ_B confirming the divalency of the Mn atom [7]. Whereas previous authors assumed that the reactive ligand is triformaldehyde trioxime in the chain form [2, 9, 10], recent work suggests that these complexes contain the hexahydrotriazine ring system in the tub form [14, 18, 22].

^1H NMR studies show that, in aqueous solutions of formaldehyde oxime or its trimer, an equilibrium exists between these two forms. The equilibrium is shifted with decreasing concentration, increasing temperature, and increasing pH to the monomer side. However, the depolymerization reaction of the trimer is slow at pH values >4 [14].

The Mn^{II} complexes in alkaline solutions of pH>9 are readily oxidized by oxygen in the air to give the intensely brown-red colored Mn^{IV} complex $Mn(CH_2NO)_6^{2-}$, e.g., [1, 2, 21]. Composition of this complex and the oxidation state of the Mn atom have been proved by numerous studies, for example spectrophotometric and ion exchange, e.g., [1, 2], conductometric [10], polarographic [5], and electrophoretic investigations [3, 4, 10]. A stability constant of log K_2=9.8 was calculated from the electrophoretic results [4]. The effective magnetic moment of the complex is 3.82 μ_B (spin-only value for a d^3 complex 3.87 μ_B) [7]. The electronic absorption spectrum shows a broad maximum at 455 nm with ε=11200 $L \cdot mol^{-1} \cdot cm^{-1}$ [1, 6, 11, 15]. On the basis of IR studies (p. 233) and previous work, particularly [10], it is assumed that the ligand is coordinated as in the solid complex, i.e., in its monomeric form and through the N atoms [18].

$Mn(CH_2NO)_6^{2-}$ is rather stable in solutions with pH>12, decomposes only slightly at pH 10 to 12, and is totally decomposed at pH 6 [9]. Strong complex formers, e.g., citrate, tartrate, cyanide, and ethylenediaminetetraacetate, in high excess, do not decompose the complex [11]. It is also stable against reductants, as are ascorbic acid, hydroxylamine, and sulfite [11, 16], but it is decomposed by special oxidants, e.g., hypoiodite [9]. In weakly alkaline solution (0.04 N NaOH) it can be heated at 90°C for 15 min without losing any of its color intensity [11, 16]. The intense color reaction observed on formation of the $Mn(CH_2NO)_6^{2-}$ complex was used for the colorimetric or spectrophotometric determination of manganese in low concentration in the presence of other metals [8 to 12, 16, 17, 19, 20]. A review on the older literature (up to 1960) about this color reaction is given in [16]. A spot test for the analytical detection of formaldehyde in the presence of hydroxylamine and Mn^{2+} ions is based on this reaction [13].

Isolated Compounds. To prepare the Mn^{IV} complex $\mathbf{Na_2Mn^{IV}(CH_2NO)_6}$, a concentrated aqueous solution of sodium methanolate was mixed with $C_3H_9N_3O_3 \cdot HCl$ and aqueous $Mn(CH_3COO)_2 \cdot 4H_2O$ and finally diluted with a large quantity of methanol. The solution became dark by air oxidation, and the compound separated in the form of small, dark violet hygroscopic platelets within 16 h [7].

The IR spectrum of $Na_2Mn(CH_2NO)_6$ was recorded from KCl, KBr, CsCl, or CsBr disks (4000 to 200 cm^{-1}) or from adamantane or polyethylene pellets (400 to 33 cm^{-1}). The important bands were assigned as follows (assignment of the infrared active fundamentals, 15 of species A_u and 15 of species E_u, was based mainly on isotopic frequency shifts (H/D, $^{14}N/^{15}N$) assuming the complex ion to have S_6 symmetry) [18]:

$\bar{\nu}_{max}$ in cm^{-1}	1422 m, 1414 sh	1161 s, 1154 sh	971 m, br	946 s, br, 934 sh
assignment	CH_2 bending	NO stretching	CH_2 rocking	CH_2 wagging
$\bar{\nu}_{max}$ in cm^{-1}	798 m	745 w, 647 s	527 vs, br, 510 sh	450 s
assignment	CNO bending	CNO wagging	Mn–N stretching	CNO rocking

Six bands between 387 and 125 cm^{-1} include skeletal deformation and lattice vibration bands. The IR results indicate that in $Mn(CH_2NO)_6^{2-}$, the formaldehyde oxime is coordinated in its monomeric form through the N atom. The IR spectrum is similar to the spectra of the Fe and Ni compounds. This indicates that the molecular structures of the complex ions are similar [18].

A complex of composition $\mathbf{Mn(CH_2NO)_3 \cdot 2H_2O}$ with Mn in the alleged three oxidation state was prepared by reaction of manganese(II) acetate with triformaldehyde trioxime in methanol as well. It formed black-brown rectangular platelets and deflagrated at temperatures above 220°C [21]. However, a manganese(III) complex with formaldehyde oxime could not be confirmed by polarographic studies [5].

References:

[1] Marczenko, Z., Minczewski, J. (Zh. Analit. Khim. **17** [1962] 23/7; J. Anal. Chem. [USSR] **17** [1962] 19/24, 21).
[2] Okáč, A. (Theory Struct. Complex Compounds Papers Symp., Wroclaw 1962 [1964], pp. 167/80, 175, 177; C.A. **63** [1965] 12654).
[3] Bečka, J., Jokl, J. (Collection Czech. Chem. Commun. **36** [1971] 2467/73, 2469).
[4] Bečka, J., Jokl, V. (Collection Czech. Chem. Commun. **36** [1971] 3263/74, 3272).
[5] Bartušek, M., Okáč, A. (Collection Czech. Chem. Commun. **26** [1961] 52/8, 54, 56).
[6] Marczenko, Z., Minczewski, J. (Roczniki Chem. **35** [1961] 1223/35, 1224, 1226; C.A. **56** [1962] 15137).
[7] Bartušek, M., Okáč, A. (Collection Czech. Chem. Commun. **26** [1961] 883/7).
[8] Bernal Nievas, J., Aznárez Alduán, J. (Rev. Acad. Cienc. Exact. Fis. Quim. Nat. Zaragoza [2] **26** [1971] 403/6; C.A. **76** [1972] No. 41671).
[9] Okáč, A., Bartušek, M. (Z. Anal. Chem. **178** [1960] 198/201).
[10] Bartušek, M., Okáč, A. (Collection Czech. Chem. Commun. **26** [1961] 2174/88, 2188).

[11] Marczenko, Z. (Anal. Chim. Acta **31** [1964] 224/32, 225).
[12] Dick, T. A. (Methods Exam. Waters Assoc. Mater. **1977** 1/15, 5; C.A. **90** [1979] No. 28789).
[13] Jungreis, E. (Chemist-Analyst **49** [1960] 14).
[14] Jensen, K. A., Holm, A. (Mat. Fys. Medd. Kgl. Danske Videnskab. Selskab. **40** No. 1 [1978] 1/23, 4, 9/12, 19/20, 22; C.A. **89** [1978] No. 110480).
[15] Marczenko, Z. (Bull. Soc. Chim. France **1964** 939/44).
[16] Marczenko, Z. (Chem. Anal. [Warsaw] **6** [1961] 477/88).

[17] Marczenko, Z. (Chem. Anal. [Warsaw] **5** [1960] 747/62).
[18] Andersen, F. A., Jensen, K. A. (J. Mol. Struct. **79** [1982] 357/60).
[19] Denigès, G. (Compt. Rend. **194** [1932] 895/7).
[20] Gottlieb, A., Hecht, F. (Mikrochem. Ver. Mikrochim. Acta **35** [1950] 337/45).

[21] Hofmann, K. A., Ehrhardt, U. (Ber. Deut. Chem. Ges. **46** [1913] 1457/66, 1464).
[22] Andersen, F. A., Jensen, K. A. (J. Mol. Struct. **60** [1980] 165/71).

27.1.2 With Butyraldehyde Oxime $CH_3CH_2CH_2CH{=}NOH$ (= C_4H_9NO)

$Mn(C_4H_9NO)_4Cl_2$. To prepare the compound, dried $MnCl_2$ was added in small portions to the ligand, and the reaction mixture was kept at 5°C. After all the chloride had been dissolved by warming the mixture to 25°C, it was filtered and left at 5°C or below. Colorless rather labile crystals separated which were collected, dried at $<10^{-2}$ Torr, and recrystallized from ethanol, butanol, or petroleum ether; m.p. 54°C. The IR spectrum of the compound (in KBr disks) shows the following characteristic bands (in cm^{-1}; shift with respect to the liquid ligand in parentheses): $\nu(OH)_{assoc}$, 3330 (−20); ν(C=N), 1661 (+11), 1613; δ(OH), 1429; γ(OH) or ν(NO), 931. The IR spectrum indicates that the ligand is coordinated in the form of a six-membered cyclic dimer containing two intramolecular hydrogen bonds OH····N. A structure is possible in which the four O atoms of the two oxime dimers are equatorially coordinated and four five-membered chelate rings are formed (the Cl atoms being axially coordinated), or a structure in which the $MnCl_2$ is sandwiched by the two cyclic dimers.

The compound is insoluble in water, and easily decomposed by it. $Mn(C_4H_9NO)_4Cl_2$ dissolves easily in ethanol as a nonelectrolyte, but the solution is not stable. The complex reacts with solvents which contain chlorine, for example with $CHCl_3$, $CHCl_2CHCl_2$, and CCl_4.

Reference:

Masui, M., Hotta, K. (Chem. Pharm. Bull. [Tokyo] **12** [1964] 564/9, 565; C.A. **61** [1964] 3901).

27.1.3 With the Polymer of 2-Propenal Oxime $\{CH_2CH(CH{=}NOH)\}_n$ (= Acroleine Oxime = $[C_3H_5NO]_n$)

The stability constant of $[Mn(C_3H_4NO)_2]_n$, which was formed over a wide range of pH, has been determined by pH titration to be log $\beta_2 = 11.6$ at 25°C and I = 0.1 M, Muto, N., Komatsu, T., Nakagawa, T. (Nippon Kagaku Zasshi **92** [1971] 43/6 from C.A. **75** [1971] No. 21372).

27.1.4 With Oximes Derived from Hydroxyaldehydes and Hydroxyketones

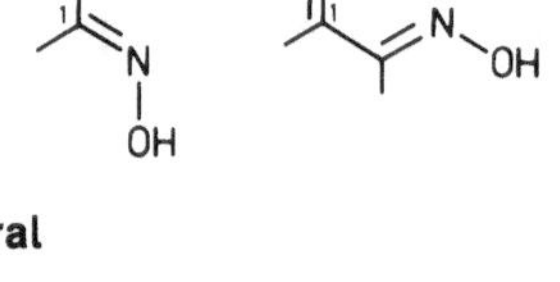

General

These ligands having a hydroxy group in 2-position with respect to the hydroxyimino or hydroxyiminomethyl group are weak dibasic acids H_2L. But on complexation, in most cases only the phenolic or alcoholic hydroxy group is deprotonated ($pK_{a1} \sim 9$ to 12 in 75% aqueous dioxane; $pK_{a2} > 12$).

In aqueous-organic solution, the species $Mn(HL)^+$ and $Mn(HL)_2$ exist. Intensely yellow or brown colored complex solutions are suitable for the spectrophotometric determination of manganese in low concentrations.

Solid compounds of the type $Mn^{II}(HL)_2 \cdot nH_2O$, $Mn^{III}(HL)_3$ and $Mn^{II}L \cdot 2H_2O$ have been prepared in aqueous ethanol. Pyridine adducts $Mn^{II}L \cdot 2py$ were obtained by refluxing the hydrates with pyridine. In the IR spectra, bands associated with the intramolecular hydrogen bond OH···N for the free ligands have vanished, indicating deprotonation of the 2-hydroxy group. Coordination of the phenolate oxygen is assumed. Shifts of the ν(C=N) bands to lower wave numbers and of the ν(N–O) bands to higher wave numbers indicate coordination of the hydroxyimino nitrogen, with the formation of five- or six-membered chelate rings. The spectra of the complexes $Mn(HL)_2 \cdot nH_2O$ are consistent with the formation of new hydrogen bonds, as shown below. The green complexes $Mn^{II}(HL)_2$ are trans-square planar, the dark brown complex $Mn^{II}(HL)_2$ with a derivative of pentyl 2-hydroxyphenyl ketone oxime is tetrahedral, while the brown complexes $Mn^{II}(HL)_2 \cdot 2H_2O$ and $Mn^{III}(HL)_3$ are monomeric octahedral. For the complexes $Mn^{II}L \cdot 2H_2O$ and $Mn^{II}L \cdot 2py$, a two-dimensional polymeric structure with bridging hydroxyimino and phenolic oxygen atoms is assumed, the water or pyridine molecules completing the octahedral coordination sphere (as shown below for the pyridine adduct of the complex with salicylaldehyde oxime). The electronic spectra, reported only for some complexes with alkyl 2-hydroxyphenyl ketone oximes, are consistent with these structures.

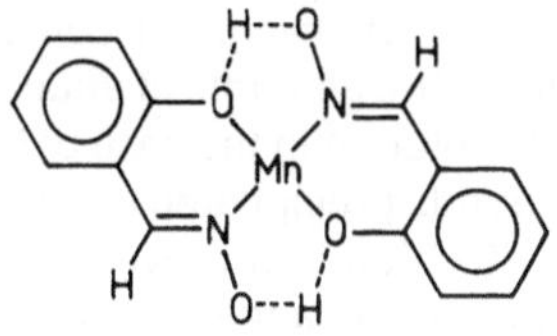

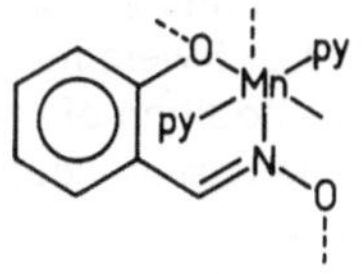

Most of the compounds are stable to air. They are insoluble in water and are nonelectrolytes in DMF or nitrobenzene.

27.1.4.1 With 2-Hydroxybenzaldehyde Oxime (= Salicylaldehyde Oxime) or Derivatives

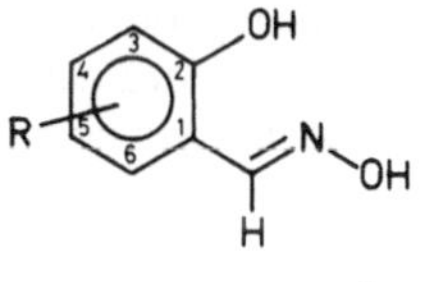

No.	1	2	3	4	5
R	H	5-CH_3	5-NO_2	5-Cl	3-OCH_3
ligand (= H_2L)	$C_7H_7NO_2$	$C_8H_9NO_2$	$C_7H_6N_2O_4$	$C_7H_6ClNO_2$	$C_8H_9NO_3$

Complexes in Solution. The stability constants of the species $Mn(HL)^+$ and $Mn(HL)_2$, $K = [Mn(HL)^+]/[Mn^{2+}][HL^-]$ and $K' = [Mn(HL)_2]/[Mn(HL)^+][HL^-]$, have been determined potentiometrically with a glass electrode in 75 vol% aqueous dioxane at 20°C and I = 0.1 M ($NaClO_4$) under hydrogen [1, 2]:

R	H	5-CH_3	5-NO_2	5-Cl
log K	5.8 ± 0.2	6.14 ± 0.2	4.42 ± 0.05	4.8 ± 0.1
log K'	6.1 ± 0.2	6.14 ± 0.2	3.90 ± 0.05	5.7 ± 0.1

For the complex with ligand 1, $Mn(C_7H_6NO_2)^+$, log K = 3.01 was determined in 10 vol% aqueous dioxane at 26°C by conductometric and spectrophotometric measurements [4]. For the same complex in butanol, log K = 5.36 was determined by spectrophotometry [3]. Only one complex was found by [3, 4], in contrast to the results of [1, 2].

The unusual relationship $K' \geqq K$ can be explained by the formation of hydrogen bonds between the hydroxyimino and the phenolate oxygen atoms (see structure on p. 235). These hydrogen bridges particularly stabilize the 1:2 complexes [1, 2]. Electronic spectra and magnetic properties of 1:2 complexes in aqueous solution are given on p. 237.

The complex with ligand 1, $Mn(C_7H_6NO_2)^+$, is quantitatively extractable into butanol from aqueous alkaline solutions (optimum pH 9.2) [3, 5]. The yellowish brown [5] butanolic layer has an absorption maxium at 420 nm ($\varepsilon = 3140 \pm 15\ L \cdot mol^{-1} \cdot cm^{-1}$) [3]. Although the butanolic solution is unstable and its color deepens slowly with time, the analytical determination of Mn^{II} is possible if the photometric measurements are completed within 30 min. The complex can also be extracted quantitatively with 5% pyridine-toluene at pH 8 to 10 (but not with toluene alone). The yellow solution ($\lambda_{max} = 420$ nm at pH 10) is stable for several weeks [5]. In a previous publication, extraction of 70% Mn from aqueous solution into a solution of ligand 1 in benzene at pH 10.5 is reported [6]. The complex is also extractable from aqueous solution into chloroform (brown solution) [7].

$Mn(C_7H_6NO_2)_2$ and **$Mn(C_8H_8NO_3)_2$**. For preparation of the green compounds, ethanolic solutions of ligands 1 or 5 were mixed with stoichiometric amounts of aqueous manganese(II) acetate. On dropwise addition of sodium hydroxide solution in the case of ligand 1 or sodium acetate solution in the case of ligand 5, the compounds precipitated. They were washed with water, recrystallized from ethanol and dried at ~60°C [8]. In a previous publication, a light green product of the same composition was obtained by adding aqueous salicylaldehyde oxime in slight excess and then ammonia to an aqueous Mn^{II} salt [9]. In aqueous solutions containing Mn^{2+} and $C_7H_6NO_2^-$ ions, a green-brown precipitate was observed at pH 9 [10, 11].

The room-temperature magnetic moments in the solid state in ethanol or pyridine are 4.51, 4.55, or 5.73 μ_B for the complex with ligand 1, and 4.35, 4.23, or 5.95, respectively, for that with ligand 5. The values of the compounds in the solid state and in ethanol suggest partial spin pairing, i.e., $S = 3/2$ for the Mn atom, while the values for the complexes in pyridine are those of high-spin ($S = 5/2$) complexes (spin-only value 5.92), indicating axial coordination of two pyridine molecules. The higher experimental values of the former complexes compared with the spin-only value of 3.87 μ_B may be explained by an orbital contribution introduced by spin-orbit coupling. These findings are in contradiction to the ESR spectra of polycrystalline samples which give broad signals around g = 2.00, indicating that there is little orbital contribution. The ESR spectra of the complexes in ethanol, chloroform, benzene, or pyridine show extremely broad, almost vanishing signals. A trans-square planar geometry of the chelate compounds with hydrogen bonds between the hydroxyimino and phenolate oxygens forming two additional chelate rings (see the left structure on p. 235) is proposed [8]. Characteristic Ir bands of $Mn(C_7H_6NO_2)_2$ were assigned as follows (in Nujol; bands in cm^{-1}; assignments by use of deuterated free ligand 1 and the deuterated Ni complex with ligand 1; shifts in parentheses): δ(OH), 1633(+10); ν(C=N), 1553(−24); ν(NO), 1202; ν(CO), 1021(+31); ν(NO), 912. The assignment of the ν(NO) bands is tentative only [12].

The complex with ligand 1 is unstable against atmospheric oxygen. Especially in the presence of traces of NH_3, it is oxidized by air to give a brown product [9]. Both complexes are insoluble in water but soluble in all common organic solvents [8] and in excess aqueous sodium hydroxide or ammonia [9 to 11].

Na[Mn(HL)L]. Compounds with ligands 1 to 4 (ligand Nos. see p. 235), were prepared by mixing an ethanolic solution of the ligand with an aqueous Mn^{II} salt solution, adjusting the pH value to about 8, and then diluting with water. The precipitates formed were centrifuged, washed with water, and dried at 100°C in vacuum. Bands in the IR spectra of the compounds in KBr disks or hexachlorobutadiene were assigned as follows (wave numbers in cm^{-1}):

No.	R	complex	ν(OH)	ν(C=N)	bands of the conjugated system
1	H	$Na[Mn(C_7H_6NO_2)(C_7H_5NO_2)]$	~3200	1625	1601, 1586, 1560, 1545
2	5-CH_3	$Na[Mn(C_8H_8NO_2)(C_8H_7NO_2)]$	~3250	1632	1618, 1582, 1550
3	5-NO_2	$Na[Mn(C_7H_5N_2O_4)(C_7H_4N_2O_4)]$	3300	1629	(2555), 1605, 1560
4	5-Cl	$Na[Mn(C_7H_5ClNO_2)(C_7H_4ClNO_2)]$	3250	1629	1600, 1585, 1540

Compared with the free ligands, the ν(OH) band has shifted to higher wave numbers, while the ν(C=N) band is not noticeably changed. Extremely broad ν(OH) bands (800 to 1000 cm^{-1}) of low intensity indicate the existence of stable intramolecular hydrogen bonds for all complexes.

The electronic absorption spectrum of a 10^{-4} M aqueous solution of pH 12 of the complex with ligand 1 shows bands at 43500, 37500, 31800, and 28500 cm^{-1}. The solution spectra of the other complexes are similar except for the complex with ligand 3, which exhibits additional bands. A cis-square planar geometry of the chelate compounds is suggested, with formation of one additional chelate ring by a hydrogen bond between the oxygen atoms of the two hydroxyimino groups [13, 14].

$Mn(C_8H_8NO_3)Cl \cdot H_2O$ was prepared by reacting equimolar amounts of $MnCl_2$ and ligand 5 in ethanol. When the pH of the solution was adjusted to 4, the compound separated as a yellow-green precipitate, which was washed with acetone and ether and dried in vacuum.

The magnetic moment is $\mu_{eff} = 6.35\ \mu_B$ indicating five unpaired electrons. The IR spectrum shows that the ν(C=N) band has shifted from ~1630 to 1600 cm^{-1} on complexation. The ν(OH) bands have shifted from 3300 and 3000 cm^{-1} to ~3410 or 3100 cm^{-1}, respectively, while the ν(NO) band (960 to 970 cm^{-1}) has shifted only slightly. The IR results suggest an ionic bond between the Mn atom and the phenolate O atom and a coordinate Mn–N bond to form six-membered chelate rings.

The compound has a high decomposition point (>280°C). It is insoluble in the common organic solvents [15].

$Mn(C_7H_5NO_2) \cdot 2H_2O$ and **$Mn(C_7H_5NO_2) \cdot 2py$.** The brown hydrate was prepared by adding an ethanol solution of $MnCl_2$ to an alkaline aqueous-ethanol solution of ligand 1 in the stoichiometric amount. The separated product was washed with water and ethanol and dried in vacuum. The brown pyridine adduct was obtained by refluxing the hydrate with pyridine for 10 to 30 min. Unreacted pyridine was removed by distillation. The solid obtained was washed with petroleum ether and dried in the air.

The magnetic moments, 4.51 μ_B for the hydrate and 5.96 μ_B for the pyridine adduct, suggest partial spin-pairing with contribution from spin-orbit coupling for the first complex. In the IR spectra, the δ(OH) band at 1618 cm^{-1} for the free ligand is replaced by a weak intensity band indicating that the hydroxyimino group is deprotonated on complexation. A band at 1260 cm^{-1} for the free ligand due to ν(CO) vibrations is shifted to higher energy indicating bonding through phenolate oxygen. A shift to lower energy (15 to 60 cm^{-1}) of the ν(C=N) band (free ligand 1575 cm^{-1}) and a shift to higher energy (100 to 125 cm^{-1}) of the ν(NO) band (free ligand

900 cm^{-1}) indicate bonding through the hydroxyimino nitrogen. The high shift of the latter band is evidently due to the deprotonation of this group. Nonligand bands in the 380 to 320 cm^{-1} or 465 to 270 cm^{-1} regions were assigned to ν(Mn–N) and ν(Mn–O) modes, respectively. Other bands are due to the coordinated water or pyridine molecules. The hydrate splits off the coordinated water at 160 to 190°C. The complexes are insoluble in water and common organic solvents, such as chloroform, ethanol, methanol, acetone, etc. From composition and physicochemical studies, a polymeric structure with bridging oxygen atoms, as shown by the right structure on p. 235, is proposed for both complexes [16].

References:

[1] Burger, K., Egyed, I. (Magy. Kem. Folyoirat **71** [1965] 143/9, 146; C.A. **63** [1965] 2364).
[2] Burger, K., Egyed, I. (J. Inorg. Nucl. Chem. **27** [1965] 2361/70, 2365, 2367).
[3] Sankara Reddy, P. B., Brahmaji Rao, S. (Current Sci. [India] **47** [1978] 84/5).
[4] Kachhawaha, M. S., Bhattacharya, A. K. (Z. Anorg. Allgem. Chem. **325** [1963] 321/4).
[5] Sankara Reddy, P. B., Brahmaji Rao, S. (J. Indian Chem. Soc. **57** [1980] 346/7).
[6] Dahl, I. (Anal. Chim. Acta **41** [1968] 9/14).
[7] Gorbach, G., Pohl, F. (Mikrochim. Acta **1951** 258/67, 264).
[8] Rani, I., Pandeya, K. B., Singh, R. P. (J. Indian Chem. Soc. **58** [1981] 73/4).
[9] Ephraim, F. (Ber. Deut. Chem. Ges. **64** [1931] 1215/8).
[10] Bhaduri, A. S. (Z. Anal. Chem. **151** [1956] 109/18, 111).

[11] Flagg, J. F., Furman, N. H. (Ind. Eng. Chem. Anal. Ed. **12** [1940] 529/31).
[12] Ramaswamy, K. K., Jose, C. I., Sen, D. N. (Indian J. Chem. **5** [1967] 156/9).
[13] Burger, K., Ruff, F., Ruff, I., Egyed, I. (Acta Chim. Acad. Sci. Hung. **46** [1965] 1/21, 7, 10, 18).
[14] Burger, K., Ruff, F., Ruff, I., Egyed, I. (Magy. Kem. Folyoirat **71** [1965] 282/91, 283, 290; C.A. **63** [1965] 14347).
[15] Patil, B. K. (Acta Ciencia Indica Ser. Chem. **7** [1981] 39/43; C.A. **97** [1982] No. 32642).
[16] Aggarwal, R. C., Singh, N. K., Singh, R. P. (Syn. Reactiv. Inorg. Metal-Org. Chem. **14** [1984] 637/50).

27.1.4.2 With 4-[4-Hydroxy-3-(hydroxyiminomethyl)phenylazo]benzenesulfonic Acid

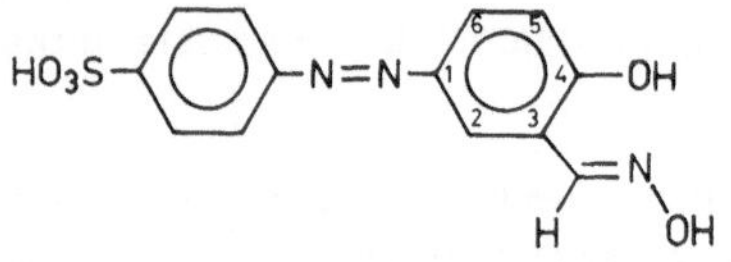

($=C_{13}H_{11}N_3O_5S$)

Manganese(II) forms 1:1 and 1:2 complexes in ethanol-water medium of pH < 7.4 with the monosodium salt of the deprotonated ligand. At pH > 7.4 a precipitate was observed. The stability constants determined potentiometrically (glass electrode) are log $K_1 = 3.52$ and log $K_2 = 3.04$ at 25°C and I = 0.2 M ($NaClO_4$) in 50 vol% ethanol-water, Mohan Das, P. N., Sunar, O. P., Trivedi, C. P. (J. Inorg. Nucl. Chem. **35** [1973] 316/9).

27.1.4.3 With 2-Hydroxy-1-naphthaldehyde Oxime

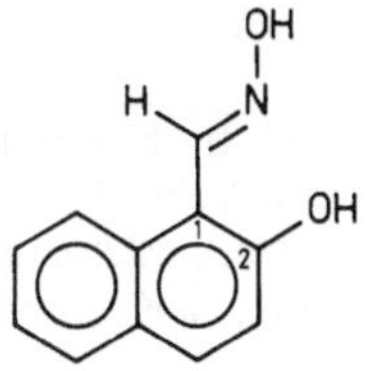

($= C_{11}H_9NO_2$)

Complexes in Solution. The stability constants of the complex species $Mn(C_{11}H_8NO_2)^+$ and $Mn(C_{11}H_8NO_2)_2$ in 50 vol% aqueous dioxane were determined potentiometrically (glass electrode): log $K_1 = 5.55$ and log $K_2 = 5.10$ at 28°C and $I = 0.05$ M (KCl) [1].

$Mn(C_{11}H_8NO_2)_2$ precipitated on mixing ethanol solutions of $MnCl_2$ and the ligand in stoichiometric amounts and adjusting the pH value to ~5.8 (NH_4Cl–NH_4OH buffer). The brown complex was washed with water and ethanol and dried in vacuum [3]. A compound of this composition was also obtained by dehydration of the monohydrate, $Mn(C_{11}H_8NO_2)\cdot H_2O$ [1].

The magnetic moment of the complex, $\mu_{eff} = 6.0\ \mu_B$, is indicative of its high-spin nature. In the IR spectrum, the ν(OH) band at 3580 cm^{-1} for the free ligand has disappeared, and the band of the intramolecular hydrogen bonding at 3310 cm^{-1} has shifted to 3265 to 3100 cm^{-1} on complexation. Other bands (in cm^{-1}) were assigned as follows (shifts in parentheses): δ(OH), 1616 to 1609 (−9 to −16); ν(C=N), 1565 to 1520 (−15 to −60); ν(CO), 1330 to 1285 (+15 to 70); ν(NO), 956 to 945 (+15 to 26). The IR spectrum is consistent with the formation of hydrogen bonds between the phenolate and the hydroxyimino oxygen atoms [3].

$Mn(C_{11}H_8NO_2)_2$ prepared as described above is reported to melt at 290°C [3]. The compound obtained by dehydration of the monohydrate decomposed at ~260°C to form Mn_3O_4 [1]. $Mn(C_{11}H_8NO_2)_2$ is insoluble in water and common organic solvents, such as ethanol, acetone, and chloroform, but is soluble in DMF. The molar conductance in DMF indicates its nonelectrolytic nature in this solvent [3].

$Mn(C_{11}H_8NO_2)_2\cdot H_2O$. The dark green compound precipitated on addition of a ligand solution in ethanol to an Mn^{II} salt solution at pH 8 to 9. The precipitate was washed with hot water and dried in vacuum [1, 2].

The magnetic moment of the hydrate, 4.32 μ_B, suggests a square planar structure of the complex with partial spin pairing. The IR spectrum of the complex in KBr reveals no change of the strong ligand band at 1630 cm^{-1} ascribed to the δ(OH) vibration. The ν(C=N) band (coupled with ν(C=C)) of the free oxime (1590 cm^{-1}) is shifted to 1540 cm^{-1} on complexation. Strong bands at 1250 and 940 cm^{-1} (shifts +15 and +10 cm^{-1}) were assigned to ν(NO) vibrations, and a strong band at 1025 cm^{-1} (shift +20 cm^{-1}) to ν(CO) vibrations. The ν(Mn–N) band was located at 520 cm^{-1}, and the ν(Mn–O) band at 490 cm^{-1}. The IR spectrum gives evidence for the presence of lattice water in the hydrate [2].

The monohydrate is stable in air [1, 2]. On heating, it loses the water molecule at ~200°C to give the anhydrous compound [1]. The complex is insoluble in water and soluble in various organic solvents. It is a nonelectrolyte in nitrobenzene and DMF [2].

$Mn(C_{11}H_7NO_2)\cdot 2H_2O$ and **$Mn(C_{11}H_7NO_2)\cdot 2py$.** The brown compounds were prepared as described on p. 237 for the complexes with salicylaldehyde oxime, but using 2-hydroxy-1-naphthaldehyde oxime as the ligand. They show properties similar to those of the salicylaldehyde oxime complexes [3].

References:

[1] Ramesh, A., Naidu, R. S., Naidu, R. R. (Indian J. Chem. A **19** [1980] 1133/5).
[2] Naidu, R. S., Ramesh, A., Naidu, R. R. (Proc. Indian Acad. Sci. Chem. Sci. **89** [1980] 417/23, 419; C.A. **94** [1981] No. 24233).
[3] Aggarwal, R. C., Singh, N. K., Singh, R. P. (Syn. Reactiv. Inorg. Metal-Org. Chem. **14** [1984] 637/50).

27.1.4.4 With 7-Hydroxy-4-methyl-*2H*-1-benzopyrane-8-carbaldehyde Oxime

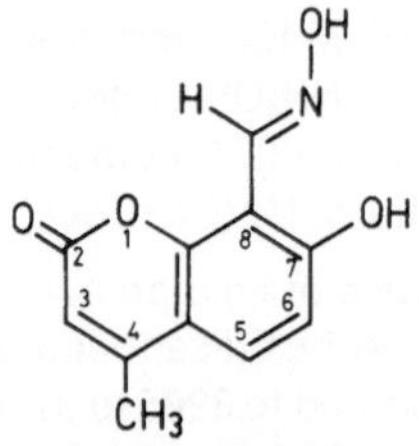

$(=C_{11}H_9NO_4)$

Stability constants log $K_1 = 3.99$ for $Mn(C_{11}H_8NO_4)^+$ and log $K_2 = 2.96$ for $Mn(C_{11}H_8NO_4)_2$ in 70 vol% aqueous methanol were determined potentiometrically (glass electrode) at 35°C and $I = 0.1$ M $NaClO_4$, Charyulu, K. J., Ettaiah, P., Omprakash, K. L., et al. (Indian J. Chem. A **23** [1984] 668/70).

27.1.4.5 With Oximes Derived from Alkyl 2-Hydroxyphenyl Ketones

Manganese(II) Complexes in Aqueous-Organic Solution

Table 6 presents stability constants for the species $Mn(HL)^+$ and $Mn(HL)_2$, $K = [Mn(HL)^+]/[Mn^{2+}][HL^-]$, $K' = [Mn(HL)_2]/[Mn(HL)^+][HL^-]$ and $\beta' = [Mn(HL)_2]/[Mn^{2+}][HL^-]^2$. The stability constants were determined potentiometrically with a glass electrode in the presence of $NaClO_4$. Because of the low solubility of the ligands and/or chelates in water, measurements on the complexes in aqueous-organic solvents were made, especially in dioxane-water.

Table 6

Stability Data for Manganese(II) Complexes with Alkyl 2-Hydroxyphenyl Ketone Oximes in Aqueous Dioxane.

No.	R	R′	R″	formula	dioxane vol%	t in °C	I in mol/L	log K	log K′	log β′	$-\Delta G_{\beta'}$	Ref.
	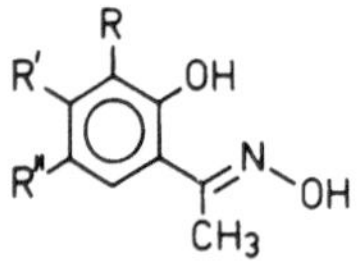											
1	H	H	H	$C_8H_9NO_2$	75	30	~0.1	10.26	7.92	—	—	[1]
					75	30	~0.1	7.57	7.35	—	—	[2]
					50	30	0.1	5.90	—	—	—	[3]
2	CH_3	H	H	$C_9H_{11}NO_2$	75	30	~0.1	8.67	—	—	—	[1]
3	H	H	CH_3	$C_9H_{11}NO_2$	75	30	~0.1	7.86	—	—	—	[1]

Table 6 (continued)

No.	R	R′	R″	formula	dioxane vol%	t in °C	I in mol/L	log K	log K′	log β′	−ΔG$_{β′}$	Ref.
4	CH_3	H	CH_3	$C_{10}H_{13}NO_2$	75$^{a)}$	—	—	—	6.73	—	9.35	[4]
5	H	CH_3	CH_3	$C_{10}H_{13}NO_2$	75	30	0.1	6.39	5.51	—	—	[5]
					75	30	→0	—	—	12.70	17.63	[5]
6	H	OH	H	$C_8H_9NO_3$	60	27	—	6.85	4.85	—	—	[8]
7	H	OCH_3	H	$C_9H_{11}NO_3$	50	28	—	5.67	5.08	—	—	[9]
8	OH	OH	H	$C_8H_9NO_4$	50	—	—	7.00	5.65	—	—	[10]
9	H	H	NO_2	$C_8H_8N_2O_4$	75	30	~0.1	6.17	—	—	—	[1]
10	Cl	H	H	$C_8H_8ClNO_2$	75	30	~0.1	7.78	—	—	—	[1]
11	H	H	Cl	$C_8H_8ClNO_2$	75	30	~0.1	9.06	6.54	—	—	[1]
12	H	CH_3	Cl	$C_9H_{10}ClNO_2$	75	27	0.1	6.47	5.90	—	—	[11]
					75	27	→0	—	—	13.10	18.01	[11]
13	Cl	H	Cl	$C_8H_7Cl_2NO_2$	75	27	0.1	6.30	5.82	—	—	[12]
					75	27	→0	—	—	12.81	17.61	[12]
14	Br	H	H	$C_8H_8BrNO_2$	75	30	~0.1	9.66	—	—	—	[1]
15	H	H	Br	$C_8H_8BrNO_2$	75	30	~0.1	9.30	7.65	—	—	[1]
16	Br	H	CH_3	$C_9H_{10}BrNO_2$	25$^{a)}$	—	0.1	—	7.89	—	—	[6]
17	Br	OH	Br	$C_8H_7Br_2NO_3$	75	27	0.1	4.73	4.42	—	—	[13]
					75	27	→0	5.72	5.46	10.18	15.37$^{b)}$	[13]
18	H	H	I	$C_8H_8INO_2$	75	30	~0.1	8.31	—	—	—	[1]
	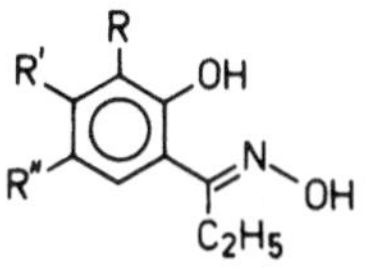											
19	H	H	NO_2	$C_9H_{10}N_2O_4$	75	40	~0.1	5.35^{c}	3.64$^{c)}$	—	—	[20]
20	H	OH	Cl	$C_9H_{10}ClNO_3$	75	27	0.1	4.69$^{c)}$	3.90$^{c)}$	—	—	[21]
					75	27	→0	5.0	4.30	—	12.78$^{d)}$	[21]
21	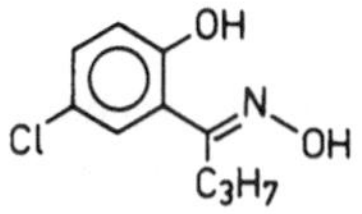			$C_{10}H_{12}ClNO_2$	75	40	0.1$^{e)}$	8.92	—	—	—	[23]
22	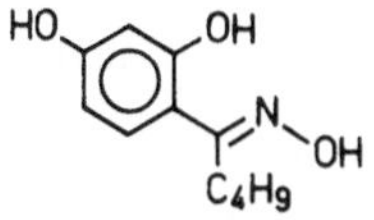			$C_{11}H_{15}NO_3$	75	27	0.1	6.17	5.70	—	—	[24]
					75	27	→0	7.00	6.12	—	17.53$^{f)}$	[24]

$^{a)}$ In ethanol; spectrophotometrically. – $^{b)}$ $ΔG_1 = 7.86$, $ΔG_2 = 7.51$ kcal/mol. – $^{c)}$ Average values from two calculation methods. – $^{d)}$ $ΔG_1 = -6.87$, $ΔG_2 = -5.91$ kcal/mol. – $^{e)}$ 0.1 M KNO_3. – $^{f)}$ $ΔG_1 = -9.12$, $ΔG_2 = -8.41$ kcal/mol.

Other values for stability constants at various ionic strengths of complexes in dioxane-water are reported by [5, 7, 11 to 13, 21, 24], in ethanol-water by [6, 11, 12, 24], in acetone-water by [8 to 12, 24], and in 2-ethoxyethanol-water by [8 to 10]. A 1:3 complex was found for ligand 1 only. The stability constant log $K_3 = 7.13$ at 30°C for $I = \sim 0.1$ M ($NaClO_4$) in 75 vol% aqueous dioxane was determined potentiometrically for this complex [2].

Deeply brown-colored solutions of 1:2 chelates were observed on reaction of Mn^{II} with the following ligands applied in 6- to 40-fold excess in aqueous ethanol at pH 9 to 10.5; absorption maxima and extinction coefficients (in parentheses) are given: ligand 4, 405 (log $\varepsilon = 3.11$) [4]; ligand 17, 395 (3.72) [17]; ligand 20, 405 (3.80) [22]. Procedures for the spectrophotometric determination of manganese in low concentrations based on these color reactions are given in [4, 17, 22]. Absorption maxima of 1:2 chelates in chloroform are reported on p. 243.

Isolated Manganese(II) Compounds

Solid complexes **$Mn(HL)_2$** or **$Mn(HL)_2 \cdot 2H_2O$** with the ligands listed in the table below were obtained by mixing ethanol solutions of the ligands with aqueous solutions of an Mn^{II} salt and adjusting the mixtures to the appropriate pH values [4, 6, 14, 15, 18, 25]. In the cases of ligands 3, 4, and 6, the reaction mixtures were stirred for 5 to 6 h, concentrated under vacuum, and allowed to stand for several days [14, 25]. In the case of ligand 2, the precipitate was digested on a water bath for about 30 min [18]. The compounds were washed with aqueous ethanol and dried in vacuum [14, 25] or in the air at 110 to 120°C [15, 18]. The formation of a greenish yellow precipitate, which is soluble in excess NaOH solution, on addition of excess 2-hydroxyacetophenone oxime to an Mn^{II} salt solution at pH 7.6, was observed in an analytical study [19]. The table lists the pH values of the aqueous-ethanolic preparation media, the effective magnetic moments (in μ_B), and the decomposition points (in °C) of the complexes (ligand numbers refer to those in Table 6, pp. 240/1):

No.	R	R′	R″	formula	pH	complex	color	μ_{eff}	structure	Ref.
	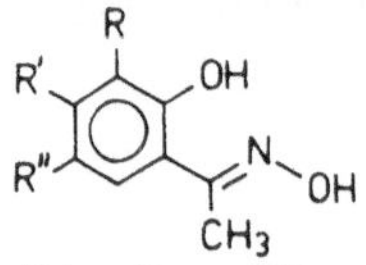									
4	CH_3	H	CH_3	$C_{10}H_{13}NO_2$	9 to 10	$Mn(C_{10}H_{12}NO_2)_2$	—	—	?	[4]
5	H	CH_3	CH_3	$C_{10}H_{13}NO_2$	5 to 6	$Mn(C_{10}H_{12}NO_2)_2$	dark green	4.30 [18]	square planar	[6, 18]
12	H	CH_3	Cl	$C_9H_{10}ClNO_2$	8	$Mn(C_9H_9ClNO_2)_2 \cdot 2H_2O$	dark brown	5.83	octa-hedral	[14]
13	Cl	H	Cl	$C_8H_7Cl_2NO_2$	8	$Mn(C_8H_6Cl_2NO_2)_2 \cdot 2H_2O$	dark brown	5.83	octa-hedral	[14]
16	Br	H	CH_3	$C_9H_{10}BrNO_2$	4 to 5	$Mn(C_9H_9BrNO_2)_2$	dark green	4.40 [15]	square planar	[6, 15]
22	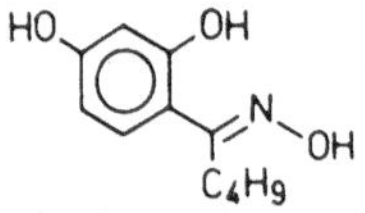			$C_{11}H_{15}NO_3$	7.5 to 8.5	$Mn(C_{11}H_{14}NO_3)_2$	dark brown	5.83	tetra-hedral	[25]

The magnetic moments of the complexes with ligands 12, 13, and 22 indicate high-spin d^5 systems [14, 25], while the low values for the complexes with ligands 5 and 16 indicate partial spin-paired systems ($S = 3/2$) [15, 18]. The ESR spectrum of the complex with ligand 16 was recorded on a sample in polycrystalline form. A broad signal centered around $g = 2$ was observed. This g value indicates that there is little orbital contribution [15].

The IR spectra of the compounds show characteristic maxima at the listed wave numbers (in cm^{-1}). Different assignments are given in the 1630 to 900 cm^{-1} range for the complexes with ligands 5 to 13 or ligands 16 and 22, respectively.

No.	ν(OH)	ν(OH)	ν(CH)	δ(OH)	ν(C=N)	ν(C=C)	ν(NO)	ν(Mn–N)	ν(Mn–O)	Ref.
5	—	—	2950	1625	1555	—	1250	570	525	[18]
12, 13	~3430	~3200	~3080	—	~1535	—	~1250	~510	~460	[14]
16	—	—	—	—	~1610	—	~1030	—	—	[15]
22[a)]	3500	3280	2875	—	1615	1565 1450	980	505[b)]	555[b)]	[25]

[a)] In KBr disks. – [b)] Coupled with ligand bands.

A band at ~1240 cm^{-1} is assigned to (δ)OH vibrations by [15]. The ν(C=N) bands are shifted by 10 to 35 cm^{-1} and the ν(NO) bands by 3 to 10 cm^{-1} to lower energy with respect to free ligand bands [14, 15, 18, 25]. Absorption bands at 3380 cm^{-1} assigned to ν(OH) vibrations of the intramolecularly bonded phenolic OH group disappeared on complexation [18, 25]. The electronic spectra show bands at about 15700 and 20000 cm^{-1} for the complexes with ligands 5 and 16 which were assigned to the $^4A_{1g} \rightarrow {}^3E_g$ and $^4A_{1g} \rightarrow {}^4B_{2g}$ transitions, respectively, in the square planar ligand field of Mn^{II} [15, 18]. Six bands (in cm^{-1}) for the complexes with ligand 12 and 13 were assigned as follows (excited states of the transitions from the ground state $^6A_{1g}$ of Mn^{II} in the octahedral ligand field given): ~19600, $^4T_{1g}$(G); ~22000, $^4T_{2g}$(G); ~24200, ^4E_g(G); ~27300, $^4T_{2g}$(D); ~29000, ^4E_g(D); ~31300, $^4T_{1g}$(P). The crystal field parameters $\Delta = 10\,Dq =$ 9100 and 8900 cm^{-1}, respectively, were calculated [14]. Six bands observed for the complex with ligand 22 as well were assigned to the same transitions, but are shifted to lower wave numbers with respect to the complexes with ligands 12 and 13: 17400, 18975, 20000, 21300, 22600, 27800. Molar extinction coefficients in the ranges between 1 to 10 $L \cdot mol^{-1} \cdot cm^{-1}$ indicate tetrahedral geometry for the complex with ligand 22 [25].

The complexes with ligands 5, 16, and 22 decompose at 210 [18], 260 [15], and 165°C [25], respectively. Those with ligands 5 and 16 can be extracted with chloroform, carbon tetrachloride, ethyl acetate, or other organic solvents. The absorption maxima of the extracts in chloroform are located at 560 nm for both chelates with log $\varepsilon = 3.05$ and 2.78, respectively [6].

The solution of the complex with ligand 16 in dimethylformamide exhibits the molar conductance of a nonelectrolyte [15].

Isolated Manganese(III) Compound

The complex $Mn(C_9H_{10}NO_2)_3$ was obtained by reaction of 2-hydroxy-5-methylacetophenone oxime (= $C_9H_{11}NO_2$) with Mn^{III} at pH 6 to 8. The reaction can be used for the gravimetric determination of Mn^{III}. The brown chelate has the magnetic moment $\mu_{eff} = 4.90\ \mu_B$, which is consistent with an octahedral structure of a $3d^4$ complex. Cryoscopic determination of the molecular weight shows that the complex is monomeric. The compound is quite stable in air. It is insoluble in water, but soluble in organic solvents. It can be extracted with chloroform [16].

References:

[1] Ingle, D. B., Khanolkar, D. D. (Indian J. Chem. A **14** [1976] 596/8).
[2] Kabadi, M. B., Venkatachalam, K. A. (Current Sci. [India] **27** [1958] 337/8).
[3] Unny, V. K. P., Vartak, D. G. (Indian J. Chem. A **21** [1982] 493/7).

[4] Jetley, U. K., Singh, J., Rastogi, S. N. (Acta Ciencia Indica Ser. Chem. **5** [1979] 169/72).
[5] Mittal, M., Lal, K., Gupta, S. P. (Rev. Roumaine Chim. **26** [1981] 1335/40; C.A. **96** [1982] No. 12185).
[6] Mittal, M., Malhotra, S. R., Lal, K., Gupta, S. P. (Natl. Acad. Sci. Letters [India] **2** [1979] 381/2).
[7] Seshagiri, V., Brahmaji Rao, S. (Australian J. Chem. **20** [1967] 2783/4).
[8] Seshagiri, V., Brahmaji Rao, S. (J. Inorg. Nucl. Chem. **36** [1974] 353/6).
[9] Babu, V. S., Raju, D. U., Naidu, R. R. (Indian J. Chem. A **18** [1979] 87/9).
[10] Reddy, K. A., Reddy, Y. K., Rao, S. B. (J. Inorg. Nucl. Chem. **43** [1981] 1933/5).

[11] Lal, K., Gupta, S. P. (Rev. Chim. [Bucharest] **27** [1976] 657/60; C.A. **86** [1977] No. 34864).
[12] Lal, K., Gupta, S. P. (Indian J. Chem. A **14** [1976] 260/3).
[13] Bhuee, G. S., Sharma, K. N., Rastogi, S. N., Srivastava, S. K., Singh, J. (Chem. Era **19** [1983] 73/6; C.A. **100** [1984] No. 216603).
[14] Lal, K., Malhotra, S. R., Singh, J., Gupta, S. P. (Acta Ciencia Indica Ser. Chem. **7** [1981] 161/2).
[15] Lal, K., Malhotra, S. R. (Anales Quim. B **79** [1983] 56/8).
[16] Dave, L. D., Patel, K. R. (Current Sci. [India] **40** [1971] 547).
[17] Bhuee, G. S., Singh, J., Rastogi, S. N. (Chim. Acta Turc. **12** [1984] 133/5).
[18] Mittal, M., Lal, K., Gupta, S. P. (J. Iraqi Chem. Soc. **1982** 81/95).
[19] Poddar, S. N. (Z. Anal. Chem. **154** [1957] 254/9).
[20] Patel, B. H., Shah, J. R., Patel, R. P. (J. Indian Chem. Soc. **52** [1975] 998/9; Current Sci. [India] **47** [1978] 625/6).

[21] Sharma, K. N., Bhuee, G. S., Rastogi, S. N., Singh, J. (Chim. Acta Turc. **10** [1982] 85/92, 88).
[22] Singh, J., Sharma, K. N., Jetley, U. K., Rastogi, S. N., Bhuee, G. S. (Chem. Era **18** [1982] 218/9; C.A. **99** [1983] No. 98456).
[23] Shah, J. R., Patel R. P. (Indian J. Chem. **11** [1973] 607/8).
[24] Singh, J., Lal, K., Gupta, S. P. (Acta Ciencia Indica Ser. Chem. **5** [1979] 20/5).
[25] Singh, J., Lal, K. Gupta, S. P. (South African J. Sci. **75** [1979] 62/5).

27.1.4.6 With 2-Hydroxybenzophenone Oxime and Its Derivatives

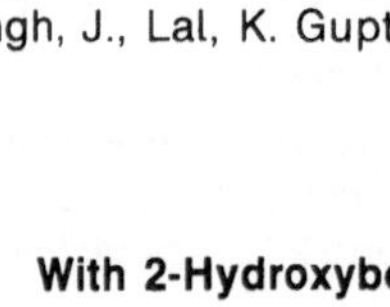

(=HL)

Stability constants of the 1:1 complexes of Mn^{II} with the following ligands in 50 vol% aqueous dioxane were determined at 30°C and I = 0.1 M ($NaClO_4$) potentiometrically with a glass electrode:

No.	R	R'	R''	formula	log K_1 [a]
1	H	H	H	$C_{13}H_{11}NO_2$	5.80
2	H	H	CH_3	$C_{14}H_{13}NO_2$	6.28
3	H	OH	H	$C_{13}H_{11}NO_3$	7.51
4	H	OCH_3	H	$C_{14}H_{13}NO_3$	5.66
5	H	H	NO_2	$C_{13}H_{10}N_2O_4$	4.77
6	H	OH	NO_2	$C_{13}H_{10}N_2O_5$	7.19
7	H	H	Br	$C_{13}H_{10}BrNO_2$	5.09

[a] Accuracy ±0.05 log unit.

A linear relationship between log K_1 and the pK_{a1} value of the ligands (dissociation of the phenolic proton) is approximately fulfilled (except for ligands 2 and 6), Unny, V. K. P., Vartak, D. G. (Indian J. Chem. A **21** [1982] 493/7).

27.1.4.7 With 1-Hydroxy-2-propionaphthone Oxime

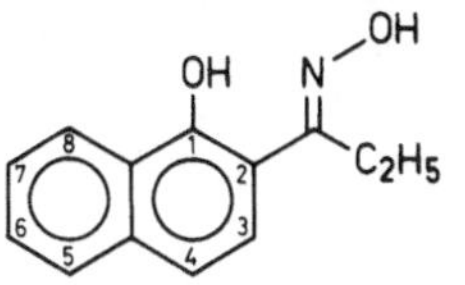

$(=C_{13}H_{13}NO_2)$

$Mn(C_{13}H_{12}NO_2)_2$. The dark green compound was prepared by adding an ethanolic solution of the oxime in excess to an aqueous Mn^{II} salt solution and adjusting the pH to between 8.0 and 8.5. The precipitate was washed with 50% aqueous ethanol and dried at 105 to 110°C. The quantitative precipitation in this pH range allows its application for gravimetric determination of Mn^{II}. The compound is stable to heat up to 140°C. It is soluble in dioxane, and sparingly soluble in ethanol, diethyl ether, acetone, chloroform, and benzene, Patkar, D. N., Merchant, R. N. (J. Indian Chem. Soc. **56** [1979] 194/5).

27.1.4.8 With 2-Hydroxy-3-methyl-1,4-naphthoquinone 1-Oxime and Its *O*-Acetyl Derivative

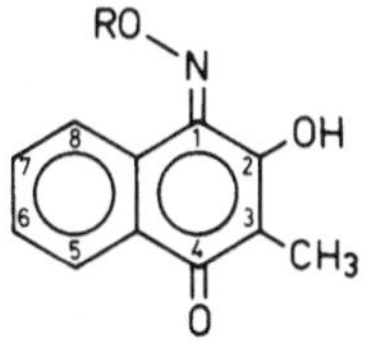

ligand 1 R = H; $(=C_{11}H_9NO_3)$
ligand 2 R = CH_3CO; $(=C_{13}H_{11}NO_4)$

The stability constant of $Mn(C_{11}H_8NO_3)^+$ was determined potentiometrically (glass electrode) in 75 vol% aqueous dioxane at 30°C with $NaClO_4$ as a supporting electrolyte (ionic strength in parentheses): log K_1 = 3.51(0.2); 3.70(0.1); 3.92(0.01); 4.34(0.005) [1].

$Mn^{III}(C_{11}H_8NO_3)_3$ was prepared by mixing ligand 1 in methanol with an aqueous solution of Mn^{II} acetate (?) in stoichiometric amounts and adjusting the pH to between 6 and 7. On cooling for several hours, dark green crystals separated. The yield increased from 15 to 54% when a mixture of $KMnO_4$ and $MnSO_4$ was used. The precipitate was washed with water and ether and dried under vacuum at 70°C [2].

The magnetic moment of the compound, $\mu_{eff} = 4.90\ \mu_B$ at room temperature, indicates the presence of Mn^{III}. The IR spectrum (Nujol and hexachlorobutadiene mulls) shows the decrease of the ν(C=N) frequency from 1583 to 1575 cm^{-1} and upward shifts of the ν(NO) frequencies from 1053 to 1099 and 925 to 943 cm^{-1}. A shift of the ν(CO) band from 1210 to 1235 cm^{-1} suggests partial double bond character of the C–O bond owing to the enhanced conjugation in the chelate ring by the establishment of a donor π-bond. The electronic spectrum shows a charge-transfer band at ~11490 to 12740 cm^{-1} (eg* → π*, parity forbidden), which causes the dark green color, and a strong band in the region 18180 to 21370 cm^{-1} corresponding to the $^5E_g \rightarrow {}^5T_{2g}$ transition for Mn^{III} in octahedral environment. The complex melts at 240°C with decomposition [2].

$Mn^{II}(C_{13}H_{10}NO_4)_2 \cdot 2H_2O$. The orange-red manganese(II) complex with the acetylated oxime was obtained by mixing deaerated solutions of aqueous Mn^{II} acetate and methanolic ligand 2 in appropriate amounts under inert atmosphere and adjusting the pH to 4. The precipitate was washed with water and ether and dried under vacuum at 70°C.

The magnetic moment, $\mu_{eff} = 5.80\,\mu_B$, is typical for a high-spin Mn^{II} complex. The IR spectrum shows the shift of the ν(NO) band from 1087 to 1031 cm^{-1} on complexation, while the ν(C=N) band at 1580 cm^{-1} is not affected. Coordination through the hydroxyimino oxygen is therefore assumed. The two coordinated water molecules give rise to a broad absorption band between 3300 and 3100 cm^{-1}. On heating, the two water molecules are lost along with one acetyl group at about 180°C. At 180 to 240°C total decomposition of the chelate takes place [2].

References:

[1] Kamini Sindhwani, S. K., Singh, R. P. (Indian J. Chem. A **20** [1981] 1040/2).
[2] Padhye, S. B., Rane, S. Y., Gupta, S. G. (Inorg. Nucl. Chem. Letters **14** [1978] 83/6).

27.1.4.9 With 5-Hydroxy-3-methyl-1-phenyl-*1H*-pyrazole-4-carbaldehyde Oxime and a Related Ketone Oxime

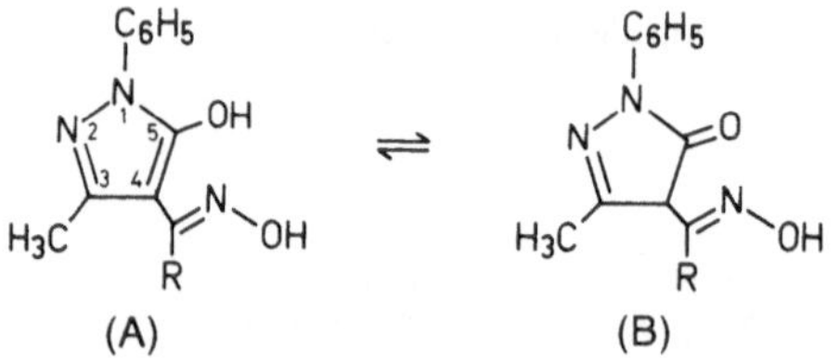

ligand 1 R = H; (= $C_{11}H_{11}N_3O_2$)
ligand 2 R = C_6H_5; (= $C_{17}H_{15}N_3O_2$)

The IR spectra of the free ligands indicate that they exist in their enol forms (A) rather than in their keto forms (B) [1, 3]. (Structure represented for ligand 1 in [1] erroneously given with R = CH_3.)

$Mn(C_{11}H_{10}N_3O_2)_2 \cdot 2H_2O$. The chelate complex with ligand 1 was prepared by adding sodium acetate to a refluxing solution of Mn^{II} nitrate and the ligand (mole ratio 1:2) in absolute ethanol. The separated solid was washed with water, ethanol, and ether [1].

The effective magnetic moment is 6.04 μ_B. The important bands in the IR spectra were assigned as follows (shifts in parentheses): ν(OH), 3400 to 3200; $\nu(C{=}N)_{cycl}$, ~1632; δ(OH), ~1600; $\nu(C{=}N)_{oxime}$, ~1546 (~50); ν(CO), ~1180; ν(NO), ~1016. The diffuse reflectance spectrum shows bands at 17391, 20408, and 25000 cm^{-1} which were assigned to the $^6A_{1g} \rightarrow {}^4T_{1g}$, $\rightarrow {}^4T_{2g}$, and $\rightarrow {}^4E_g$, $^4A_{1g}$ transitions, respectively, for Mn^{II} in the octahedral ligand field. The complex is a nonelectrolyte in DMF [1].

$Mn(C_{17}H_{14}N_3O_2)_2 \cdot 2H_2O$. The chelate complex with ligand 2 was prepared by adding sodium acetate to an ethanol solution of Mn^{II} nitrate and the ligand in the mole ratio 1:2 and digesting the mixture on a water bath for 30 min. The precipitate was filtered and washed with ethanol [2]. In a previous paper, preparation of the pale yellow compound from Mn^{II} acetate and the ligand in ethanol is reported [3].

In the IR spectrum of the complex, the broad free ligand band in the 3200 to 2200 cm^{-1} region which is associated with strong hydrogen bonding has dissappeared. The $\nu(C{=}N)_{oxime}$

band has shifted from ~1530 to ~1490 cm^{-1}, but the ν(NO) band is observed at almost the same position as that of the free ligand, ~1000 cm^{-1}. The diffuse reflectance spectrum of the complex exhibits bands at 11900, 15620, and 21050 cm^{-1}. The effective magnetic moment of the complex is 5.84 μ_B. It behaves as a nonelectrolyte in ethanol [2].

References:

[1] Patel, Y. M., Shah, J. R. (Indian J. Chem. A **24** [1985] 800/2).
[2] Rana, A. K., Shah, J. R. (Indian J. Chem. A **20** [1981] 142/4).
[3] Suenaga, E. (Yakugaku Zasshi **79** [1959] 207/10; C.A. **1959** 13136).

27.1.4.10 With *(Z)*-Furoin Oxime

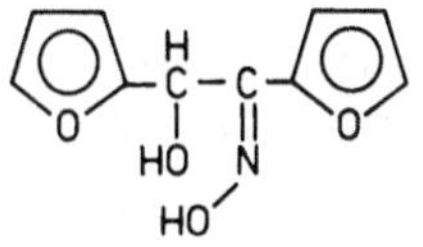

$(= C_{10}H_9NO_4)$

Mn^{2+} ions form a brown precipitate with the ligand in alcoholic solution. A spot test for the detection of microgram amounts of manganese based on this reaction is described by Armeanu, V., Camboli, D. (Rev. Chim. [Bucharest] **10** [1959] 529/30).

27.1.5 With Monooximes Derived from 1,2-Diketones or Polyketones

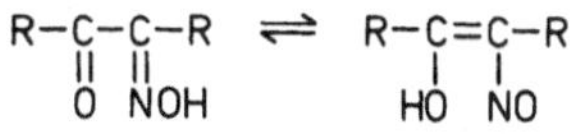

General

Mn^{2+} ions form soluble MnL^+ and MnL_2 complexes with the ligands HL in aqueous-organic medium. Solid compounds of the type $MnL_2 \cdot nH_2O$ with all the ligands HL and of the type $Mn(HL)X_2$ and $Mn(HL)_2X_2$ with benzil monooxime (see p. 249), were prepared by reacting an Mn^{II} salt and the ligand in aqueous methanol or ethanol. A dinuclear Mn^{IV} compound with 1,3-dimethylvioluric acid (see p. 252), was prepared in aqueous solution.

In the IR spectra of the anhydrous complexes MnL_2, vanishing of the free ligand ν(OH) band indicates deprotonation of the hydroxyimino group. For all the complexes, shifts of the ν(CO), ν(C=N), and ν(NO) bands indicate coordination of the carbonyl oxygen and hydroxyimino nitrogen forming five-membered chelate rings. Several authors discuss the conversion of the hydroxyimino ketone form of the free ligands to the nitroso enol form on complexation.

For the complexes MnL_2 with benzil monooxime a tetrahedral structure is assumed, whereas the complexes MnL_2 with the pyrazoline derivatives are presumably polymeric octahedral with one ring nitrogen being involved in complexation.

The complexes are insoluble in water, but soluble in organic solvents. They are nonelectrolytes. Together with other transition metal complexes, they were studied with a view to analytical application.

27.1.5.1 Manganese(II) Complexes in Solution

Ligands:

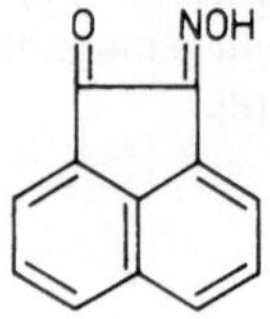

1)

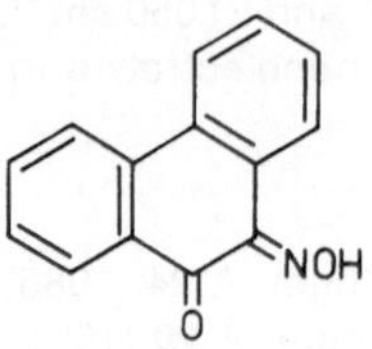

2)

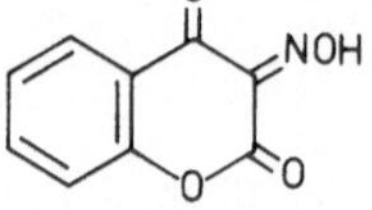

3)

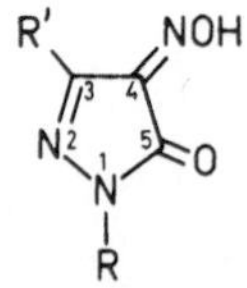

4) $R = H, R' = CH_3$
5) $R = H, R' = C_6H_5$
6) $R = C_6H_5, R' = CH_3$

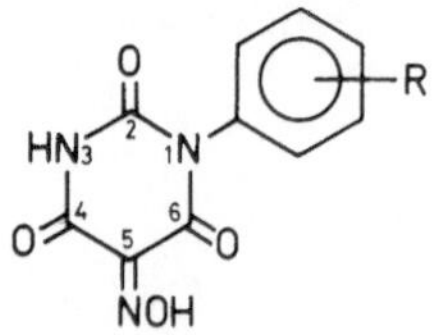

7) R = H
8) $R = 2\text{-}CH_3$
9) $R = 3\text{-}CH_3$
10) $R = 4\text{-}CH_3$

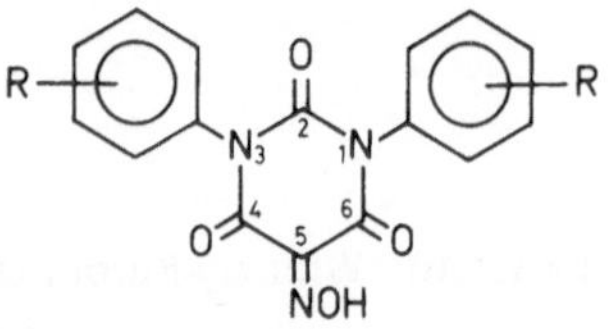

11) R = H
12) $R = 2\text{-}CH_3$
13) $R = 3\text{-}CH_3$
14) $R = 4\text{-}CH_3$

Table 7 lists stability constants of the species MnL^+ and MnL_2 formed by reaction of manganese(II) with the ligands HL presented. Measurements were performed potentiometrically with a glass electrode. The pK_{a1} values given are presumably those of the nitrosoenol forms of the ligands, which are the stronger acids.

Table 7

Stability Constants of Manganese(II) Complexes with Monooximes Derived from 1,2-Diketones in Aqueous-Organic Medium.

No.	formula	organic solvent in vol%	t in °C	I in mol/L	pK_{a1}	log K_1	log K_2	Ref.
1	$C_{12}H_7NO_2$	75% dioxane	30	0.1($NaClO_4$)	9.30	6.00	—	[1]
2	$C_{14}H_9NO_2$	75% dioxane	30	0.1($NaClO_4$)	10.16	5.28	4.04	[2]
3	$C_9H_5NO_4$	40% ethanol	25	0.1($NaClO_4$)	3.50	3.1 a)	—	[3]
4	$C_4H_5N_3O_2$	50% ethanol	20	0.08 KNO_3	6.46	2.33	2.67	[4]
			30		6.36	2.36	2.37	[4]
5	$C_9H_7N_3O_2$	50% ethanol	20	0.08 KNO_3	6.46	2.57	2.38	[4]
			30		6.43	2.87	2.30	[4]
6	$C_{10}H_9N_3O_2$	50% ethanol	20	0.08 KNO_3	6.24	2.50	2.22	[4]
			30		6.14	2.35	2.40	[4]
		75% dioxane	40	0.1($NaClO_4$)	6.98	3.65	—	[7]
7 b)	$C_{10}H_7N_3O_4$	50% ethanol	18	0.1($NaClO_4$)	5.02	5.30	4.48	[5]
			31		4.76	4.95	4.21	[5]
8 b)	$C_{11}H_9N_3O_4$	50% ethanol	18	0.1($NaClO_4$)	4.60	5.08	4.12	[5]
			31		4.42	4.68	3.95	[5]
9 b)	$C_{11}H_9N_3O_4$	50% ethanol	18	0.1($NaClO_4$)	4.65	5.13	4.38	[5]
			31		4.46	4.80	4.12	[5]

Table 7 (continued)

No.	formula	organic solvent in vol%	t in °C	I in mol/L	pK_{a1}	log K_1	log K_2	Ref.
10[b)]	$C_{11}H_9N_3O_4$	50% ethanol	18	0.1($NaClO_4$)	5.34	5.42	4.63	[5]
			31		5.07	5.14	4.46	[5]
11	$C_{16}H_{11}N_3O_4$	75% dioxane	30	0.1($NaClO_4$)	7.48	3.10	2.88	[6]
12	$C_{18}H_{15}N_3O_4$	75% dioxane	30	0.1($NaClO_4$)	6.38	2.75	2.70	[6]
13	$C_{18}H_{15}N_3O_4$	75% dioxane	30	0.1($NaClO_4$)	6.08	2.60	2.70	[6]
14	$C_{18}H_{15}N_3O_4$	75% dioxane	30	0.1($NaClO_4$)	5.56	2.24	2.71	[6]

[a)] By spectrophotometry. – [b)] For these ligands, pK_{a2} values are also given ranging from 10.5 to 11 at 18°C.

Additional values for stability constants at 42°C are reported in [5].

Thermodynamic data of formation are given for the 1:2 complexes with ligands 4 to 6 in 50% ethanol; all data in kcal/mol [4]:

ligand No.	complex	$-\Delta G_{\beta 2}$ 20°C	$-\Delta G_{\beta 2}$ 30°C	$-\Delta H_{\beta 2}$ 20 to 30°C	$T\Delta S$ 20°C	$T\Delta S$ 30°C
4	$Mn(C_4H_4N_3O_2)_2$	6.49	6.63	2.43	4.06	4.20
5	$Mn(C_9H_6N_3O_2)_2$	6.68	6.84	2.02	4.66	4.82
6	$Mn(C_{10}H_8N_3O_2)_2$	6.41	6.56	2.02	4.39	4.54

The blueish green 1:1 complex with ligand 3 shows an absorption maximum at 625 nm with $\varepsilon = 50\ L \cdot mol^{-1} \cdot cm^{-1}$ in 40 vol% aqueous ethanol. In this solution, the complex is stable for ~10 min [3].

References:

[1] Sindhwani, S. K., Singh, R. P. (Indian J. Chem. **9** [1971] 1000/1).
[2] Trikha, K. C., Dutt, Y., Singh, R. P. (Indian J. Chem. **6** [1968] 376/8).
[3] Manhu, G. S., Bhat, A. N., Jain, B. D. (Current Sci. [India] **37** [1968] 461/3).
[4] Shah, N. R., Shah, J. R. (J. Inorg. Nucl. Chem. **43** [1981] 1583/90).
[5] Singh, B. R., Gosh, R. (Indian J. Chem. A **21** [1982] 320/2).
[6] Mathur, P., Goel, D. P., Singh, R. P. (Monatsh. Chem. **109** [1978] 839/45, 841).
[7] Shah, J. R., Patel, R. P. (J. Prakt. Chem. **315** [1973] 980/2).

27.1.5.2 Isolated Manganese(II) and Manganese(IV) Compounds

With *(E)*-Benzil Monooxime

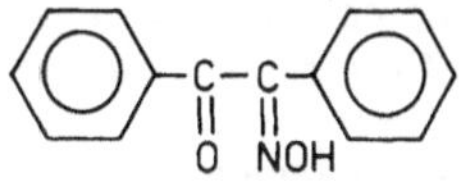

(= $C_{14}H_{11}NO_2$)

$Mn^{II}(C_{14}H_{10}NO_2)_2$ was prepared by dissolving stoichiometric amounts of $MnCl_2$ and the ligand in methanol, adjusting the solution to pH 8.3, and concentrating it on a water bath. The yellow precipitate, which separated on cooling, was washed with water, dilute aqueous NaOH and water again, and dried in vacuum. The effective magnetic moment is 5.66 μ_B corresponding to a high-spin $Mn^{II}(d^5)$ complex. Characteristic bands (in cm^{-1}) in the IR spectrum (KBr) were assigned as follows (shifts in parentheses): ν(C=N), 1605(−5); ν(CO), 1320(+10); ν(NO), 970(−15). The ν(OH) bands of the free ligands, at 3060 cm^{-1} (enol group of the tautomeric ligand form) or at 2900 cm^{-1} (hydroxyimino group), have disappeared on complexation. The complex is insoluble in water and dilute alkali solution, and sparingly soluble in organic solvents. The absorption spectrum in the visible and UV region of solutions in methanol or chloroform exhibits a charge-transfer band at 36000 cm^{-1} and a band at 24500 cm^{-1} assigned to a d-d transition of Mn^{II}. A tetrahedral structure is assumed for the dissolved complex. The molar conductance of a solution in nitrobenzene indicates a nonelectrolyte [1].

$Mn^{II}(C_{14}H_{11}NO_2)X_2$ (X = NO_3, Cl) and **$Mn(C_{14}H_{11}NO_2)_2(ClO_4)_2$**. Manganese(II) nitrate, chloride, or perchlorate were reacted in ethanol with the ligand in a 1:2 proportion. The resulting solutions were refluxed for 30 min to 1 h, then concentrated and cooled. If necessary, some petroleum ether was added to precipitate the light yellow crystals, which were washed with absolute ethanol and ether and dried in vacuum.

The magnetic moments of the compounds are ~5.9 μ_B. The main bands in the IR spectra were located at the following wave numbers (in cm^{-1}; shifts in parentheses): ν(OH), ~3230 (~−70); ν(C=O), 1630(−30); ν(C=N), 1590(~−20); ν(Mn–O), 460; ν(Mn–N), ~320. The nitrate shows bands at 1410 and 1280 cm^{-1} indicating monodentate coordination of the NO_3 group. A broad hump at ~1100 cm^{-1} for the perchlorate indicates the presence of ionic ClO_4 groups. In the visible electronic spectra of 10^{-2} M solutions of the chloride and perchlorate in chloroform, bands were observed around 20000, 22400, and 23400 cm^{-1}, which were assigned to the lowest energy d-d transitions of Mn^{II} in the cubic ligand field. Values for the molar extinction coefficient of 1 to 2 $L \cdot mol^{-1} \cdot cm^{-1}$ indicate tetrahedral geometry for the compounds. The molar conductance of 10^{-3} M solutions of the nitrate and chloride in acetone (~11 $cm^2 \cdot \Omega^{-1} \cdot mol^{-1}$) indicates nonelectrolytes, while that of the perchlorate (238 $cm^2 \cdot \Omega^{-1} \cdot mol^{-1}$) corresponds to a 1:2 electrolyte [2].

With Derivatives of 1-*H*-Pyrazole-4,5-dione 4-Oxime (= HL)

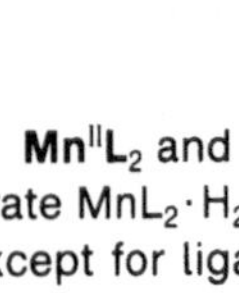

1) R = H, R′ = CH_3; (= $C_4H_5N_3O_2$)
2) R = H, R′ = C_6H_5; (= $C_9H_7N_3O_2$)
3) R = C_6H_5, R′ = H; (= $C_9H_7N_3O_2$)
4) R = C_6H_5, R′ = CH_3; (= $C_{10}H_9N_3O_2$)
5) R = R′ = C_6H_5; (= $C_{15}H_{11}N_3O_2$)
6) R = 4-$C_6H_4NO_2$, R′ = CH_3; (= $C_{10}H_8N_4O_4$)
7) R = 4-$C_6H_4NO_2$, R′ = C_6H_5; (= $C_{15}H_{10}N_4O_4$)
8) R = 4-C_6H_4Cl, R′ = CH_3; (= $C_{10}H_8ClN_3O_2$)

$Mn^{II}L_2$ and **$Mn^{II}L_2 \cdot nH_2O$**. Complexes of composition MnL_2 with all the ligands, a monohydrate $MnL_2 \cdot H_2O$(?) with ligand 1, and dihydrates $MnL_2 \cdot 2H_2O$ of complexes with all the ligands except for ligand 2 and 5 are described.

All the hydrates and the anhydrous complexes with ligands 2 and 5 were prepared by adding hot freshly prepared solutions of the respective ligand in hot aqueous 50 or 96% ethanol to hot $Mn(NO_3)_2$ or $MnSO_4$ solutions (mole ratio 2:1). Precipitation occurred immediately (in the case of ligand 5) or on addition of solid [3] or aqueous sodium acetate (5 or 10%). Some 1 N HNO_3 had to be added to this mixture in the case of ligand 7 [7]. The complex with ligand 4 was also prepared by digesting a mixture of aqueous Mn^{II} acetate and a methanolic or ethanolic ligand solution on a water bath for 20 min [8]. The green complexes with ligands 1 to 4, 6, 8 and yellow compounds with ligands 5 and 7 were washed with water and ethanol [3, 8] and dried in air [3] or in vacuum over $CaCl_2$ [8]. The complex MnL_2 with ligand 5 and the dihydrates with ligands 6 and 7 had to be dried at 105 to 110°C to attain weight constancy [7]. Previous preparations for the complexes with ligands 2 and 5 are given in [9], with ligand 4 in [10]. The anhydrous compounds were obtained by heating the dihydrates at 150 to 210°C for ligand 1 [3, 7] and at 135 to 170°C for ligand 3.

Magnetic measurements at 294 K yielded effective magnetic moments of 5.56 to 5.60 μ_B for the anhydrous complexes and of 5.78 to 5.82 μ_B for the dihydrated complexes with ligands 1 to 3 indicating high-spin $3d^5$ complexes in all cases [12]. The lowered magnetic moments of the anhydrous compounds are presumably due to some metal-metal interaction [3, 12]. In the IR spectra, the absence of free ligand ν(OH) bands for all the complexes suggests the deprotonation of the hydroxyimino group. Lowering of the $\nu(C{=}N)_{cycl}$ band suggests participation of ring nitrogen in complexation. The ν(NO) band is higher compared with those of the free ligands, indicating hydrogen bonding for the latter [3]. Far-IR bands at 635 and 425 cm^{-1} observed for the complex with ligand 4 were assigned to Mn–ligand bands [8]. Magnetic moments (in μ_B) at 303 K, characteristic IR bands (in Nujol or hexachlorobutadiene [3, 8]; shifts in parentheses), and decomposition temperatures for some of the complexes are tabulated below:

No.	complex	μ_{eff}	$\nu(CO)^{a)}$	$\nu(C{=}N)_{cycl}$	$\nu(C{=}N)_{oxime}$	ν(NO)	t_{dec} in °C	Ref.
1	$MnL_2 \cdot H_2O$	5.42	1660(−40)	1620(+5)	1580(−10)	1060(+30)	330[b)]	[3]
2	MnL_2	5.41	1650(−57)	1640(−25)	1500(−110)	1150(+108)	350	[3]
4	MnL_2	5.35	1636(−69)	1595(−17)	1518(−72)	1122(+82)	320	[3]
4	$MnL_2 \cdot 2H_2O$	—	1640(−45)	—	—	1250[c)]	—	[8]
5	MnL_2	5.53	1635(−47)	1592(−23)	1495(−97)	1095(+40)	325	[3]

a) Assignment to ν(CO) + ν(C=N) by [8]. – b) Decomposition point of the anhydrous compound. – c) This band does not appear in the IR spectrum of the free ligand [8].

The diffuse reflectance spectra of the anhydrous complexes with ligands 1, 2, 4, and 5 show three bands in the ranges normally expected for octahedral Mn^{II} complexes. The bands were assigned as follows (wave numbers in cm^{-1}): ~16000, $^6A_{1g} \rightarrow {}^4T_{1g}(G)$; ~23000, $^6A_{1g} \rightarrow {}^4T_{2g}(G)$; ~25000, $^6A_{1g} \rightarrow {}^4E_g$, $^4A_{1g}(G)$ [3]. For the dihydrate with ligand 4, two bands at 6000 and 15750 cm^{-1} were observed [8].

Thermogravimetric analyses show that the dihydrates start to lose their water at temperatures between 150 and 200°C and that the anhydrous complexes decompose at temperatures between 250 and 350°C [3, 8], from the thermogram in [7]. The dihydrated complex with ligand 4 is somewhat soluble in hot water [10], soluble in DMF, but practically insoluble in benzene and other noncoordinating solvents [8]. The anhydrous complexes with ligands 1, 2, 4, and 6 are nonelectrolytes in DMF [3]. The physical properties suggest a monomeric octahedral structure for the dihydrates and a polymeric distorted-octahedral structure for the anhydrous compounds [3, 12].

$MnL_2 \cdot 2py$ was obtained by adding pyridine to the hot aqueous ethanol solution of $MnSO_4$ and ligand 2 with exclusion of air. The pyridine was removed at 150°C [7].

With Derivatives of Violuric Acid and a Related Compound

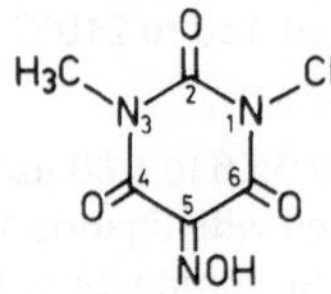
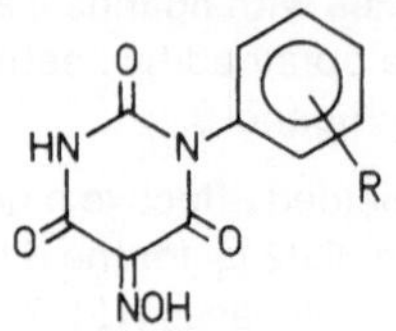
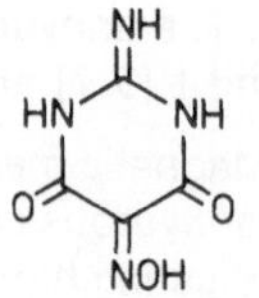

1) 1,3-Dimethylvioluric Acid ($=C_6H_7N_3O_4$)

Arylvioluric Acids
2) R=H; ($=C_{10}H_7N_3O_4$)
3) R=2-CH$_3$; ($=C_{11}H_9N_3O_4$)
4) R=3-CH$_3$; ($=C_{11}H_9N_3O_4$)
5) R=4-CH$_3$; ($=C_{11}H_9N_3O_4$)

6) 2-Imino-4,5,6-(*1H,3H*)-pyrimidinetrione 5-Oxime ($=C_4H_4N_4O_3$)

A yellow compound of composition $\mathbf{Mn_2^{IV}(C_6H_6N_3O_4)_2(C_6H_7N_3O_4)\cdot 6OH}$**(?)** was obtained by concentrating an aqueous solution containing equimolar amounts of $Mn(NO_3)_2$ and 1,3-dimethylvioluric acid (ligand 1). The color of the solution changed from rose to yellow in three or four months. From the yellow solution, crystals separated, which were gathered, washed with cold water, and dried in air. The mentioned change of color occurred within 15 d when the Mn:ligand ratio was 4:1. The magnetic moment of the compound is 3.73 μ_B. Therefore an oxidation state of four is suggested for the Mn atoms (spin-only value for a d^3 complex 3.87 μ_B). Bands in the IR spectrum were assigned as follows (shifts on complexation in parentheses; wave numbers in cm^{-1}): ν(OH), 3400, 3310; ν(C=O), 1712(−28), 1695(−45), 1635(−40); ν(C=C)+ν(C=N), 1590(−10); ν(C–N), 1280(−2), 1250(−5), 1195(−15); δ(OH), 1135(−10); ν(C=O), 1060(−10), 1030(−10). The shifts suggest that complex formation involves the nitrogen of the hydroxyimino group and the oxygen at the 6-position. TG and DSC studies show an endothermic effect at 170 to 230°C and an exothermic effect at 260 to 390°C corresponding to the dehydration and decomposition of the compound, respectively. An enthalpy of dehydration of 70.5 kJ/mol H_2O was evaluated [4].

Highly colored compounds of composition $\mathbf{Mn^{II}L_2\cdot nH_2O}$ with ligands 2 through 5 were prepared by adding concentrated Mn^{II} salt solutions to saturated aqueous solutions of the NH_4 salts of the ligands. In the IR spectra of the compounds, the ν(OH) band of the hydroxyimino group has vanished, and a band in the 1425 to 1395 cm^{-1} region for the free ligands, which was associated with ν(NO) vibrations, has shifted to lower frequencies. New bands in the 520 to 510 cm^{-1} and 420 to 410 cm^{-1} regions indicate the existence of Mn–O and Mn–N bonds, respectively. The compounds split off water at 180 to 210°C. They have no sharp melting points, but start to decompose at 200°C. They are insoluble in water and soluble in ethanol, acetone, dioxane, and other organic solvents [5].

An undefined dark brown crystalline compound with ligand 6 (34.5% Mn; 23.8% N) was obtained by combining solutions of an Mn^{II} salt and the K salt of the ligand. It melts with decomposition at 258°C and is insoluble in all organic solvents [6].

References:

[1] Bhargava, P. P., Bembi, R., Tyagi, M. (J. Indian Chem. Soc. **60** [1983] 214/7).
[2] Mohapatra, B. K., Mishra, R. C., Panda, D. (J. Indian Chem. Soc. **58** [1981] 1154/6).
[3] Shah, N. R., Shah, J. R. (J. Indian Chem. Soc. **58** [1981] 851/4).
[4] Romero Molina, M. A., Salas Peregrin, J. M., Lopez Gonzales, J. D., Valenzuela Calahorro, C. (Anales Quim. B **79** [1983] 383/7).
[5] Singh, B. R., Ghosh, R. (Indian J. Chem. A **21** [1982] 320/2).
[6] Handa, R. L., Dutt, S. (J. Indian Chem. Soc. **27** [1950] 647/50).
[7] Hovorka, V., Šůcha, L. (Collection Czech. Chem. Commun. **29** [1964] 983/92, 984, 989).

[8] Rustagi, S. C., Arora, H. C., Rao, G. N. (Proc. Chem. Symp., Aligarh, India, 1972, Vol. 2, pp. 49/54; C.A. **81** [1974] No. 180503).
[9] Hovorka, V., Sýkora, V. (Chem. Listy **35** [1941] 89/93; C **1942** I 2040).
[10] Hovorka, V., Sýkora, V. (Collection Trav. Chim. Tchecoslovaquie **11** [1939] 70/6, 75).

[11] Hovorka, V., Šůcha, L. (Collection Czech. Chem. Commun. **25** [1960] 55/9).
[12] Šůcha, L. (Collection Czech. Chem. Commun. **29** [1964] 993/8, 996).

27.1.6 With Nitrosophenols and Related Compounds

General

The o-nitrosophenols and -naphthols are known to be tautomeric with the corresponding o-quinone monooximes:

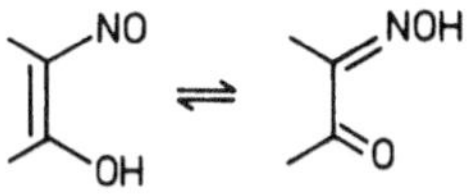

Spectral studies indicate that the quinone oxime form exists in the solid state and in acid solution, while in basic solution the nitrosophenol form predominates. These ligands are therefore closely related to the 1,2-diketone monooximes.

Strongly colored compounds of the type MnL_2 were prepared in aqueous medium with the isomeric nitrosonaphthols. The complexes are sparingly soluble in aqueous solution and more soluble in various organic solvents, forming colored solutions which were studied with reference to their analytical interest. Green or brown ternary complexes of the types $Mn^{II}LA$, $Mn^{II}(HL)A_2$, and $Mn^{III}L_2A$ derived from nitrosonaphthols HL and a second ligand HA were also synthesized. For all the solid compounds with nitrosonaphthols, IR data indicate that coordination occurs through hydroxyimino nitrogen and carbonyl oxygen, i.e., the coordinated ligands exist in their quinone oxime form. The ternary complexes $Mn^{II}(HL)A_2$ and $Mn^{III}L_2A$ are monomeric. $Mn^{II}LA$ is polymeric octahedral.

27.1.6.1 With Derivatives of o-Nitrosophenol

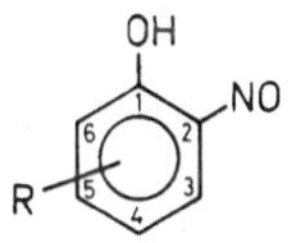

No.	R	formula
1	5-N(CH$_3$)$_2$	$C_8H_{10}N_2O_2$
2	4-CH$_3$	$C_7H_7NO_2$
3	5-OCH$_3$	$C_7H_7NO_3$
4	5-NHCOCH$_3$	$C_8H_8N_2O_3$

Green to brown colored solutions of Mn^{II} complexes in dimethylformamide are formed on reaction of manganese(II) chloride with ligands 1 to 4 and cyclohexylamine or triethylamine (mole ratio 1:2:2). The electronic spectra reveal absorption maxima at 450 nm and in the range from 820 to 842 nm. The highest specific absorption coefficients ($\alpha = 63\ L\cdot g^{-1}\cdot cm^{-1}$ at 450 nm and 13.5 $L\cdot g^{-1}\cdot cm^{-1}$ at 842 nm) were observed for the complex with ligand 1 and cyclohexylamine. A dark brown solid 1:2 Mn^{II} complex, obtained by reacting aqueous $MnCl_2\cdot 4H_2O$ with ligand 1 in hot aqueous HCl and raising the pH to 6.5 shows a similar spectral curve in dimethylformamide. Because of its high absorption in the near-IR region the compound was proposed as a component of IR filter media, Coleman, R. A., Rodgers, J. L. (U.S. 2971921 [1959/61]).

27.1.6.2 With Nitrosonaphthols

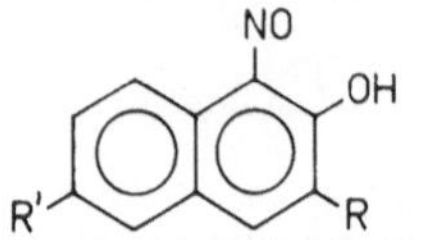

1) $R = R' = H$; ($= C_{10}H_7NO_2$)
2) $R = R' = SO_3H$; ($= C_{10}H_7NO_8S_2$)
3) $R = H$, $R' = SO_2NH_2$; ($= C_{10}H_8N_2O_4S$)

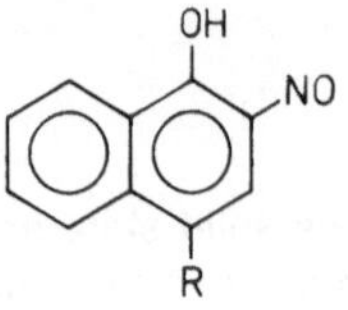

4) $R = H$; ($= C_{10}H_7NO_2$)
5) $R = SO_3H$; ($= C_{10}H_7NO_5S$)

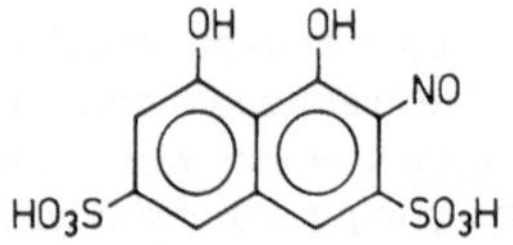

6) ($= C_{10}H_7NO_9S_2$)

Complexes in Aqueous or Aqueous-Organic Solutions. The table lists stability constants of 1:1 and 1:2 Mn^{II} chelates with ligands 1, 2, 4, and 5, determined potentiometrically with a glass electrode [1 to 5, 13] or spectrophotometrically [14] in aqueous [2,5,13,14] or aqueous-organic medium [1,3,4]:

No.	H_nL	solvent (vol%)	t in °C	I in mol/L	log K_1	log K_2	Ref.
1	HL	75% dioxane	30	0 corr	7.52	6.25	[1]
2	H_3L	—	25	→0	3.73	—	[2]
	H_3L	—	25	0.1 (KCl)	2.7	—	[13]
4	HL	75% dioxane	30	0 corr	6.78	5.42	[1]
		75% dioxane	30	0 corr	~7.1	~5.5	[3]
		50% acetone	35	0.1 (KNO_3)	3.91	3.53	[4]
5	H_2L	—	25	→0	2.07	—	[2,14]
		—	35	0.1 (KNO_3)	3.69	2.84	[5]

Yellow-green solutions of Mn^{II} complexes in dimethylformamide were obtained by reacting $MnCl_2$ with ligands 1 and 3 in the presence of cyclohexylamine (mole ratio 1:2:2). The solutions show absorption maxima (in nm) at the following wavelengths (specific absorption coefficient α in $L \cdot g^{-1} \cdot cm^{-1}$ in parentheses): 820 (6.5) and 415 (16.2) for ligand 1; 818 (3.2) for ligand 3. For the complex with ligand 3, $\alpha = 7.7$ at 450 nm (no maximum, peak below 400 nm) [6].

The Mn^{II} complex with 1-nitroso-2-naphthol in chloroform was separated from other metal complexes of this ligand by thin-layer chromatography. The R_f value was 0.19 for a 1:1 mixture of silica and alumina as the sorbent and benzene as the solvent [9]. The same complex is catalytically active in the chemiluminescent oxidation of luminol by H_2O_2. A method for the determination of small amounts of 1-nitroso-2-naphthol was based on this activity [10]. On reaction of Mn^{II} with 2-nitrosochromotropic acid (ligand 6), a pale violet coloration was observed at pH 5.6 [11] and a pink coloration in alkaline medium [12].

$Mn(C_{10}H_6NO_2)_2$. The dark brown complex with 1-nitroso-2-naphthol (ligand 1) was obtained by treating a 0.1 M aqueous solution of $MnCl_2$ with an excess of the ligand (2% in 50% aqueous alcohol) [7]. In another procedure, an Mn^{II} salt and ligand 1 were reacted in the mole ratio 1:2 in aqueous medium. The complex formed immediately on adjusting the pH to between 2 and 6 [8]. In a third procedure, aqueous $MnCl_2 \cdot 4H_2O$ was reacted with a suspension of ligand 1 in aqueous sodium acetate with heating. The separated product was washed with water and ether [6] and dried over $CaCl_2$. Excess ligand was removed by Soxhlet extraction with petroleum ether [8]. The black complex with 2-nitroso-1-naphthol (ligand 4) was prepared from $MnCl_2$ and the ligand in sodium acetate solution [7].

Characteristic IR absorption bands (cm^{-1}) of the complex with ligand 1 in KBr pellets were assigned as follows (free ligand bands in parentheses): ν(C=N), 1485 (1656); ν(C=O), 1315 (1612); ν(N–O), 1190 (1078). The spectrum of the complex with ligand 4 shows so many bands in the 1500 to 1200 cm^{-1} region that the pertinent bands could not be located with certainty. On the basis of the IR spectra, coordination of the carbonyl oxygen and the oxygen (?) of the deprotonated hydroxyimino group is suggested; i.e., for both complexes in the solid state, the quinone oxime structure is assumed [7]. In the near-IR region, a solution of the complex with ligand 1 in dimethylformamide shows a maximum at 840 nm with the specific absorption coefficient $\alpha = 16.4\ L \cdot g^{-1} \cdot cm^{-1}$ [6]. The UV-visible spectra of both complexes in ethanol show maxima at 410 nm for the complex with ligand 1 and at 450 nm for that with ligand 4 [7]. The comparison with the spectra of the free ligands in acidic or alkaline ethanol solutions suggests a resonance between the quinone oxime structure and the nitrosonaphthol structure for the complex with ligand 1, but the predominance of the nitrosonaphthol structure for the complex with ligand 4 [7].

The microcrystalline compounds decompose before melting [7]. The complex with 1-nitroso-2-naphthol is soluble in chloroform [9] and dimethylformamide [6]. All the compounds are only slightly soluble in water, ethanol, ether, or acetone [7]. Because of its high absorption in the near-IR region, the complex with ligand 1 was proposed as a component in IR filter media [6].

References:

[1] Callahan, C. M., Fernelius, W. C., Block, B. P. (Anal. Chim. Acta **16** [1957] 101/8).
[2] Mäkitie, O. (Maatalouden Tutkimuskeskus Maantutkimuslaitos No. 79 [1961] 1/61, 52; C.A. **56** [1962] 2035).
[3] Van Uitert, L. G., Fernelius, W. C. (J. Am. Chem. Soc. **76** [1954] 375/9).
[4] Lingaiah, P., Sundaram, E. V. (Current Sci. [India] **45** [1976] 51/2).
[5] Lingaiah, P., Sundaram, E. V. (Indian J. Chem. **10** [1972] 670/1).
[6] Coleman, R. A., Rodgers, J. L. (U.S. 2971921 [1959/61]).
[7] Gurrieri, S., Siracusa, G. (Inorg. Chim. Acta **5** [1971] 650/4).

[8] Patil, S. V., Raju, J. R. (Current Sci. [India] **41** [1972] 117/8).
[9] Senf, H.-J. (J. Chromatog. **27** [1967] 331/3).
[10] Pilipenko, A. T., Kalinichenko, I. E., Matveeva, E. Ya. (Zh. Analit. Khim. **33** [1978] 1612/7; J. Anal. Chem. [USSR] **33** [1978] 1257/61).
[11] Datta, S. K., Saha, S. N. (Mikrochim. Acta **1961** 361/5).
[12] Datta, S. K., Ghose, P. (Naturwissenschaften **45** [1958] 515/6).
[13] Mäkitie, O. (Suomen Kemistilehti B **40** [1967] 27/31).
[14] Mäkitie, O. (Suomen Kemistilehti B **39** [1966] 218/20).

27.1.6.3 With 1-Nitroso-2-naphthol or 2-Nitroso-1-naphthol and Other Ligands

$Mn^{II}(C_{10}H_6NO_2)(acac)\cdot nH_2O$ (n = 0, 1) and **$Mn^{II}(C_{10}H_7NO_2)(acac)_2$**. The greenish hydrated complex with the deprotonated ligands was prepared by adding a mixture of 1-nitroso-2-naphthol (= $C_{10}H_7NO_2$) in methanol and of KOH in water to the solution of Mn^{II} acetylacetonate in methanol and refluxing the mixture for 3 to 4 h. The brownish green complex with non-deprotonated 1-nitroso-2-naphthol, $Mn(C_{10}H_7NO_2)(acac)_2$, was prepared by stirring the same reaction mixture at room temperature. The complexes were collected, washed with methanol, and dried in vacuum. The yellowish green anhydrous compound, $Mn(C_{10}H_6NO_2)(acac)$, was obtained by heating the hydrate at 130 to 150°C [1].

The magnetic moments are 5.22, 5.43, and 4.99 μ_B for $Mn(C_{10}H_6NO_2)(acac)$, its hydrate, and $Mn(C_{10}H_7NO_2)(acac)_2$, respectively. Characteristic IR bands (in cm^{-1}) were observed in the following regions (free ligand bands in parentheses): ν(C=O), 1610 to 1600, 1590 to 1580 (1629); ν(C=N), 1560 to 1555 (1568); ν(NO), 1210 to 1190 (1083). Bands of the ν(OH) vibrations, between 3600 and 3200 cm^{-1}, were observed only in the spectra of $Mn(C_{10}H_6NO_2)(acac)\cdot H_2O$ and $Mn(C_{10}H_7NO_2)(acac)_2$. For the hydrate, the presence of the ρ(OH) band at 820 cm^{-1} indicates the coordinated nature of the water molecule. For this compound, a band at 340 cm^{-1} was associated with the ν(Mn–N) vibration, and a band at 290 cm^{-1} with the ν(Mn–O) vibration [1].

The compounds do not melt, but decompose at 240 to 270°C. They are soluble only in coordinating solvents, such as DMF, DMSO, etc. They are nonelectrolytes in DMF. On the basis of the studies, a monomeric octahedral structure is suggested for $Mn(C_{10}H_7NO_2)(acac)_2$ and a polymeric octahedral structure with bridging oxygen atoms of the naphthol group for $Mn(C_{10}H_6NO_2)(acac)$ and its hydrate [1].

$Mn^{II}(C_{10}H_6NO_2)(C_7H_6NO_2)$. The ternary complex of 1-nitroso-2-naphthol (= $C_{10}H_7NO_2$) and 2-aminobenzoic acid (= $C_7H_7NO_2$) was precipitated by adding aqueous $MnCl_2$ (5 mmol) to a mixture of the nitroso compound (5 mmol) in the minimum amount of acetone and 2-aminobenzoic acid (5 mmol) and NaOH (5 mmol) in water. It was washed with acetone-water mixture and ethanol and dried in vacuum. The pyridine adduct, **$Mn^{II}(C_{10}H_6NO_2)(C_7H_6NO_2)\cdot 2py$**, was obtained by refluxing an ethanolic suspension of the complex with excess pyridine for 30 min. After cooling of the reaction mixture and stirring for 2 h, the adduct was precipitated by adding excess ether. It was filtered, washed with ether, and dried in air [2].

The magnetic moment of $Mn(C_{10}H_6NO_2)(C_7H_6NO_2)$ is $\mu_{eff} = 5.27\ \mu_B$ indicating the presence of five unpaired electrons. In the IR spectrum, characteristic IR bands due to 1-nitroso-2-naphthol appear in the following regions (free ligand bands in parentheses): ν(C=O), 1620 to 1610 (1629); ν(C=N), 1560 to 1545 (1568); ν(N–O), ~1210 (1083 cm^{-1}). The shifts are in accord with the coordination of this ligand through the carbonyl oxygen and the hydroxyimino nitrogen of the ligand in its quinone oxime form. Strong negative shifts of the ν(NH) bands of 2-aminobenzoic acid for the complex suggest coordination of the amino nitrogen, whereas positive shifts of the $\nu(COO^-)$ bands indicate bridging bidentate behavior of the COO^- group. The ν(Mn–N) and ν(Mn–O) bands are in the regions 380 to 330 and 320 to 280 cm^{-1}, respectively. For the pyridine adduct, the $\nu(Mn\text{–}N)_{pyridine}$ band is located at ~270 cm^{-1}. A polymeric octahedral structure with *trans*-square planar arrangement of the nitrogens and carbonyl oxygens and axial coordination of bridging carbonyl oxygens is proposed for the ternary complex and a monomeric octahedral structure for the pyridine adduct, as shown below:

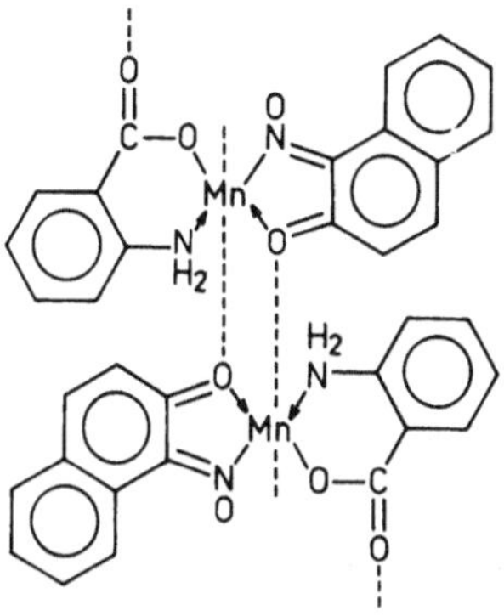

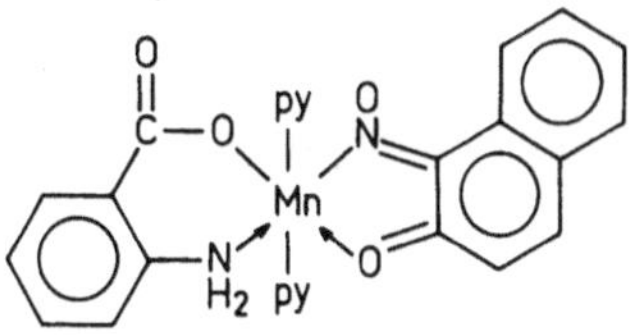

$Mn(C_{10}H_6NO_2)(C_7H_6NO_2)$ decomposes at temperatures above 200°C. It is insoluble in common organic solvents, but slightly soluble in coordinating solvents like DMF, DMSO, etc. It is a nonelectrolyte in DMF [2].

$Mn^{III}(C_{10}H_6NO_2)_2(CH_3COO)$. The complexes with 1-nitroso-2-naphthol or 2-nitroso-1-naphthol (= $C_{10}H_7NO_2$) were prepared by combining methanolic solutions of the ligands and $Mn(CH_3COO)_3 \cdot 2H_2O$. The brownish black complexes, which separated instantaneously, were filtered, washed with methanol, and dried over P_2O_5 in vacuum [3].

The effective magnetic moments of the complexes are 3.26 and 3.72 μ_B, respectively. These values, which are considerably lower than the spin-only value expected for high-spin Mn^{III} complexes (4.90 μ_B), may be due to metal-metal interactions. The most important bands in the IR spectra (KBr) were assigned as follows (in cm^{-1}; shifts in parentheses): ν(C=O), ~1590(−25); ν(C=N), ~1510(−10); ν(N–O), 1090(+10). The shifts indicate that the ligands are coordinated in their quinone oxime form by the hydroxyimino nitrogen and the carbonyl oxygen atoms, forming five-membered chelate rings. Two medium intensity bands around 1610 and 1395 cm^{-1} can be assigned to $\nu_{as}(COO)$ and $\nu_s(COO)$ vibrations, respectively, of the acetate group. The solid state electronic spectrum of the complex with 1-nitroso-2-naphthol shows two bands, at 450 and 900 nm. The solution spectra of the complexes in benzene show a band at 790 nm and two shoulders at 450 and 590 nm for the complex with 1-nitroso-2-naphthol, and a band at 830 nm and two shoulders at 550 and 710 nm for the complex with 2-nitroso-1-naphthol. These bands can be assigned to the split components of the $^5E_g \rightarrow {}^5T_{2g}$ transition for manganese(III) in the octahedral ligand field [3].

Thermogravimetric analysis shows that both complexes decompose in two stages. The residues of the first stage are the acetate-free complexes, that of the second stage is Mn_3O_4. Decomposition temperatures are 231 and 375°C for the complex with 1-nitroso-2-naphthol and

221 and 379°C for the complex with the 2-nitroso-1-naphthol. The complexes are soluble in benzene, nitrobenzene, acetonitrile, etc., but are insoluble or sparingly soluble in water, methanol, etc. [3].

References:

[1] Aggarwal, R. C., Bala, R., Prasad, R. L. (Syn. Reactiv. Inorg. Metal-Org. Chem. **14** [1984] 171/84; C.A. **101** [1984] No. 162350).

[2] Aggarwal, R. C., Bala, R., Prasad, R. L. (Indian J. Chem. A **22** [1983] 955/8).

[3] Sarasukutty, S., Sunder Ram, A. N., Prabhakaran, C. P. (Indian J. Chem. A **15** [1977] 914/6).

27.1.6.4 With Nitrosoquinolinols

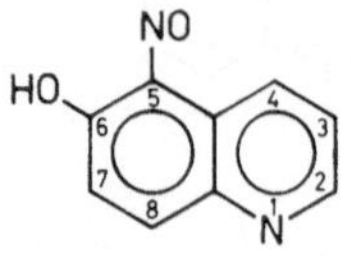

ligand 1 (= $C_9H_6N_2O_2$)

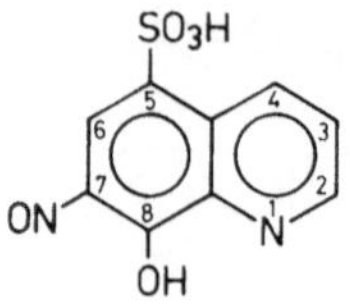

ligand 2 (= $C_9H_6N_2O_5S$)

Mn^{2+} ions form 1:1 and 1:2 complexes with ligand 2 in aqueous medium. Stability constants of the $Mn(C_9H_5N_2O_5S)^+$ and $Mn(C_9H_5N_2O_5S)_2$ species in 0.1 M KNO_3 were determined by potentiometric titration with the glass electrode: log K_1 = 4.02 and log K_2 = 3.50 at 25 ± 0.5°C. The complex solution shows an absorption maximum at 410 nm [1].

An undefined brown-green Mn^{II} complex with ligand 1 was obtained by reacting an Mn^{II} salt with the sodium salt of the ligand in hot aqueous solution. The compound is only slightly soluble in water or alcohol, but readily soluble in benzene or chloroform [2].

A yellow-orange chelate complex of composition **$Mn(C_9H_4N_2O_5S)\cdot 2H_2O$** was obtained by reacting the disodium salt of ligand 2 with $MnCl_2$ in aqueous solution under reflux for 30 min. The compound, which precipitated on cooling, was washed with water and ethanol and dried over P_4O_{10}. Characteristic IR bands were assigned as follows (in Nujol; shifts on complexation in parentheses): ν(C=O), 1605 (−5); ν(C=N), 1590 (+10); ν(C=C), 1570; ν(NO), 1540 (+10). Bands in the 3600 to 3100 cm^{-1} region and around 1650 cm^{-1} were associated with vibrations of the coordinated water molecules. The electronic spectrum of the complex in aqueous solution shows peaks at 440, 305, and 235 nm, the first of which was shifted by +10 nm on complexation. The complex decomposes at 262 to 268°C. It is insoluble in cold water and in organic solvents [3].

References:

[1] Choi, Q. W., Lee, D. H., Oh, J. S., Lee, K. W. (Daehan Hwahak Hwoejee **12** [1968] 81/4 from C.A. **70** [1969] 41356).

[2] Travagli, C. (Ann. Chim. [Rome] **58** [1968] 625/9).

[3] Aly, M. M., Issa, I. M., Allam, M. G., El-Haty, M. T. (J. Inorg. Nucl. Chem. **35** [1973] 2080/3).

27.1.6.5 With a Derivative of 10-Nitroso-9-phenanthrol

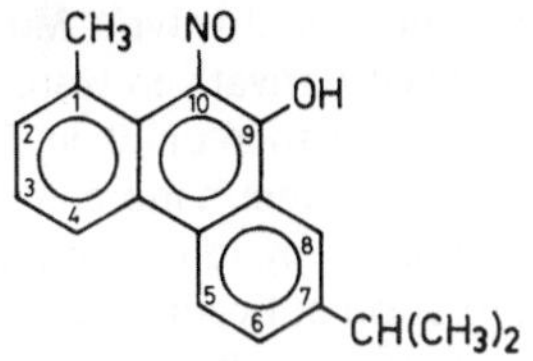

(tautomeric with Retenequinone Monooxime $= C_{18}H_{17}NO_2$)

A solution of an Mn^{II} complex with retenequinone monooxime was prepared by reacting $MnCl_2$ with the ligand in the presence of cyclohexylamine (mole ratio 1:2:2) in DMF. The yellow solution shows an absorption maximum at 790 nm with the specific absorption coefficient $\alpha = 0.7\ L \cdot g^{-1} \cdot cm^{-1}$. At 450 nm (no peak), $\alpha = 8.0\ L \cdot g^{-1} \cdot cm^{-1}$, Coleman, R., Rodgers, J. L. (U.S. 2971921 [1959/61]).

27.1.7 With *(Z)*-2-Furancarbaldehyde Oxime

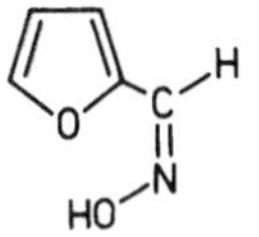

(= anti-furfuraldoxime $= C_5H_5NO_2$)

Formation of only the $Mn(C_5H_4NO_2)^+$ species was observed in solution. Its stability constant was determined by pH measurement (glass electrode) at 20°C and I = 0.1 M $NaClO_4$ in various dioxane-water mixtures: log $K_1 = 2.72$ for 40 vol%, log $K_1 = 3.27$ for 50 vol%, and log $K_1 = 4.24$ for 60 vol% dioxane [1].

$Mn(C_5H_5NO_2)_2Cl_2$ was prepared by dropwise addition of the stoichiometric amount of the ligand dissolved in absolute ethanol to a hot solution of $MnCl_2 \cdot 4H_2O$ in absolute ethanol. After refluxing for 2 h, part of the ethanol was removed. The yellow-brown precipitate was filtered, washed with cyclohexane and acetone, and dried in vacuum [2].

The magnetic moment of the compound is $\mu_{eff} = 5.96\ \mu_B$ at 24°C [2]. Characteristic IR bands (in cm^{-1}) of the complex in CsI were assigned as follows (free ligand bands in parentheses): ν(OH), 3375, 3140, 3030(3160, 3040); ν(CN), 1650(1640); ν(NO), 970(970), $\nu(Mn-O)_{oxime}$, 657, 632; $\nu(Mn-O)_{furan}$, 470. From the shifts of the ν(OH) and ν(CN) bands and the appearance of new bands in the 660 to 630 cm^{-1} region and around 450 cm^{-1} on complexation, coordination through the hydroxyimino oxygen and furan ring oxygen is inferred [3]. The compound melts at 202°C with decomposition [2]. It is a nonelectrolyte in ethanol and nitrobenzene [3].

References:

[1] Gupta, V. K., Bhat, A. N. (Indian J. Chem. A **18** [1979] 342/6).
[2] Sen, B., Pickerell, M. E. (J. Inorg. Nucl. Chem. **35** [1979] 2573/5).
[3] Gupta, V. K., Bhat, A. N. (Z. Naturforsch. **32b** [1977] 225/8).

27.1.8 With Oximes Derived from Aldehydes and Ketones Containing Chelating N-Heterocyclic Groups

General

In this chapter complexes with oximes of 2-pyridinecarbaldehyde and related aldehydes or ketones are described. For complexes with oximes of other N-heterocyclic compounds, see pp. 246, 248, 250, 252, 258, and 264.

Solid compounds of the type **$Mn(HL)_2X_2$** with all the ligands HL were prepared in ethanol or in mixtures of ethanol and 2,2-dimethoxypropane. Compounds of the type **$MnL_2L'_2$** with HL = pyridine-2-carbaldehyde oxime and L' = pyridine or its alkyl derivatives were obtained by reacting $Mn(HL)_2Cl_2$ with the respective pyridine base in the presence of alkali. For all the compounds, shifts of free ligand ν(C=N) and ν(N–O) bands in the IR spectra indicate coordination of the hydroxyimino nitrogen atoms. Shifts and splittings of pyridine ring vibrations indicate coordination of the pyridine nitrogen atoms with formation of five-membered chelate rings. All the complexes are octahedral, most of them monomeric. The halogeno complexes are dimeric except for those with pyridine-2-carbaldehyde oxime and its 6-methyl derivative. In these compounds, two bridging halogen atoms are shared by the two coordination octahedra, whereas the other two halogen atoms are ionic. Splitting of the ν(OH) band in the IR spectra indicates H bonding (intermolecular or intramolecular) for the complexes of the oximes with *E*-configuration. Large positive shifts of the ν(C=N) and ν(CO) bands give evidence of strong H bonding in the free ligands. All the compounds are stable at room temperature. Most of them are insoluble in water or nonpolar solvents. Most of them are non-electrolytes in methanol or ethanol.

27.1.8.1 Complexes in Solution

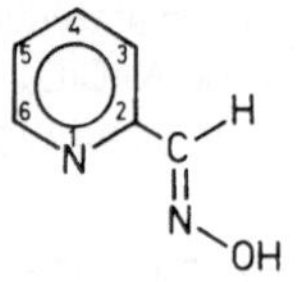

ligand 1

(= $C_6H_6N_2O$)

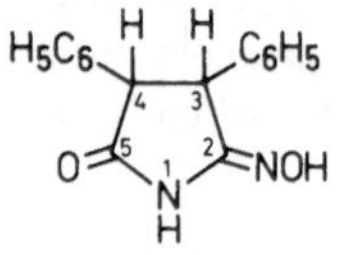

ligand 2

(= $C_{16}H_{14}N_2O_2$)

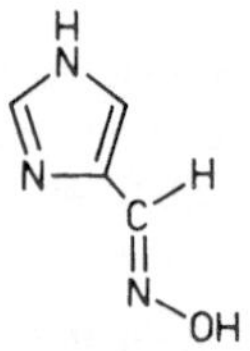

ligand 3

(= $C_4H_5N_3O$)

Stability constants of the complex species $Mn(C_6H_5N_2O)^+$ and $Mn(C_6H_5N_2O)_2$, which formed with ligand 1 in aqueous solution, have been determined potentiometrically (glass electrode) to be $\log K_1 = 5.2 \pm 0.2$ and $\log K_2 = 3.9 \pm 0.2$ at 25°C and I = 0.3 M ($NaClO_4$). In the UV absorption spectrum, complex formation is indicated by a small shift of the ligand bands at ~42000 and ~34000 cm^{-1} to higher wave numbers and by a significant decrease in the extinction coefficients calculated from ligand concentration [1].

By mixing an ethanolic solution of ligand 2 with an aqueous Mn^{II} solution and adjusting the pH to between 10 and 11.5 with ammonia, a soluble violet complex (Mn^{III} ?) was obtained. It exhibits an absorption maximum at 490 nm, the absorbance being practically constant in the cited pH range. The formation of the complex allows the spectrophotometric determination of Mn. In NaOH solution, the oxime forms a reddish brown complex with manganese ions [2].

Mn^{2+} ions form a yellow-brown 1:1 complex with ligand 3 in alkaline aqueous solution. The complex is anionic, probably a hydroxo species. It shows an absorption maximum at 350 nm with $\log \varepsilon = 3.89$ at pH 11.8. The color reaction can be utilized for the spectrophotometric determination of manganese in low concentrations. The effects of the reaction conditions and other metal ions were studied [3].

References:

[1] Burger, K., Egyed, I., Ru, I. (J. Inorg. Nucl. Chem. **28** [1966] 139/45, 143, 145).
[2] Mongay, C., Rodriquez, E., Borull, F., Cerda, V. (Anales Quim. B **78** [1982] 247/51).
[3] Fernandez Pereira, C., Gasch Gomez, J. (Anal. Letters **18** [1985] 2219/27).

27.1.8.2 Isolated Manganese(II) Compounds

With Oximes Derived from Pyridine-2-carbaldehyde or Its 6-Methyl Derivative

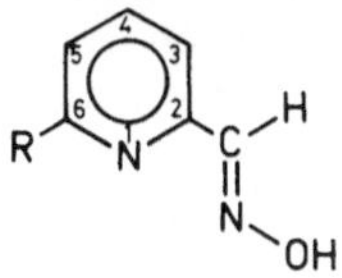

R = H; $(=C_6H_6N_2O)$
R = CH_3; $(=C_7H_8N_2O)$

$[Mn(C_6H_6N_2O)_2X_2]$ and **$[Mn(C_7H_8N_2O)_2X_2]$** (X = NO_3, Cl, Br, I, ½SO_4, CH_3COO, NCS). The yellow compounds were prepared by mixing aqueous solutions (for X = ½SO_4 and CH_3COO) or ethanol solutions (for the other anions) of the respective Mn^{II} salt hydrates with hot ethanol solutions of the ligands. Most of the compounds crystallized immediately. The reaction mixtures were stirred for several hours in the case of X = NO_3. They were refluxed for 30 min in the case of X = NCS or CH_3COO. All the compounds were washed with ethanol and ether and dried in vacuum over P_4O_{10} [1].

The magnetic moments of the complexes at room temperature are in the range 5.80 to 5.90 μ_B expected for high-spin $3d^5$ complexes. The magnetic susceptibility down to liquid nitrogen temperature obeys the Curie-Weiss law with a relatively small value of the Weiss constant. Pertinent IR bands (in cm^{-1}) of the complexes in Nujol or CsI were assigned as follows (ligand bands in parentheses): ν(OH), ~3250 (several bands from 3200 to 2790); $\nu(CN)_{oxime}$, ~1660(1520); ν(NO), ~1074(950). Six anion bands for the nitrates in the range between 1300 and 720 cm^{-1} suggest unidentately coordinated nitrate groups, whereas the ν(SO) band at ~990 cm^{-1} for the sulfates indicates bidentate coordination of the sulfate groups. Anion bands of the acetates and isothiocyanates were located as follows: $\nu_{as}(COO)$, ~1650; $\nu_s(COO)$, ~1390; ν(CN), ~2045; ν(CS), 810; δ(NCS), ~485. The far-IR spectra show the $\nu(Mn–N)_{ligand}$ bands at ~250 and ~215(sh) cm^{-1}. The ν(Mn–X) bands were located at the following wave numbers: $\nu(Mn–O)_{nitrate}$, ~320; $\nu(Mn–O)_{acetate}$, ~350; ν(Mn–Cl), ~270; $\nu(Mn–N)_{NCS}$, ~260; ν(Mn–Br) and ν(Mn–I), <200 cm^{-1}. The diffuse reflectance spectra in the UV and visible region show six (for the complexes with X = Cl, Br) or five maxima (for the other complexes) in the 16000 to 30000 cm^{-1} range (maxima reported in detail for each complex) which were associated with the transitions $\rightarrow {}^4T_{1g}(G)$, $\rightarrow {}^4T_{2g}(G)$, $\rightarrow {}^4A_{1g}(G)$, ${}^4E_g(G)$, $\rightarrow {}^4T_{2g}(D)$, and $\rightarrow {}^4E_g(D)$ from the ${}^6A_{1g}$ ground state of the coordinated Mn^{2+} ion. In the 30000 to 47000 cm^{-1} range, two to five charge-transfer bands were observed. Crystal field parameters Dq, B, C, and β were calculated for all the complexes, Dq ranging from 1024 for $[Mn(C_7H_8N_2O)SO_4]$ to ~1140 for $[Mn(C_6H_6N_2O)(NCS)_2]$. The physical properties indicate distorted octahedral molecular structures with the unidentate anions in *trans*-axial positions [1].

The complexes are stable at room temperature. They are soluble in most common organic solvents except for the compounds with X = ½SO_4 and CH_3COO. From the solutions in acetone, acetonitrile, or nitromethane, they can be recovered unchanged. In aqueous solutions, they are decomposed [1].

$[Mn(C_6H_5N_2O)_2L'_2]$ and **$[Mn(C_7H_7N_2O)_2L'_2]$** (L′ = pyridine, 2-, 3-, or 4-methylpyridine, or 2-, 3-, or 4-ethylpyridine). To prepare the compounds, anhydrous K_2CO_3 and some NaOH were added to a boiling mixture of $[Mn(C_6H_6N_2O)_2Cl_2]$ or $[Mn(C_7H_8N_2O)_2Cl_2]$ (see above) and the respective pyridine base. The reaction mixture was heated for 2 to 3 h. On cooling, crystals separated which were purified by dissolving them in chloroform and pouring the solution into a large volume of the pyridine base. The crystals formed were washed with the pyridine base and dried over P_4O_{10}.

The complexes are monomeric in formamide. Their magnetic moment is independent of temperature: values between 5.87 and 5.92 μ_B have been determined for the range between 78 and 300 K. The Weiss constants are between −2 and −8 K. The IR spectra of the complexes in CsI exhibit the $\nu(C{=}N)_{oxime}$ band at ~1500 cm^{-1} (−20) and the ν(NO) bands at ~1200 cm^{-1} (+215; shifts in parentheses). The ν(OH) bands, in the 3200 to 2800 cm^{-1} range for the free ligands, vanish on complexation. Bands at 260, 250, and 230 cm^{-1}, which were assigned to $\nu(Mn{-}N)_{oxime}$ and $\nu(Mn{-}N)_{pyridine}$ vibrations, suggest coordination of the hydroxyimino and pyridine nitrogen atoms in *trans* positions. The coordination of the pyridine nitrogen atoms is also indicated by shifting and splitting of the pyridine ring vibrations [2].

The diffuse reflectance spectra at room temperature exhibit weak absorption bands in the 18200 to 20000, 22500 to 25500, and 23000 to 26300 cm^{-1} ranges corresponding to the $^6A_{1g}\rightarrow{}^4T_{1g}(G)$, $^4A_{1g}\rightarrow{}^4T_{2g}(G)$, and $^6A_{1g}\rightarrow{}^4A_{1g}, {}^4E_g(G)$ transitions, respectively, for octahedrally coordinated manganese(II) (bands reported in detail for each compound). Ligand field parameters (Dq, B, C, β) were calculated from the bands: Dq ranging from 700 for $[Mn(C_6H_5N_2O)_2py_2]$ to 820 for $[Mn(C_7H_7N_2O)_2(\gamma\text{-}C_7H_9N)_2]$ ($\gamma\text{-}C_7H_9N$ = 4-ethylpyridine).

The compounds are insoluble in water, slightly soluble in nonpolar solvents and soluble in polar organic solvents. They are nonelectrolytes in ethanol [2].

With 2-Quinoline- or 3-Isoquinolinecarbaldehyde Oximes

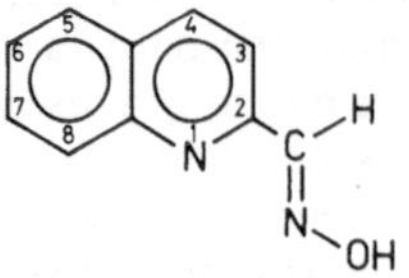

(= $C_{10}H_8N_2O$)

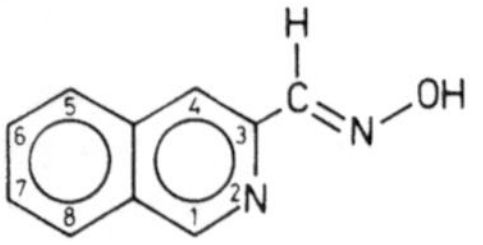

$[Mn(C_{10}H_8N_2O)_2(NO_3)](NO_3)$ and **$Mn(C_{10}H_8N_2O)_2X_2$** (X = Cl, Br, I, ½SO_4, CH_3COO, NCS, NCSe). Compounds with each of the ligands were prepared by mixing boiling solutions of the ligands in a 1:1 (v/v) mixture of ethanol and 2,2-dimethoxypropane with boiling solutions of the respective Mn^{II} salt hydrate (mole ratio 2:1) in the same solvent or in a small amount of water (the latter in the case of X = ½SO_4 and CH_3COO), and refluxing the mixture for 2 to 6 h. On cooling, polycrystalline solids separated, which were filtered, washed with water (in the case of X = ½SO_4 and CH_3COO), ethanol, and ether, and dried over P_2O_5 [3].

The compounds with X = NO_3, ½SO_4, CH_3COO, NCS, and NCSe are monomeric in formamide. X-ray powder diffraction studies show that the compounds with X = CH_3COO, NCS, and NCSe are isostructural and those with X = NO_3 and ½SO_4 are structurally similar to the former compounds, but that those with X = Cl, Br, and I differ in structure from the other compounds. The complexes with X = NO_3, ½SO_4, CH_3COO, NCS, and NCSe show temperature-independent values of μ_{eff} = ~5.9 μ_B. The magnetic moments of the halo complexes decrease from this value at room temperature to ~5.8 for X = Cl, Br and to ~5.6 for X = I at 78 K. From the susceptibility data of the latter compounds, values for the exchange coupling constant of J = −0.16 cm^{-1} for X = Cl, −0.15 cm^{-1} for X = Br, and −0.8 cm^{-1} for X = I were obtained, indicating a slight antiferromagnetic interaction. For all the compounds, g = 2 was calculated [3].

Characteristic IR bands (in cm^{-1}) of the compounds in CsI were assigned as follows (shifts on complexation in parentheses): $\nu(OH)_{free}$, 3480 to 3400; $\nu(OH)_{H\ bonded}$, 3370 to 3360; ν(CH), 3200 to 3090; $\nu(OH)_{coupled}$, 3050 to 3040; $\nu(C{=}N)_{overtone}$, 2850 to 2840; $\nu(C{=}N)_{oxime}$, ~1620(+100); ν(NO), ~1070(+90). Five anion bands in the 1320 to 700 cm^{-1} range for the complexes with X = NO_3 indicate the presence of bidentate coordinated nitrate groups (C_{2v}). The appearance of bands at 1350 and 830 cm^{-1} also indicates ionic nitrate groups (D_{3h}) in these complexes. For the complexes with X = ½SO_4, six anion bands were observed in the 1180 to 450 cm^{-1} range suggesting bidentate sulfate groups, whereas the presence of monodentate acetate groups in the acetato complexes is confirmed by the appearance of $\nu_{as}(COO)$ at ~1650 cm^{-1} and $\nu_s(COO)$ at ~1425 cm^{-1}. Splitting of the ν(CN) bands (~2070 and ~2020 cm^{-1}) indicates *cis*-configuration for the complexes with X = NCS or NCSe. In the far-IR spectrum, $\nu(Mn–N)_{ligand}$ bands were observed at ~375, ~350, ~302, and ~280 cm^{-1}. Bands of Mn–X vibrations were assigned as follows: ν(Mn–Cl), ~235, ~215; ν(Mn–Br), ~145, ~115; $\nu(Mn–N)_{NCS}$, ~325; ν(Mn–O), ~240, ~200, and ~170 for X = NO_3, CH_3COO, or $\frac{1}{2}SO_4$, respectively [3].

The UV-visible reflectance spectra of the complexes show three intense intra-ligand bands in the 46500 to 30000 cm^{-1} region, indicating that the *E*-configuration of the free ligands exists also in the solid complexes. Very intense maxima at ~26000 and ~19500 cm^{-1} were assigned to charge-transfer transitions [3].

All the complexes are quite stable under normal conditions. The compounds are insoluble in water and nonpolar solvents, but are soluble in polar organic solvents except for the halo complexes. Solutions of the nitrato complexes in methanol (~10^{-3} M) show the molar electrical conductance of 1:1 electrolytes (~95 $cm^2 \cdot \Omega^{-1} mol^{-1}$). The complexes with X = $\frac{1}{2}SO_4$, CH_3COO, NCS, or NCSe are nonelectrolytes in methanol. The physical properties indicate a dimeric *cis*-octahedral structure of the type $[(HL)_2MnX_2Mn(HL)_2]X_2$ for the complexes with X = Cl, Br, I, with two common halogen atoms of the two coordination octahedra. For the remaining complexes a monomeric *cis*-octahedral structure is assumed [3].

With *(E)*-Methyl or *(E)*-Phenyl 2-Pyridyl Ketone Oximes

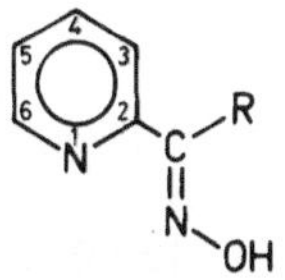

1) R = CH_3; (= $C_7H_7N_2O$ = HL)
2) R = C_6H_5; (= $C_{12}H_{10}N_2O$ = HL)

$[Mn(HL)_2(NO_3)]NO_3$ and **$Mn(HL)_2X_2$** (HL = ligand 1 or ligand 2; X = Cl, Br, I, $\frac{1}{2}SO_4$, CH_3COO, NCS, NCSe). All the compounds were prepared by the procedure given for the complexes with quinoline- and isoquinolinecarbaldehyde oximes except for using ligand 1 or 2 and a 3:2 (v/v) mixture of ethanol and 2,2-dimethoxypropane as the solvent [4]. The compound with ligand 2 and X = Cl was also prepared in ethanol alone. The yellow crystals obtained were recrystallized from ethanol and dried in vacuum and over P_2O_5 at 110°C [5].

Physical and chemical properties of these compounds show the same features as those of the complexes with the quinoline- and isoquinolinecarbaldehyde oximes. Analogous structures are therefore assumed. Details differing somewhat for the complexes treated here compared to those on p. 262 are outlined in the following: Magnetic moments varying between 5.80 and 6.00 μ_B are temperature-independent between 78 and 300 K (however, exchange coupling constants are of the same order of magnitude). Differing assignments are given for the IR spectra of the complex with ligand 2 and X = Cl: $\nu(OH)_{free}$, ~3250; $\nu(OH)_{H\ bonded}$, 3130; ν(CH), 3060; $\nu(OH)_{coupled}$, 2870 (same frequencies for mulls in hexachlorobutadiene and solutions in chloroform). Additional bands of the solution spectrum at 2860 and 1420 cm^{-1} were assigned to $\nu(CN)_{overtone}$ and ν(CN) vibrations, respectively. The IR spectra indicate strong H bonding between the hydroxyimino oxygen and π-orbitals of the phenyl ring. The UV-visible spectrum of $Mn(C_{12}H_{10}N_2O)_2Cl_2$ in Nujol shows a band beginning at 22000 cm^{-1} which merges into a strong band starting at 32000 cm^{-1} with shoulders at 28400, 28800, and 31200 cm^{-1}. A 10^{-3}M solution of this compound in dry methanol shows the molar electric conductance 106 $cm^2 \cdot \Omega^{-1} \cdot mol^{-1}$ [5].

References:

[1] Mohan, M., Malik, W. U., Dutt, R., Srivastava, A. K. (Monatsh. Chem. **111** [1980] 1273/85, 1274, 1276, 1278).
[2] Mohan, M., Varshney, P. K. (Acta Chim. [Budapest] **108** [1981] 147/59, 150, 154; C.A. **96** [1982] No. 154368).
[3] Mohan, M., Kumar, M. (Syn. Reactiv. Inorg. Metal-Org. Chem. **13** [1983] 331/48, 333, 336, 339).
[4] Mohan, M., Paramhans, B. D. (Indian J. Chem. A **19** [1980] 759/65).
[5] Sen, B., Malone, D. (J. Inorg. Nucl. Chem. **34** [1972] 3509/16, 3511, 3513).

27.1.9 With Carboxamide Oximes RC(NH$_2$)=NOH

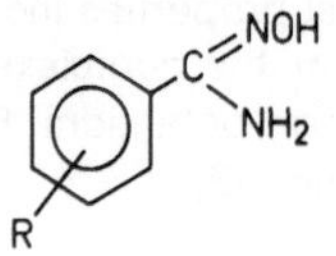

ligand 1 R=H;
(=$C_7H_8N_2O$)

ligand 2 R=4-$CH_3(CH_2)_8O$;
(=$C_{16}H_{26}N_2O_2$)

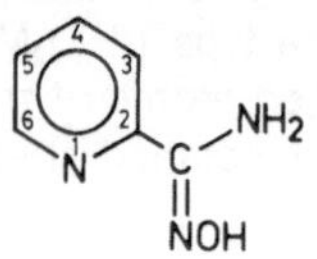

ligand 3
(=$C_6H_7N_3O$)

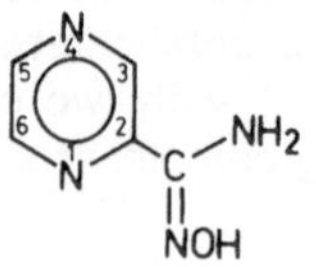

ligand 4
(=$C_5H_6N_4O$)

The stability constants of the 1:1 and 1:2 complexes with ligand 1 and 2 in 70 vol% aqueous dioxane determined by potentiometric titration with the glass electrode are: log $K_1 = 7.62 \pm 0.24$ and log $K_2 = 6.65 \pm 0.24$ for ligand 1 and log $K_1 = 7.43 \pm 0.36$ and log $K_2 = 7.12 \pm 0.36$ for ligand 2 at 22°C and I = 0.1 M (KNO_3) [1].

The formation of colored complexes was observed on reaction of Mn^{II} with ligand 3 in neutral and basic aqueous medium: a yellow soluble complex at pH 7, which is destroyed by acids, a brown-red soluble complex at pH 7 to 8 which is decolorized by acids, and a dark brown precipitate at pH 8 to 10, which is soluble in hot water, HNO_3, and 2N HCl. The latter complex can be utilized for the gravimetric determination of manganese [2].

The light yellow complex with ligand 4, **$Mn(C_5H_6N_4O)_2Cl_2$**, was prepared by adding $MnCl_2 \cdot 4H_2O$ (1 mmol) in solid form to a solution of the ligand (2.5 mmol) in dry ethanol and refluxing the mixture for 0.5 h. The separated complex was filtered hot, washed with the solvent, and dried over fused $CaCl_2$. The yield was 70% [3].

The room temperature magnetic moment, $\mu_{eff} = 6.04\ \mu_B$, indicates a high-spin d^5 complex. In the IR spectrum, the ν(NO) band is shifted from 952 cm^{-1} for the free ligand to wave numbers between 1000 and 1030 cm^{-1} for the transition metal complexes indicating coordination through hydroxyimino nitrogen. Participation of the heterocyclic ring nitrogen in complex formation is indicated by positive shifts of the in-plane and out-of-plane ring deformation bands (at 620 to 640 and 410, 440 cm^{-1}, respectively, for the free ligand) [3].

The complex melts at 282°C. It seems to be amorphous in nature and is insoluble in solvents like ethanol, acetone, nitromethane, etc. In aqueous medium, the complex is decomposed. The molar conductance of a 10^{-3} M solution in methanol, 156.3 $L \cdot \Omega^{-1} \cdot mol^{-1}$, corresponds to a 1:2 electrolyte presumably generated through solvolysis [3].

References:

[1] Nematov, I., Petrukhin, O. M., Martirosov, A. E., Talipov, Sh. T. (Zh. Neorgan. Khim. **23** [1978] 1299/302; Russ. J. Inorg. Chem. **23** [1978] 715/7).
[2] Vicente Perez, S., Losado del Barrio, J., Lorenzo Abad, E. (Anales Quim. **81** [1985] 75/81, 81).
[3] Sanyal, G. S., Modak, A. B. (Syn. Reactiv. Inorg. Metal-Org. Chem. **16** [1986] 113/25).

27.1.10 With Other Monooximes

A complex of composition $Mnpy_4[C(CN)_2NO]_2$ with **(hydroxyimino)propanedinitrile** $HON{=}C(CN)_2$ (= nitrosodicyanmethanide) is described in "Manganese" D 3, 1982, p. 90.

With 2-(Hydroxyimino)-N-phenylacetamide (= Isonitrosoacetanilide) **and Derivatives**

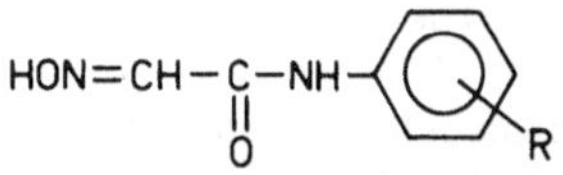

The ligand with R = H (isonitrosoacetanilide = $C_8H_8N_2O_2$) gives a precipitate with Mn^{2+} ions in aqueous ammonia [1]. Green colorations were observed on the reactions of Mn^{2+} ions with the N-methyl (= $C_9H_{10}N_2O_2$) or N-ethyl (= $C_{10}H_{12}N_2O_2$) derivatives of isonitrosoacetanilide [8]. Yellow or yellow-brown colors are reported for the following ligands:

R	formula	Ref.	R	formula	Ref.
2- or 4-Cl	$C_8H_7ClN_2O_2$	[2]	3- or 4-COOH	$C_9H_8N_2O_4$	[5,6]
4-I	$C_8H_7IN_2O_2$	[3]	4-$COOC_2H_5$	$C_{11}H_{12}N_2O_4$	[5]
2- or 4-OCH_3	$C_9H_{10}N_2O_3$	[4,7]	4-$COOC_4H_9$	$C_{13}H_{16}N_2O_4$	[5]
2- or 4-OC_2H_5	$C_{10}H_{12}N_2O_3$	[4]			
3- or 4-COOH					

References:

[1] Buscarons, F., Buscarons Jr., F., (Anal. Chim. Acta **32** [1965] 568/74, 569).
[2] Buscarons, F., Julve, E. (Anales Real. Soc. Espan. Fis. Quim. [Madrid] B **59** [1963] 51/8, 56).
[3] Buscarons, F., Julve, E. (Chim. Anal. [Paris] **46** [1964] 72/6).
[4] Buscarons, F., Mena, R. (Chim. Anal. [Paris] **45** [1963] 72/9, 78).
[5] Duñach, J., Buscarons, F. (Analusis **2** [1973] 510/3).
[6] Buscarons, F., Duñach, J. (Chim. Anal. [Paris] **43** [1961] 457/60).
[7] Buscarons, F., Mena, R. (Anales Real Soc. Espan. Fis. Quim. [Madrid] B **57** [1961] 495/502, 500/1).
[8] Buscarons, F., Izquierdo, A. (Anales Real Soc. Espan. Fis. Quim. [Madrid] B **58** [1962] 601/6).

27.2 Complexes with Dioximes

General

This section initially treats complexes with *vic*-dioximes (see p. 266) comprising mainly substituted glyoximes, like the well-known dimethylglyoxime. These ligands form 1:1 and 1:2 complexes of high stability with Mn^{II} in aqueous-organic medium. They are bidentate and form five-membered chelate rings by coordination of the hydroxyimino nitrogen atoms. The high stability, particularly of the 1:2 complexes with *vic*-dioximes, results from their unique structure involving two five-membered and two six-membered chelate rings, the latter being formed by H bonds between the hydroxyimino oxygens of the two coordinated ligands (see structure on p. 267).

The remaining dioximes having the NOH groups on opposite sides of the ligand molecule or in *meta* position generally contain additional donor atoms or groups (heterocyclic nitrogen or hydroxy groups) offering the possibility of chelate formation. Intensely colored soluble complexes are formed on reaction of Mn^{2+} ions with the dioximes of heterocyclic nitrogen compounds in alkaline aqueous solution. These ligands having the NOH groups at adjacent positions to the heterocyclic nitrogen are tridentate and form two chelate rings on coordination. Some of these complexes are believed to contain Mn^{III} (formed by air oxidation of Mn^{II} at high pH values (~10)). All these soluble complexes are of analytical interest.

Solid Mn^{II} compounds of the type $[MnL(H_2O)_2]_n$ with dioximes derived from bis(hydroxyaldehydes) or bis(hydroxyketones) were prepared in ethanol or DMF. The complexes with these tetradentate ligands bearing the chelating groups on opposite sides of the molecule are chain-polymers, water molecules completing the coordination octahedron of the Mn atom. Yellow or brown complexes of the trigonal bipyramidal type $[Mn(H_2L)X_2]$ and of the octahedral type $[Mn(H_2L)_2]X_2$ with tridentate 2,6-diacetylpyridine dioxime and nitrate, halide or pseudohalide anions X, were prepared in hot methanol. Polymeric complexes again of the type MnL $1.5H_2O$ and $Mn(HL)_2$ with H_2L = p-benzoquinone dioxime were isolated from aqueous solution. For all the complexes, coordination of the hydroxyimino nitrogen is indicated by shifts of the ν(CN) band to lower or higher energy and of the ν(NO) band to higher energy. For the complexes with p-benzoquinone dioxime, additional coordination of the hydroxyimino oxygen is indicated by unusually large shifts and by splittings of the ν(NO) band on coordination. These complexes presumably consist of layers which are connected by H bonds in the case of the hydrate or by Mn–O bonds in the case of $Mn(HL)_2$. For all the polymeric complexes, magnetic measurements indicate metal-metal interactions.

27.2.1 With *vic*-Dioximes

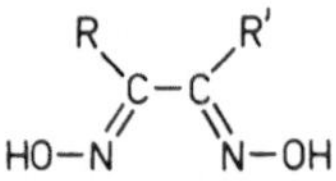

The table presents stability constants of 1:1 and 1:2 complexes of Mn^{II} with the ligands HL in aqueous-organic medium determined by potentiometric titration with a glass electrode:

No.	R (or ligand)	R'	formula	organic solvent (vol%)	t in °C	I in mol/L	log K_1	log β_2	Ref.
1	H	C_6H_5	$C_8H_8N_2O_2$	methanol	30	0.25(LiCl)	5.85	9.65	[1]
2	H	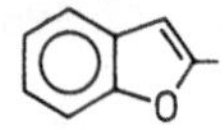	$C_{10}H_8N_2O_3$	75% methanol	30	0.25($NaClO_4$)	5.78	8.88	[2]
3	CH_3	CH_3	$C_4H_8N_2O_2$	50% dioxane	25	0.3($NaClO_4$)	8.6	17.2	[3]
4	C_6H_5	C_6H_5	$C_{14}H_{12}N_2O_2$	50% dioxane	25	0.3($NaClO_4$)	7.9	15.3	[3]
5	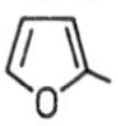		$C_{10}H_8N_2O_4$	75% dioxane	25	0.1($NaClO_4$)	6.4 ± 0.1	11.7 ± 0.1	[4]
6	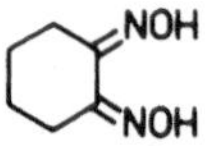 NOH / NOH		$C_6H_{10}N_2O_2$	50% dioxane	25	0.3($NaClO_4$)	8.2	15.4	[5]
7	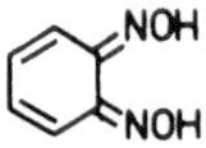 NOH / NOH		$C_6H_6N_2O_2$	75% dioxane	25	0.1($NaClO_4$)	$K_1 \ll K_2$	8.8	[6]

Thermodynamic data of formation are reported for the 1:2 complex with ligand 5, $Mn(C_{10}H_7N_2O_4)_2$: $\Delta H_{\beta2} = -10 \pm 4$ kcal/mol, $\Delta G_{\beta2} = -15.9 \pm 2$ kcal/mol, and $\Delta S_{\beta2} = 20 \pm 9$ cal·mol^{-1}·K^{-1} [4]. Equilibrium constants K for $MnL_2(H_2O)_2 \rightleftharpoons MnL_2(H_2O)(OH)^- + H^+$ and K′ for $MnL_2(OH)(H_2O)^- \rightleftharpoons MnL_2(OH)_2^{2-} + H^+$ were determined potentiometrically with a glass electrode at 25°C and I = 0.3 M($NaClO_4$): log K = 4.8, log K′ = 3.9 for ligand 3 [3]. For the reaction $MnL_2 + 2OH^- \rightleftharpoons MnL_2(OH)_2^{2-}$, log β = 7.0 was found for ligand 6 at the same conditions [5].

From the stability and thermodynamic data it is concluded that the nonreleased proton of each of the coordinated ligands participates in a strong intramolecular hydrogen bond which stabilizes the 1:2 compared to the 1:1 complexes [7], especially those with ligand 7 where $K_2 \gg K_1$ [6]. Therefore a square bipyramidal structure was proposed for the 1:2 complexes with a square planar arrangement of the two ligands (as shown below) and the two water molecules (or OH groups in hydroxy complexes) occupying the axial positions at the Mn atom [3, 7]:

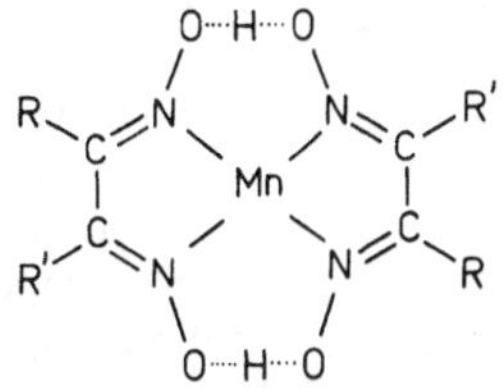

The complexes are of analytical interest [3 to 7]. The Mn^{II} complex with dimethylglyoxime (ligand 3) is catalytically active in the chemiluminescent oxidation of luminol by H_2O_2. A method for the determination of small amounts of dimethylglyoxime based on this activity was developed [8].

References:

[1] Verma, H. S., Saxena, R. C., Gupta, S. L. (Acta Ciencia Indica **3** [1977] 97/8; C.A. **87** [1977] No. 173564).

[2] Verma, H. S., Saxena, R. C., Rastogi, S. C. (J. Indian Chem. Soc. **56** [1979] 828/9).

[3] Burger, K., Ruff, I. (Talanta **10** [1963] 329/38, 332).

[4] Burger, K., Papp-Molnár, E. (Acta Chim. Acad. Sci. Hung. **53** [1967] 111/20, 115, 118; Magy. Kem. Folyoirat **73** [1967] 77/81; C.A. **66** [1967] No. 121634).

[5] Burger, K., Ruff, I., Papp-Molnár, E. (Ann. Univ. Sci. Budapest Rolando Eotvos Nominatae Sect. Chim. **7** [1965] 49/54; C.A. **65** [1965] 6378).

[6] Burger, K., Ruff, I. (Acta Chim. Acad. Sci. Hung. **49** [1966] 1/9, 5).

[7] Burger, K. (in: Flaschka, H. A., Barnard Jr., A. J., Chelates Anal. Chem. **2** [1969] 179/212, 186/7).

[8] Pilipenko, A.T., Kalinichenko, I. E., Matveeva, E. Ya. (Zh. Anal. Khim. **33** [1978] 1612/7).

27.2.2 With Dioximes Derived from Bis(hydroxyaldehydes) or Bis(hydroxyketones)

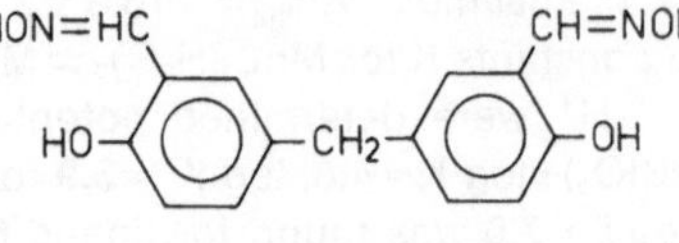

1) ($=C_{15}H_{14}N_2O_4$)

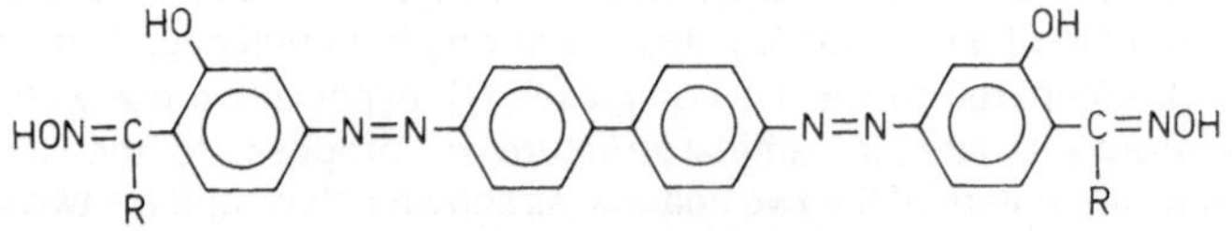

2) R = H; ($=C_{26}H_{18}N_6O_4$)
3) R = CH_3; ($=C_{28}H_{24}N_6O_4$)

$[MnL(H_2O)_2]_n$ (with H_2L = ligands 1 and 3) and **$\{[MnL(H_2O)_2]\cdot H_2O\}_n$** (with H_2L = ligand 2). For preparation, solutions of an Mn^{II} salt and the respective ligand in the mole ratio 1:1 were mixed and the pH adjusted by addition of sodium acetate to the refluxing mixture. Solvents used were: absolute alcohol for preparation of the complex with ligand 1 [1]; water for the Mn salt (chloride) and DMF for preparation of the complex with ligand 2 [2]; and DMF for preparation of the complex with ligand 3 [3]. The complex with ligand 1 precipitated immediately. It was filtered off and dried at 70°C. It was purified by extraction with water, then with alcohol, and was dried in vacuum at 50°C [1]. The reaction mixtures containing the complexes with ligands 2 and 3 were refluxed for 2 to 4 h, filtered off, washed with hot DMF, hot water, finally with ethanol and dried at 80°C [2, 3]. The table below lists the color and characteristic bands with assignments in the IR and diffuse reflectance spectra:

No.	complex (color)	characteristic IR data[a)] bands in cm^{-1}	electronic transitions from $^6A_{1g}$	$\bar{\nu}_{max}$ in cm^{-1}	Ref.
1	$[Mn(C_{15}H_{12}N_2O_4)(H_2O)_2]_n$ (dark green)	$\nu(OH)_{H_2O}+\nu(OH)_{NOH}$ 3575 $\nu(C{=}N)$, 1600(−40) $\nu(CO)$(?), 1025(+20)	$\rightarrow {}^4T_{1g}$ $\rightarrow {}^4T_{2g}$ $\rightarrow {}^4A_{1g}, {}^4E_g$	16560 21050 24960	[1]
2	$\{[Mn(C_{26}H_{18}N_6O_4)(H_2O)_2]\cdot H_2O\}_n$ (brown)	$\nu(OH)_{H_2O}+\nu(OH)_{NOH}$ $+\nu(OH)_{COH}$, 3200 $\nu(C{=}N)+\nu(N{=}N)$ 1620, 1570(−10) $\nu(CO)$, 1320(+20)	$\rightarrow {}^4T_{1g}$ $\rightarrow {}^4T_{2g}$	19010 22220	[2]
3	$[Mn(C_{28}H_{22}N_6O_4)(H_2O)_2]_n$ (brown)	$\nu(OH)_{H_2O}+\nu(OH)_{NOH}$ 3450 to 3200(br) $\nu(C{=}N)$, 1625(−31) $\nu(N{=}N)$, 1600 $\delta(OH)_{H_2O}$, 1300 $\nu(CO)$, 1136(−4) $\nu(NO)$, 1058(+33)	$\rightarrow {}^4T_{1g}$ $\rightarrow {}^4T_{2g}$	19230 21739	[3]

a) In Nujol in the case of ligand 1, in KBr in the case of ligand 2, not stated for ligand 3. – b) Diffuse reflectance spectra; two additional bands in the spectrum of the complex with ligand 1 (18180 and 23810 cm^{-1}) were not assigned.

The effective magnetic moments (5.43, 5.58, and 5.87 μ_B for the complexes with ligands 1, 2, or 3, respectively), which are lower than the spin-only values for $3d^5$ complexes (5.92 μ_B), suggest the presence of metal-metal interactions [1 to 3]. Negative shifts of the ν(C=N) bands and positive shifts of the ν(NO) bands in the IR spectra indicate coordination through the hydroxyimino nitrogen, while the ν(N=N) band in the azo compounds is unaffected. The temperatures of dehydration and decomposition determined by thermogravimetric analyses are: 250 and 372°C for the complex with ligand 1 [1], 200 and 330°C for that with ligand 2 [2], and 200 and 299 for that with ligand 3 [3]. From the TG results an energy of activation for the decomposition reaction of ~23 kcal/mol and the order of reaction ~2 were obtained for the complex with ligand 1 [1]. This complex is soluble in water and ethanol [1]. All the complexes are insoluble in common organic solvents [1 to 3]. The complex with ligand 3 is a nonelectrolyte in DMF [3]. The physical properties indicate a polymeric octahedral structure for all the compounds involving the oxygen of the deprotonated phenolic hydroxyl group and the hydroxyimino nitrogen from each of two ligand molecules in complex formation. Two water molecules complete the coordination octahedron of the Mn atom [1 to 3].

References:

[1] Karampurwala, A. M., Patel, R. P., Shah, J. R. (J. Macromol. Sci. Chem. A **15** [1981] 431/8).
[2] Rana, A. K., Shah, N. R., Patil, M. S., et al. (Makromol. Chem. **182** [1981] 3387/95).
[3] Suthar, H. B., Shah, J. R. (Angew. Makromol. Chem. **130** [1985] 147/53).

27.2.3 With Dioximes Derived from 2,6-Pyridinedicarbaldehyde or from a Related Diketone

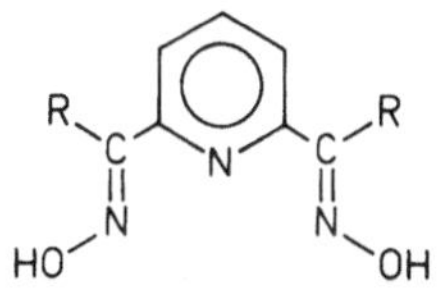

1) R = H; $(=C_7H_7N_3O_2)$
2) R = CH_3; $(=C_9H_{11}N_3O_2)$

Complexes in Solution. Stability constants of the complex species $Mn(C_7H_6N_3O_2)^+$ and $Mn(C_7H_6N_3O_2)_2$ with the mono-deprotonated ligand 1 in aqueous solution, log K_1 = 4.4 and log K_2 = 4.1, have been determined by potentiometric titration with a glass electrode at 25°C and I = 0.1 M ($NaClO_4$). Aqueous solutions of the $Mn(C_7H_6N_3O_2)_2$ species are blue-violet in the pH region 4 to 9. The color is due to an absorption maximum at 590 nm with log ε = 2.48 at pH 3.8 and log ε = 3.05 at pH 7.8. Chelate formation involving the pyridine nitrogen and the deprotonated hydroxyimino group is suggested [1].

In higher pH regions, at pH 9.5 to 11.5, Mn^{2+} ions form intensely green soluble chelate complexes with both ligands. The solutions appear red in a thick layer or at high concentrations of manganese. The color is due to an absorption maximum at 598 nm with log ε = 3.61 for the complex with ligand 2. The color is stable for several hours. While tartaric acid or triethanolamine do not affect the complexes, they are destroyed by reducing agents such as ascorbic acid, hydroxylammonium chloride, or hydrazinium sulfate. Therefore, they are believed to contain manganese(III) formed by air oxidation. The complexes can be utilized for the spectrophotometric determination of manganese [2].

$[Mn(C_9H_{11}N_3O_2)X_2]$ (X = NO_3, Cl, Br, I, NCS, NCSe). The compounds with X = NO_3, Cl, or Br precipitated on slowly adding a hot methanol solution of ligand 2 to a boiling methanol solution of the respective hydrated Mn^{II} salt in the mole ratio 1:1. The mixture was allowed to stand for 30 min, then heated to boiling and filtered hot. The compound with X = I was obtained by refluxing combined methanol solutions of freshly prepared MnI_2 and the ligand for 30 min, then allowing the mixture to stand for 30 min. The compounds with X = NCS or NCSe were prepared by refluxing a suspension of $[Mn(C_9H_{11}N_3O_2)Cl_2]$ in methanol with an excess of solid KSCN or KSeCN for about 2 h. All the compounds were washed with methanol and ether and dried over P_2O_5 in vacuum. Rapid isolation was necessary to preserve the 1:1 stoichiometry of the complexes. Longer reaction times resulted in a rearrangement to form the 1:2 complexes (p. 271).

The table lists some physical data of the yellow complexes (magnetic moments μ_{eff} in μ_B, wave numbers of far-IR bands and bands in the UV-visible region in cm^{-1}, molar conductance Λ in $cm^2 \cdot \Omega^{-1} \cdot mol^{-1}$):

complex	μ_{eff}	far-Ir bands[a]	UV-visible bands[b]	Λ
$[Mn(C_9H_{11}N_3O_2)(NO_3)_2]$	5.94	$\nu(Mn\text{-}N_{py})$ 250; $\nu(Mn\text{-}N)_{oxime}$ 220	18000, 25000	1.8
$[Mn(C_9H_{11}N_3O_2)Cl_2]$	6.10	$\nu_s(Mn\text{-}Cl)$ 288; $\nu_{as}(Mn\text{-}Cl)$ 265; $\nu(Mn\text{-}N_{py})$ 248; $\delta(Mn\text{-}Cl)$ 236; $\nu(Mn\text{-}N)_{oxime}$ 220	15600, 21150, 22800, 24700, 26500	30.2
$[Mn(C_9H_{11}N_3O_2)Br_2]$	6.00	$\nu(Mn\text{-}N_{py})$ 250; $\nu_s(Mn\text{-}Br)$ 235; $\nu(Mn\text{-}N_{oxime})$ 220; $\nu_{as}(Mn\text{-}Br)$ 208; $\delta(Mn\text{-}Br)$ 196	15200, 20980, 22800, 24500, 26500	20.4
$[Mn(C_9H_{11}N_3O_2)I_2]$	6.00	$\nu(Mn\text{-}N_{py})$ 245; $\nu(Mn\text{-}N_{oxime})$ 218	15300, 21150, 22760, 24250, 26500	52.8
$[Mn(C_9H_{11}N_3O_2)(NCS)_2]$	5.89	$\nu(Mn\text{-}N_{NCS})$ 267; $\nu(Mn\text{-}N_{py})$ 250; $\nu(Mn\text{-}N)_{oxime}$ 227; $\nu(Mn\text{-}S)_{NCS}$ 210	17500, 24900	2.0
$[Mn(C_9H_{11}N_3O_2)(NCSe)_2]$	5.89	$\nu(Mn\text{-}N_{NCSe})$ 265; $\nu(Mn\text{-}N_{py})$ 245; $\nu(Mn\text{-}N)_{oxime}$ 225	17700, 25000	2.0

[a] In CsI. – [b] Diffuse reflectance spectra. Similar bands were observed in ethanol with ε from ~1 to ~8 $L \cdot mol^{-1} \cdot cm^{-1}$.

The magnetic moments are typical for high-spin configurations. The Curie-Weiss law is obeyed with relatively small values of the Weiss constant (2 K for the complex with X = NO_3, ~ −5 K for the other complexes). The IR spectra in the 4000 to 900 cm^{-1} range show bands in the following regions, which were assigned as follows (shifts in parentheses): $\nu(OH)_{free}$, 3450 to 3300; $\nu(OH)_{H\,bonded}$, 3200 to 3150; $\nu(CH)$, 3050 to 3000; $\nu(OH)_{coupled}$, 2920 to 2750; $\nu(C{=}N)_{NOH}$, ~1600 (+40); $\nu(NO)$, ~1050(+80). The ring vibrations (not given in the publication) are also shifted, indicating coordination of the pyridine nitrogen. Four anion bands in the 1500 to 1200 cm^{-1} region for the complex with $X = NO_3$ indicate the presence of bidentate and monodentate nitrato groups. Two ν(CN) bands near 2100 cm^{-1} for the complexes with X = NCS and NCSe indicate the presence of both bridging and terminal isothiocyanato and isothioselenato groups for these compounds. The electronic spectra of the complexes with X = Cl, Br, I differ from those of octahedral Mn^{II} complexes, but are similar to those of trigonal bipyramidal symmetry. The bands at ~18000 and ~25000 cm^{-1} for the complexes with $X = NO_3$, NCS, or NCSe were assigned to ligand → Mn or Mn → ligand charge-transfer transitions, respectively.

The complexes are quite stable at room temperature. They are insoluble in water and nonpolar organic solvents, but, except for the complexes with X = NCS and NCSe, they are soluble in polar solvents. The molar conductance values indicate partial solvolytic displacement of coordinated ions for the complexes with X = Cl, Br, or I. The physical properties indicate a monomeric trigonal bipyramidal structure for the complexes with X = Cl, Br, or I, a monomeric octahedral structure for those with $X = NO_3$, and a polymeric octahedral structure with bridging NCS or NCSe groups for the complexes with these anions [3].

$[Mn(C_9H_{11}N_3O_2)_2]X_2$ ($X = NO_3$, Cl, Br, I, NCS, or NCSe). The 1:2 complexes with $X = NO_3$, Cl, Br, or I were prepared by slowly adding a hot methanol solution of $MnX_2 \cdot 4H_2O$ or MnI_2 (2 mmol) to a boiling methanol solution of the ligand (4.5 mmol). On cooling the solutions, yellow crystalline solids deposited. The dark brown compounds with X = NCS or NCSe were prepared by refluxing a suspension of $[Mn(C_9H_{11}N_3O_2)_2]Cl_2$ with excess KSCN or KSeCN (mole ratio 1:4) for 1 h. All the compounds were washed with methanol and ether and dried over P_2O_5 in vacuum.

Magnetic moments (5.89 to 5.96 μ_B) and spectral data are those expected for octahedral $3d^5$ complexes. In the far-IR spectrum, the $\nu(Mn–N)_{py}$ bands were located at 240 to 250 cm^{-1} and the $\nu(Mn–N)_{oxime}$ bands at 220 to 230 cm^{-1} (other IR and electronic spectral data are not given in detail). The molar conductance values are those of 1:2 electrolytes. The complexes are quite stable at room temperature. Their solubility properties are similar to those of the 1:1 complexes (p. 270). The physical properties suggest a monomeric octahedral structure with the hydroxyimino and pyridine nitrogen atoms as the coordination sites [3].

References:

[1] Bag, S. P., Fernando, Q., Freiser, H. (Anal. Chem. **35** [1963] 719/22).
[2] Hartkamp, H. (Angew. Chem. **72** [1960] 349).
[3] Mohan, M., Kumar, M. (Transition Metal Chem. [Weinheim] **10** [1985] 255/8).

27.2.4 With Other Dioximes Derived from N-Heterocyclic Dioxo Compounds

Mn^{2+} ions give characteristic color reactions in ammoniacal or sodium hydroxide solutions, respectively, with the following ligands: brown, greenish yellow with 2,5-pyrrolidinedione dioxime ($= C_4H_7N_3O_2$) [1]; yellow, green with 3,4-diphenyl-2,5-pyrrolidinedione dioxime ($= C_{16}H_{15}N_3O_2$) [2]; rose-colored, yellow with 1,3-isoindolinedione dioxime ($= C_8H_7N_3O_2$) [1]; violet, blue-violet with 2,6-piperidinedione dioxime ($= C_5H_9N_3O_2$) [3]. With 3-methylene-2,6-piperidinedione dioxime ($= C_6H_9N_3O_2$), green precipitates were observed in both media [4]. The colored complexes are of analytical interest [1 to 4].

References:

[1] Buscaróns, F., Abelló, J. (Anales Real Soc. Espan. Fis. Quim. [Madrid] B **58** [1962] 591/600, 597/8).

[2] Buscaróns, F., Caralt, E. (Inform. Quim. Anal. [Madrid] **27** [1973] 138/46, 146; C.A. **80** [1974] No. 22280).

[3] Buscaróns, F., Iturriaga, H. (Anales Real Soc. Espan. Fis. Quim [Madrid] B **63** [1967] 95/100).

[4] Rius, J., Mongay, C., Cerdá, V. (Afinidad **38** [1981] 235/40).

27.2.5 With 1,4-Benzoquinone Dioxime

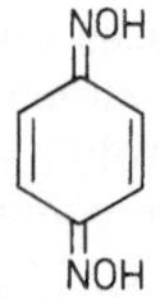

$(=C_6H_6N_2O_2)$

$Mn(C_6H_4N_2O_2)\cdot 1.5H_2O$ was prepared by adding the stoichiometric amount of $Mn(NO_3)_2$ to an aqueous solution of $Na_2C_6H_4N_2O_2$ heated to 50°C. The microcrystalline, colored compound (no color given) has the magnetic moment $\mu_{eff}=2.80$, 3.49, or 3.53 μ_B at 78, 300, and 381 K, respectively.

The IR spectrum recorded from KBr disks (3800 to 400 cm^{-1}) or liquid-paraffin mulls (400 to 30 cm^{-1}) shows that the ν(C=N) band of the free ligand at 1350 cm^{-1} has split and shifted to lower wave numbers on coordination (1315 and 1298 cm^{-1}). The ν(NO) band at 1000 cm^{-1} also has split but shifted to higher frequency (1260 and 1070 cm^{-1}). Bands at 396 and 320 cm^{-1} presumably are due to the ν(Mn–N) or ν(Mn–O) vibrations, respectively. Broad bands in the 3600 to 2400 cm^{-1} region are due to water molecules participating in hydrogen bonds. The IR spectrum, particularly the positions and intensities of the ν(NO) bands compared to those of the 1:2 complex (see below), suggest that both the nitrogen and oxygen atoms are involved in coordination. The physical properties suggest a polymeric layer structure as shown in **Fig. 25** with *trans*-square planar arrangement of the N and O donor atoms and H bonds through the water molecules between the layers.

The complex is insoluble in water and common organic solvents such as ethanol, benzene, or DMF.

$Mn(C_6H_5N_2O_2)_2$ was obtained by reacting stoichiometric amounts of $Mn(NO_3)_2$ and $Na_2C_6H_4N_2O_2$ in aqueous solution at 50°C. The microcrystalline, colored precipitate (no color given) was washed with water and dried at 100°C. Magnetic measurements at 78, 300, and 396 K yielded the effective magnetic moments 3.81, 4.79, or 4.92 μ_B, respectively. The important bands in the IR spectrum were assigned as follows (free ligand bands in parentheses): ν(C=N), 1318, 1300(1350); ν(NO), 1270, 1175, 1095(1000); ν(Mn–N)?, 400; ν(Mn–O)?, 320. The splits and shifts of the ν(C=N) band suggest coordination of the hydroxyimino nitrogen atom. The ν(NO) band is also split, but significantly shifted to higher wave numbers, this being accompanied by considerable weakening. A broad diffuse band in the 2500 to 1700 cm^{-1} range characterizes the formation of chelate hydrogen bonds. The diffuse reflectance spectrum (not given in the publication) indicates an octahedral symmetry for the 1:2 complex which is insoluble in water and the common organic solvents. All the physical properties are consistent with a polymeric three-dimensional structure. It is suggested that the Mn atom is equatorially coordinated by four N atoms of four hydroxyimino groups two of them deprotonated and that the axial positions are occupied by O atoms of nondeprotonated hydroxyimino groups from adjacent layers. H bonds between the hydroxyimino O atoms within the layers are assumed. The complex reacts with Mn^{2+} ions to give the 1:1 complex.

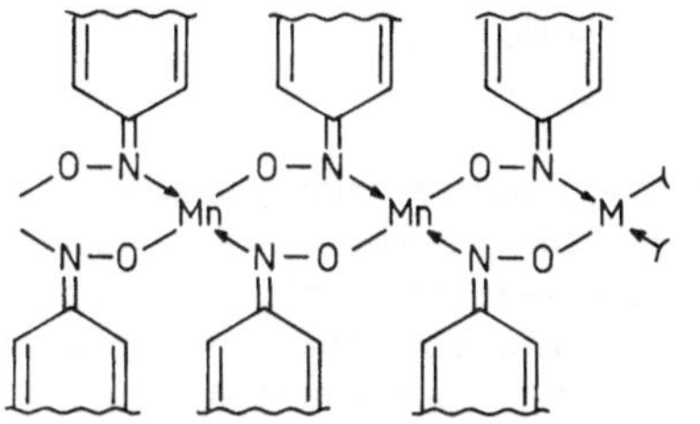

Fig. 25. Structure proposed for the 1:1 complex with 1,4-benzoquinone dioxime, $Mn(C_6H_4N_2O_2)\cdot1.5H_2O$, (water molecules omitted).

Reference:

Titov, N. M., Kalinichenko, I. I., Purtov, A. I., Nikonenko, E. A. (Zh. Neorgan. Khim. **26** [1981] 2153/8; Russ. J. Inorg. Chem. **26** [1981] 1159/62).

27.2.6 With Dioximes Derived from Dicarboxamides

1) $HN[CH_2C(NH_2)=NOH]_2$ $(=C_4H_{11}N_5O_2)$
2) $HN[CH_2CH_2C(NH_2)=NOH]_2$ $(=C_6H_{15}N_5O_2)$
3) $O[CH_2CH_2C(NH_2)=NOH]_2$ $(=C_6H_{14}N_4O_3)$

The complex with ligand 1, **$[Mn(C_4H_{11}N_5O_2)_2]Cl_2$**, was prepared by dropwise addition of an alcohol solution of $MnCl_2\cdot4H_2O$ to an aqueous solution of the ligand (mole ratio 1:2). The tan precipitate, which formed immediately, was washed with absolute ethanol, acetone, and dry ether, and air-dried. The yield was 76% based on $MnCl_2\cdot4H_2O$. On heating, the complex darkens and decomposes at about 200°C. The compound is slightly soluble in water and insoluble in alcohol, acetone, and ether. The aqueous solution of the complex exhibits an absorption maximum at 670 nm (extinction coefficient $\varepsilon\approx4$ $L\cdot mol^{-1}\cdot cm^{-1}$). Measurements of the electric conductance on a 10^{-3} M aqueous solution indicate dissociation into three ions [1].

Characteristic yellow-brown or yellow color reactions were observed on reaction of Mn^{2+} ions with ligands 2 or 3 in neutral aqueous solution as well as in acetic acid and sodium acetate media [2].

References:

[1] Eddy, L. P., Levenhagen, W. W., McEwen, S. K. (Inorg. Syn. **11** [1968] 89/93).
[2] Mantecón, A., Cádiz, V., Cerdá, V. (Afinidad **38** [1981] 137/42).

27.3 Complexes with Derivatives of N-Nitrosohydroxylamine

27.3.1 With N-Nitroso-N-phenylhydroxylamine (Ammonium Salt = Cupferron)

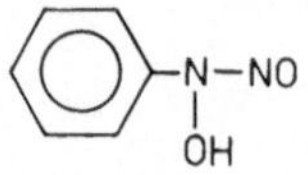

$(=C_6H_6N_2O_2)$

$Mn(C_6H_5N_2O_2)_2$. The formation of the sparingly soluble complex from $MnCl_2$ and cupferron in aqueous solution was revealed by refractometric measurements (method of continuous variation) [10]. The orange precipitate prepared by reacting an Mn^{II} salt with cupferron in water [1, 2, 9] was washed with water and dried under a stream of dry air at 60°C [1, 2] or washed with water and alcohol and air-dried for at least 24 h at room temperature [3]. The yield was 80% [2].

IR spectra were recorded from KBr disks [1, 2]. Characteristic bands were assigned as follows (shifts from those of cupferron in parentheses): $\nu(C{=}C)_{ring}$, 1605(+5), 1490; ν(CN), 1360(+20); ν(NN), 1310(−12), 1298(−24); ν(N–O), 1220, 1190, 1162, 932(+17); ν(Mn–O), 545, 490, 405 cm^{-1}. Two ν(N=O) bands, at 1460 and 1275 cm^{-1}, for cupferron have disappeared on complexation [1]. In another publication, the ν(N–O) bands were located at 1288 and 927 cm^{-1} [2]. The number of three ν(Mn–O) bands suggests a tetrahedral structure of the complex [1], as was also assumed on the basis of the X-ray diffraction and magnetic studies [9].

The complex is monomeric in chloroform. Its X-ray powder patterns are similar to those of the Fe^{II}, Ni^{II}, or Zn complexes, but differ from those of the Cu^{II} complex [9]. The effective magnetic moment measured on a powdered sample at room temperature is 4.92 μ_B. The value is lower than that expected for a $3d^5$ complex (5.92 μ_B). The difference may be attributed to a residual orbital angular momentum or a distortion of the crystal field [8]. Varied temperature magnetic measurements (no values given) indicate no association in the solid state [9].

Varied temperature measurements of the electrical resistivity ϱ indicate that the complex is a semiconductor. The resistivity obeys the relation $\varrho = \varrho_0 \exp(E/kT)$ with $\varrho_0 = 1.2 \times 10^9\ \Omega \cdot cm$ and the activation energy of electrical conduction E = 1.15 eV. This energy was compared with the energy of 4.30 eV corresponding to a charge-transfer band at 307 nm in dioxane [5].

The compound decomposes at 192°C under an atmosphere of dry nitrogen, as determined thermogravimetrically at a heating rate of 4 to 5°C/min [2]. When the complex has been heated in vacuum at 10°C/min, the DTA curve shows an endothermic peak at ~170°C followed by an exothermic one at ~250°C and a second endothermic peak at ~280°C [3]. The compound is not strongly hygroscopic but contains small amounts of adsorbed water [2].

The complex is extractable with various organic solvents from the aqueous medium [6, 11, 12], for example with chloroform in the pH region 4.5 to 9 to an extent of 16% from aqueous 0.1 M sodium perchlorate solution [12]. Extraction studies with isopentyl alcohol from various aqueous mineral acids show that the complex is quantitatively extractable above pH 5. The composition of the solvent adducts is 1:2:2 (Mn:ligand:isopentyl alcohol). The extraction constant of the complex from perchloric acid solution was evaluated [6]. The flame photometric determination of manganese(II) using solutions of the chelate in chloroform, 4-methyl-2-pentanone, pentyl acetate, or toluene was investigated [11].

Absorption spectra in the UV, visible and near-IR regions were recorded from solutions of the complex in various solvents [1, 4, 7]. An absorption maximum at 580 nm with $\varepsilon = 2\ L \cdot mol^{-1} \cdot cm^{-1}$ of the complex in dioxane was assigned to the $^6A_1(S) \rightarrow {}^4T_1(G)$ transition of Mn^{II}. In the 750 to 2000 nm region, solutions in methanol and dioxane show five bands, and solutions in ethanol or isopropanol show two bands with ε from 1 to 100 [4]. In the UV region

(200 to 300 nm) four bands with $\varepsilon = (3\ \text{to}\ 4) \times 10^4\ L \cdot mol^{-1} \cdot cm^{-1}$ were observed in the same solvents [1, 7]. The assignments of these bands [1, 7] and solvent effects on all the absorption bands are discussed [1, 4, 7].

$Mn(C_6H_5N_2O_2)_2$ reacts with pyridine to give the ternary complex **$Mn(C_6H_5N_2O_2)_2 \cdot 2py$**. This complex is monomeric in solution and has presumably an octahedral structure [9].

$M[Mn(C_6H_5N_2O_2)_3] \cdot nH_2O$ (M = Na or K). The compounds were prepared by adding the Na or K salt of the ligand to a hot alcohol solution of $MnCl_2$ and warming the mixture. After removing the NaCl or KCl, which had precipitated, the solution was evaporated to a small volume. The complexes separated after long standing [13].

References:

[1] Abou El Ela, A. H., Abdel-Kerim, F. M., Afifi, H. H., Aly, H. F. (Z. Naturforsch. **28b** [1973] 610/4).
[2] Bottei, R. S., Schneggenburger, R. G. (J. Inorg. Nucl. Chem. **32** [1970] 1525/45, 1527, 1535, 1542).
[3] Wendlandt, W. W., Iftikhar Ali, S., Stembridge, C. H. (Anal. Chim. Acta **31** [1964] 501/8, 504).
[4] Abou El Ela, A. H., Afifi, H. H. (Z. Naturforsch. **30b** [1975] 215/8).
[5] Aly, H. F., Abdel-Kerim, F. M., Afifi, H. H. (Z. Naturforsch. **31a** [1976] 675/6).
[6] Nadezhda, A. A., Ivanova, K. P., Gorbenko, F. P. (Ukr. Khim. Zh. **46** [1980] 1315/20; Soviet Progr. Chem. **46** No. 12 [1980] 76/9).
[7] Abou El Ela, A. H., Afifi, H. H. (Z. Naturforsch. **29a** [1974] 719/24).
[8] Abou El Ela, A. H., Afifi, H. H. (Z. Naturforsch. **29b** [1974] 524/6).
[9] Charalambous, J., Haines, L. I. B., Harris, N. J., et al. (J. Chem. Res. S **1984** 220/1).
[10] Gyunner, E. A., Belykh, N. D. (Zh. Neorgan. Khim. **11** [1966] 386/91; Russ. J. Inorg. Chem. **11** [1966] 210/3).

[11] Eshelman, H. C., Armentor, J. (Develop. Appl. Spectrosc. **3** [1963] 190/5).
[12] Starý, J., Smižanská, J. (Anal. Chim. Acta **29** [1963] 545/51).
[13] Iinuma, H. (Res. Rept. Fac. Eng. Gifu Univ. **2** [1952] 77/8 from C.A. **1955** 11487).

27.3.2 With N,N'-Dinitroso-1,4-phenylenedihydroxylamine
(Diammonium Salt = Dicupferron)

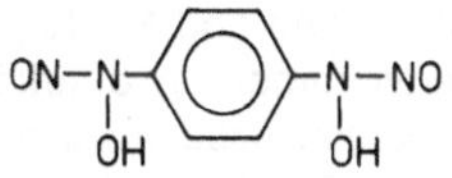

$(= C_6H_6N_4O_4)$

$Mn(C_6H_4N_4O_4) \cdot H_2O$ was prepared from $MnSO_4$ and the ligand in aqueous solution. The microcrystalline brown precipitate was air-dried at 110°C. The yield was 94%. X-ray powder patterns indicate that it is not isomorphous with other metal compounds of the ligand (d values listed). In the IR spectrum, a band at 1265 cm^{-1} was assigned to ν(N=O) vibrations and a band at 921 cm^{-1} to ν(N–O) vibrations, the latter shifted by +18 cm^{-1} compared with the disodium salt of the ligand.

The compound is not strongly hygroscopic, but contains small amounts of adsorbed water which could not be removed by drying at 110°C. It does not melt and does not lose the water molecule until decomposition. At 225°C, it decomposes explosively. The presumably polymeric complex is insoluble in water, ethanol, ether, dioxane, benzene, dimethylformamide, and tetrahydrofuran.

Reference:

Bottei, R. S., Schneggenburger, R. G. (J. Inorg. Nucl. Chem. **32** [1970] 1525/45, 1526, 1528, 1531, 1535).

27.3.3 With Cyclohexyl or 1-Naphthyl Derivatives of N-Nitrosohydroxylamine

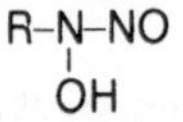

1) R = cyclohexyl; (= $C_6H_{12}N_2O_2$)
2) R = 1-naphthyl; (= $C_{10}H_8N_2O_2$)

Mn^{2+} ions form a colorless complex with ligand 1 in sodium acetate or neutral aqueous solution [1, 2]. The complex can be extracted into chloroform at pH values higher than 5, the extraction reaching a maximum between pH 7.5 and 9.6. The amounts of the extracted ion were 53.4% after shaking for 15 min and 86.4% after 60 min. The extraction behavior as a function of pH changed in the presence of reductants or oxidants. This, together with the different color in the organic phase (green), suggests an oxidation process of Mn^{II} followed by extraction of a new complex. An analytical application of the method was proposed [2].

The Mn^{II} complex with ligand 2 was proposed as a component in fungicidal products [3].

References:

[1] Buscaróns, F., Canela, J. (Anal. Chim. Acta **67** [1973] 349/55).
[2] Buscaróns, F., Canela, J. (Anal. Chim. Acta **70** [1974] 113/9).
[3] Leitner, G. J. (U.S. 2951008 [1960]).

27.4 Complexes with Other Nitroso Compounds

27.4.1 With 1-Nitrosoacetaldehyde Oxime (= Ethylnitrosolic Acid = $C_2H_4N_2O_2$)

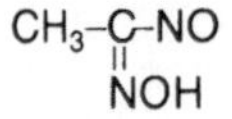

$[(C_4H_9)_4N][Mn(C_2H_3N_2O_2)_3]$ and **$[(C_6H_5)_4P][Mn(C_2H_3N_2O_2)_3]$**. The compounds were prepared by two methods: 1) An aqueous solution of potassium ethylnitrosolate and $MnSO_4$ in a 2:1 mole ratio was filtered to remove a white solid. Addition of tetrabutylammonium or tetraphenylphosphonium bromide gave dark brown solids which were recrystallized from methanol. 2) N-Hydroxyacetamide oxime hydrochloride, $CH_3C(=NOH)NHOH \cdot HCl$ (15 mmol), was added to an aqueous solution of $KMnO_4$ (5 mmol). After a white precipitate had been removed, the preparation was carried out as above. The compounds were air-dried and kept in vacuum over P_2O_5.

The magnetic moment of the tetrabutylammonium compound, $\mu_{eff} \approx 2\mu_B$, is consistent with a low-spin d^5 octahedral configuration. In the IR spectra of the compounds (KBr or CsI pellets), bands between 1550 and 1350 cm^{-1} were associated with the $\nu_{as}(NCN)$ vibrations and strong bands at 1220 and at about 1150 to 1100 cm^{-1} with the $\nu(N{=}O)$ and $\nu(N{-}O)$ vibrations, respectively. Another strong absorption at 1360 to 1300 cm^{-1} was not assigned. The far-IR region shows strong bands at 378 and 338 cm^{-1} which were assigned to $\nu(Mn{-}N)$ and $\nu(Mn{-}O)$

vibrations. The IR spectrum is similar to that of $Cs[Fe(C_2H_3N_2O_2)_3]$ for which an X-ray diffraction study was made. The results indicate for the complexes a facial geometry with each of the three ligands bound by one nitrogen atom and one oxygen atom. The electronic solution spectrum (in acetonitrile), shows intense absorption maxima at ~270, ~580, and ~780 nm. Of these, the first two were assigned to $\pi \rightarrow \pi^*$ and $n \rightarrow \sigma^*$ transitions of the ligand. Voltammetric studies of the complex in acetonitrile indicate two species of different oxidation states linked by a one-electron transfer: $Mn^{II}(C_2H_3N_2O_2)_3^- + e^- \rightleftharpoons Mn^{I}(C_2H_3N_2O_2)_3^{2-}$. On further reduction, the ligand is reduced, forming N-hydroxyacetamide oxime. All attempts to obtain the corresponding Mn^{III} species by oxidation of the Mn^{II} species resulted in the decomposition of the Mn^{II} complex.

Reference:

Gouzerh, P., Jeannin, Y., Rocchiccioli-Deltcheff, C., Valentini, F. (J. Coord. Chem. **6** [1979] 221/3).

27.4.2 With p-Nitrosodimethylaniline

$(=C_8H_{10}N_2O)$

$Mn(C_8H_{10}N_2O)Cl_2$ precipitated when an excess of a concentrated ether solution of the ligand was added to an alcohol solution of $MnCl_2$. The colored complex was dried at 80°C [1].

The enthalpy of formation, $\Delta H = -8.85 \pm 0.15$ kcal/mol at 25°C, for the crystalline complex from $MnCl_2$ and the ligand both in the crystalline states was determined calorimetrically [2].

The compound is stable in air. Thermogravimetric and differential thermal analyses in static air atmosphere revealed a primary loss of the NO group followed by decomposition at 400 to 500°C. The complex is rapidly hydrolyzed in water and insoluble in most common solvents [1]. The enthalpy of solution of the complex in an excess of 1 M aqueous HCl is $\Delta H = -8.519 \pm 0.066$ at 25°C [2].

References:

[1] Condorelli, G., Gurrieri, S., Musumeci, S. (Boll. Sedute Accad. Gioenia Sci. Nat. Catania [4] **8** [1966] 791/8 from C.A. **69** [1968] No. 113031).
[2] Gurrieri, S., Cali, R., Siracusa, G. (J. Chem. Eng. Data **18** [1973] 22/3).

28 Complexes with Azo Compounds

General Remarks

Manganese(II) complexes with azo compounds R–N=N–R′ are known, where R and R′ are aromatic or heterocyclic groups. As shown by IR data, the nitrogen atoms of the azo groups are coordinated to manganese if additional donor atoms in sterically suited positions permit the formation of fused chelate rings with the metal ion:

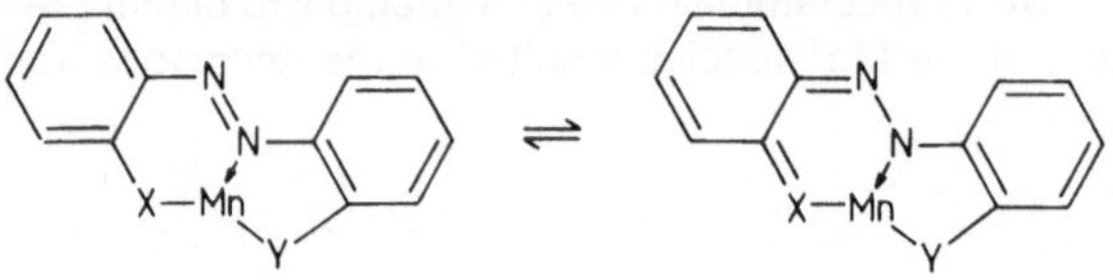

X, Y = O,O or O,N

In the pyridylazo compounds, the pyridine nitrogen atom is coordinated to manganese forming a five-membered chelate ring:

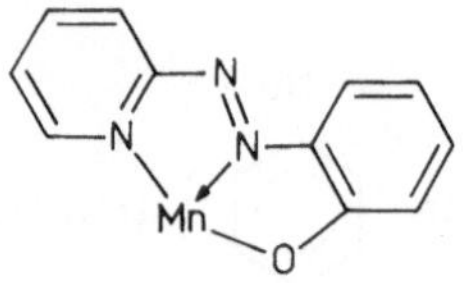

The azo group is always coordinated as the trans =*(E)*-isomer. Stabilization of the complexes by the hydrazo-keto tautomeric form was observed, especially for the pyrazolylazo complexes. Included also are complexes where the azo group is not involved in coordination, e.g. chelates of phenylazo derivatives of salicylaldehyde or salicylic acid.

Generally, 1:1 and 1:2 complexes of manganese(II) are formed with the ligand anions in mixed aqueous-organic solution at pH 8 to 12. The Irving-Williams order $Zn^{II} < Cu^{II} > Ni^{II} > Co^{II} > Mn^{II}$ was shown to be valid for the complex stabilities. The manganese complexes are more stable than the corresponding compounds of alkaline earth metals.

Most of the complexes are insoluble in water, but are soluble in polar organic solvents. Because of their intensely red to purple colors, the complexes are used for the analytical determination of manganese, especially those with pyridylazo compounds. This section is confined to complexes of known composition or of definite physical data whereas mere color reactions which are frequently observed with azo compounds in the presence of manganese ions are not described.

28.1 With Arylazo Compounds

28.1.1 With o,o′-Dihydroxyazobenzene and Related Compounds

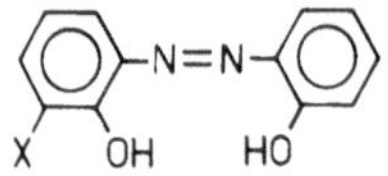

No.	X	formula
1	H	$C_{12}H_{10}N_2O_2$
2	Cl	$C_{12}H_9ClN_2O_2$
3	Br	$C_{12}H_9BrN_2O_2$
4	I	$C_{12}H_9IN_2O_2$
5	CH_3	$C_{13}H_{12}N_2O_2$

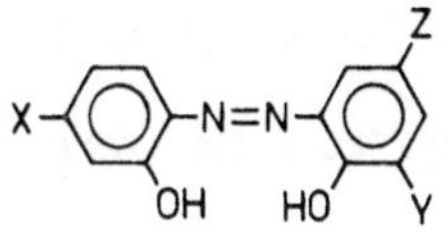

No.	X	Y	Z	formula
6	NH_2	H	H	$C_{12}H_{11}N_3O_2$
7	OH	H	H	$C_{12}H_{10}N_2O_3$
8	OH	H	SO_3H	$C_{12}H_{10}N_2O_6S$
9	OH	SO_3H	SO_3H	$C_{12}H_{10}N_2O_9S_2$
10	OH	Cl	SO_3H	$C_{12}H_9ClN_2O_6S$
11	OH	SO_3H	Cl	$C_{12}H_9ClN_2O_6S$

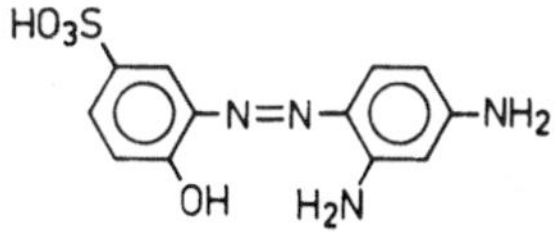

ligand 12 (= $C_{12}H_{12}N_4O_4S$)

Complexes in Solution. The formation of a 1:2 complex with ligand 1 at pH ~ 10 was established by a broad absorption band in the range 450 to 550 nm and by Job's method of continuous variation. This complex can be extracted into methyl isobutyl ketone, most effectively in the presence of Na_2SO_4 [1]. The formation of 1:1 complexes was assumed in buffered aqueous solution on reaction of Mn^{2+} ions with the doubly deprotonated ligands 7 to 11. The pH values of maximum complexation are 6.8, 6.7, 6.3, 6.8, and 10.6, respectively. Absorption maxima observed are shown below; extinction coefficients, ε_{max} in 10^3 $L \cdot mol^{-1} \cdot cm^{-1}$, are given in parentheses:

ligand	7	8	9	10	11
λ_{max} (ε_{max})	500 (24.4)	— (15.6)	500 (20.0)	490 (22.8)	520 (17.0)

The values were compared with those of corresponding complexes of other transition metal cations and related with the cation hydrolysis constants and ionic radii. An approximate function was derived from the experimental data relating pH_{max} with pK_{a1}, the first dissociation constant of the ligands [2].

A manganese(II) complex is formed in dimethylformamide solution with excess sodium salt of ligand 12 in the presence of triethylamine. Two absorption maxima occurred in its visible spectrum at $\lambda = 490$ and 700 nm [3].

MnL complexes with ligands 1 to 7 were precipitated on reacting hot ligand solutions in methanol or appropriate solvents with $Mn(CH_3COO)_2$, as described in [4, 5] for azo complexes of other metals [6]. The polarographic behavior of the different complexes in DMF was investigated and the influence of the substituents on the electroreduction of the complexes was studied. Peak potentials $E_p = E_{1/2} - 1.109$ RT/nF (in V versus SCE) were determined oscillopolarographically in DMF at 25 ± 0.2°C, for the first reduction step of the individual complexes. The E_p values are more negative for Mn^{II} complexes than for corresponding Ni^{II} and Zn^{II} complexes. The introduction of meta and para substituents into a benzene ring of ligand 1 alters the values of E_p but does not appreciably influence the nature of the electroreduction process. Formation of a delocalized coordinated radical anion from the azo ligand upon reduction is discussed rather than a lowered oxidation state of the manganese ion [4].

References:

[1] Kikuchi, S., Manabe, M., Okinaka, H. (Niihama Kogyo Koto Senmon Gakko Kiyo Rikogakuhen **20** [1984] 41/5; C.A. **101** [1984] No. 47779).

[2] Salikhov, V. D., Shuklina, G. I. (Zh. Obshch. Khim. **48** [1978] 2578/83; J. Gen. Chem. [USSR] **48** [1978] 2341/5).

[3] Coleman, R. A., Rodgers, J. L., American Cyanamide Co. (U.S. 3042624 [1959/62] 1/3; C.A. **58** [1963] 3129).
[4] Andreeva, M. A., Stepanov, B. I. (Zh. Obshch. Khim. **28** [1958] 2966; J. Gen. Chem. [USSR] **28** [1958] 2995/6).
[5] Zhuchenko, T. A., Kuznetsova, L. I., Kogan, V. A., Garnovskii, A. D., Osipov, O. A., Gorelik, M. V., Gladysheva, T. Kh., Alekseenko, V. A., Zayakina, T. A. (Zh. Neorgan. Khim. **16** [1971] 2169/72; Russ. J. Inorg. Chem. **16** [1971] 1157/9).
[6] Budnikov, G. K., Maistrenko, V. N., Toropova, V. F. (Zh. Obshch. Khim. **46** [1976] 1589/93; J. Gen. Chem. [USSR] **46** [1976] 1548/51).

28.1.2 With Nitrophenylazo Derivatives of 1,2-Benzenediol

1) $R_1 = NO_2$, $R_2 = H$ $(= C_{12}H_9N_3O_4)$
2) $R_1 = H$, $R_2 = NO_2$ $(= C_{12}H_9N_3O_4)$
3) $R_1 = R_2 = NO_2$ $(= C_{12}H_8N_4O_6)$

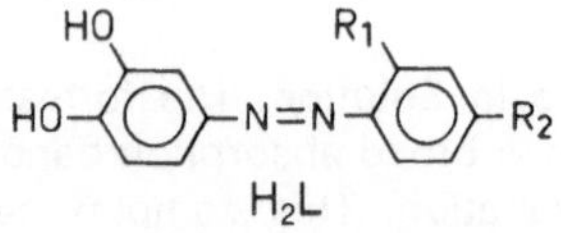

H_2L

ESR spectrometric studies reveal the formation of 1:1 and 1:2 complexes in aqueous ethanol solutions of $MnCl_2$ and ligand 3 buffered with ammonium acetate. The stability constants $K_1 = 1.75 \times 10^{-4}$ L/mol and $\beta_2 = 7.94 \times 10^9$ L^2/mol^{-2} were determined at room temperature [1]. No binary manganese(II) complexes with ligands 1 to 3 were obtained in aqueous ethanol at pH 8 to 11 [1, 2].

Ternary complexes with N-heterocyclic diamines of composition $[Mn\,bpy_2(HL)_2]$ in solution with either of the ligands 1 to 3 and 2,2′-bipyridyl were found in the chloroform extract of an ethanol-water solution containing Mn^{II} ions, the azo ligand, and 2,2′-bipyridyl, on adjusting the mixture to the pH range 8 to 11 with acetate-ammonia buffer. The optimum complex concentration in the $CHCl_3$ extract was detected spectrophotometrically at pH ~10 where one phenolic proton is released per azo ligand. Based on the specific distribution of reactants and complexes between the organic and the aqueous phase, the stability constants β^* are calculated for the equilibrium $Mn\,bpy_2^{2+} + 2HL^- \rightleftharpoons Mn\,bpy_2(HL)_2$. Extinction curves of the azo ligand mono anions as well as of their ternary complexes show that the ligand bands are preserved with a bathochromic shift for the complexes [2]:

ligand No.	$\lambda_{max(HL^-)}$ in nm	$\beta^* \times 10^{-9}$	$\lambda_{max\,(complex)}$ in nm	$\varepsilon \times 10^{-4}$ in $L \cdot mol^{-1} \cdot cm^{-1}$
1	470	1.5±0.18	500	2.3±0.54
2	400	1.7±0.64	510	3.0±0.81
3	450	4.2±0.18	525	5.8±0.59

The complex with ligand 3 exhibits an additional absorption band doubtfully assigned to the undissociated form of this acid ligand. The complexes may be of interest for the spectrophotometric determination of manganese [2].

Solid ternary complexes with ligand 3 and phenanthroline or bathophenanthroline (= 4,7-diphenylphenanthroline = $C_{24}H_{16}N_2$) were isolated from evaporated and dehydrated chloroform extracts of buffered solutions containing $MnCl_2 \cdot 4H_2O$, the azo ligand, and the diamine in ethanol-water, when the reactants were mixed at least in 1:100:10 mole ratios [1, 3].

Spectrophotometric studies on chloroform or toluene extracts exhibit pH-dependent absorption maxima. On account of the observed shifts (the corresponding λ_{max} values of the free ligand are given in parentheses) the formation of the following complexes was suggested:

complex	$Mnphen_2L$	$Mn(C_{24}H_{16}N_2)_2L$	$Mnphen_2(HL)_2$	$Mn(C_{24}H_{16}N_2)_2(HL)_2$
pH of formation	11	10	8	8
λ_{max} in nm	640 (600)	600 (600)	535 (510)	580 (510)

The IR spectrum of the complex $Mnphen_2L$ in KBr exhibits a broad band at 3490 to 3370 cm^{-1}, attributed to a 5-membered chelate ring formed on complexation by the doubly deprotonated benzenediolate group of the ligand. Bands assigned to ν(CO) and δ(OH) of the ligand at 1435, 1395, 1285, 1265, 1230, and 1070 cm^{-1} are displaced to 1452, 1295, and 1042 cm^{-1} in the complex or disappear. Additional coordination of both phenanthroline molecules through their N atoms is demonstrated by the shift of the ν(N=C–C=N) and ν(C=C) bands from 1625, 1590, 1560, 1510, and 1420 cm^{-1} for phen to 1655, 1588, and 1568 cm^{-1} for the complex. Other vibrations with lower wave numbers are also displaced or have disappeared on complexation. Bands of the complex occurring at 515 and 495 cm^{-1} in the far-IR may be caused by Mn–O and $Mn–N_{phen}$ stretching vibrations [1].

Thus, it is assumed that in the complexes obtained at pH 8 manganese is coordinated to the nitrogen atoms of two bidentate phen or bathophenanthroline molecules in equatorial positions and to the deprotonated oxygens of two monodentate HL^- ligands in the axial position of an octahedron. In the complexes obtained at pH 10, the two oxygen atoms of only one bisdeprotonated ligand are fixed in two equatorial *cis*-positions of an octahedron. The remaining sites are occupied by the four nitrogen atoms of the diamine ligands [1].

References:

[1] Zeinalova, S. A., Guseinov, I. K., Marov, I. N., Rustamov, N. Kh., Kalinichenko, N. B. (Zh. Neorgan. Khim. **27** [1982] 3095/100; Russ. J. Inorg. Chem. **27** [1982] 1752/5).
[2] Zeinalova, S. A., Guseinov, I. K., Rustamov, N. Kh. (Zh. Analit. Khim. **38** [1983] 241/4; J. Anal. Chem. [USSR] **38** [1983] 187/91).
[3] Guseinov, I. K., Zeinalova, S. A. (Azerb. Khim. **1980** No. 2, pp. 130/4; C.A. **94** [1981] No. 149606).

28.1.3 With Phenylazo Derivatives of Salicylaldehyde

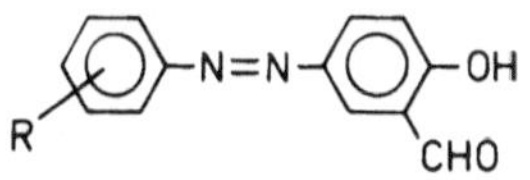

ligands 1 to 6

ligand No.	R	formula
1	3-Cl	$C_{13}H_9ClN_2O_2$
2	4-Cl	$C_{13}H_9ClN_2O_2$
3	$2-CH_3$	$C_{14}H_{12}N_2O_2$
4	$3-CH_3$	$C_{14}H_{12}N_2O_2$
5	$4-CH_3$	$C_{14}H_{12}N_2O_2$
6	$4-SO_3H$	$C_{13}H_{10}N_2O_5S$

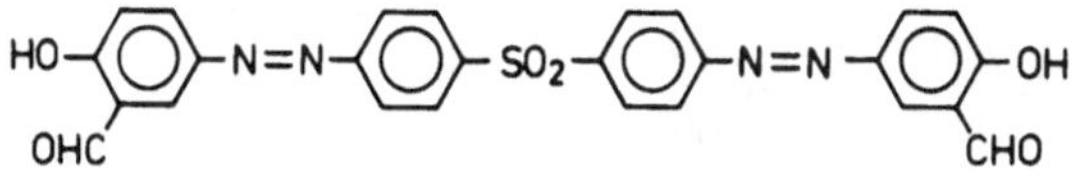

ligand 7 (= $C_{26}H_{18}N_4O_6S$)

Solid complexes were prepared at pH 5 to 6, as described in [4] for the corresponding Co^{II} complexes, from Mn^{II} salts and the ligands 1, 3, 4, 5 in ethanol, and ligand 2 in toluene or benzene-ethanol. D.c. conductivities have been measured on complex pellets at 312.5 and 416.6 K [1].

A semiconductive behavior of the complexes is deduced for the temperature range studied, the analytical composition (not given in the paper) remaining essentially unchanged [1].

A complex of composition $Mn(C_{13}H_9N_2O_5S)_2 \cdot 7.5H_2O$ was obtained from manganese(II) acetate and ligand 6 (mole ratio 1:2) in heated aqueous solution. Yellow needles result after recrystallization from water. A structure similar to the chelate structure of salicylaldehyde complexes was concluded from the IR spectrum. The ν(CO) band of the ligand at 1650 cm^{-1} is shifted to a lower wave number [2].

A dark colored amorphous complex was formed on reaction of a manganese(II) salt in aqueous solution with ligand 7 in dimethylformamide. The mixture was buffered with CH_3COONa and refluxed for 4 h. The precipitate was washed with hot water, hot DMF, and ethanol, then dried at 80°C. The IR spectrum (KBr disks), exhibits a strong broad absorption band at 3250 cm^{-1} ascribed to a coupling of ν(OH) water and ν(OH) phenol modes. Strong bands of ν(CO) at 1600 (ligand 1660) cm^{-1} and of the phenolic ν(C–O) around 1300 (ligand 1280) cm^{-1}, as well as the disappearance of the ligand δ(OH) band at 1380 cm^{-1}, indicate chelation of the ligand to Mn through the carbonyl and the deprotonated phenolic oxygen atoms. The electronic spectrum shows two absorption bands at 23260 and 20830 cm^{-1} assigned to $^6A_{2g} \rightarrow ^4T_{2g}$ and $^6A_{2g} \rightarrow ^4T_{1g}$ electron transitions, respectively, suggesting an octahedral structure [3].

Two water molecules are assumed to be in the coordination sphere. The compound with a mole ratio Mn:ligand:H_2O = 2:3:4 is probably a polymer. The low magnetic moment $\mu_{eff} = 3.79\ \mu_B$ resulting from susceptibility measurements (Gouy balance) at 30°C suggests metal-metal interactions. The complex is insoluble in all common organic solvents [3].

References:

[1] Pardeshi, L., Rasheed, A., Bhobe, R. A. (J. Indian Chem. Soc. **57** [1980] 388/90).
[2] Tanaka, M. (Bull. Chem. Soc. Japan **37** [1964] 1210/6, 1214).
[3] Patil, M. S., Shah, N. R., Rana, A. K., Karampurwala, A. M., Shah, J. R. (J. Macromol. Sci. Chem. A **16** [1981] 737/43).
[4] Deshmukh, K. G., Bhobe, R. A. (Current Sci. [India] **46** [1977] 67/9).

28.1.4 With Phenylazo Derivatives of Salicylic Acid and Related Compounds

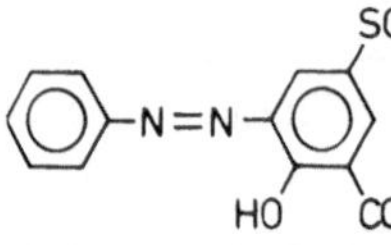

ligand 1 (= $C_{13}H_{10}N_2O_6S$)

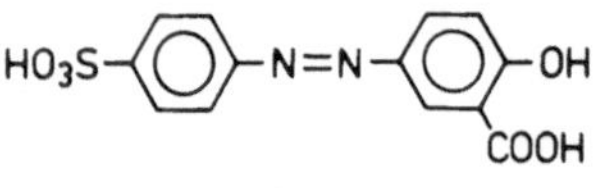

ligand 2 (= $C_{13}H_{10}N_2O_6S$)

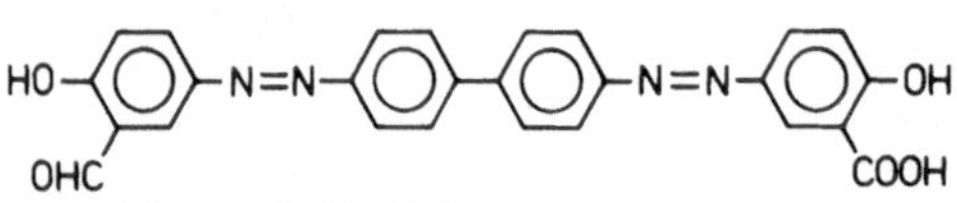

ligand 3 (= $C_{26}H_{18}N_4O_5$)

Complexes in Solution. The formation of 1:1 complexes with the sodium sulfonate form of ligand 1 was shown by spectrophotometric measurements. A mild break in the titration curve near pH 8 was considered to show chelation of both the deprotonated phenolic and carboxylic oxygen atoms. A chelate with coordinated but undissociated phenolic group is expected to be formed at lower pH values. The stability constants of both species were determined potentiometrically at 28°C and I = 0.1 M ($NaClO_4$): log K = 2.99 and 3.64, respectively [1].

With ligand 2 as its monosodium salt, the formation of a 1:1 and a 1:2 complex was evaluated potentiometrically (glass electrode) on aqueous solutions containing Mn^{2+} ions at $25 \pm 0.1°C$ and $I = 0.1 M$ (KNO_3). For the pH range from 7 to 9.3, where hydrolytic effects could be excluded, only the carboxyl proton of the ligand is involved in complex formation ($pK_{a1} = 2.38$, $pK_{a2} = 11.04$). Stability constants were determined: $\log K_1^* = 4.94 \pm 0.02$ referring to the equilibrium $Mn^{2+} + HL^{2-} \rightleftharpoons MnHL$ and $\log K_2^* = 3.5 \pm 0.1$ referring to $MnHL + HL^{2-} \rightleftharpoons [Mn(HL)_2]^{2-}$. Bidentate chelation, with the phenolic group coordinated, is presumed. The stability of the 1:1 complex slightly decreased, compared with those of manganese 5-sulfosalicylate and the salicylate, due to the greater electron withdrawing effect of the para-sulfophenylazo group [2].

$[Mn(C_{26}H_{16}N_4O_5)(H_2O)_2]_n$. The dark colored polychelate, erroneously formulated in the paper as $[Mn(C_{26}H_{14}N_4O_5)(H_2O)_2]$, was prepared from equimolar amounts of ligand 3 in refluxing dimethylformamide and manganese(II) chloride in ethanol. After addition of CH_3COONa, the mixture was refluxed for 2 h, and the solid obtained was washed several times with hot water, with DMF, finally with ethanol, and dried at 45°C. Its IR spectrum displays characteristic bands in the regions 3560 to 2800 cm^{-1} due to ν(OH) vibrations, at 1630 to 1575 cm^{-1} due to ν(CO), and at 1112 cm^{-1} due to further OH vibrations. Alteration of these bands with respect to those of the free ligand is indicative of chelation by phenolate and carboxylate anions of the ligand, of aldehyde and carboxylate oxo groups additionally coordinated, and of coordinated water. The electronic reflectance spectrum shows bands at 15150 and 20410, assigned to $^6A_{1g} \rightarrow {}^4T_{1g}$ and $^6A_{1g} \rightarrow {}^4T_{2g}$, respectively, and at 23810 cm^{-1} presumably owing to a ligand band. Octahedral geometry is implied from these observations and also by the magnetic moment (5.76 B.M.). The compound is insoluble, or poorly soluble, in common organic solvents. A low specific conductance $\lambda = 2.9 \times 10^{-6}\ \Omega^{-1} \cdot cm^{-1}$ was observed in DMF [3].

References:

[1] Saxena, H. B., Saxena, M. C. (Proc. Natl. Acad. Sci. India A **47** [1977] 41/5, 43; C.A. **89** [1978] No. 153535).
[2] Murakami, Y., Tagaki, M. (Bull. Chem. Soc. Japan **37** [1964] 268/72).
[3] Deshpande, U. G., Shah, J. R. (J. Macromol. Sci. Chem. A **23** [1986] 97/104).

28.1.5 With Other Derivatives of Phenylazobenzene

ligand 1
($= C_{14}H_{12}N_2O_3 = H_2L$)

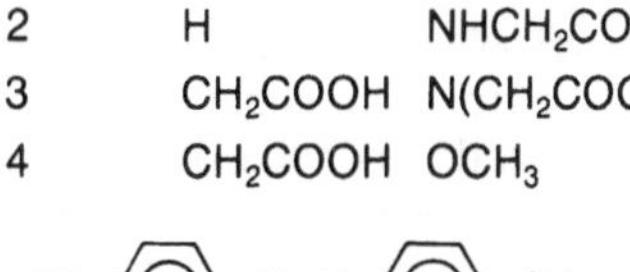

ligand No.	R_1	R_2	R_3	R_4	formula
2	H	$NHCH_2COOH$	H	H	$C_{16}H_{16}N_4O_4$
3	CH_2COOH	$N(CH_2COOH)_2$	H	H	$C_{20}H_{20}N_4O_8$
4	CH_2COOH	OCH_3	CH_3	NO_2	$C_{18}H_{18}N_4O_7$

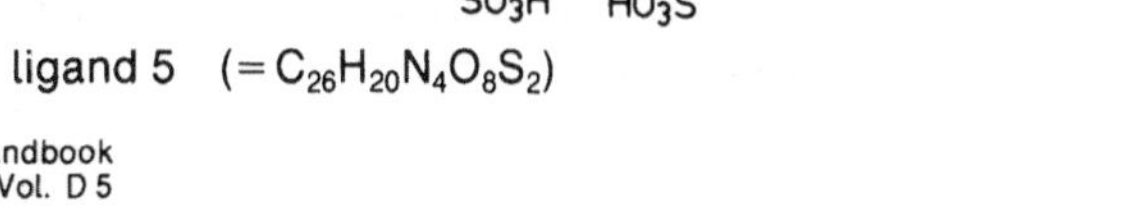

ligand 5 ($= C_{26}H_{20}N_4O_8S_2$)

Complexes in Solution. The formation constant of the complex $Mn(C_{14}H_{10}N_2O_3)$, with ligand 1 formed by the reaction $Mn^{2+} + C_{14}H_{12}N_2O_3 \rightleftharpoons Mn(C_{14}H_{10}N_2O_3) + 2H^+$, was determined potentiometrically (glass electrode) in 75/25 (v/v) dioxane-water mixtures at 30°C under N_2: log K = 10.6 [1].

The formation of manganese(II) complexes with the ligands 2 to 4 in buffered aqueous solutions at pH 9.35 was concluded in a polarographic study from a lowering of the reduction half-wave potentials $E_{1/2}$ observed for the ligands ($\Delta E_{1/2} = -0.07$ to -0.10 V). No composition of the complexes was given. At pH 2.5 there is no evidence of complex formation [2].

A solid complex, with the possible composition $Na_2Mn(C_{26}H_{16}N_4O_8)$, is formed by manganese(II) salts and ligand 5 as its disodium salt (= Brilliant Yellow indicator) in acidic aqueous solution. The yellow nonhygroscopic crystalline compound can be dried at 120°C and is stable to CH_3COOH and even to 0.1N HCl. Because of its low solubility it may serve in gravimetric analysis of manganese [3].

References:

[1] Snavely, F. A., Fernelius, W. C., Douglas, B. E. (J. Soc. Dyers Colour. **73** [1957] 491/5), also see Sillén, L. G., Martell, A. E. (Chem. Soc. [London] Spec. Publ. No. 17 [1964] 688).

[2] Lastovskii, R. P., Dyatlova, N. M., Kolpakova, I. D., Krinitskaya, L. V. (Zh. Obshch. Khim. **37** [1967] 121/8; J. Gen. Chem. [USSR] **37** [1967] 109/14).

[3] Pushimov, Yu. V., Kazakhov, B. I. (Fiz. Khim. Metody Anal. Kontrolya Proizvod. Mater. 4th Konf. Rab. Vuzov Zavodsk. Lab. Yugo-Vostoka SSSR, Makhachkala, USSR, 1972, Vol. 3, pp. 105/6; C.A. **80** [1974] No. 907).

28.1.6 With 1-(Phenylazo)-2-naphthols

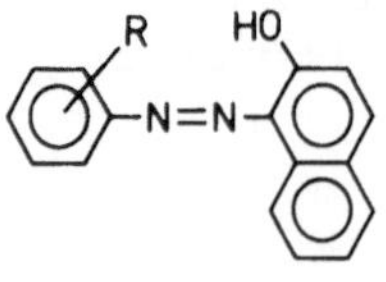

ligands 1 to 15

(R = H, OH, OCH_3, NO_2, Cl, Br, I, SO_3H, CH_3, COOH)

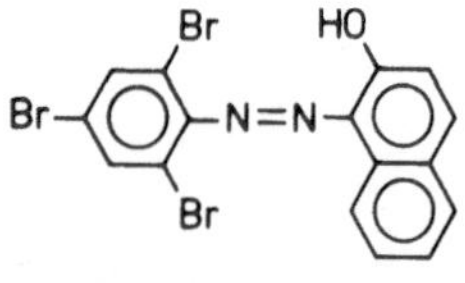

ligand 16

Formation in Solution. Formation constants $K = [MnL^+]/[Mn^{2+}][L^-]$ for complexes with monodeprotonated ligands 1 to 6, 8 to 13, 15, or 16 were determined potentiometrically (glass electrode) in a 3:1 (v/v) acetone-water medium at 25±1°C and ionic strength I = 0.100 M (KNO_3). Under these conditions, only the sulfonated ligand 14 is doubly deprotonated in its manganese(II) complex ($K_{14} = [MnL]/[Mn^{2+}] \cdot [L^{2-}]$) [1]. The log K values, see p. 285, decrease in the order $OCH_3 > CH_3 > H > OH > Cl > NO_2$ for *ortho*-substituted ligands and $CH_3 > OCH_3 > H > OH > I > Br > Cl > NO_2 > SO_3H$ for *para*-substituted ligands; this order corresponds to the sequence of most proton dissociation constants, pK_a, determined in the solvent mixture. The slightly greater stability of the chelates with *ortho*-substituted ligands 2 and 3, without response to the pK_a values of the ligands, was attributed to a tridentate ligation of these ligands forming two fused chelate rings upon additional coordination of the phenyl O atom. Bidentate ligation by the ionized O atom of the naphthol group and one coordinated azo N atom was proposed for the remaining complexes [1].

No.	R	ligand	log K	No.	R	ligand	log K
1	H	$C_{16}H_{12}N_2O$	7.32±0.04	9	4-OCH_3	$C_{17}H_{14}N_2O_2$	7.76±0.07
2	2-OH	$C_{16}H_{12}N_2O_2$	7.24±0.04	10	4-NO_2	$C_{16}H_{11}N_3O_3$	4.19±0.03
3	2-OCH_3	$C_{17}H_{14}N_2O_2$	8.03±0.04	11	4-Cl	$C_{16}H_{11}ClN_2O$	6.04±0.03
4	2-NO_2	$C_{16}H_{11}N_3O_3$	4.07±0.03	12	4-Br	$C_{16}H_{11}BrN_2O$	6.29±0.03
5	2-Cl	$C_{16}H_{11}ClN_2O$	5.66±0.05	13	4-I	$C_{16}H_{11}IN_2O$	6.68±0.04
6	2-CH_3	$C_{17}H_{14}N_2O$	7.35±0.05	14	4-SO_3H	$C_{16}H_{12}N_2O_4S$	3.92±0.03
7	2-COOH	$C_{17}H_{12}N_2O_3$	—	15	4-CH_3	$C_{17}H_{14}N_2O$	7.84±0.04
8	4-OH	$C_{16}H_{12}N_2O_2$	6.96±0.04	16	—	$C_{16}H_9Br_3N_2O$	6.02±0.04

$Mn(C_{16}H_{10}N_2O_2)$ or **$Mn(C_{17}H_{10}N_2O_3)$** with doubly deprotonated ligands 2 or 7, respectively, were prepared analogously to the method described earlier for Cu and Ni complexes, as cited in [2]. An Mn^{II} salt and ligand 2 ($=C_{16}H_{12}N_2O_2$) were boiled in ethanol whereas ligand 7 ($=C_{17}H_{12}N_2O_3$) was reacted in water, aqueous ammonia, or in boiling ethanol.

Classical and oscillographic polarography yields two reversible one-electron waves of the solids dissolved in DMF containing 0.1M $(C_2H_5)_4NClO_4$ as the supporting electrolyte at 25±0.2°C. In classical polarograms the half-wave potentials of the first reduction step are shifted to $E_{1/2}=-1.04$ V for $Mn(C_{16}H_{10}N_2O_2)$ and −0.98 V for $Mn(C_{17}H_{10}N_2O_3)$, compared with −0.57 and −0.47 V for the free ligands [2]. Likewise, the oscillopolarographic peak potential of the complex with ligand 2 is situated at −1.03 V [3]. In the presence of benzoic acid as a proton donor, the first reduction wave is shifted cathodically by about 0.34 or 0.30 V, with respect to $E_{1/2}$ of the free ligands. An additional wave at −1.8 V corresponds to the reduction of the solvated protons, liberated in complex formation [2]. The somewhat more negative shift of the complex with ligand 2, as compared to ligand 7, is proposed to reflect the individual electron density at the azo groups which is reduced. On addition of the first electron to the complex, a relatively stable radical anion results, proved by its ESR spectrum. By use of an oscillopolarographic method the heterogeneous rate constants of charge transfer, $K_s=0.011$ and 0.009 cm/s, respectively, were determined. A second wave at $E_{1/2}=-1.47$ V is not substantially shifted from the positions of the free ligands and, being due to the reduction of the azo group, merges with the reduction wave of the Mn^{2+} ion. Manganese complexes are reduced at more negative potentials than aluminum complexes also studied [2]. Based on the high negative shifts of the first, and on the constancy of the second reduction potentials, a structure is discussed for the neutral complexes with two fused chelate rings of quasiaromatic character. It is assumed that the metal is coordinated to the phenolic oxygens and one of the azo nitrogens of the ligands [2, 3].

References:

[1] Manku, G. S., Chadha, R. C., Nayar, N. K., Sethi, M. S. (J. Inorg. Nucl. Chem. **34** [1972] 1091/4).

[2] Toropova, V. F., Budnikov, G. K., Maistrenko, V. N. (Zh. Obshch. Khim. **42** [1972] 1207/11; J. Gen. Chem. [USSR] **42** [1972] 1203/6).

[3] Budnikov, G. K., Maistrenko, V. N., Toropova, V. F. (Zh. Obshch. Khim. **46** [1976] 1589/93; J. Gen. Chem. [USSR] **46** [1976] 1548/51).

28.1.7 With Derivatives of (2-Hydroxyphenyl)azonaphthol

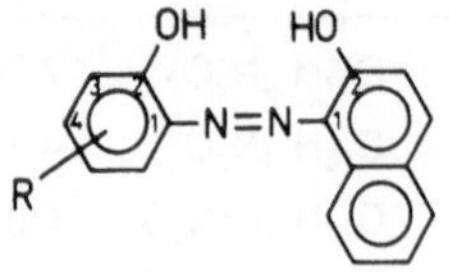

ligand No.	R	formula
1	4-OH	$C_{16}H_{12}N_2O_3$
2	4-NH_2	$C_{16}H_{13}N_3O_2$
3	3-Cl	$C_{16}H_{11}ClN_2O_2$
4	5-Cl	$C_{16}H_{11}ClN_2O_2$
5	3-Br	$C_{16}H_{11}BrN_2O_2$
6	3-I	$C_{16}H_{11}IN_2O_2$
7	3-CH_3	$C_{17}H_{14}N_2O_2$
8	4-SO_3H	$C_{16}H_{12}N_2O_5S$
9	5-SO_3H	$C_{16}H_{12}N_2O_5S$
10	5-SO_2NH_2	$C_{16}H_{13}N_3O_4S$
11	R' = H	$C_{16}H_{12}N_2O_8S_2$
12	R' = Cl	$C_{16}H_{11}ClN_2O_8S_2$

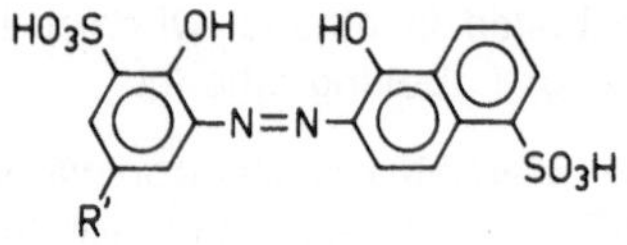

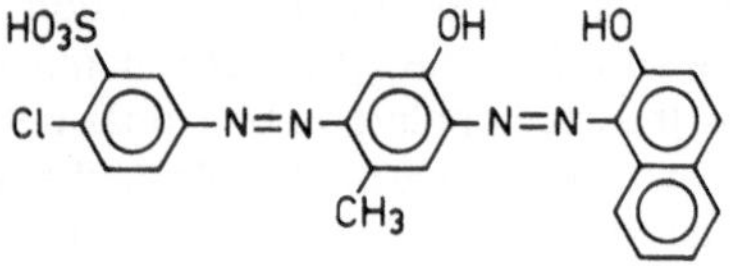

ligand 13 (= $C_{23}H_{17}ClN_4O_5S$)

Complexes in Solution. A formation constant, log K_1 = 14.19, was determined potentiometrically (glass electrode) at 30°C in 3 : 1 (v/v) dioxane-water for the complex $Mn(C_{16}H_9ClN_2O_2)$ with ligand 4 (= H_2L) [1]. The pH potentiometric studies on aqueous solutions containing a manganese(II) salt and Solochrome Violet RS (the monosodium salt of ligand 9) reveal the formation of 1:1 and 1:2 complexes with the stability constants log K_1 = 11.45 and log β_2 = 23.1 at 25°C and $I = 0_{corr}$ [2].

Spectrophotometric studies indicate complex formation in DMF solution of $MnCl_2$ with ligand 9 or 10 in the presence of triethylamine or cyclohexylamine, respectively [3]. In aqueous solution, colored complexes are observed at pH 11 with Acid Chrome Blue Black, presumably the disodium salt of ligand 11 [4], and with the monosodium salt of ligand 13 (= Solochrome Green V) [10]. At pH 8.5 a complex is formed with the disodium salt of ligand 12 (= Solochrome Fast Navy 2RS or Acid Chrome Blue Black K) [5].

Evidence of complex formation with the sulfonated ligands 8 or 9 was also provided by polarographic studies in alkaline aqueous solutions [2, 6 to 8]. The half-wave potential of Solochrome Violet RS (monosodium salt of ligand 9), $E_{1/2} = -0.573$ V, observed at pH 9.2 and 30°C vs. SCE, is shifted on complexation by $\Delta E_{1/2} = -0.145$ V. This value is independent of pH and metal concentration and is linearly related to the ratio of the log K values of the azo and the corresponding hydrazo complexes formed in the polarographic process and to the pK_a values of the respective ligands. The electroreduction is performed on the azo group coordinated to the manganese(II) ion without dissociation of the metal. The second wave, due to the reduction $Mn^{II} \rightarrow Mn^0$ with $E_{1/2} = -1.479$ V, is not affected by the ligand [2].

Isolated complexes MnL, obtained with the ligands 1 to 3 or 5 to 7 (= H_2L), were prepared analogously to the corresponding o,o'-dihydroxyazobenzene chelates, see p. 279. Based on equal behavior in their polarographic reduction the same quasiaromatic two-ring chelate structures are discussed as for those complexes [9].

References:

[1] Snavely, F. A. (Investigations of the Coordination Tendencies of o-Substituted Arylazo Compounds, Pennsylvania State College 1952 from Sillén, L. G., Martell, A. E., Chem. Soc. [London] Spec. Publ. No. 17 [1964] 711).

[2] Florence, T. M., Belew, W. L. (J. Electroanal. Chem. Interfacial Electrochem. **21** [1969] 157/67, 160/2; C.A. **70** [1969] No. 102488).

[3] Coleman, R. A., Rodgers, J. L., American Cyanamide Co. (U.S. 3042624 [1959/62] 1/3; C.A. **58** [1963] 3129).

[4] Butenko, G. A., Grzhegorzhevskii, A. S., Korzh, V. P. (Vopr. Khim. Khim. Tekhnol. No. 28 [1973] 60/4; C.A. **80** [1974] No. 43712).

[5] Abd El Raheem, A. A., El-Sabban, M. Z., Dokhana, M. M. (Z. Anal. Chem. **188** [1962] 96/109, 104/6).

[6] Dean, J. A., Bryan, H. A. (Anal. Chim. Acta **16** [1957] 87/93, 93).

[7] Dean, J. A., Bryan, H. A. (Anal. Chim. Acta **16** [1957] 94/100, 98).

[8] Palmer, S. M., Reynolds, G. F. (Z. Anal. Chem. **216** [1966] 202/7, 204).

[9] Budnikov, G. K., Maistrenko, V. N., Toropova, V. F. (Zh. Obshch. Khim. **46** [1976] 1589/93; J. Gen. Chem. [USSR] **46** [1976] 1548/51).

[10] Amin, A. M., Khalifa, H., Moustafa, A. S. (Z. Anal. Chem. **173** [1960] 138/48, 139/40, 143).

28.1.8 With 1-(Naphthylazo)- and 1-(Hydroxynaphthylazo)-2-naphthols

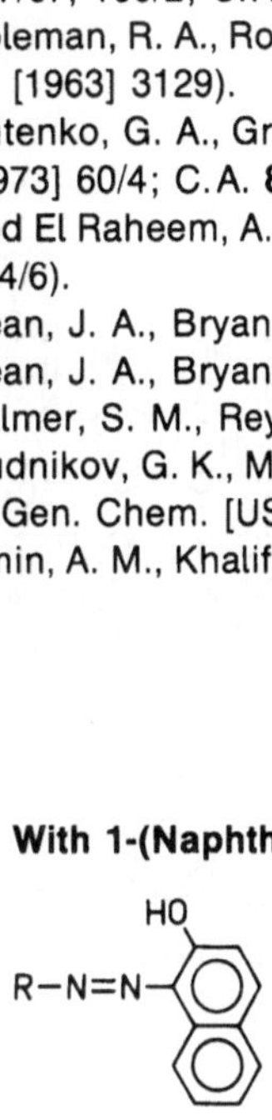

ligand 1 R = 1-naphthyl
ligand 2 R = 2-naphthyl } (= $C_{20}H_{14}N_2O$)

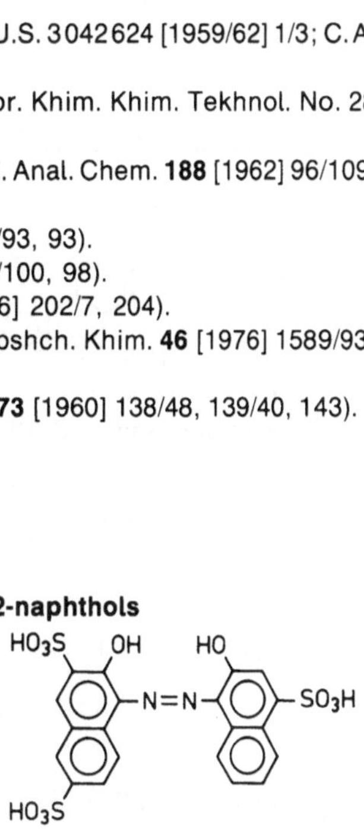

ligand 3 (= $C_{20}H_{14}N_2O_{11}S_3$)

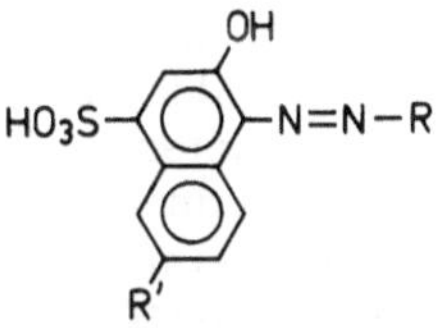

ligand 4 R = 1-hydroxy-2-naphthyl, R' = H (= $C_{20}H_{14}N_2O_5S$)
ligand 5 R = 2-hydroxy-1-naphthyl, R' = H (= $C_{20}H_{14}N_2O_5S$)
ligand 6 R = 1-hydroxy-2-naphthyl, R' = NO_2 (= $C_{20}H_{13}N_3O_7S$)

Formation constants log $K_1 = 7.02 \pm 0.04$ and 7.27 ± 0.05 for complexes with ligands 1 or 2, respectively, were determined potentiometrically (glass electrode) in a 3:1 (v/v) acetone-water solvent at 25 ± 1°C and ionic strength I = 0.100 M (KNO_3). The slightly lower stabilities, compared with the 1-(phenylazo)-2-naphthol complex, are attributed to steric hindrance [1]. Formation of a 1:1 complex with ligand 3 (trisodium salt = hydroxy naphthol blue) was revealed spectrophotometrically in aqueous phosphate buffered solution at pH 6.85 and 23°C, using Job's method of continuous variation. A stability constant log K = 4.42 ± 0.02 was established by extinction measurement. The blue color, exhibited by the ligand mono anion, $C_{20}H_{13}N_2O_{11}S_3^-$ (λ_{max} = 650 nm), decreased on complexation [2], also see [3].

Even in very low concentrations, complexes with the ligands 4 to 6 were detected in aqueous solutions at pH 10, by their waves in a.c. voltammetry, when the complexes were enriched by adsorption on carbon paste electrodes [4].

References:

[1] Manku, G. S., Chadha, R. C., Nayar, N. K., Sethi, M. S. (J. Inorg. Nucl. Chem. **34** [1972] 1091/4).
[2] Brittain, H. G. (Anal. Letters A **11** [1978] 355/62; C.A. **89** [1978] No. 84115).
[3] Brittain, H. G. (Anal. Chim. Acta **96** [1978] 165/70).
[4] Narayanan, A., Neeb, R. (Z. Anal. Chem. **269** [1974] 344/8).

28.1.9 With Azo Derivatives of Chromotropic Acid

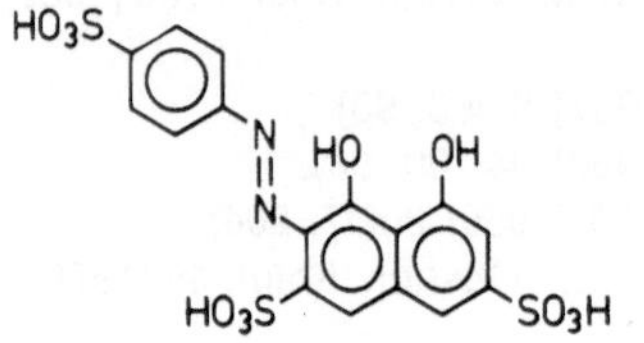

1) $(=C_{16}H_{12}N_2O_{11}S_3)$
trisodium salt = SPADNS = Na_3H_2L

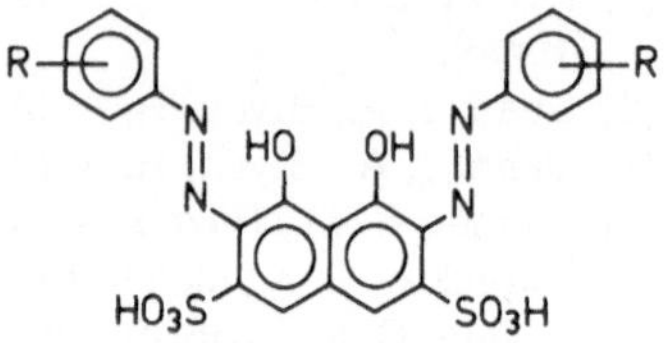

2) R = 3-SO_3H $(=C_{22}H_{16}N_4O_{14}S_4)$
3) R = 4-SO_3H $(=C_{22}H_{16}N_4O_{14}S_4)$
4) R = 4-COOH $(=C_{24}H_{16}N_4O_{12}S_2)$

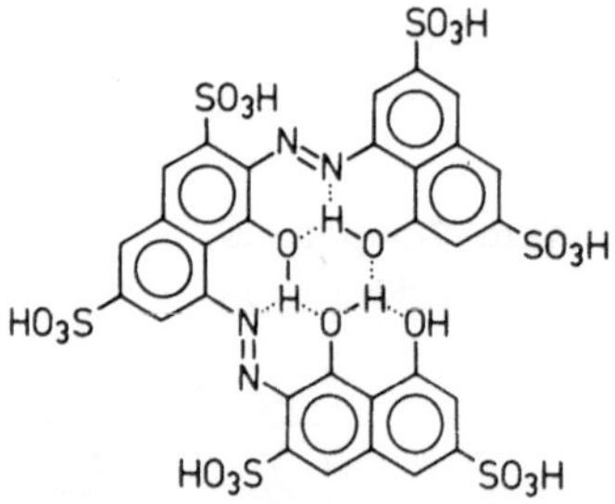

5) $(=C_{30}H_{20}N_6O_{22}S_6)$

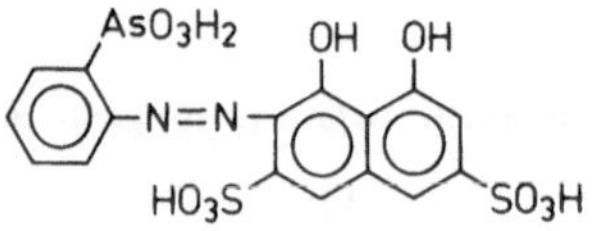

6) $(=C_{16}H_{13}AsN_2O_{11}S_2)$

The formation of a 1:1 complex in aqueous solutions of $Mn(ClO_4)_2$ with ligand 1 at 25°C, pH 8.0 to 9.3, and ionic strength I = 0.1M ($NaClO_4$) is indicated by the shift of the absorption maximum from $\lambda = 500$ nm (free ligand) to $\lambda = 600$ nm ($\varepsilon_{mol} = 6.7 \times 10^3$ L·mol^{-1}·cm^{-1}). The ligand exchange reaction with ethylenediaminetetraacetate (= L'), MnL + L' → MnL' + L (charges and protons omitted) at 25°C and pH 8.0 to 9.3 was monitored by increased transmittance at 600 nm on disappearance of the binary complex. Rapid formation of a mixed ligand intermediate [L-Mn^{II}-L'] was evaluated in kinetic studies with rate-determining release of the ligand L to yield an Mn^{II} ethylenediaminetetraacetate complex. A high overall rate constant $k \geqq 5 \times 10^6$ L·mol^{-1}·s^{-1} was determined for the analogous Mn, Cd, and Zn complexes compared to those of less reactive Co, Ni, and Cu complexes [1].

Complexes of Mn^{2+} ions with the isomeric ligands 2 or 3 or with ligand 4 are revealed in organic solutions of low acidity. An optimum yield of the complex with ligand 2 is indicated in propanol containing 4 to 5 vol% of water. It shows two absorption maxima at 590 and 645 nm; the latter one with $\varepsilon_{mol} = 6.5 \times 10^4$ L·mol^{-1}·cm^{-1} is shifted 104 nm from λ_{max} of the ligand [2]. Complex formation with the para-substituted ligand 3 in acetone is also reported in [5, 6]. Acetone containing 3.2 to 4.0 vol% of water provides the most favorable medium to form a 1:2 complex with ligand 4. λ_{max} values were observed at 645 and 720 nm. The latter maximum with

$\varepsilon_{max} = 15 \times 10^4$ L·mol^{-1}·cm^{-1} is shifted to longer wavelengths by 180 nm in comparison to the free ligand. This reaction does not proceed in pure aqueous propanol [2]. Different complex types and chelate structures are postulated for reactions in aqueous propanol [2], also see [3, 5], or in aqueous acetone [2, 5], or dioxane [5].

A 1:1 complex established spectrophotometrically by the method of continuous variation is formed with calcichrome, the hexasodium salt of ligand 5 in aqueous solutions at pH 8 to 12. At pH 10.5 two absorption maxima are exhibited at ~308 and 525 nm. The absorption intensity at 590 nm ($\varepsilon = 1.5 \times 10^4$ L·mol^{-1}·cm^{-1}) was measured for quantitative determination of manganese [4].

The formation of a 1:1 complex with ligand 6 in aqueous solution at pH 9 to 10.5 and mole ratios 1:1 to 1:4 is shown by an absorption maximum at $\lambda = 520$ nm [7, 8]. Susceptibility studies yield magnetic moments between 5.80 and 5.95 μ_B, indicative of a spin-free Mn^{II} complex. Coordination of the two deprotonated phenolic oxygen atoms is assumed [8].

References:

[1] Mentasti, E. (Anal. Chim. Acta **111** [1976] 177/85, 179, 182).
[2] Savvin, S. B., Petrova, T. V., Dzherayan, T. G. (Zh. Analit. Khim. **30** [1975] 2092/7; J. Anal. Chem. [USSR] **30** [1975] 1761/4).
[3] Petrova, T. V., Savvin, S. B., Dzherayan, T. G., Muk, A. A. (Zh. Analit. Khim. **29** [1974] 1067/83; J. Anal. Chem. [USSR] **29** [1974] 914/8).
[4] Ishii, H., Einaga, H. (Bunseki Kagaku **15** [1966] 1124/9; C.A. **67** [1967] No. 39849).
[5] Savvin, S. B., Petrova, T. V., Dzherayan, T. G. (Zh. Analit. Khim. **35** [1980] 1485/94; J. Anal. Chem. [USSR] **35** [1980] 955/62).
[6] Petrova, T. V., Savvin, S. B., Dzherayan, T. G. (Zh. Analit. Khim. **28** [1973] 1888/93; J. Anal. Chem. [USSR] **28** [1973] 1678/82).
[7] Lyutsedarskii, V. A., Lozanovskaya, I. N., Kaplina, E. P. (Tr. Novocher-Kassk. Politekh. Inst. No. 220, [1969], 58/66; C.A. **75** [1971] No. 155430).
[8] Lozanovskaya, I. N., Lyutsedarskii, V. A., Kaplina, E. P. (Tr. Novocher-Kassk. Politekh. Inst. No. 220 [1969] 67/73; C.A. **76** [1972] No. 18396).

27.1.10 With Azo Derivatives of 3-Hydroxy-2-naphthoic Acids

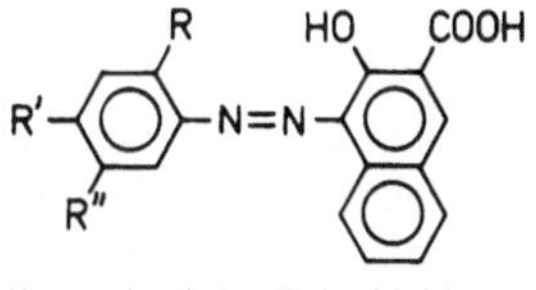

ligands 1 to 5 (= H_2L)

ligand	R	R'	R''	formula
1	SO_3H	CH_3	Cl	$C_{18}H_{13}ClN_2O_6S$
2	SO_3H	Cl	CH_3	$C_{18}H_{13}ClN_2O_6S$
3	COOH	H	Cl	$C_{18}H_{11}ClN_2O_5$
4	SO_3H	CH_3	H	$C_{18}H_{14}N_2O_6S$
5	COOH	H	$C(O)NHC_6H_5$	$C_{25}H_{17}N_3O_6$

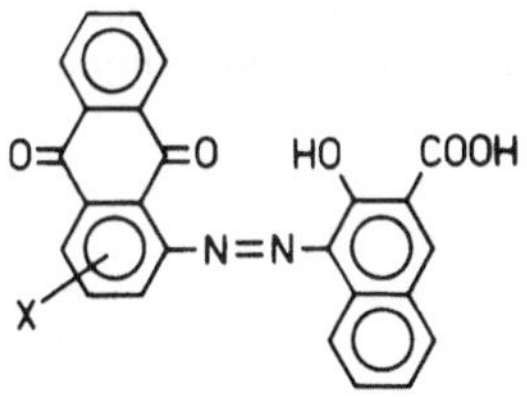

ligand 6 X = H (= $C_{25}H_{14}N_2O_5$)
ligand 7 X = Cl (= $C_{25}H_{13}ClN_2O_5$)

Pigments of the general composition MnL are produced on an industrial scale by precipitation with Mn^{2+} ions of the water-soluble ligands 1 to 5 or similar ones, frequently used as their disodium salts. The general name, Manganese BON Reds or BON Maroons, is derived from the azo coupling component BON = β-oxynaphthoic acid as an essential group and from the splendid colorizing effects of the manganese lakes or toners. These compounds exhibit clean, intense tones of higher lightfastness and durability compared to their alkaline earth analogs. Thus they are widely used in printing inks and lacquers, but less in plastic industries, due to the catalytic properties of manganese favoring decomposition reactions. The extensive literature published on this field of manganese compounds, of importance to technical problems, cannot be cited in detail; see patents and other technical publications. Pigments like the following, best known and approved ones, are compiled in handbooks, e.g. [1 to 3], and the references therein. The compound $Mn(C_{18}H_{11}ClN_2O_5S)$ obtained with ligand 1 and called Pigment Red 48:4, Permanent Red 2B, or Manganese BON Red, Color Index No. 15865, has long been applied in automotive and other high-quality industrial finishes as it offers a supreme and durable masstone lightfastness and high bleed and fair bake resistance. The isomeric compound with ligand 2 (Pigment Red 52 = Lithol Red 2 G) supplies durable red lakes. The very lightfast $Mn(C_{18}H_9ClN_2O_5)$ produced with ligand 3 (Pigment Red 55 = Yellow BON Maroon) with high solvent bleed resistance is utilized for automotive and other outdoor enamels.

A red pigment with the ligand 6 or 7 was proposed for large scale production of printing inks and of lacquers for Soft-PVC [4].

References:

[1] Fytelson, M. (in: Kirk-Othmer Encycl. Chem. Technol. 3rd Ed. **17** [1982] 841, 851/2).
[2] Ullmanns Enzykl. Tech. Chem. 4th Ed. **18** [1979] 674/5.
[3] The Society of Dyers and Colourists, Bradford, Yorksh., Eng., and American Association of Textile Chemists and Colorists, Lowell, Mass., Colour Index, 3rd Edition, 1971/1975, Suppl. Vol. 4.
[4] Dimroth, P., Henning, G., BASF A.-G. (Ger. Offen 2060557 [1972]).

28.1.11 With 4-Hydroxy-3-(phenylazo)coumarin

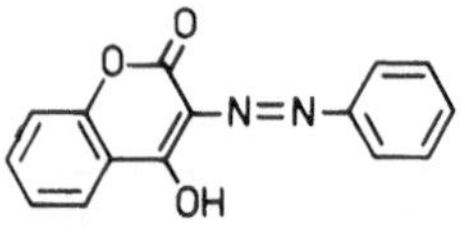

or tautomers ($=C_{15}H_{10}N_2O_3$)

$Mn(C_{15}H_9N_2O_3)_2$ precipitated from mixed equimolar solutions of Mn^{2+} ions in water and of the ligand in methanol at room temperature. After recrystallization from dioxane the air-stable compound melts at 170°C. Its IR spectrum shows an absorption band at 1660 cm^{-1} due to ν(CO) vibrations shifted on complexation from 1740 cm^{-1} for the free ligand, whereas the ν(OH) band of the ligand at 3300 cm^{-1} has disappeared, and a band assigned to ν(N=N) is located at 1550 cm^{-1}. A complex structure is assumed with two bidentate ligands forming 6-membered chelate rings with manganese bound by the enolate oxygen atoms in 4-position of the coumarin system and the azo nitrogen atom nearest to the phenyl ring. Involvement of the additional carbonyl group in chelation was not considered. UV bands, which are only slightly blue shifted (in comparison to the ligand) to 252 and 424 nm, show increased intensities on complex formation. The compound is soluble in nonpolar solvents and nonconducting in methanol or dioxane solution, Bharat Kumar, B., Krishna Rao, K. S. R. M., Ganorkar, M. C. (Current Sci. [India] **42** [1973] 461/3; C.A. **79** [1973] No. 61034).

28.2 With Pyridylazo Compounds

General Reference:

Shibata, S. (in: Flaschka, H. A., Barnard, A. J., Chelates in Analytical Chemistry, Vol. 4, London, New York 1972, pp. 1/232).

28.2.1 With 2-(2-Pyridylazo)phenols

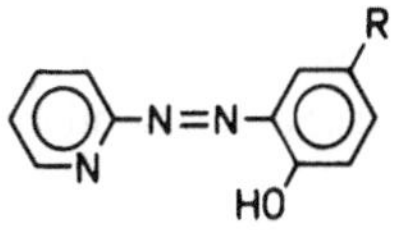

1) R = H (= $C_{11}H_9N_3O$)
2) R = CH_3 (= $C_{12}H_{11}N_3O$)

Complexes in Solution. Potentiometric titrations (glass electrode) on solutions of a manganese(II) salt and ligand 1 in aqueous methanol (50 vol%) at 25°C, I = 0.1 M ($NaClO_4$) under N_2 reveal the formation of a 1:1 and a 1:2 complex with the stability constants log K_1 = 5.6 and log K_2 = 7.0. The pH meter readings are corrected for the effect of the organic solvent [1]. A value of log β_2 = 10.52 ± 0.02 was determined spectrophotometrically in 5% (v/v) aqueous ethanol for the complex $Mn(C_{11}H_8N_3O)_2$ at 23 to 25°C and I = 0.1 M ($NaClO_4$) in the pH range from 6 to 9. The molar extinction coefficient in this solution is $\varepsilon = 2.1 \times 10^4$ $L \cdot mol^{-1} \cdot cm^{-1}$ observed at λ_{max} = 530 nm. At pH values >9 the absorbance decreases, and a precipitate is formed which is presumed to be a hydroxy form of the chelate [2].

A blue-violet complex with the methyl substituted ligand 2 was evaluated spectrophotometrically in weakly alkaline aqueous solution. Maximum absorbance $\varepsilon = 1.8 \times 10^4$ $L \cdot mol^{-1} \cdot cm^{-1}$ at λ_{max} = 560 nm develops at pH ~ 9 [5].

$Mn(C_{11}H_8N_3O)_2$. Mixed equal volumes of 0.005 M solutions of a manganese(II) salt in water and of ligand 1 in ethanol were heated with a few drops of aqueous ammonia. The crystalline complex, which precipitated on standing, was washed with water and dried in a vacuum [3]. The IR spectrum of complex pellets in CsI shows many strong and broad absorption bands due to skeletal vibrations of the ligand in the 1600 to 900 cm^{-1} region which do not allow analysis of complex-induced bands. A strong band at 636 cm^{-1} (sh) is ascribed to ν(Mn–O) but is superimposed with bands caused by shifted pyridine C–H deformations and ring vibrations. Bands of medium intensity at 243 (sh) and 222 cm^{-1} assigned to ν(Mn–N) indicate pyridine ring coordination of the anionic chelating ligand [4]. A chelate molecule is postulated with the Mn^{2+} ion coordinated by two terdentate ligand molecules each one forming two five-membered rings by nitrogen atoms from the pyridine ring and the azo groups or from azo nitrogen and the phenolato oxygen atom [3].

The mass spectrum recorded at 380°C resembles that of the analogous zinc complex. It revealed initial formation of the 1:1 complex. The major fragmentation pattern is displayed as: $Mn(C_{11}H_8N_3O)_2^+$ (m/e = 451) → $Mn(C_{11}H_8N_3O)^+$ (m/e = 253) → $Mn(C_5H_4N)^+$ (m/e = 133) → Mn^+ (m/e = 55) where a reduction of manganese to the oxidation state +1 is implied. The large peak at m/e = 170 is apparently due to ligand fragmentation by loss of one hydrogen and the azo group atoms [3].

References:

[1] Anderson, R. G., Nickless, G. (Anal. Chim. Acta **39** [1967] 469/77, 475).
[2] Betteridge, D., John, D. (Analyst [London] **98** [1973] 390/411, 401/3, 410).
[3] Betteridge, D., John, D. (Talanta **15** [1968] 1227/40, 1232, 1236).
[4] Betteridge, D., John, D. (Analyst [London] **98** [1973] 377/89, 383/4).
[5] Nakagawa, G., Wada, H. (Nippon Kagaku Zasshi **83** [1962] 1098/102, A 70; C.A. **59** [1963] 12143).

28.2.2 With 4-(2-Pyridylazo)-1,3-benzenediol

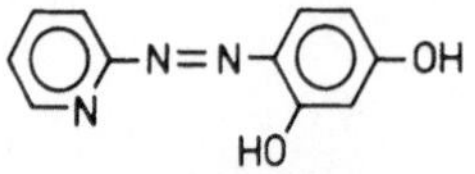

$(=C_{11}H_9N_3O_2=H_2L)$

Complexes in Solution. Various types of distinct manganese(II) complexes and of complex equilibria, listed in the table, were evaluated spectrophotometrically (sp) and potentiometrically (glass electrode = gl) at 25 ± 0.1°C:

equilibrium	method	medium	I in mol/L	log K	Ref.
(1) $Mn^{2+}+2L^- \overset{\beta_2}{\rightleftharpoons} MnL_2^{2-}$	sp	water	0.1($NaNO_3$)	log β_2 = 15.6	[1]
(2) $Mn^{2+}+2HL^- \overset{K_{eq}}{\rightleftharpoons} MnL_2^{2-}+2H^+$	sp	water	0.1($NaNO_3$)	K_{eq} = −8.99	[1]
(3) $Mn^{2+}+HL^- \overset{K_1^*}{\rightleftharpoons} MnHL^+$	gl	a)	<0.01($HClO_4$)	K_1^* = 9.79 b)	[3]
	gl	a)	~0.005($HClO_4$)	9.7	[4]
	gl	a)	0 corr	10.02 b)	[3]
(4) $MnHL^+ + HL^- \overset{K_2^*}{\rightleftharpoons} Mn(HL)_2$	gl	a)	<0.01($HClO_4$)	K_2^* = 9.13 b)	[3]
	gl	a)	0.005($HClO_4$)	9.2	[4]
	gl	a)	0 corr	9.36 b)	[3]
(5) $Mn(HL)_2 \overset{K_1'}{\rightleftharpoons} Mn(HL)L^- + H^+$	gl	a)	~0.005($HClO_4$)	pK_1' = 8.8	[5]
(6) $Mn(HL)L^- \overset{K_2'}{\rightleftharpoons} MnL_2^{2-} + H^+$	gl	a)	~0.005($HClO_4$)	pK_2' = 10.3	[5]

a) 50 vol% dioxane-water. – b) Values are based on pH meter readings corrected for the organic solvent.

The value of $\beta_2 = K_{eq}/K_{a3}^2$ for the overall reaction concerning equation (1) is calculated from the equilibrium constant K_{eq} of reaction (2) studied in the pH range 11.2 to 11.7. The value for dissociation of the *ortho*-hydroxy group, pK_{a3} = 12.31, of the free ligand H_3L^+ is taken from literature cited in [1]. A value of $K_{inst} = 3.90 \times 10^{-12}$ reported without further information for a 1:2 complex in water was determined spectrophotometrically at pH 10 and λ = 490 nm [2]. Potentiometric titration curves of Mn^{2+} ions with the ligand in 50 vol% dioxane exhibit two buffer regions. The first region at pH < 7 corresponds to the release of the proton from the o-hydroxy group with formation of $Mn(HL)^+$ and $Mn(HL)_2$; see equations (3) and (4) [5]. Despite the more acidic nature of the *para*-hydroxy group (pK_{a2} = 6.9 [4], 6.87 for I < 0.01 M and 7.08 for I = 0 (corr) [3]), the weakly acidic proton (pK_{a3} = 12.4 [4], 13.42 for I < 0.01 M, 13.70 for I = 0 (corr) [3]) in the *ortho*-position of the terdentate ligand is the first to be replaced by manganese whereas the OH groups in *para*-position are retained. The constants K_1^* and K_2^* are calculated on the assumption that the dissociation of the proton in *ortho*-position is independent of a preceding dissociation of the proton in *para*-position [3]. The protons in *para*-position are released only in the second, more alkaline buffer region according to equations (5) and (6). The acid dissociation constants pK'_1 and pK'_2 demonstrate the acid strengthening effect on the *para*-OH group as a result of complex formation. This effect is still higher according to the Irving-Williams order for analogous Cu, Ni, or Zn complexes [3, 5].

The formation of the orange-red MnL_2^{2-} complex at pH 9 to 10.7 was confirmed by spectrophotometric studies [1, 2, 6, 8 to 11]. Absorption maxima were observed in the range 490 to 510 nm with molar extinction coefficients of ~4×10^4 to 8×10^4 $L \cdot mol^{-1} \cdot cm^{-1}$. Maximum absorption of the free ligand in weakly alkaline solutions is observed at 410 [2] or 415 nm [7]. The absorbance of MnL_2^{2-} decreases according to the conditions, after ½ h [1], 4 h [11], several days [10], or on heating the solution above 40°C [2]. Reducing agents like $H_2NOH \cdot HCl$ [7] or L-ascorbic acid [1, 6, 8, 11] prevent oxidation of manganese(II) by atmospheric oxygen. One isosbestic point in the absorbance curves obtained at the pH range 6.10 to 11.70 proves only one complex species to be formed with highest sensitivity at pH 11.2 to 11.7 [1]. Obviously erroneously, a 1:3 mole ratio has been established at the pH range 10.3 to 11.2 for an orange-red complex with identical absorption properties [7].

Separation of the orange complex MnL_2^{2-} from the yellow ligand was performed on an anion exchange column, using 0.1 N NaOH for complete elution. The Mn^{2+} is assumed to be hexacoordinated by two tridentate ligand anions [10, 11]. Extraction into organic solvents is performed by addition of diphenylguanidine. The base forms a soluble guanidinium associate of MnL_2^{2-} [12].

The complex is decomposed upon addition of ethylenediaminetetraacetic acid [6, 11].

$Mn(C_{11}H_8N_3O_2)_2 \cdot 2H_2O$. For preparation, the ligand dissolved in dilute aqueous HCl was added to an aqueous solution of $Mn(ClO_4)_2$, the mole ratio slightly exceeding 2:1. Dilute aqueous ammonia was added slowly with stirring until a precipitate formed. Digestion on a water bath accelerated coagulation. The heat-sensitive complex was washed with ethanol followed by water and was dried in a vacuum at room temperature [4].

References:

[1] Nonova, D., Evtimova, B. (Talanta **20** [1973] 1347/51).

[2] Tataev, O. A., Anisimova, L. G. (Zh. Analit. Khim. **26** [1971] 184/7; J. Anal. Chem. [USSR] **26** [1971] 166/8).

[3] Stanley, R. W., Cheney, G. E. (Talanta **13** [1966] 1619/29, 1620/5).

[4] Corsini, A., Mai-Ling Yih, I., Fernando, Q., Freiser, H. (Anal. Chem. **34** [1962] 1090/3).

[5] Corsini, A., Fernando, Q., Freiser, H. (Inorg. Chem. **2** [1963] 224/6).

[6] Yotsuyanagi, T., Goto, K., Nagayama, M., Aomura, K. (Bunseki Kagaku **18** [1969] 477/81; C.A. **71** [1969] No. 56313).

[7] Ueda, K., Yamamoto, Y., Ueda, Sh. (Nippon Kagaku Zasshi **90** [1969] 903/7, A 49; C.A. **71** [1969] No. 119327).

[8] Ahrland, S., Herman, R. G. (Anal. Chem. **47** [1975] 2422/6, 2424).

[9] Tokar, L. V., Per'kov, I. G., Tyvetshang, S. J. (Vestn. Kharkov. Univ. No. 215 [1981] 43/5; C.A. **95** [1981] No. 226619; Ref. Zh. Khim. **1981** No. 21 V 146).

[10] Herman, R. G., Norman, L. J. (Inorg. Nucl. Chem. Letters **14** [1978] 183/8).

[11] Nagarkar, S. G., Eshwar, M. C. (Chem. Era **11** No. 4 [1975] 1/3; C.A. **84** [1976] No. 115451).

[12] Pyatnitskii, I. V., Grigalashvili, K. I., Mamuliya, S. G. (Mater. Mezhfak. Konf. Tbilis. Inst. Estestv. Naukam Khim. Biol. Geogr. Geol., Tbilisi 1984, pp. 19/21 from Ref. Zh. Khim. **1985** No. 19 V 228).

28.2.3 With 4-(2-Pyridylazo)benzenamines

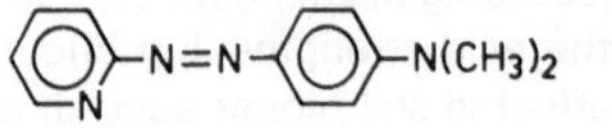

1) $(=C_{13}H_{14}N_4=L)$

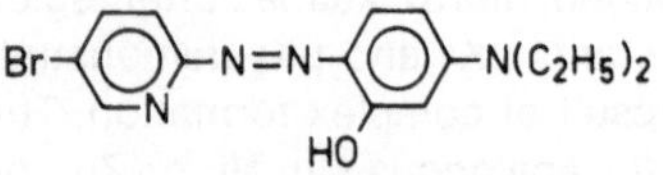

2) $(=C_{15}H_{17}BrN_4O=HL)$

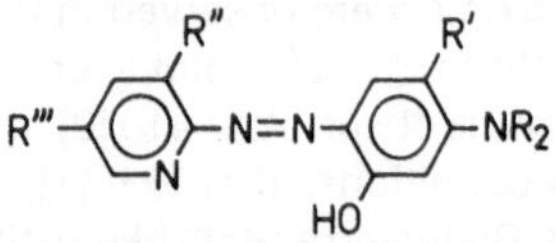

3) $R=CH_3$, $R'=R''=R'''=H$ $(=C_{13}H_{14}N_4O)$
and further ligands: $R=H$, CH_3, C_2H_5;
$R'=H$, CH_3; $R''=H$, Br; $R'''=H$, Cl, Br, I

Complexes in Solution. A very small stability constant, $K_1=5\pm0.1$ L/mol, was determined spectrophotometrically for the complex MnL^{2+} with ligand 1 in aqueous solution at pH 6.0 to 6.5 and 25°C, I = 0.15 M ($NaNO_3$), obviously due to hydrolytic or other effects [1]. Absorbance measurements using equal conditions at the pH range between 8.0 and 8.8 (glycine buffer) yield $K_1=2.1$ L/mol for the same complex. A formation rate constant $k_1=1.5\times10^6$ $L^3\cdot mol^{-1}\cdot s^{-1}$ has been fixed under pseudo first order conditions, with $[Mn^{2+}]\gg$ [ligand 1] by the continuous flow method. A reaction model is established kinetically, including the replacement of water from the inner hydration sphere by one ligand donor atom entering first and the rapid closure of a 5-membered chelate ring thereafter. The rate of complex formation is strongly increased on addition of sodium dodecyl sulfate in about its critical micelle concentration of 2.6×10^{-3} M. With this concentration of the anionic surfactant a maximum rate enhancement of 400-fold has been calculated from a comparison of the Mn^{II} concentrations required to give the same amount of complex in the presence and absence of sodium dodecyl sulfate. The reaction enhancement has been explained quantitatively by an enrichment of the reactants in the region of the micelle surface [2]. The visible spectrum of the complex shows an absorption maximum at $\lambda=540$ nm with the molar extinction coefficient $\varepsilon=3.27\times10^4$ $L\cdot mol^{-1}\cdot cm^{-1}$ at pH 6. A five-membered chelate ring is assumed to be formed by coordination of the pyridine nitrogen and one of the azo nitrogen atoms [1].

The stability constant $\beta_2=(7.0\pm0.3)\times10^{11}$ $L^2\cdot mol^{-2}$ was determined spectrophotometrically for the complex $Mn(C_{15}H_{16}BrN_4O)_2$ with ligand 2 in aqueous ethanol at pH = 8.5. An extremely high extinction ($\varepsilon=1.27\times10^5$ $L\cdot mol^{-1}\cdot cm^{-1}$) was observed at $\lambda_{max}=575$ nm [3]. Values of $\lambda_{max}=558$ nm and $\varepsilon=1.06\times10^5$ $L\cdot mol^{-1}\cdot cm^{-1}$ for the purple complex at pH 7 to 10 were reported in [4]. At pH 10, two maxima were observed in 50% (v/v) ethanol at $\lambda=525$ and 560 nm, with $\varepsilon=8.8\times10^4$ $L\cdot mol^{-1}\cdot cm^{-1}$ each. Bathochromic shifts of 80 and 115 nm, respectively, were observed, relative to the uncomplexed ligand. The noncharged complex can be extracted into organic solvents [5]. Pure ethanol instead of the water mixture increases the solubility and stability of the complex [3].

Appearance of characteristic red to violet colors was also reported on reaction of Mn^{2+} ions in weakly acid [6] or weakly alkaline [7] aqueous ethanol with ligand 3 and the further ligands of this type, see above.

$Mn(C_{13}H_{13}N_4O)_2$ was prepared by mixing an ethanolic solution of ligand 3 with an acidic solution of manganese(II) chloride. Its visible spectrum is characterized by the same double peak in the 500 to 560 nm range observed for the complex solution with ligand 2. The core electron binding energies for $N_{azo}(1s)$ and O(1s) of the ligand 3 donor atoms are determined by X-ray photoelectron spectroscopy of the powdered chelate and are discussed with respect to the chelate stability. The extremely high binding energy of the phenolate oxygen O(1s) to Mn^{II} reflects the preferred binding tendency of this metal to oxygen in contrast to Ni^{II}, Cu^{II}, and Zn^{II} ions in corresponding complexes. A chelate structure is suggested with two N, N, O-terdentate ligand anions, each one forming two fused 5-membered rings with the six-coordinate manganese(II) ion [8].

References:

[1] Klotz, I. M., Loh Ming, W.-C. (J. Am. Chem. Soc. **75** [1953] 4159/62).
[2] Holzwarth, J., Knoche, W., Robinson, B. H. (Ber. Bunsenges. Physik. Chem. **82** [1978] 1001/5).
[3] Wei, Fusheng, Qu, Peihua, Zhu, Yurei (Fenxi Huaxue **9** [1981] 345/7; C.A. **95** [1981] No. 231307).
[4] Qiu, Xing-Chu, Zhang, Yu-Sheng, Zhu Ying-Quan (Soil Sci. **138** [1984] 432/5; C.A. **102** [1985] No. 44943), Qiu, Xing-chu, Zhanf, Yu-sheng, Zhu, Ying-Quan (Chem. Anal. [Warsaw] **30** [1985] 127/30; C.A. **104** [1986] No. 81090).
[5] Johnson, D. A., Florence, T. M. (Talanta **22** [1975] 253/65, 259/60).
[6] Gusev, S. I., Shchurova, L. M. (Zh. Analit. Khim. **21** [1966] 1042/9; J. Anal. Chem. [USSR] **21** [1966] 927/33, 930).
[7] Shibata, S., Furukawa, M., Toei, K. (Anal. Chim. Acta **66** [1973] 397/409, 401).
[8] Kudo, Y., Yoshida, N., Fujimoto, M., Tanaka, K., Toyoshima, I. (Chem. Letters **1985** 1573/6; C.A. **104** [1986] No. 12571).

28.2.4 With Isomeric (2-Pyridylazo)naphthols and Their Derivatives

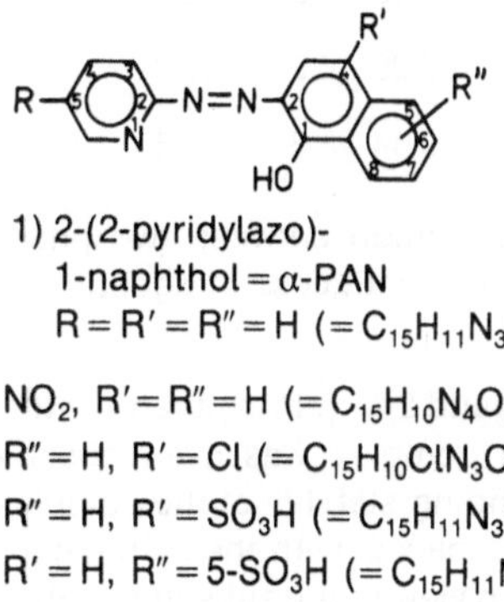

1) 2-(2-pyridylazo)-1-naphthol = α-PAN
R = R' = R" = H (= $C_{15}H_{11}N_3O$)

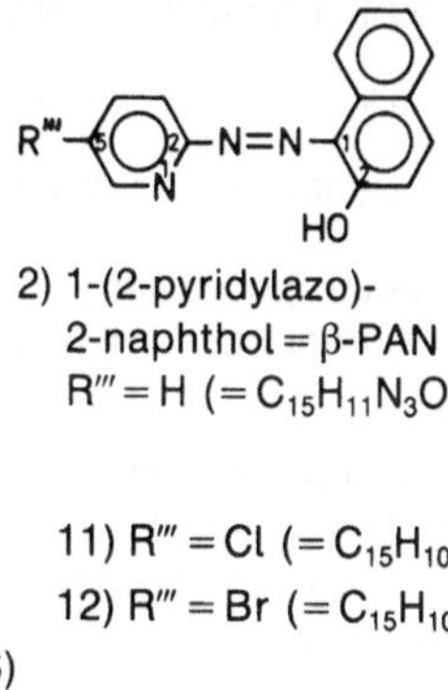

2) 1-(2-pyridylazo)-2-naphthol = β-PAN
R''' = H (= $C_{15}H_{11}N_3O$)

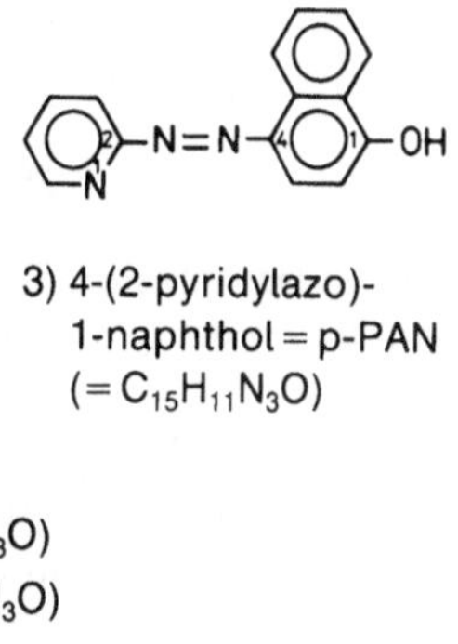

3) 4-(2-pyridylazo)-1-naphthol = p-PAN
(= $C_{15}H_{11}N_3O$)

4) R = NO_2, R' = R" = H (= $C_{15}H_{10}N_4O_3$)
5) R = R" = H, R' = Cl (= $C_{15}H_{10}ClN_3O$)
6) R = R" = H, R' = SO_3H (= $C_{15}H_{11}N_3O_4S$)
7) R = R' = H, R" = 5-SO_3H (= $C_{15}H_{11}N_3O_4S$)
8) R = R' = H, R" = 6-SO_3H (= $C_{15}H_{11}N_3O_4S$)
9) R = R' = H, R" = 7-SO_3H (= $C_{15}H_{11}N_3O_4S$)
10) R = R' = H, R" = 8-SO_3H (= $C_{15}H_{11}N_3O_4S$)

11) R''' = Cl (= $C_{15}H_{10}ClN_3O$)
12) R''' = Br (= $C_{15}H_{10}BrN_3O$)

28.2.4.1 Manganese(II) Complexes in Solution

Spectrophotometric evidence has been obtained for the formation of red to purple complexes in weakly alkaline solutions of aqueous Mn^{2+} ions and the ligands 1 to 3 [1, 2, 5 to 8, 12, 15, 16], 4 [1], 11, and 12 [13], dissolved in various organic solvents. Because of their higher solubilities the complexes are frequently enriched by extraction into the organic medium. The mole ratio 1:2 is determined by the continuous variation and mole ratio methods [5, 7, 12 to 14]. Oxidation of Mn^{II} to Mn^{III} was prevented by addition of $NH_2OH \cdot HCl$ [15] or ascorbic acid [5] to β-PAN complex solutions in the pH region 8.8 to 9.6.

Stepwise and overall stability constants were determined for 1:1 and 1:2 complexes with the chelating ligands 1, 2, and 6 to 10 listed below together with the pertinent pK values for dissociation of the phenolic proton of the ligands (pK_{a2}):

ligand	method[a)]	t in °C	solvent	I in L/mol	pK_{a2}	log K_1	log β_2	Ref.
1= α-PAN	sp	30	50% dioxane	0.1	10.37	7.18	14.68	[1]
	sp	24±1	40% ethanol	0.1($NaClO_4$)	10.20	—	13.54±0.06[b)]	[2]
	sp	24±1	40% ethanol	0.1($NaClO_4$)	10.20	—	13.27±0.05[c)]	[2]
	dis	30	water-CCl_4	0.10($NaClO_4$)	9.63	—	13.30	[2]
2= β-PAN	sp	30	50% dioxane	0.1	12.3	7.90	16.33	[1]
	gl	25	50% dioxane	0.01($HClO_4$)	12.3	8.5	16.4	[4]
	sp	24±1	40% ethanol	0.1($NaClO_4$)	12.20	—	15.77±0.06[b)]	[2]
	sp	24±1	40% ethanol	0.1($NaClO_4$)	12.20	—	15.69±0.08[c)]	[2]
	dis	30	water-CCl_4	0.10($NaClO_4$)	11.62	—	16.13	[2]
	dis	31±2	water-CCl_4	0.1($NaClO_4$)	11.2	—	15.3±0.5	[3]
	ex	25	water-methanol	0.10($NaClO_4$)	12.3	—	16.8	[5, 17]
6	gl	25	50% methanol	0.1($NaClO_4$)	8.63	5.9	11.9	[9, 10]
7	gl	25	50% methanol	0.1($NaClO_4$)	9.11	6.0	12.5	[9]
8	gl	25	50% methanol	0.1($NaClO_4$)	9.13	6.2	12.7	[9]
9	gl	25	50% methanol	0.1($NaClO_4$)	9.09	6.2	12.0	[9]
10	gl	25	50% methanol	0.1($NaClO_4$)	10.44	8.3	14.5	[9]

[a)] gl = glass electrode, sp = spectrophotometry, dis = distribution between two phases, ex = extraction using a nonionic surfactant. – [b)] Excess of Mn. – [c)] Excess of ligand.

Stability is higher for manganese complexes with ligand 2 (= β-PAN) than for the isomer with ligand 1 (= α-PAN) [1, 2]. In chloroform, the 1:2 complex was even observed with substoichiometric amounts of ligand 2 [11]. The outstanding stability of the complexes with ligand 10, a linear relationship between log K of 1:1 complexes with the sulfonated ligands No. 6 to 10, and the respective pK_{a2} values are discussed in view to the substituent effects [9].

With ligand 3 = p-PAN a more intensely colored complex is formed than with β-PAN = ligand 2. It was presumed, from the close similarity of their spectra and in analogy to the more thoroughly studied zinc complexes, that the *para*-hydroxy substituted ligand anion reacts with manganese in a quinoid resonance form which is stabilized by bidentate nitrogen chelation involving a negatively charged pyridine nitrogen atom [12].

For the uncharged complexes MnL_2 with ligand 1 or 2 (= HL), extracted into carbon tetrachloride from their water solution, a distribution ratio $K_D = [MnL_2]_{CCl_4}/[MnL_2]_{H_2O} = 10^4$ is inferred. Formation constants log β_2 based thereon, as well as the pH values for half extraction, 7.30 and 7.85, respectively, are in good agreement with the values predicted by spectrometry [2, 3]. Distribution studies on solutions of Mn^{2+} ions and ligand 2 in the borate-buffered systems H_2O-$CHCl_3$ and H_2O-CCl_4 at 22°C and I = 0.01 M establish the 1:2 complex with the extraction constants $\log K_{ex} = [MnL_2]_{org} \cdot [H^+]^2_{H_2O} \cdot [Mn^{2+}]^{-1}_{H_2O} \cdot [HL]^{-2}_{org} = -11.0$ for $CHCl_3$ and −9.8 for CCl_4 [6] based on K_D values of $10^{5.4}$ and 10^4 from [3], respectively.

A red-violet complex with ligand 1 and a blue solution with the nitro-substituted ligand 4 resulted from reactions with Mn^{2+} ions in 50% aqueous dioxane [1]. Characteristic data for the absorption spectra of several complexes in ethanol-water, $CHCl_3$, and ether:

ligand	1	2	2	3	11	12
solvent	ethanol-water	ethanol-water	ethanol-water	ethanol-water	$CHCl_3$	ether
v/v ratio	2:3	2:3	7:3	7:3	—	—
pH	9 to 9.5	9 to 9.5	ammoniacal	ammoniacal	~8 to 11	~8 to 12
λ_{max} in nm	586, 546	546, 516	552, 525, 332	595, 580, 396	566	574
$\varepsilon \cdot 10^{-4}$ in $L \cdot mol^{-1} \cdot cm^{-1}$	4.09 to 4.15	4.04 to 4.07	—	—	7.2	7.2
Ref.	[2]	[2]	[12]	[12]	[13]	[13, 14]

The prominent absorption maximum (in nm, molar absorbance in $L \cdot mol^{-1} \cdot cm^{-1}$) of the complex with ligand 2 = β-PAN at pH 9.2 is displaced to 558 in n-pentyl alcohol [15], 560 (5.85×10^4) in ether [16], 562 (4.8×10^4) in $CHCl_3$ [13 to 15], 568 (4.7×10^4) in benzene [15], and 570 in CCl_4 [15]. For other values, see [6].

Optimal conditions for the analytical determination and separation of Mn^{II} from interfering ions have been investigated by extraction studies of manganese(II) pyridylazonaphtholates [1, 5 to 8, 11, 16, 19]. The highest absorptivities and selectivities are exhibited by the complexes with ligands 11 and 12 [13, 14].

Substitution of β-PAN in $Mn(C_{15}H_{10}N_3O)_2$ by ethylenediaminetetraacetate was studied kinetically by the decrease in absorbance at 554 nm in ammonia-buffered aqueous solution containing solubilizing methanol and Triton X-100. The low conditional rate constant $k = 1.13\ s^{-1}$ for the manganese complex at 25°C and pH 8.8, determined by the stopped-flow method, allowed the simultaneous determination of Mn^{2+} and Cd^{2+} ions [18].

With ligands 1 or 2 (= HL) formation of the complex MnL_2 is accompanied by the occurrence of a precipitate presumed to be a hydroxy species. For the β-PAN chelate a formation constant $K = [MnL_2(OH)^-]/[MnL_2] \cdot [OH^-] = 10^{7.57}$ L/mol^{-1} was evaluated spectrophotometrically at $\lambda = 546$ nm in 40% ethanol, in accordance with a value $K = 10^{7.64}$ L/mol^{-1} from extraction studies [2]. The formation of any hydroxy species is prevented by addition of triethanolamine and Triton X-100 [5].

References:

[1] Kawase, A. (Bunseki Kagaku **16** [1967] 569/76, 572, 575; C.A. **68** [1968] No. 18243).
[2] Betteridge, D., John, D. (Analyst [London] **98** [1973] 390/411, 397/403, 406/10).
[3] Betteridge, D., Fernando, Q., Freiser, H. (Anal. Chem. **35** [1963] 294/8).
[4] Corsini, A., Mai-Ling Yih, I. Fernando, Q., Freiser, H. (Anal. Chem. **34** [1962] 1090/3).
[5] Goto, K., Taguchi, Sh., Fukue, Y., Ohta, K., Watanabe, H. (Talanta **24** [1977] 752/3).
[6] Inczédy, J., Varga-Puchony, Z. (Acta Chim. [Budapest] **79** [1973] 63/9).
[7] Shibata, Sh. (Anal. Chim. Acta **25** [1961] 348/59).
[8] Berger, W., Elvers, H. (Z. Anal. Chem. **171** [1959] 185/93, 188/90).
[9] Anderson, R. G., Nickless, G. (Analyst [London] **93** [1968] 13/9, 17).
[10] Anderson, R. G., Nickless, G. (Talanta **14** [1967] 1221/8, 1225).

[11] Ashok Rao, K., Rangamannar, B. (J. Radioanal. Nucl. Chem. **93** [1985] 319/26).
[12] Betteridge, D., Todd, P. K., Fernando, Q., Freiser, H. (Anal. Chem. **35** [1963] 729/33, 730).
[13] Shibata, Sh., Furukawa, M., Kamata E., Goto, K. (Anal. Chim. Acta **50** [1970] 439/46, 442/4).

[14] Shibata, Sh., Goto, K., Kamata, E. (Anal. Chim. Acta **45** [1969] 279/88, 283/5).
[15] Donaldson, E. M., Inman, W. R. (Talanta **13** [1966] 489/97).
[16] Shibata, Sh. (Anal. Chim. Acta **23** [1960] 367/9).
[17] Goto, K., Taguchi, Sh., Fukue, Y., Ohta, K., Watanabe, H. (Mizu Shori Gijutsu **19** [1978] 745/8; C.A. **90** [1979] No. 33484).
[18] Nakagawa, K., Ogata, T., Haraguchi, K., Ito, S. (Bunseki Kagaku **30** [1981] 149/53; C.A. **95** [1981] No. 54132).
[19] Cheema, M. N., Qureshi, I. H., Ashraf, M., Hanif, I. (J. Radioanal. Chem. **35** [1977] 311/9).
[20] Iida, Y., Furukawa, M., Shibata, Sh. (Huaxue Shiji **6** [1984] 260/3 from C.A. **102** [1985] No. 38628).

28.2.4.2 Isolated Manganese(II) Compounds

Isomeric chelates $\mathbf{Mn(C_{15}H_{10}N_3O)_2}$ with ligand 1 = α-PAN or ligand 2 = β-PAN were prepared by heating solutions of manganese(II) salts in water and either ligand in ethanol in the presence of aqueous ammonia. The complexes precipitating on standig were washed with water and dried in vacuum [1]. For precipitation of the intensely pink $Mn(C_{15}H_{10}N_3O)_2$ with ligand 2 the pH was raised to 10 with NaOH. The mixture was refluxed on a water bath for 2 h. The complex separated on concentration. It was successively washed with water, ethanol, and acetone, then dried at 60°C; melting point 300°C [2]. According to a procedure reported in [4] for analogous Co, Cu, and Zn complexes, the manganese complex with ligand 2 was also precipitated from hot methanolic solutions of the reactants in a 1:2 mole ratio [3], or from methanol-water mixtures at pH 9 [5]. It was washed with hot methanol and dried in vacuum [4].

The IR spectra (CsI pellets) of both complexes with the isomeric ligands resemble that of $Mn(C_{11}H_8N_3O)_2$ with 2-(2-pyridylazo)phenol; see p. 291. Many broad and overlapping bands in the 1600 to 900 cm^{-1} region hamper identification of complex bonding. Broad strong bands at 1330 cm^{-1}, assigned to ν(N=N) and shifted from higher frequencies on complexation of either ligand, indicate coordination of the azo groups. Broadening of a band at ~1100 cm^{-1} in the complexes is probably caused by a shift of a ν(C–O) band. The following absorption bands (in cm^{-1}) are observed in the far-IR spectra of the chelates $Mn(C_{15}H_{10}N_3O)_2$:

assignment	*)	ν(Mn–O)	ν(Mn–N)	ν(Mn–N)
ligand 1 = α-PAN	625	420	240	—
ligand 2 = β-PAN	626	438	244	219

*) The bands near 620 cm^{-1} may be due to coupling of a vibration mode ν(Mn–O) with δ(CH) or ring vibrations of the ligands. The complex stereochemistry is determined by a terdentate bite of the ligands [6].

Susceptibility measurements of the complex with ligand 2 = β-PAN yield the magnetic moment $\mu_{eff} = 6.31\ \mu_B$ at 295.0 K, consistent with with a sextet ground state in the compound. Its EPR spectra exhibit one almost symmetrical signal centered around g = 2.00, more narrow in pyridine solution than for the powder sample, indicating a symmetric ligand field [2]. The polarographic behavior of the β-PAN complex in DMF resembles that of the chelates with o,o′-bis(hydroxyaryl)azo compounds; see p. 279. Classical polarograms (Hg pool reference) at 25 ± 0.2°C on solutions containing 0.1 M $(C_2H_5)_4NClO_4$ reveal three reduction waves with $E_{1/2} = -0.54$, -1.22, and -1.44 V compared to $E_{1/2} = -0.49$ and -1.29 V for the free ligand. The cathodic shifts $\Delta E_{1/2}$ are substantially lower than for the chelates with two hydroxy groups also studied. The first two waves correspond to reversible one-electron reduction steps with conversion of the neutral complex to the anionic species $Mn(C_{15}H_{10}N_3O)_2^-$ and $Mn(C_{15}H_{10}N_3O)_2^{2-}$. The third two-electron wave in the classical polarograms represents an irreversible reduction of the ligand liberated after the discharge of the complex. Two pairs of symmetrical cathodic-anodic peaks are displayed oscillopolarographically, permitting the calculation of the heterogeneous rate constants of charge transfer, $K_s = 0.0175$ and 0.0156 cm/s, respectively [3].

Both the complexes with isomeric ligands exhibit mass spectra similar to that observed for $Mn(C_{11}H_8N_3O)_2$; see p. 291. The parent peak of the chelate with β-PAN is smaller, however, than for the other two compounds [1]. The 1:1 complexes prevail after initial loss of one ligand molecule [6]. Either complex is insoluble in water [2, 6]. Solubility in common organic solvents such as chloroform, DMF, nitrobenzene, and pyridine is stated for the β-PAN complex [2].

With 4-(2-pyridylazo)-1-naphthol (= ligand 3) a violet precipitate was observed by spot test reactions in ethanol-water containing ammonia [7].

References:

[1] Betteridge, D., John, D. (Talanta **15** [1968] 1227/40, 1227, 1236).
[2] Bhoon, Y. K., Pandeya, K. B., Singh, R. P. (J. Indian Chem. Soc. **57** [1980] 286/8).
[3] Toropova, V. F., Budnikov, G. K., Maistrenko, V. N. (Zh. Obshch. Khim. **44** [1974] 364/8; J. Gen. Chem. [USSR] **44** [1974] 345/9).
[4] Zhuchenko, T. A., Kuznetsova, L. I., Kogan, V. A., Garnovskii, A. D., Osipov, O. A., Gorelik, M. V., Gladysheva, T. Kh., Alekseenko, V. A., Zayakina, T. A. (Zh. Neorgan. Khim. **16** [1971] 2169/72; Russ. J. Inorg. Chem. **16** [1971] 1157/9).
[5] Shibata, Sh. (Anal. Chim. Acta **23** [1960] 367/9).
[6] Betteridge, D., John, D. (Analyst [London] **98** [1973] 377/89, 377, 383/4).
[7] Betteridge, D., Todd, P. K., Fernando, Q., Freiser, H. (Anal. Chem. **36** [1963] 729/33).

28.2.5 With 1-(2-Pyridylazo)-2-naphthol N-Oxide

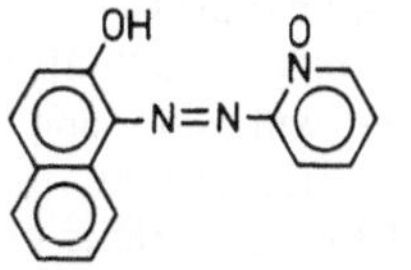

$(= C_{15}H_{11}N_3O_2)$

$Mn(C_{15}H_{10}N_3O_2)_2$. $MnCl_2 \cdot 4H_2O$ and the ligand in a 2:1 mole ratio were heated in DMF to 50 to 60°C for 1 h, to 90°C for another 1 h, and to 125 to 130°C for final 2 h to yield 55.5% of dark brown-violet crystals. After recrystallization from a 1:7 (v/v) DMF-acetone mixture the complex melts at 360°C. The IR spectrum (KBr disks) shows characteristic absorption bands at 3075, 1445, 1250, 1210, 830, and 455 cm^{-1}. The bands assigned to ν(NO) and δ(NO) are only slightly shifted from the respective ligand bands at 1260 and 1220 cm^{-1}, as a result of opposite effects on the π-bonded N-oxide group when coordinated to manganese via oxygen. The far-IR band is assigned to a ν(Mn–O) vibration [1]. The electronic spectrum shows absorption maxima at 537 [1, 2], 246, and 222 nm, and minima at 432, 344, and 275 nm in DMF, with a bathochromic shift of 82 nm in relation to the λ_{max} value of the ligand at 455 nm [1]. It is concluded from the spectra that manganese is six-coordinate. The terdentate ligand molecules chelate through the oxygen atoms of the phenolate and N-oxide groups and through one azo nitrogen atom [1]. The electrophoretic mobilities and shifts of the absorption maximum of the complex were studied in the pH range 3.5 to 8.7 [2]. The solubility of the complex in various organic solvents and the color of such solutions (in parentheses) were stated as follows: very soluble in warm and soluble in cold DMF (violet-red), soluble in warm, slightly soluble in cold ethanol, acetone, or $CHCl_3$ (red), slightly soluble in pyridine (blueish red) and also in warm methanol, ether, or CCl_4 (red), but insoluble in benzene. A violet solution of the water insoluble complex is obtained in concentrated sulfuric acid, yellow or yellow-orange solutions in ethanol on addition of 0.1 N HCl or CH_3COOH [1].

Mass spectrometric analyses confirm the complex composition. Whereas the molecular ion itself is not present, fragment peaks with high intensities are obtained at m/e = 552 ($Mn(C_{15}H_{10}N_3O_2)_2$ minus O_2), 303, 220, 130 [1, 3].

References:

[1] Koprivanac, N., Jovanović-Kolar, J., Renko, D. (Croat. Chem. Acta **54** [1981] 149/55; C.A. **94** [1981] No. 218844).

[2] Renko, D., Koprivanac, N., Jovanović-Kolar, J., Osterman, D. (Kem. Ind. [Zagreb] **28** No. 2 [1979] 53/8, 56; C.A. **92** [1980] No. 24222).

[3] Koprivanac, N., Jovanović-Kolar, J., Kramer, V. (Intern. J. Mass Spectrom. Ion Phys. **47** [1983] 531/4; C.A. **98** [1983] No. 98349).

28.2.6 With 10-(2-Pyridylazo)-9-phenanthrol

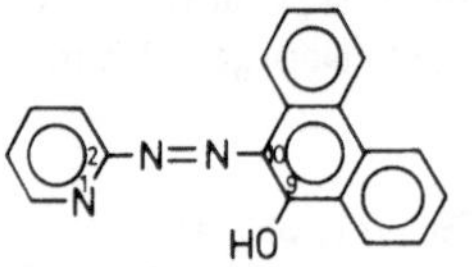

(= $C_{19}H_{13}N_3O$)

Complexes in Solution. The formation constants log K_1 = 8.80 and log K_2 = 8.25 of 1:1 and 1:2 complexes in 3:1 (v/v) dioxane-water mixtures and the formation enthalpy ΔG_1 = −12.04 kcal/mol were determined potentiometrically at 300 K and ionic strength I = 0.1 M ($NaClO_4$) [1].

$Mn(C_{19}H_{12}N_3O)_2 \cdot 2H_2O$ was prepared like the complex with 1-(2-pyridylazo)-2-naphthol, $Mn(C_{15}H_{10}N_3O)_2$ (see p. 298), in aqueous ethanolic solution at pH 9.0 [2] or 10.0 [3]. The pink complexes precipitated on concentration of the solution were washed with water and benzene and dried at 80°C [2], or with water, ethanol, and acetone and dried at 60°C [3, 4]. The compound melts above 300°C [2, 3]. The electronic spectrum in $CHCl_3$ at various pH values exhibits three absorption bands in neutral or alkaline media and complex dissociation in acidic solution [2]. In neutral solutions the absorption bands are also recorded at λ = 554, 519, 492, and 410 nm with the molar extinction coefficients ε = 43830, 36590, 27830, 13910 $L \cdot mol^{-1} \cdot cm^{-1}$, respectively [4]. Susceptibility measurements at 294.0 K yield the magnetic moment $\mu_{eff} = 6.48\ \mu_B$ consistent with a sextet ground state. The ESR spectrum with g = 2.00 and a zero-field splitting of 819.8 G (= 0.82 cm^{-1}) in the solid state suggest a very low complex symmetry. In pyridine solution (g = 2.02) no zero-field splitting was observed [3]. Whereas, from the electronic spectra, terdentate chelation of the ligands was proposed with comparatively poor involvement of the azo group and a quinone hydrazone resonance structure of the ligand [4], the authors argue in [3] for bidentate bonding of the ligand without participation of the pyridine ring due to the magnetic and ESR data. Water molecules are proposed to occupy the axial sites of the coordination sphere in the solid state of the complex. They may be replaced by pyridine when dissolved in this solvent [3].

The mass spectrum recorded at 300°C (70 eV) shows the parent peak of the molecular ion $Mn(C_{19}H_{12}N_3O_2)_2^+$ at m/e = 651 together with essential peaks of fragmentation products at m/e =

623 (assigned to $[(C_{19}H_{12}N_3O)Mn(C_{19}H_{12}NO)]^+$), 353 ($Mn(C_{19}H_{12}N_3O)^+$), 325 ($Mn(C_{19}H_{12}NO)^+$), various peaks of ligand fragmentation, 133 ($Mn(C_5H_5N)^+$), 55 (Mn^+) [5].

$Mn(C_{19}H_{12}N_3O)_2 \cdot 2H_2O$ is insoluble in water but soluble in common organic solvents such as $CHCl_3$, DMF, nitrobenzene, and pyridine [3].

References:

[1] Rishi, A. K., Garg, B. S., Singh, R. P. (Indian J. Chem. A **14** [1976] 912/3).
[2] Bhoon, Y. K., Pandeya, K. B., Singh, R. P. (J. Indian Chem. Soc. **51** [1974] 960/2).
[3] Bhoon, Y. K., Pandeya, K. B., Singh, R. P. (J. Indian Chem. Soc. **57** [1980] 286/8).
[4] Bhoon, Y. K., Pandeya, K. B., Singh, R. P. (Bull Soc. Chim. France I **1979** 104/7).
[5] Bhoon, Y. K., Singh, R. P. (Ann. Chim. [Rome] **69** [1979] 477/82).

28.2.7 With Other Pyridylazo Compounds (Derivatives of 1-Methylanabasine)

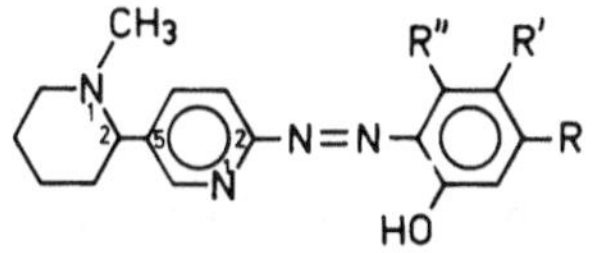

No.	R	R′	R″	formula
1	OH	H	H	$C_{17}H_{20}N_4O_2$
2	NH_2	H	H	$C_{17}H_{21}N_5O$
3	$N(C_2H_5)_2$	H	H	$C_{21}H_{29}N_5O$
4	OH	*cyclo*-C_6H_{11}	H	$C_{23}H_{30}N_4O_2$
5	OH	H	C_7H_{15}	$C_{24}H_{34}N_4O_2$

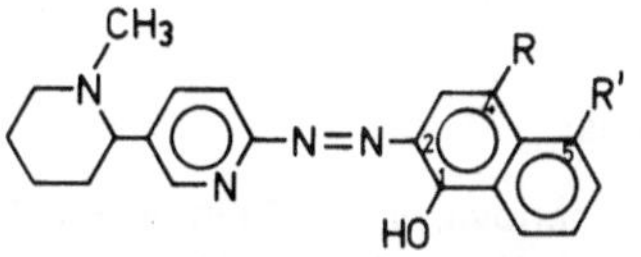

No.	R	R′	formula
6	H	H	$C_{21}H_{22}N_4O$
7	SO_3H	H	$C_{21}H_{22}N_4O_4S$
8	H	SO_3H	$C_{21}H_{22}N_4O_4S$

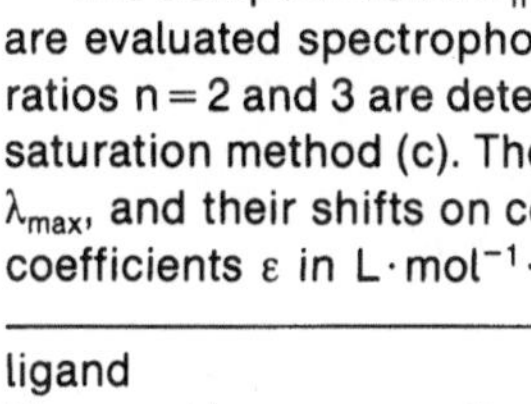

No. 9 (= $C_{21}H_{22}N_4O_2$)

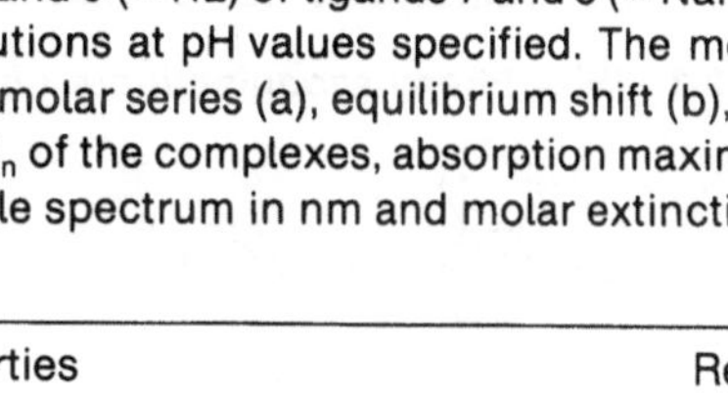

No. 10 (= $C_{29}H_{29}N_5O_3$)

The compounds MnL_n with either ligands 1 to 4, 6, and 9 (= HL) or ligands 7 and 8 (= NaHL) are evaluated spectrophotometrically in aqueous solutions at pH values specified. The mole ratios n = 2 and 3 are determined by the method of isomolar series (a), equilibrium shift (b), or saturation method (c). The overall stability constants β_n of the complexes, absorption maxima, λ_{max}, and their shifts on complexation, $\Delta\lambda$, in the visible spectrum in nm and molar extinction coefficients ε in $L \cdot mol^{-1} \cdot cm^{-1}$ are recorded below:

ligand No.	pH_{opt}	method	n	complex properties log β_n	λ_{max}	$\Delta\lambda$	ε	color	Ref.
1	10.0	a, b	2	13.60	505	—	60825	pink	[1]
2	9.75	a, b	3	18.89	520	65	75740	pink	[1,3]
3	9.75	a	3	21.74	560	—	112300	red	[1]
4	10.0	a, b	3	19.11	545	—	27860	red	[1]

ligand No.	pH_{opt}	method	n	complex properties log β_n	λ_{max}	$\Delta\lambda$	ε	color	Ref.
6	8.0	a, c	3	15.93	600 (560)	—	38020	violet	[1]
7	8.5	a, b	3	15.44	575	—	27435	raspberry red	[1]
7	8.0 to 9.0	a, b, c	3	18.35	575	55	26880	raspberry red	[4]
8	7.6	a, b	2	—	600	110	17360	—	[5]
9	9.0	a, b	2	13.54	555	—	34840	red	[1]

With the ligands 5 and 10 water-insoluble compounds are formed. The red complex with ligand 5, presumably a 1:1 chelate at pH 9.5, is sufficiently stable in aqueous ethanol or aqueous acetone and can be extracted by $CHCl_3$, higher alcohols, or ethyl acetate, or other organic solvents. An absorption maximum, $\lambda_{max}=530$ nm shifted from 410 nm for the ligand, with $\varepsilon=23080$ $L\cdot mol^{-1}\cdot cm^{-1}$, is recorded [2]. The 1:2 complex with ligand 10, extracted into $CHCl_3$, exhibits $\lambda_{max}=570$ nm and $\varepsilon=5.2\times10^4$ $L\cdot mol^{-1}\cdot cm^{-1}$ at the pH range 7.6 to 9.8 [6]. For earlier studies on complexes with similar ligands, see [7].

References:

[1] Kagramanova, N. G., Talipov, Sh. T., Dzhiyanbaeva, R. Kh. (Izv. Akad. Nauk Kaz.SSR Ser. Khim. **19** No. 2 [1969] 14/8; C.A. **71** [1969] No. 56314).

[2] Shesterova, I. P., Kostylev, N. F., Talipov, Sh. T., Dzhiyanbaeva, R. Kh. (Uzb. Khim. Zh. **10** No. 6 [1966] 7/10; C.A. **66** [1967] No. 72068).

[3] Kagramanova, N. G., Talipov, Sh. T., Dzhiyanbaeva, R. Kh. (Nauchn. Tr. Tashkent. Gos. Univ. No. 323 [1968] 28/32; C.A. **72** [1970] No. 96356).

[4] Kagramanova, N. G., Talipov, Sh. T., Dzhiyanbaeva, R. Kh. (Nauchn. Tr. Tashkent. Gos. Univ. No. 323 [1968] 11/5; C.A. **72** [1970] No. 96255).

[5] Sharipova, Sh. T., Dzhiyanbaeva, R. Kh., Talipov, Sh. T. (Nauchn. Tr. Tashkent. Gos. Univ. No. 323 [1968] 39/45, 40; C.A. **72** [1970] No. 96261).

[6] Podlipskaya, S. E., Dzhiyanbaeva, R. Kh., Talipov, Sh. T., Davranova, R. (Nauchn. Tr. Tashkent. Gos. Univ. No. 419 [1972] 57/61 from Ref. Zh. Khim. **1973** No. 1G73; C.A. **79** [1973] No. 121549).

[7] Podlipskaya, S. E., Dzhiyanbaeva, R. Kh., Talipov, Sh. T. (Nauchn. Tr. Tashkent. Gos. Univ. No. 288 [1967] 87/9; C.A. **68** [1968] No. 106053).

28.3 With Quinolylazo Compounds

28.3.1 With Phenylazoquinolinols and Their Derivatives

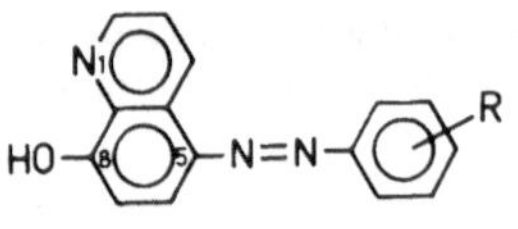

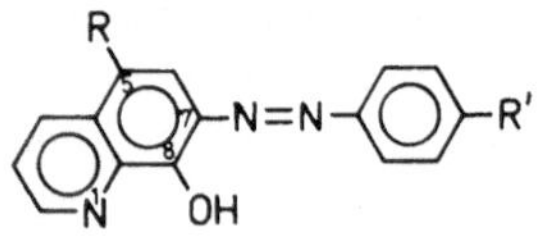

No.	R	formula
1	H	$C_{15}H_{11}N_3O$
2	2-OH	$C_{15}H_{11}N_3O_2$
3	3-OH	$C_{15}H_{11}N_3O_2$
4	4-OH	$C_{15}H_{11}N_3O_2$

No.	R	R′	formula
5	H	H	$C_{15}H_{11}N_3O$
6	SO_3H	H	$C_{15}H_{11}N_3O_4S$
7	SO_3H	NO_2	$C_{15}H_{10}N_4O_6S$

Complexes in Solution. The formation constants of MnL^+ and MnL_2 complexes with the ligands 1 to 4 in 50% (v/v) dioxane-water were determined potentiometrically (glass electrode) at 25°C and ionic strength I = 0.1 M ($NaClO_4$). Values of log K_1 and log β_2 are given below together with the acid dissociation constants of the ligands:

ligand HL	1	2	3	4
pk_2	8.84	8.51	8.84	9.15
log K_1	6.20 ± 0.1	7.11 ± 0.1	6.56 ± 0.1	6.64 ± 0.1
log β_2	12.57 ± 0.05	13.01 ± 0.05	12.52 ± 0.05	12.66 ± 0.05

The pH meter readings are corrected for effects of the organic solvent. Complexation of ligands 1 to 4 to Mn^{2+} ions takes place at pH values lower by about 0.5 to 0.8 than for 8-quinolinol [1].

Complex formation with the ligands 6 or 7 (= H_2L) was studied in alkaline aqueous solutions at 25°C and the existence of the compound MnL was also shown by spectrophotometric measurements. Chelation is suggested to occur through the phenolate oxygen and the pyridine N atom but not by the azo group [2].

Colored complexes are also observed with analogous ligands derived from p-cresol or resorcinol, e.g., violet Mn^{II} complexes in alcohol at pH 6 to 7 with 4-methyl-2-(2-quinolylazo)-phenol (= $C_{16}H_{13}N_3O$) [4] and purple complexes in water at pH ≦3 with 4-(2-quinolylazo)-1,3-benzenediol (= $C_{15}H_{11}N_3O_2$) [5].

Solid Complexes. A light brown solid supposed to be $Mn(C_{15}H_{10}N_3O)_2$ was obtained from an aqueous slurry of ligand 5 and manganese(II) acetate at pH 5, heated on a steam bath for 4 h. The precipitate was washed with water and dried at 60°C. Coordination was proposed via the deprotonated OH group, one azo and the heterocyclic nitrogen atom. The complex may serve as a moderately light-stable pigment [3].

References:

[1] Fernando, Q., Freiser, H. (Anal. Chem. **37** [1965] 1249/51).

[2] Uusitalo, E. (Suomen Kemistilehti B **31** [1958] 264/8; C.A. **1959** 2778).

[3] Potnis, S. P., Arora, K. K. (Paintindia **18** No. 6 [1968] 17/23, 52; C.A. **70** [1969] No. 30021).

[4] Rakhmatullaev, K. Z., Rakhmatullaeva, M. A., Rakhimov, Kh. R., Talipov, Sh. T. (Uzb. Khim. Zh. **14** [1970] 23/6 from C.A. **73** [1970] No. 126569).

[5] Talipov, Sh. T., Rakhmatullaev, K., Babaev, N., Kulmuratov, N. (Nauchn. Tr. Tashkent. Gos. Univ. No. 323 [1968] 46/9 from C.A. **72** [1970] No. 71258).

28.3.2 With Quinolylazo Compounds of Naphthols or Acenaphthylenol

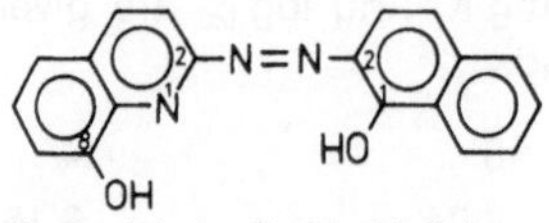

ligand 1 (= $C_{19}H_{13}N_3O_2$)

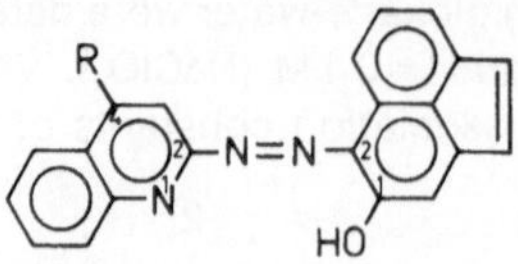

ligand 3 R = H (= $C_{21}H_{13}N_3O$)
ligand 4 R = CH_3 (= $C_{22}H_{15}N_3O$)

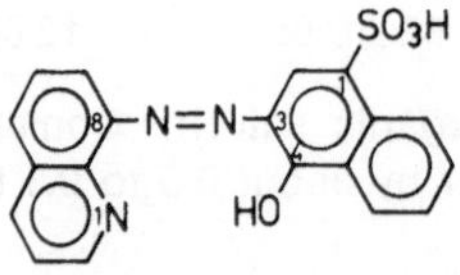

ligand 2 (= $C_{19}H_{13}N_3O_4S$)

The 1:1 and 1:2 complexes with the ligands 2, 3, or 4 in mixed solvents are evaluated by potentiometric titrations (glass electrode, pH meter readings corrected for the organic solvent). Stability constants recorded increase on decreased ionic strengths (I) maintained with $NaClO_4$. Enthalpies of formation in kcal/mol are calculated from pertinent K_1 values:

ligand No.	t in °C	solvent, vol%	I in mol/L	log K_1	log K_2	ΔG_1	Ref.
2	25	methanol, 50	0.1	8.6	7.0	—	[1]
3	30	dioxane, 75	0.2	5.95	5.10	−8.11	[2]
	30	dioxane, 75	0.1	5.98	5.88	—	[2]
	30	dioxane, 75	0.01	6.40	6.41	—	[2]
	30	dioxane, 75	0.005	8.00	6.93	—	[2]
4	30	dioxane, 75	0.2	6.36	5.73	−8.67	[2]

With ligand 1, a blue-violet chelate, presumably of 1:1 mole ratio, is evaluated spectrophotometrically in 50% (v/v) dioxane [4].

A water-insoluble red manganese complex of ligand 2 can be extracted by pentanol, ether, benzene, $CHCl_3$, and CCl_4 to give yellow solutions. It decomposes, however, upon extraction with CCl_4 [3].

A further ligand 4-hydroxy-3-[(4-methyl-2-quinolyl)azo]-1-naphthalenesulfonic acid (= $C_{20}H_{15}N_3O_4S$) of similar type as ligand 2 forms a water-soluble blue complex at pH 7.0 to 7.7, determined by the decolorizing substitution reaction with ethylenediaminetetraacetate [5].

References:

[1] Anderson, R. G., Nickless, G. (Talanta **14** [1967] 1221/8, 1225).
[2] Singh, J., Garg, B. S., Singh, R. P. (Indian J. Chem. A **17** [1979] 104/5).
[3] Mehta, Y. L., Garg, B. S., Singh, R. P. (Current Sci. [India] **43** [1974] 11/2).
[4] Ishizuki, T., Wada, H., Kodama, K., Nakagawa, G. (Anal. Chim. Acta **176** [1985] 63/70, 68).
[5] Chadha, R. C., Garg, B. S., Singh, R. P. (Proc. Indian Natl. Acad. Sci. A **45** [1979] 20/4).

28.3.4 With Phenylene- or Biphenylenebis(azoquinolinols)

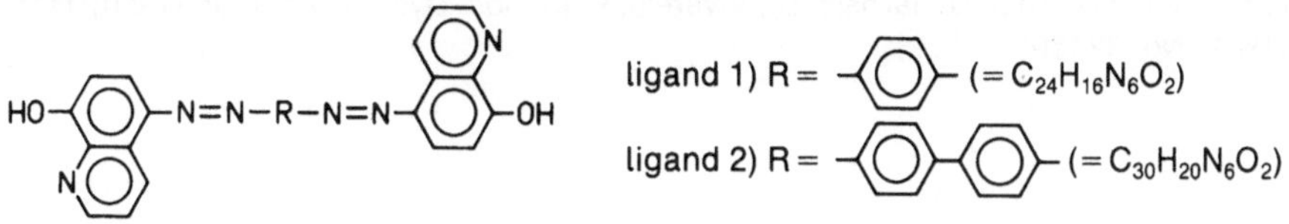

[$Mn(C_{24}H_{14}N_6O_2)_2(H_2O)_2]_n$ (I) and **[$Mn(C_{30}H_{18}N_6O_2)_2(H_2O)_2]_n$** (II) complexes were prepared by refluxing an equimolar mixture of $Mn(CH_3COO)_2 \cdot 4H_2O$ and ligand 1 for 6 h or ligand 2 for 4 h in dimethyl sulfoxide. The precipitates were washed with hot DMSO and dried [1, 2], e.g., at 140°C for $Mn(C_{30}H_{18}N_6O_2) \cdot 2H_2O$ to obtain amorphous dark brown or red-brown solids.

The IR spectra of the complexes (KBr pellets) show bands at the following wave numbers (in cm^{-1}): $\nu(OH)_{water}$ at 3400 to 3350, $\delta(OH)_{water}$ at 1640 to 1630, $\nu(C{=}N)_{py}$ at ~1590, ν(C–OMn) at 1130(I), 1120(II), $\nu(Mn{-}O)_{water}$ at ~750, $\nu(Mn{-}O)_{phenol}$ at 625(I), 625 to 620(II), and ν(Mn–N) at 460(I), 510 to 480(II). Bidentate chelation merely by the quinolinate N and O atoms of the ligand is inferred for either complex, without involvement of the azo sites, due to the vanished ligands' OH band, the new bands induced by Mn-ligand vibrations, and the shifts in $\nu(C{=}N)_{py}$. Water is proved to be coordinated additionally [1, 2]. Absorption bands in the electronic spectrum of $Mn(C_{24}H_{16}N_6O_2) \cdot 2H_2O$ with ligand 1 exhibit a band at 500 nm due to a $\pi \rightarrow \pi^*$ transition and d→d transition bands at λ = 580, 490, 440, 380, 350, and 330 nm, assigned to the electron transitions from the $^6A_{1g}$ ground state to the $^4T_{1g}(G)$, $^4T_{2g}(G)$, $^4A_{1g}(G)$, $^4E_g(G)$, $^4T_{2g}(D)$, and $^4E_g(D)$ states, respectively [1]. These data suggest an octahedral coordination sphere. A polymeric chain structure is proposed where the Mn atoms are bridged by two intermolecular quinolinate groups and are coordinated also by two axial water molecules. This structure is consistent with the insolubility of both complexes in water and common organic solvents [1, 2].

TGA studies (heating rate 10°C/min) show that both complexes gradually lose the theoretical amount of water between 80 and 260°C and decompose with an accelerated weight loss between 300 and 380°C. A thermostability order for analogous complexes of other transition metals is given [1, 2].

References:

[1] Banerjie, V., Dey, A. K. (J. Indian Chem. Soc. **59** [1982] 991/2; C.A. **97** [1982] No. 229088).
[2] Banerjie, V., Dey, A. K. (Polym. Bull. [Berlin] **1** [1978] 685/90; C.A. **92** [1980] No. 173671).

28.4 With Azo Compounds Derived from Pyrazole or Pyrazolinones

28.4.1 With 1-[(1-Phenyl-*1H*-pyrazol-3-yl)azo]-2-naphthol

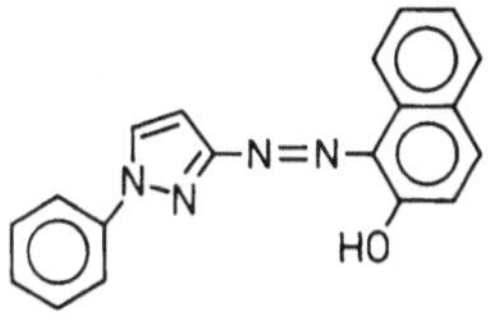

(= $C_{19}H_{14}N_4O$)

$Mn(C_{19}H_{13}N_4O)_2$ was prepared by boiling the ligand and manganese(II) acetate (mole ratio 4:1) in ethanol for 30 min. The hot mixture was filtered and the precipitate washed with hot ethanol. The complex does not melt below 300°C. The electronic spectrum of a 3×10^{-5} M acetonitrile solution shows two absorption maxima at $\lambda = 518$ and 550 nm with $\log \varepsilon = 4.44$ and 4.37, respectively, Folli, U., Iarossi, D., Vivarelli, P. (J. Soc. Dyers Colour. **96** [1980] 414/7; C.A. **94** [1981] No. 32150).

28.4.2 With 5-Amino-4-arylazo-2,4-dihydro-*3H*-pyrazol-3-ones

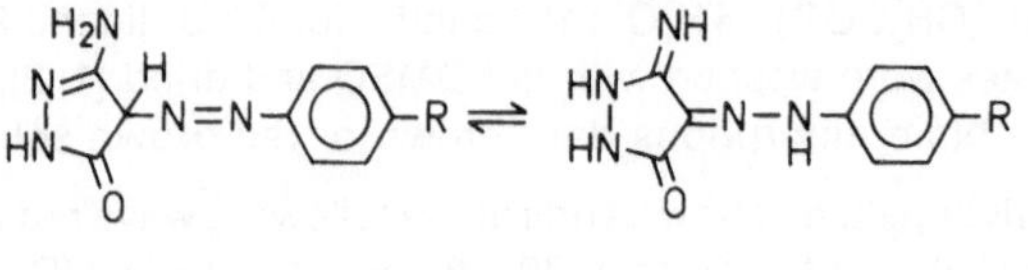

(=HL)

No.	R	formula
1	H	$C_9H_9N_5O$
2	NO_2	$C_9H_8N_6O_3$
3	OCH_3	$C_{10}H_{11}N_5O_2$
4	CH_3	$C_{10}H_{11}N_5O$

Complexes of composition $\mathbf{MnL_2}$ (with ligand 1), $\mathbf{MnL_2 \cdot 2H_2O}$ (with ligands 2 or 3), and **MnLCl** with ligands 1 to 4 were prepared by a method developed for azoimidazolate complexes of other transition metal ions: The ligands in concentrated ethanol solutions were added to manganese(II) chloride solutions in 2:1 or 1:1 mole ratios, respectively. After refluxing for 3 h and cooling of the mixture, crystals separated which were recrystallized from ethanol and dried in a vacuum over silica gel [2].

The blue shifts of the ν(CO) vibrations at 1675 to 1655 cm and $\nu(N{-}C_{phenyl})$ vibrations at 1257 to 1237 cm^{-1} observed in the IR spectra (KBr disks) on complexation lead to the assumption that the carbonyl oxygen and the α-nitrogen of the phenylazo group are the bonding sites of the bidentate ligands in the keto-hydrazone tautomeric form. Stable six-membered chelate rings are formed due to formation of a conjugate π-bonded system involving d-orbitals of the metal ion. The values of molar conductances, determined in 0.001 M solutions of the complexes in dimethylformamide at 25°C, indicate a nonelectrolyte nature of the 1:2 complexes ($\Lambda = 35$ to 40 $cm^2 \cdot \Omega^{-1} \cdot mol^{-1}$) whereas the 1:1 complexes behave as 1:1 electrolytes ($\Lambda = 62$ to 72 $cm^2 \cdot \Omega^{-1} \cdot mol^{-1}$) [1].

References:

[1] Mahmoud, M. R., Adam, F. A., Yousef, K., El-Haty, M. T. (Bull. Soc. Chim. Belges **92** [1983] 13/9).

[2] Mahmoud, M. R., Abd El-Hamide, R., Abd El-Wahab, A. A., Salman, H. M. (Bull. Soc. Chim. Belges **91** [1982] 11/7, 12).

28.4.3 With Other 4-Arylazo-2,4-dihydro-5-methyl-*3H*-pyrazol-3-ones

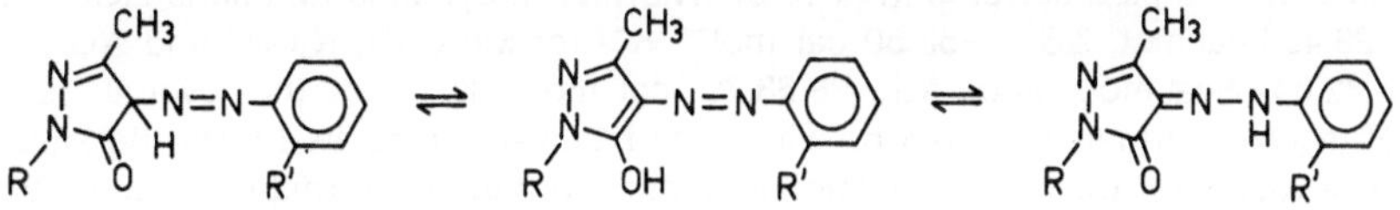

ligand No.	R	R′	formula
1	H	OH	$C_{10}H_{10}N_4O_2$
2	C_6H_5	OH	$C_{16}H_{14}N_4O_2$
3	C_6H_5	SO_3H	$C_{16}H_{14}N_4O_4S$
4	C_6H_5	OCH_2COOH	$C_{18}H_{16}N_4O_4$
5	C_6H_5	OCH_2CH_2COOH	$C_{19}H_{18}N_4O_4$

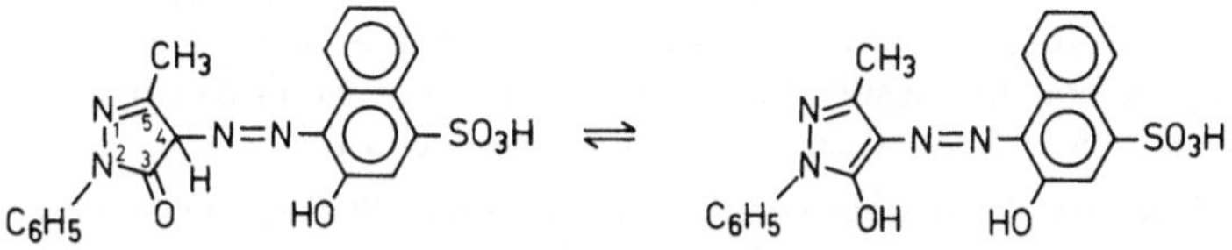

ligand 6 (= $C_{20}H_{16}N_4O_5S$)

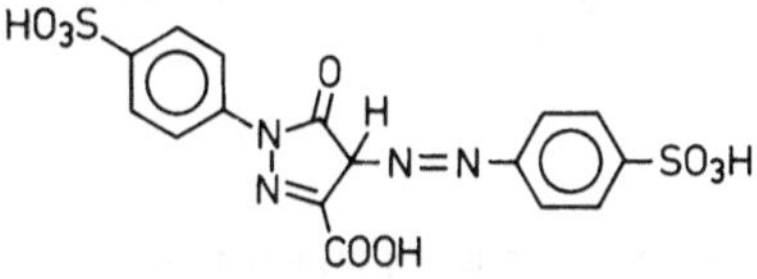

ligand 7 (= $C_{16}H_{12}N_4O_9S_2$)

Complexes in Solution. Stability constants of complexes MnL^+ and MnL_2 with the ligands 1 or 2 (=HL) and MnL with the ligands 2, 4, or 5 (= H_2L) were determined potentiometrically (mostly with the glass electrode = gl) in mixed solvents and a nitrogen atmosphere. Values of log K_1 and log K_2 are tabulated below, together with the pertinent pK_a values of the ligands:

ligand No.	method	t in °C	solvent in vol%	I in mol/L	pK_a	log K_1	log K_2	Ref.
1 (=HL)	pot	30	50% ethanol	0.8 (KNO_3)	8.69	5.06	4.65	[1]
	pot	30	50% ethanol	0.8 (KNO_3)	8.67	5.04	4.48	[1]
2 (=HL)	gl	25	75% methanol	0.1 ($NaClO_4$)	8.70	5.48	4.43	[2]
		30	75% methanol	0.1 ($NaClO_4$)	8.40	5.28	4.35	[2]
		35	75% methanol	0.1 ($NaClO_4$)	8.00	5.00	4.28	[2]
2 (= H_2L)	gl	30	75% dioxane	—	13.12	13.16	—	[3]
4 (= H_2L)	gl	30	75% dioxane	—	9.67	8.49	—	[5]
5 (= H_2L)	gl	30	75% dioxane	~0	10.78	7.80 ±0.08	—	[6]

For the equilibrium $Mn^{2+} + H_2L \rightleftharpoons MnL + 2H^+$ with ligand 2 [3, 4] or Mordant Red, the monosodium salt of ligand 6 [4] in 3:1 (v/v) dioxane-water at 30°C, values of log K_{eq} = −10.6 or

−9.8, respectively, were determined potentiometrically. The pH meter readings were corrected for the effects of mixed solvents in [4 to 6]. The thermodynamic functions $\Delta G_1 = -7.31$ and $\Delta H_1 = -23.46$ kcal/mol, $\Delta S_1 = -53.50$ cal·mol^{-1}·K^{-1} for $Mn(C_{16}H_{13}N_4O_2)^+$ and $\Delta G_{\beta_2} = -13.34$, $\Delta H_{\beta_2} = -32.64$ kcal/mol, and $\Delta S_{\beta_2} = -63.70$ cal·mol^{-1}·K^{-1} for the overall reaction to $Mn(C_{16}H_{13}N_4O_2)_2$, formed in solutions with ligand 2, were calculated from the stability constants. The values for the overall reaction indicate spontaneous formation of the 1:2 complex [2]. The stabilities of complexes with the ligands 4 and 5 are nearly equal [5] or slightly higher [6] than those of analogous thioether compounds. It is assumed that tetradentate ligands 4 and 5 are bonded to manganese through the oxygen atoms of the deprotonated OH groups of the pyrazolol tautomer and of the carboxylate group and that they are additionally coordinated by one azo nitrogen and the ether oxygen atom. A stability order is given for complexes with these ligands, with Mn^{2+} in a mean position among many divalent metal ions [5, 6].

Spectrophotometric studies on ethanolic solutions of an Mn^{2+} salt and ligand 3 reveal the formation of 1:1 and 1:2 complexes. Complex spectra recorded from ~200 to 450 nm show two bands, intensified, but not shifted from the ligand positions and two additional bands at 333 and ~420 nm due to splitting of the ligand charge transfer band at ~385 nm. Octahedral complex geometries are proposed with a tridentate O, N, O-bite of the ligand dianion coordinated as the ketohydrazone; this assumption is based on comparison with similar complexes of some other metal ions and varied substituents on the ligand [7].

A 1:1 complex is formed in aqueous solution of manganese(II) salts and ligand 7 in the pH range from 0 to 1. At ionic strength I = 0.2 M ($NaClO_4$) a stability constant log K = 5.55 was determined; free energy of formation $\Delta G = -7.61$ kcal/mol. The metal ion is expected only to replace the carboxylate proton and to form a chelate ring by coordination of one azo nitrogen atom. Precipitates occur on further titration at pH ~6.8 [8].

References:

[1] Shah, N. R., Shah, J. R. (Acta Ciencia Indica Ser. Chem. **7** No. 1 [1981] 17/9; C. A. **97** [1982] No. 29326).
[2] Jain, A. K., Goyal, R. N., Agarwal, D. D. (Thermochim. Acta **47** [1981] 243/5).
[3] Snavely, F. A. (from Sillén, L. G., Martell, A. E., Chem. Soc. [London] Spec. Publ. No. 17 [1964] 754 pp., 708).
[4] Snavely, F. A., Fernelius, W. C., Douglas, B. E. (J. Soc. Dyers Colour. **73** [1957] 491/5).
[5] Snavely, F. A., Craver, G. (Inorg. Chem. **1** [1962] 890/2).
[6] Snavely, F. A., Magen, W., Kozart, D. (J. Inorg. Nucl. Chem. **27** [1965] 679/81).
[7] El-Inany, G. A., El-Wahab, S. A., Issa, Y. M. (Egypt. J. Chem. **25** [1982] 101/14, 107/12; C. A. **99** [1983] No. 96154).
[8] Limaye, S. N., Saxena, M. C. (J. Indian Chem. Soc. **62** [1985] 192/3).

28.5 With Pyrimidylazo and Oxopyrimidinylazo Compounds

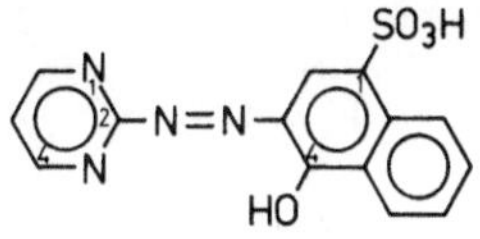

1) $(=C_{14}H_{10}N_4O_4S)$

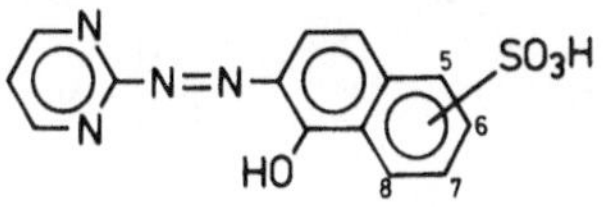

2 to 5) $(=C_{14}H_{10}N_4O_4S)$

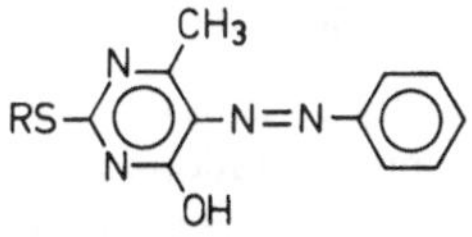

or tautomers

6) R=H $(=C_{11}H_{10}N_4OS)$
7) R=CH_3 $(=C_{12}H_{12}N_4OS)$
8) R=C_2H_5 $(=C_{13}H_{14}N_4OS)$

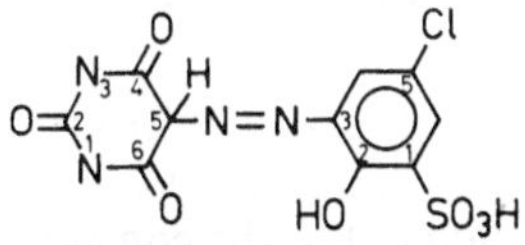

or tautomers

9) $(=C_{10}H_7ClN_4O_7S)$

Complexes in Solution. Stability constants of 1:1 and 1:2 complexes in 50 vol% aqueous methanol with the isomeric hydroxy(2-pyrimidylazo)naphthalenesulfonic acids (ligands 1 to 5) were determined potentiometrically under N_2 at 25°C and ionic strength I=0.1 M ($NaClO_4$) together with the dissociation constants of the ligand hydroxy groups, pK_{a2}:

ligand No.	position of SO_3H	pK_{a2}	log K_1	log K_2
1	—	8.41	5.3	4.9
2	5	8.87	5.3	5.3
3	6	8.91	5.8	5.2
4	7	8.66	5.7	5.3
5	8	10.10	7.5	5.7

The pK_{a2} values of the ligands and the complex stabilities are lower than those of their pyridylazo analogs (p. 296), with the same relationship to the position of the sulfonic acid group, reflecting equal steric and electronic effects [1].

A complex $[Mn(C_{10}H_6ClN_4O_7S)]^+$ with the monodeprotonated ligand 9, known as lumomagneson, in aqueous solutions is revealed spectrophotometrically (method of continuous variations) at pH 8 [2]. The formation of complexes with ligands 6 to 8 in 40% DMF solution at pH 4.5 is indicated by polarographic measurements. Shifts $\Delta E_{1/2}=-0.195$, -0.165, and -0.165 V were observed in comparison to the half wave potentials $E_{1/2}=-0.47$, -0.36, and -0.32 V (vs. SCE), respectively, of the free ligands [3].

$Mn(C_{14}H_8N_4O_4S)\cdot 2H_2O$. Aqueous solutions of a manganese(II) salt and ligand 1 were mixed in a 1:2 mole ratio and aqueous ammonia was added. The resulting precipitate was washed and dried under extremely reduced pressure [1].

References:

[1] Anderson, R. G., Nickless, G. (Analyst [London] **93** [1968] 20/5).

[2] Sychev, A. Ya., Pfannmeller, W., Isak, V. G. (Zh. Fiz. Khim. **55** [1981] 794/5; Russ. J. Phys. Chem. **55** [1981] 451/2).

[3] Jain, R. (Ann. Chim. [Rome] **74** [1984] 853/9; C.A. **102** [1985] No. 214410).

28.6 With Triazolylazo Compounds

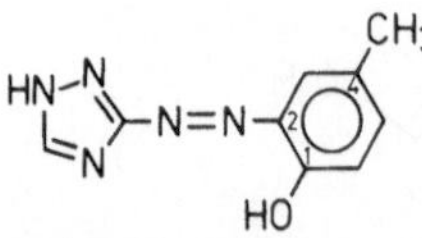

or tautomers

ligand 1 (= $C_9H_9N_5O$)

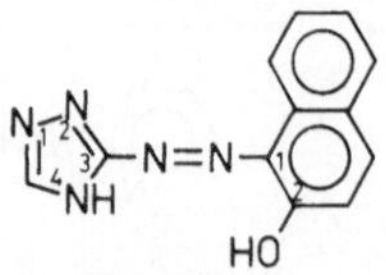

or tautomers

ligand 2 (= $C_{12}H_9N_5O$)

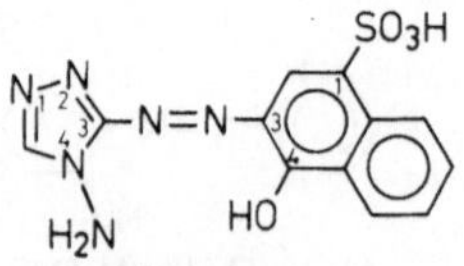

or tautomers

ligand 3 (= $C_{12}H_{10}N_6O_4S$)

Complexes in Solution. A 1:1 and a 1:2 complex with the stability constants log K_1 = 4.00 and log K_2 = 1.79 (error limits ±0.08) are evaluated by potentiometric titrations in aqueous solutions of a manganese(II) salt and the monosodium salt of ligand 3 at 30±0.1°C and ionic strength I = 0.10 M ($NaClO_4$) where the pertinent pk_a value is 5.81. The overall free energy of formation $\Delta G = -33.65$ kcal/mol, resulting from β_2, indicates spontaneous complex formation [1].

A pink-colored complex is indicated with ligand 1 in aqueous solution at pH 7 by its absorption band at λ_{max} = 510 nm [4]. Complex formation with ligand 2 in 40 vol% ethanol solution at pH 7.5 to 8.5 is suggested from spectrophotometric studies [2]. The complex can be extracted into $CHCl_3$ and methyl isobutyl ketone and is destroyed by ethylenediaminetetraacetic acid [3].

References:

[1] Garg, S. K., Mukherjee, S., Garg, B. S., Singh, R. P. (Indian J. Chem. A **20** [1981] 535/6).
[2] Cacho Palomar, J., Nerin de la Puerta, C. (Afinidad **39** No. 377 [1982] 44/6; C.A. **96** [1982] No. 228154).
[3] Cacho, J., Nerin, C. (Anal. Chim. Acta **131** [1981] 271/5).
[4] Činátlová, H., Šůcha, L. (Sb. Vys. Sk. Chem. Technol. Praze Anal. Chem. H **12** [1977] 105/17, 112; C.A. **90** [1979] No. 47796).

28.7 With Tetrazolylazo Compounds

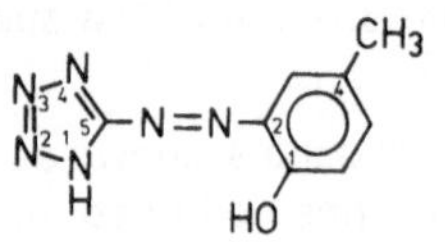

ligand 1 (= $C_8H_8N_6O$)

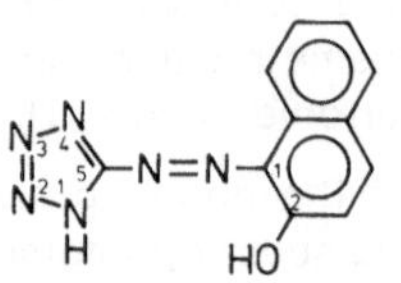

ligand 2 (= $C_{11}H_8N_6O$)

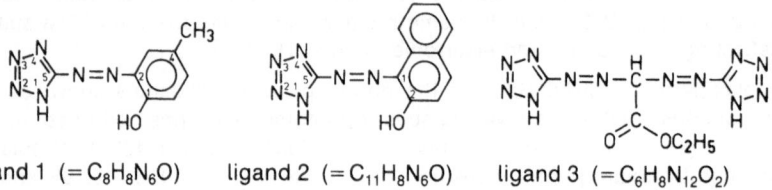

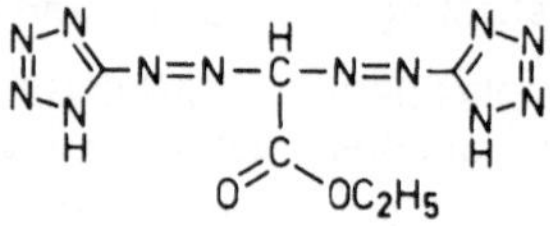

ligand 3 (= $C_6H_8N_{12}O_2$)

Complexes in Solution. A complex $Mn(C_{11}H_6N_6O)$ (= MnL) was evaluated spectrophotometrically in a 10% (by weight) ethanol solution of ligand 2 (= H_2L) and Mn^{2+} ions in the pH range 5.6 to 7.7. Stability constants log K = 5.96±0.18 were determined for the equilibrium $Mn^{2+} + H_2L \rightleftharpoons MnL + 2H^+$ and log K = −3.8 for $Mn^{2+} + HL^- \rightleftharpoons MnL + H^+$, respectively, by potentiometric titrations (glass electrode) at 20°C and ionic strength I = 0.1 M ($NaClO_4$). The complex absorption maximum is displayed at λ = 510 nm, compared to 405 nm for HL^- and to 450 and 490 nm for L^{2-}. The molar extinction coefficient is $\varepsilon = (16.9 \pm 1.2) \times 10^3\ L \cdot mol^{-1} \cdot cm^{-1}$ [1].

A red complex with λ_{max} = 515 nm is observed with ligand 1 in aqueous solution at pH 7 [2].

A 1:2 complex was indicated spectrophotometrically (Job's method of continuous variation) with the disodium salt of ligand 3 at a pH between 6 and 9 in aqueous solutions of a manganese(II) salt. The yellow-red solution exhibits an absorption maximum at λ = 435 nm [3].

References:

[1] Suchánek, M., Šůcha, L. (Collection Czech. Chem. Commun. **43** [1978] 1393/400).
[2] Činátlová, H., Šůcha, L. (Sb. Vys. Sk. Chem. Technol. Praze Anal. Chem. H **12** [1977] 105/17, 112; C.A. **90** [1979] No. 47796).
[3] Jonassen, H. B., Chamblin, V. C., Wagner, V. L., Henry, R. A. (Anal. Chem. **30** [1959] 1660/3).

28.8 With an Azo Compound of a Tetraazomacrocyclic Ligand

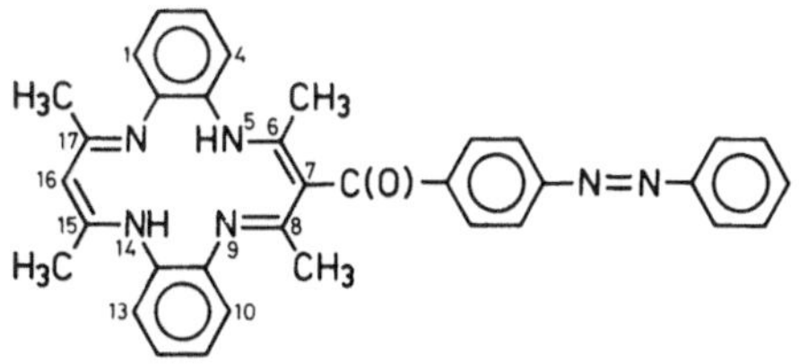

($= C_{35}H_{32}N_6O = H_2L$)

[Mn($C_{35}H_{30}N_6O$)X] with X = Cl, Br. The manganese(III) complexes were prepared as follows: 5.4 mmol of ligand was suspended in a solution of 16.3 mmol of the acetylacetonato complex $Mn(acac)_2X$ in 250 mL of dry, degassed acetonitrile and the mixture was stirred for 2 d with protection from light, filtered and the filtrate evaporated to dryness in vacuum. The residue of the chloro complex was then dissolved in benzene (150 mL), filtered and the filtrate reduced to half its volume in vacuum; hexane was added to afford crystallization overnight. The green powder was washed with hexane, redissolved in CH_2Cl_2 (20 mL) and placed on an alumina chromatography column which was eluted with acetonitrile. The middle portion was collected and its volume reduced to 100 mL. The crude bromide was washed with cold benzene (200 mL) and recrystallized from hot (60°C) acetonitrile. The deep green crystals of each complex were dried in vacuum at room temperature over P_4O_{10}, yielding 60 and 79%, respectively. The compounds melt at 227 and 233°C, respectively. Absorption bands of the IR and electronic spectra with their assignments are shown below [1]:

complex	$\bar{\nu}$ in cm^{-1}	$\bar{\nu}$ in cm^{-1}	$\bar{\nu}$ in kK			
$Mn^{III}(L)Cl$	1652	283	16.7, 17.5	21.7, 22.7, 26.1	28.6	30 33.9, 36.4
$Mn^{III}(L)Br$	1650	—	16.7, 17.4	21.6, 22.5, 25.9	28.5	30 33.8, 36.2
assignments	ν(CO)	ν(Mn–X)	charge transfer	$n \rightarrow \pi^*$	$\pi \rightarrow \pi^*$	$\pi \rightarrow \pi^*$

The magnetic moments, $\mu_{eff} = 4.95$ and $4.89\,\mu_B$, corrected for diamagnetism, are consistent with a high-spin electronic configuration of the Mn^{3+} ion. Whereas the band at ~28.5 kK is associated with an electron transition of the macrocyclic Schiff base, the bands occurring at 30 and ~20 kK are assigned to electron transitions of the azo chromophore in the complexes. Irradiation of the complex solutions with an Ar^+ ion laser induces spectral change resulting in new bands at 41.2, 36.5, and 23.5 kK for the chloride and at 41.3, 36.0, and 23.1 kK for the bromide complex. This change indicates photoisomerization of the azo linkage from the *trans*- to the *cis*-form. The quantum yields of this reaction, examined as a function of the exciting radiation wave number, reach a maximum at 19.9 and 20.5 kK (~501.7 and 488 nm) for the chloride and the bromide complexes, respectively, the region where the charge transfer and $n \rightarrow \pi^*$ transitions of the azo group are located in the complex spectra [1].

The highest m/e values observed in the mass spectra are 605 for the chloride complex corresponding to the fragment $[Mn(C_{35}H_{30}N_6O)]^+$ and 503 for the bromide indicating additional scission of the azo substituent.

Reference:

Sabbak, El-Ssaed O. A. (Diss. Syracuse Univ., Syracuse, N.Y., 1977, pp. 1/165, 118/54; Diss. Abstr. Intern. B **39** [1978] 757).

28.9 With Thiazolylazo Compounds

28.9.1 With (2-Thiazolylazo)phenols

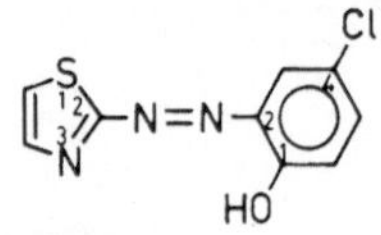

ligand 1

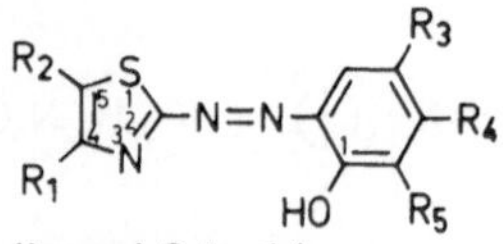

ligand 2 to 14

Complexes in Solution. A complex $Mn(C_9H_5ClN_3OS)^+$ with ligand 1 (= $C_9H_6ClN_3OS$) was established spectrophotometrically in 10% (v/v) dioxane-water mixtures. Its stability constant log K_1 = 2.86 was determined by back-titration with ethylenediaminetetraacetate at 20°C and I = 0.1 M (KNO_3) (pK_a of the ligand = 7.21). The analogous 2-(2'-thiazolylazo)phenol (not substituted by Cl) and its bromo- or iodo-substituted derivatives form violet complexes of lower stabilities [5 to 7].

Stability constants for complexes MnL^+ and MnL_2 with the ligands 2 to 14 (= HL) and their pK_a values determined potentiometrically in 6:4 (v/v) dioxane-water at 25 ± 0.1°C and I = 0.1 M ($NaClO_4$) are listed below:

No.	R_1	R_2	R_3	R_4	R_5	formula	pK_a	log K_1	log K_2	Ref.
2[1)]	H	H	CH_3	H	H	$C_{10}H_9N_3OS$	—	4.71	4.41	[1]
3	H	H	C_2H_5	H	H	$C_{11}H_{11}N_3OS$	9.02[2)]	3.73[2)]	4.35[2)]	[2]
4[1)]	H	H	CH_3	H	CH_3	$C_{11}H_{11}N_3OS$	—	5.16	5.20	[1]
5[1)]	CH_3	H	CH_3	H	H	$C_{11}H_{11}N_3OS$	—	5.06	5.26	[1]
6[1)]	CH_3	H	CH_3	H	CH_3	$C_{12}H_{13}N_3OS$	—	5.45	5.88	[1]
7	CH_3	CH_3	Cl	H	H	$C_{11}H_{10}ClN_3OS$	9.15	4.33	4.21	[1, 3]
8	CH_3	CH_3	OCH_3	H	H	$C_{12}H_{13}N_3O_2S$	10.22	5.40	5.48	[3]
9	CH_3	CH_3	CH_3	H	H	$C_{12}H_{13}N_3OS$	10.55	5.22	5.41	[1, 3]
10	CH_3	CH_3	C_2H_5	H	H	$C_{13}H_{15}N_3OS$	10.47	5.28	5.31	[1, 3]
11	CH_3	CH_3	C_6H_5	H	H	$C_{17}H_{15}N_3OS$	9.94	5.00	4.87	[1, 3]
12	CH_3	CH_3	H	$N(CH_3)_2$	H	$C_{13}H_{16}N_4OS$	11.58	7.11	7.27	[1, 3]
13	CH_3	CH_3	CH_3	H	OCH_3	$C_{13}H_{15}N_3O_2S$	10.00	—	10.60[3)]	[1, 3]
14	CH_3	CH_3	CH_3	H	CH_3	$C_{13}H_{15}N_3OS$	11.02	5.74	6.10	[1, 3]

[1)] Position of substituents established according to [8]. – [2)] In 30% (v/v) dioxane-water. – [3)] Overall stability constant β, determined by the mid-point method.

Generally, the complex stabilities are increased on stepwise enhancing of the number of electron-donating methyl groups in the ligand [1 to 3]. Titrations of solutions of an Mn^{II} salt and ligand 2 in 50% (v/v) methanol at 25 ± 0.5°C and I = 0.1 M ($NaClO_4$) indicate the 1:2 complex with the overall stability constant log β_2 = 7.6 ± 0.3 [4].

Ligand 3 appears to form mononuclear complexes only in solutions containing the ligand and Mn^{2+} ions in a 3:1 or higher mole ratio, whereas at lower mole ratios the formation curves suggest the presence of binuclear or polynuclear complexes [2].

The complex with ligand 1 exhibits an absorption maximum at 550 n with $\varepsilon = 6.38 \times 10^3$ L·mol^{-1}·cm^{-1}, compared to $\lambda_{max} = 535$ nm observed for the free ligand [5 to 7]. In contrast, complexes MnL^+ derived from the ligands 7 to 14 are reported to display their absorption maxima in the range 620 to 650 nm only, shifted from 540 to 570 nm for the ligand anions [3].

References:

[1] Kai, F., Takeshita, H., Sukimoto, S., Tamaoku, K. (J. Inorg. Nucl. Chem. **43** [1981] 3013/5).
[2] Kai, F., Egoshi, H. (Anal. Letters **8** [1975] 575/84, 581).
[3] Kai, F., Akashi, W. (Anal. Letters **13** [1980] 1451/64; C.A. **94** [1981] No. 113766).
[4] Nickless, G., Pollard, F. H., Samuelson, T. J. (Anal. Chim. Acta **39** [1967] 37/46, 43).
[5] Kai, F., Izumi, H. (Anal. Letters **3** [1970] 307/14).
[6] Kai, F., Izumi, H. (Nippon Kagaku Zasshi **91** [1970] 850/4, A48; C.A. **74** [1971] No. 49261).
[7] Kai, F., Izumi, H. (Kagaku No Ryoiki **25** [1971] 248/50; C.A. **74** [1971] No. 134425).
[8] Kawase, A. (Bunseki Kagaku **11** [1962] 621/8, 623).

28.9.2 With (2-Thiazolylazo)benzenediols

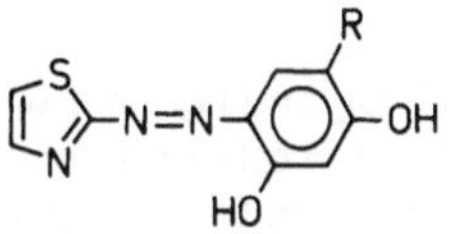

ligand 1 R = H (= $C_9H_7N_3O_2S$ = H_2L)
ligand 2 R = COOH (= $C_{10}H_7N_3O_4S$)

Complexes in Solution. Various complex equilibria with ligand 1 in aqueous organic media, studied potentiometrically (glass electrode) are listed in the table below; pH meter readings were corrected for the organic solvent.

equilibrium	medium	I in mol/L	constants	Ref.
(1) $Mn^{2+} + HL^- \overset{K_1^*}{\rightleftharpoons} MnHL^+$	50 vol% dioxane	<0.01 ($HClO_4$)	log $K_1^* = 9.43 \pm 0.02$	[1]
	50 vol% dioxane	0_{corr}	log $K_1^* = 9.66$	[1]
(2) $MnHL^+ + HL^- \overset{K_2^*}{\rightleftharpoons} Mn(HL)_2$	50 vol% dioxane	<0.01 ($HClO_4$)	log $K_2^* = 8.6 \pm 0.2$	[1]
	50 vol% dioxane	0_{corr}	log $K_2^* = 8.8$	[1]
	20 vol% butyl alcohol	0.1 (KNO_3)	log $K_2^* = 9.32 \pm 0.22$ [a]	[2]
(3) $Mn^{2+} + 2HL^- \overset{\beta_2^*}{\rightleftharpoons} Mn(HL)_2$	50 vol% methanol	0.1 ($NaClO_4$)	log $\beta_2^* = 13.1 \pm 0.2$	[3]
(4) $Mn(HL)_2 \rightleftharpoons Mn(HL)L^- + H$	50 vol% dioxane	<0.01 ($HClO_4$)	pK = 7.88 ± 0.05	[1]
(5) $Mn(HL)L^- \rightleftharpoons MnL_2^{2-} + H$	50 vol% dioxane	<0.01 ($HClO_4$)	pK = 9.38 ± 0.11	[1]

[a] Determined spectrophotometrically.

The assignment of the constants in [1, 3] is based on the structural similarity of ligand 1 and 2-pyridylazobenzenediol (p. 292). The *ortho*-hydroxy hydrogen of ligand 1 is replaced by manganese(II) in the first part of the titration with sodium hydroxide according to equations (1), (2), or (3), in spite of the normally more acidic nature of the *para*-hydroxy group.

The constants K_1^*, K_2^*, and β_2 are calculated on the assumption that the retained *para*-hydroxy proton does not influence the dissociation constant pk_{a3} of the chelating *ortho*-hydroxy group (pk_{a3} of the free ligand H_3L^+: 10.76 for I = 0.1 M ($NaClO_4$) in 50 vol% methanol-water, 12.80 ± 0.04 for I < 0.01 M ($HClO_4$), and 13.07 for I = 0 corr in 50 vol% dioxane-water) [1, 3]. The protons in *para*-position are released from the complex only in the second, more alkaline region of the NaOH titration, according to equations (4) and (5). The acid dissociation constants (pK) demonstrate the acid-strengthening effect on the *para*-hydroxy group as a result of complex formation. The probable origin of this effect is discussed. It is still higher, according to the Irving-Williams order, for analogous Cu, Ni, or Zn complexes but lower than for the pyridylazo complex [1]. Stability constants of the complex MnL with ligand 1 were also determined spectrophotometrically in aqueous-organic media at pH values ranging from 8.0 to 10.3 and at $I \to 0$. They are based on the equilibrium $Mn^{2+} + HL^- \rightleftharpoons MnL + H^+$. The values of log K vary from 5.8 to 6.3 in 10 to 50 vol% dimethylformamide or from 6.5 to 8.8 in corresponding mixtures of acetone, dioxane, or ethanol with water [4]. Formation of a deep red complex MnL_2^{2-} with ligand 1 is suggested in aqueous solution at pH 8 (λ_{max} = 530 nm and $\varepsilon = 2.5 \times 10^4$ $L \cdot mol^{-1} \cdot cm^{-1}$). The species was absorbed on an anion exchanger and eluted after the free ligand anion HL^- [5]. A red complex with λ_{max} = 540 nm and ε = 41760 $L \cdot mol^{-1} \cdot cm^{-1}$, observed in 20 vol% butyl alcohol, was formulated as $Mn(HL)_2$ [2]. Octahedral complex structures were proposed with two ligands chelating by the thiazole N, one azo N and the deprotonated oxygen atom in *ortho*-position to the azo group [2, 5].

A red complex with ligand 2 in aqueous solution is indicated at pH 6.8 by its absorption maximum at λ = 520 nm, $\varepsilon = 1.4 \times 10^4$ $L \cdot mol^{-1} \cdot cm^{-1}$ [6].

$Mn(C_9H_6N_3O_4S)Cl$ was prepared from ethanolic solutions of $MnCl_2$ and ligand 1 with a few drops of aqueous ammonia. The green-black powder (λ_{max} = 527 nm in ethanol) decomposes at 180°C. Susceptibility measurements indicated five unpaired electrons for manganese [1].

References:

[1] Stanley, W., Cheney, G. E. (Talanta **13** [1966] 1619/29).
[2] Gaokar, U. G., Eshwar, M. C. (Mikrochim. Acta **1982** II 247/52).
[3] Nickless, G., Pollard, F. H., Samuelson, T. J. (Anal. Chim. Acta **39** [1967] 37/46, 39/42).
[4] Rudometkina, T. F., Ivanov, V. M., Busev, A. I. (Zh. Analit. Khim. **32** [1977] 1674/9; J. Anal. Chem. [USSR] **32** [1977] 1329/33).
[5] Herman, R. G., Norman, L. J. (Inorg. Nucl. Chem. Letters **14** [1978] 183/8).
[6] Kasiura, K., Szczygielska, M. (Chem. Anal. [Warsaw] **16** [1971] 671/6, 673/4; C.A. **75** [1971] No. 147392).

28.9.3 With (2-Thiazolylazo)naphthols

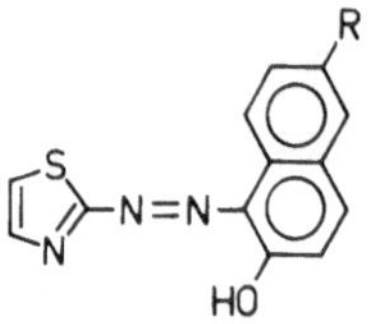

ligand 1 R = H (= $C_{13}H_9N_3OS$)
ligand 2 R = SO_3H (= $C_{13}H_9N_3O_4S_2$ = H_2L)

Complex in Solution. Stability constants were determined potentiometrically (glass electrode) for 1:1 and 1:2 complexes with ligand 2 at 25°C and I = 0.1 M ($NaClO_4$): log K_1 = 4.3 and log K_2 = 3.3 in water (pK_a = 8.38) and log K_1 = 4.9 and log K_2 = 4.0 in 50% aqueous methanol (pK_a = 8.44) [1].

$Mn(C_{13}H_8N_3OS)_2$ was precipitated by mixing, in a 1:2 mole ratio, a warm methanol solution of $Mn(CH_3COO)_2$ with the ligand 1 in an organic solvent, similarly as for analogous Cu, Ni, and Co complexes [2]. The water-insoluble complex also separated in aqueous solution [3]. Chloroform extracts at pH 9.2 to 9.8 show absorption maxima at 565 to 585 nm, $\varepsilon = 3.8 \times 10^4$ L·mol^{-1}·cm^{-1} [4] or 570 nm, $\varepsilon = 4.1 \times 10^4$ L·mol^{-1}·cm^{-1} [3]. The complex is also extracted into CCl_4 [4] or benzene at pH>8 [4, 5]. Polarographic measurements (Hg pool reference) in dimethylformamide solution, as for the complex $Mn(C_{15}H_{10}N_3O)_2$ with 1-(2-pyridylazo)-2-naphthol (see p. 298), reveal three reduction waves with the half-wave potentials $E_{1/2} = -0.41$, -1.17, and -1.45 V. The first two waves correspond to reversible one-electron reduction steps. The third wave indicates an irreversible two-electron reduction. The characteristic features of the oscillopolarogram (with two pairs of symmetrical cathodic-anodic peaks) are essentially the same as those of $Mn(C_{15}H_{10}N_3O)_2$ and may be interpreted in the same way [6].

References:

[1] Nickless, G., Pollard, F. H., Samuelsson, T. J. (Anal. Chim. Acta **39** [1967] 37/46, 43/4).
[2] Zhuchenko, T. A., Kuznetsova, L. I., Kogan, V. A., Garnovskii, A. D., Osipov, O. A. (Zh. Neorgan. Khim. **16** [1971] 2169/72; Russ. J. Inorg. Chem. **16** [1971] 1157/9).
[3] Nakagawa, G., Wada, H. (Nippon Kagaku Zasshi **83** [1962] 1185/9, 1187, A 76/7; C.A. **59** [1963] 9289).
[4] Grzegrzolka, E. (Chem. Anal. [Warsaw] **22** [1977] 303/9, 307; C.A. **87** [1977] No. 193129).
[5] Navratil, O. (Sb. Ref. Celostatni Radiochem. 3rd Konf., Liblice, Czech., 1964, pp. 112/8, 114; C.A. **64** [1966] 16712).
[6] Toropova, V., Budnikov, G. K., Maistrenko, V. N. (Zh. Obshch. Khim. **44** [1974] 364/8; J. Gen. Chem. [USSR] **44** [1974] 345/9).

28.9.4 With Thiazolylazo Compounds of 8-Hydroxy-5-quinolinesulfonic Acid

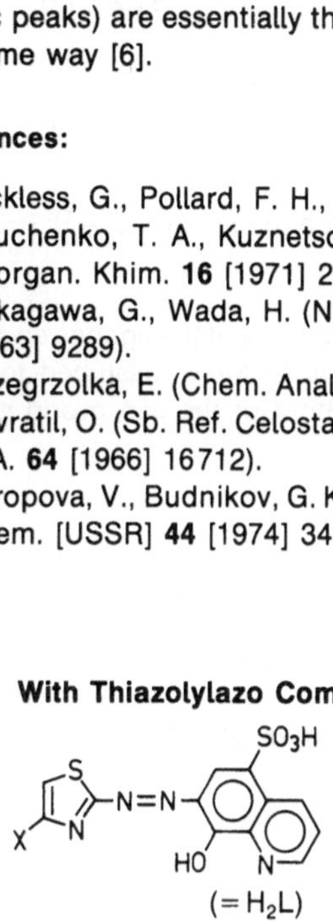

ligand 1 X = H (= $C_{12}H_8N_4O_4S_2$)
ligand 2 X = Br (= $C_{12}H_7BrN_4O_4S_2$)

Complexes in Solution. Complex formation with the ligands 1 or 2 was established spectrophotometrically in dimethylformamide-water or in mixed solvents containing 0.8 vol% DMF, water, and various concentrations of dioxane, acetone, or ethanol. Whereas 1:1 complexes predominate in most solvent mixtures, 1:2 complexes are formed only in solvents with certain contents of acetone or ethanol. Overall stability constants β_1 and β_2 were calculated from the equilibrium constants, K_{eq}, which were determined, at optimum pH values from 5.6 to 7.0 and I→0, for the equations: $Mn^{2+} + HL^- \rightleftharpoons MnL + H^+$ and $Mn^{2+} + 2HL^- \rightleftharpoons MnL_2^{2-} + 2H^+$. Selected values are presented together with the wavelengths, λ_{max} in nm, and molar extinction, ε in L·mol^{-1}·cm^{-1}, of maximum absorption:

complex	solvent	vol%	pk_{a2}	pK_{eq}	log β_n	λ_{max}	$\varepsilon \times 10^{-3}$	Ref.
$Mn(C_{12}H_6N_4O_4S_2)$	DMF	12.0	6.80	1.71	5.09	510	29.4 ± 0.7	[1]
	DMF	50.0	6.78	1.42	5.36	510	38.9 ± 2.1	[1]
	dioxane[a)]	15.2	6.98	1.67	5.31	510	29.5 ± 2.6	[1]

complex	solvent	vol%	pk_{a2}	pK_{eq}	log β_n	λ_{max}	$\varepsilon \times 10^{-3}$	Ref.
$Mn(C_{12}H_6N_4O_4S_2)$	dioxane	50.0	7.40	1.31	6.09[b]	510	39.0±4.3	[1 to 3]
	acetone[a]	28.3	6.80	1.60	5.20	510	36.2±1.4	[1]
	acetone	50.0	6.80	1.26	5.54	510	39.6±1.9	[1, 3]
	ethanol[a]	21.1	6.70	1.68	5.02	510	32.5±0.4	[1]
$Mn(C_{12}H_6N_4O_4S_2)_2^{2-}$	acetone[a]	35.2	6.85	1.95	11.75	510	77.2±3.1	[1]
	acetone	43.8	6.70	1.72	11.68	510	79.6±2.1	[1]
	ethanol[a]	30.0	6.75	2.08	11.42	510	73.8±3.1	[1]
	ethanol	50.0	6.90	1.61	12.19	510	81.2±1.6	[1]
$Mn(C_{12}H_5BrN_4O_4S_2)$	dioxane[a]	50.0	7.15	—	6.38	525	37.0±0.7	[3]
$Mn(C_{12}H_5BrN_4O_4S_2)_2^{2-}$	acetone[a]	40.0	6.30	—	6.23	530	35.2±1.3	[3]
	acetone	50.0	6.45	—	12.15	530	62.7±1.5	[3]

[a] Each solvent mixture also contains 0.8 vol% DMF and water added to 100%. – [b] Erroneous value for β in [3].

The influence of the solvent polarities on the tautomeric equilibria of the ligands and on the complex stabilities is discussed. Coordination of the ligands to Mn is assumed to occur through the deprotonated phenolic oxygen, one nitrogen atom of the azo group and the thiazole nitrogen atom [1, 3].

References:

[1] Rudometkina, T. F., Ivanov, V. M., Busev, A. I. (Zh. Analit. Khim. **31** [1976] 877/83; J. Anal. Chem. [USSR] **31** [1976] 715/20).

[2] Rudometkina, T. F., Ivanov, V. M., Busev, A. I. (Zh. Neorgan. Khim. **22** [1977] 142/5; Russ. J. Inorg. Chem. **22** [1977] 77/9).

[3] Rudometkina, T. F., Ivanov, V. M., Busev, A. I. (Zh. Analit. Khim. **32** [1977] 446/ 51; J. Anal. Chem. [USSR] **32** [1977] 354/8).

28.9.5 With Benzothiazolylazo Compounds

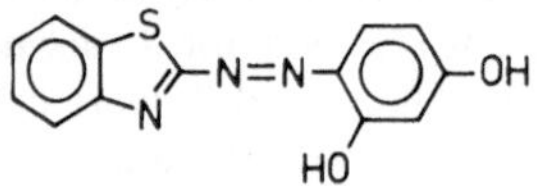

ligand 1 (= $C_{13}H_9N_3O_2S$)

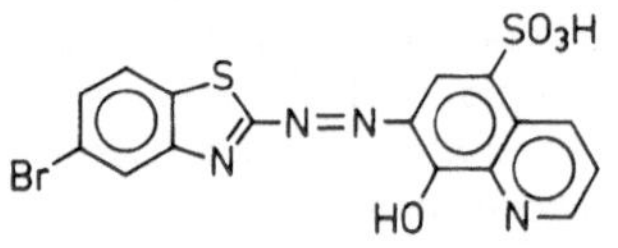

ligand 2 (= $C_{16}H_9BrN_4O_4S_2$)

Complexes in Solution. Complexes are formed with either ligand (= H_2L) according to a reaction $Mn^{2+} + HL^- \rightleftharpoons MnL + H^+$ at suitable pH in mixed solvents containing water, 1 vol% of dimethylformamide, and other organic solvents. Stability constants are calculated from the equilibrium constants K_{eq} which were determined spectrophotometrically at the ionic strength $I \rightarrow 0$ at $\lambda_{max} = 510$ nm and pH 7.9 for ligand 1 and $\lambda_{max} = 530$ to 545 nm and pH 4.9 to 5.1 for ligand 2, respectively. Values for selected conditions are listed together with the pertinent molar absorptivities ε in $L \cdot mol^{-1} \cdot cm^{-1}$.

complex	solvent[a)]	vol%	pk_{a1}	pk_{a2}	pK_{eq}	log K	$\varepsilon \times 10^{-3}$	Ref.
$Mn(C_{13}H_7N_3O_2S)$	dioxane	6.2	5.95	9.80	2.89	6.90	2.37	[1]
	dioxane	40.0	6.40	11.75	2.52	9.23	2.22	[1]
	acetone	8.4	6.65	10.15	3.12	7.03	2.40	[1]
	acetone	35.4	6.05	10.85	3.75	7.10	2.40	[1]
	ethanol	8.8	5.58	9.40	3.23	6.10	2.34	[1]
	ethanol	41.9	5.90	11.15	3.88	7.26	2.25	[1]
$Mn(C_{16}H_7BrN_4O_4S)$	dioxane	50.0	1.7	6.1	0.82	5.28	3.73 ± 0.19	[2]

[a)] Plus 1 vol% DMF, made up to 100% with water.

The roles of solvent nature and dielectric constant on the complex stability are discussed for $Mn(C_{13}H_7N_3O_2S)$ [1]. Solubility and stabilities of $Mn(C_{16}H_7BrN_4O_4S_2)$ are compared to those of the complexes with the thiazolyl- and 4-bromothiazolylazo compounds of 8-hydroxy-5-quinolinesulfonic acid [2].

References:

[1] Rudometkina, T. F., Ivanov, V. M., Busev, A. I. (Zh. Analit. Khim. **30** [1975] 2322/8; J. Anal. Chem. [USSR] **30** [1975] 1951/6, 1953).

[2] Rudometkina, T. F., Ivanov, V. M., Busev, A. I. (Zh. Neorgan. Khim. **22** [1977] 142/5; Russ. J. Inorg. Chem. **22** [1977] 77/9; Zh. Analit. Khim. **32** [1977] 446/51; J. Anal. Chem. [USSR] **32** [1977] 354/8).

28.9.6 With a Thiadiazolylazo Compound

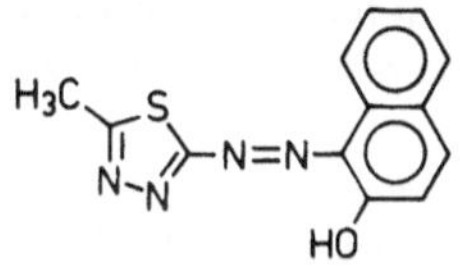

(= 1-(5-Methyl-1,3,4-thiadiazolyl-2-azo)-2-naphthol $= C_{13}H_{10}N_4OS$)

The formation of a manganese(II) complex is indicated spectrophotometrically at pH 8 and 9 in aqueous solution on addition of the ligand in ethanol. Chloroform extracts exhibit $\lambda_{max} = 578$ nm and $\varepsilon = 3.7 \times 10^4$ $L \cdot mol^{-1} \cdot cm^{-1}$ at pH 11.20, the value of optimum formation, Cacho Palomar, J., Nerin de la Puerta, C. (Afinidad **38** No. 374 [1981] 345/7; C.A. **95** [1981] No. 180138).

29 Complexes with Triazenes

29.1 With 3-Hydroxytriazenes or Tautomeric Triazene 1-Oxides

Triazene = diazoamine compounds forming complexes with manganese belong to the acid type Ar–N=N–N(OH)–R or its N-oxide tautomer Ar–N(H)–N=N$^{(+)}$(R)–O$^{(-)}$, both derived from N-substituted hydroxylamines. Only that ligand tautomer is depicted in this section that is most relevant to the manganese complexes. Chelation occurs by metal substitution of the proton and formation of further coordinate bonds.

Intensely red-colored solutions developed by hydroxytriazenes, frequently observed in the presence of manganese, are not considered here if definite complexes have not been established.

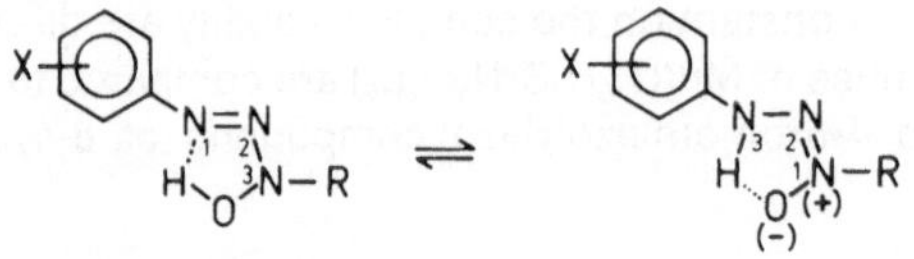

ligands 1 to 12

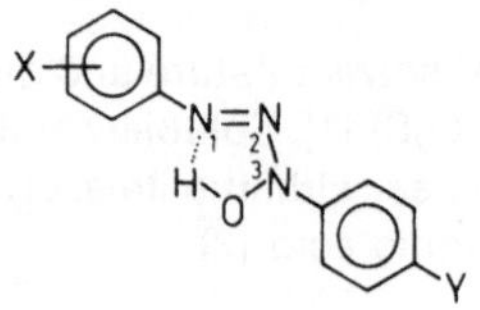

ligands 13 to 23

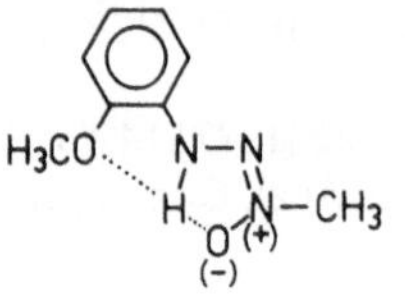

ligand 24 (= $C_8H_{11}N_3O_2$ = HL)

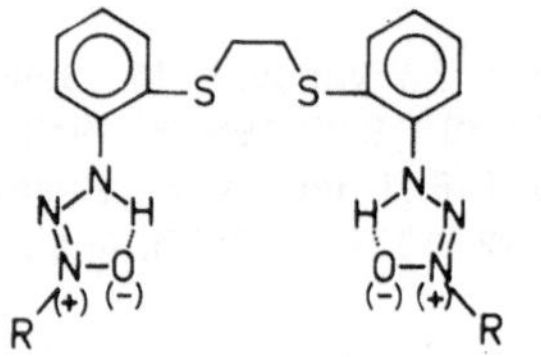

ligand 25 R = CH_3 (= $C_{16}H_{20}N_6O_2S_2$ = H_2L)
ligand 26 R = C_2H_5 (= $C_{18}H_{24}N_6O_2S_2$ = H_2L)

Complexes in Solution. Stability constants of complexes MnL^+ and MnL_2 with the ligands 1 to 23 in 70% (v/v) dioxane-water mixtures were determined potentiometrically (glass electrode), at 25 ± 0.5°C and I = 0.1 M (KCl). Values of log K_1 and log K_2 are tabulated below together with the pK_a values of the ligands.

No.	R	X	ligand HL	pK_a	log K_1	log K_2	Ref.
	With 3-alkyl-1-aryl-3-hydroxytriazenes:						
1	CH_3	2-Cl	$C_7H_8ClN_3O$	11.31	5.50	3.86	[1]
2	CH_3	4-Cl	$C_7H_8ClN_3O$	11.52	5.72	4.07	[1]
3	CH_3	4-CH_3O	$C_8H_{11}N_3O_2$	12.69	7.39	5.60	[2]
4	CH_3	4-CH_3	$C_8H_{11}N_3O$	12.52	7.01	5.57	[2]
5	C_2H_5	H	$C_8H_{11}N_3O$	12.28	6.54	4.90	[3]
6	C_2H_5	2-Cl	$C_8H_{10}ClN_3O$	11.44	5.79	3.99	[4]
7	C_2H_5	4-Cl	$C_8H_{10}ClN_3O$	11.56	5.94	4.20	[4]
8	C_2H_5	4-CH_3O	$C_9H_{13}N_3O_2$	12.95	7.39	5.60	[3]
9	C_2H_5	4-CH_3	$C_9H_{13}N_3O$	12.67	7.01	5.57	[3]
10	C_3H_7	H	$C_9H_{13}N_3O$	12.39	6.06	5.59	[5]
11	C_3H_7	4-Cl	$C_9H_{12}ClN_3O$	11.74	6.09	4.35	[5]
12	C_3H_7	4-CH_3	$C_{10}H_{15}N_3O$	—	7.18	5.64	[5]

No.	R	X	ligand HL	pK_a	log K_1	log K_2	Ref.
	With 1,3-diaryl-3-hydroxytriazenes ($R = C_6H_4Y$):						
13	C_6H_5	2-Cl	$C_{12}H_{10}ClN_3O$	10.519	5.359	4.419	[6]
14	C_6H_5	4-Cl	$C_{12}H_{10}ClN_3O$	10.72	5.83	4.67	[15]
15	C_6H_5	4-Br	$C_{12}H_{10}BrN_3O$	10.86	5.90	4.95	[7]
16	C_6H_5	$4-HO_3S$	$C_{12}H_{11}N_3O_4S$	9.988	4.825	3.376	[8]
17	C_6H_5	$4-CH_3O$	$C_{13}H_{13}N_3O_2$	11.95	7.24	6.10	[9]
18	C_6H_5	$4-CH_3C(O)NH$	$C_{14}H_{14}N_4O_2$	11.663	6.793	5.880	[10]
19	C_6H_5	$2-CH_3$	$C_{13}H_{13}N_3O$	11.55	6.68	5.39	[11]
20	C_6H_5	$4-CH_3$	$C_{13}H_{13}N_3O$	11.78	7.02	5.72	[11]
21	C_6H_5	$4-CH_3C(O)$	$C_{14}H_{13}N_3O_2$	10.962	6.094	4.801	[12]
22	$4-C_6H_4Cl$	H	$C_{12}H_{10}ClN_3O$	10.648	5.6	4.6	[13]
23	$4-C_6H_4CH_3$	$4-CH_3$	$C_{14}H_{15}N_3O$	12.175	7.820	7.050	[14]

The preponderance of different tautomers for alkylaryl-substituted ligands on one hand and diaryl-substituted ligands on the other hand is assumed to affect nature and stabilities of the corresponding complexes [3]. Electron-withdrawing chloro substituents in the phenyl groups of ligands 1, 2, 6, 7, 11 lower the stabilities of manganese chelates corresponding to their decreased ligand basicities [1, 4, 5], whereas inverse effects were observed for donating methyl groups of ligands 4, 9, and 12 [2, 3, 5], or for methoxy substituents of ligands 3 or 8 [2, 3], due to electronic and steric reasons discussed by the authors. Substituent effects for some diarylhydroxytriazenes were also discussed [11]. A stability order Pd > Cu > Ni > Zn > Mn results for complexes of these metal(II) ions with each ligand [1 to 15].

$Mn(C_8H_{10}N_3O_2)_2$ was obtained as a crystalline solid on reacting manganese(II) acetate with ligand 24. The compound is high-spin octahedral with the magnetic moment $\mu_{eff} = 6.27\,\mu_B$. On account of its dipole moment (<1 Debye), *trans*-orientation of the ligand was suggested [16].

$Mn(C_{16}H_{18}N_6O_2S_2)$ and **$Mn(C_{18}H_{22}N_6O_2S_2)$**. Mixed solutions containing 1:1 mole ratios of $Mn(CH_3COO)_2$ and ligand 25 or 26 in ethanol were refluxed for ~½ h. Upon concentration the red crystalline complexes separated. They were washed with hot water and finally with alcohol and dried over P_4O_{10} in vacuum.

The visible spectrum of the complexes shows a strong charge transfer absorption with a shoulder at ~19000 cm^{-1}. Magnetic moments, corrected for diamagnetism, $\mu_{eff} = 6.02\,\mu_B$ for $Mn(C_{16}H_{18}N_6O_2S_2)$ and 6.11 μ_B for $Mn(C_{18}H_{22}N_6O_2S_2)$, indicate neutral high-spin Mn^{II} complexes with hexadentate ligands suggested to coordinate through two deprotonated N atoms and through their O and S atoms additionally. The complexes are insoluble in water and ethanol but soluble in acetone and chloroform [17].

References:

[1] Dugar, S. M., Jaimni, J. P. C., Sogani, N. C. (Bull. Acad. Polon. Sci. Ser. Sci. Chim. **14** [1966] 535/9, 537; C.A. **66** [1967] No. 22763).

[2] Dugar, S. M., Sogani, N. C. (J. Indian Chem. Soc. **47** [1970] 479/82).

[3] Dugar, S. M., Sogani, N. C. (J. Inst. Chem. [India] **42** [1970] 184/9; C.A. **74** [1971] No. 63648).

[4] Dugar, S. M., Sogani, N. C. (J. Indian Chem. Soc. **45** [1968] 646/8).

[5] Dugar, S. M., Sogani, N. C. (J. Inst. Chem. [India] **41** [1969] 231/4; C.A. **72** [1970] No. 104548).
[6] Purohit, D. N., Sogani, N. C. (Z. Anorg. Allgem. Chem. **331** [1964] 220/4).
[7] Purohit, D. N., Sogani, N. C. (J. Proc. Inst. Chem. [India] **37** [1965] 212/5; C.A. **64** [1966] 7425).
[8] Purohit, D. N., Sogani, N. C. (Z. Anal. Chem. **203** [1964] 97/101).
[9] Purohit, D. N., Sogani, N. C. (Indian J. Chem. **3** [1965] 58/9).
[10] Purohit, D. N., Dugar, S. M., Sogani, N. C. (Proc. Natl. Acad. Sci. India A **34** [1964] 233/8; C.A. **62** [1965] 3459).

[11] Purohit, D. N., Sogani, N. C. (Bull. Chem. Soc. Japan **37** [1964] 1727/9).
[12] Purohit, D. N., Sogani, N. C. (J. Indian Chem. Soc. **41** [1964] 20/4).
[13] Purohit, D. N., Sogani, N. C. (Bull. Acad. Polon. Sci. Ser. Sci. Chim. **12** [1964] 85/9; C.A. **60** [1964] 15205).
[14] Purohit, D. N., Sogani, N. C. (J. Proc. Inst. Chem. [India] **36** [1964] 219/24; C.A. **62** [1965] 69).
[15] Purohit, D. N., Sogani, N. C. (Bull. Chem. Soc. Japan **37** [1964] 476/8).
[16] Zacharias, P. S., Chakravorty, A. (Proc. 1st Chem. Symp., Chandigarh, India, 1969 [1970], Vol. 2, pp. 49/52; C.A. **74** [1971] No. 60335).
[17] Mukkanti, K., Bhoon, Y. K., Pandeya, K. B., Singh, R. B. (J. Indian Chem. Soc. **59** [1982] 830/2).

Ligand Formula Index

The organic ligands treated in this volume are arranged in the index according to the system of Hill, A. (J. Am. Chem. Soc. **22** [1900] 478/94). In this system, the first criterion for the location of a ligand is the number of carbon atoms, the second, the number of hydrogen atoms. Then comes the number of atoms of the other elements, set in alphabetical order.

The first column contains the empirical formulas of the ligands. The second column shows their linearized structure formulas. Additional ligands occurring in mixed ligand complexes and in adducts are placed as subheadings. Thus, the complex $Mn_2(C_4H_5NO_4)_2(edta)^{4-}$, derived from iminodiacetic acid and ethylenediamine-N,N,N',N'-tetraacetic acid, see p. 54, may be found under the main entry $C_4H_7NO_4$ and the subheading "and $C_{10}H_{16}N_2O_8$" (= H_4edta). But it is also listed under the main entry $C_{10}H_{16}N_2O_8$ and the subheading "and $C_4H_7NO_4$". Additional ligands which are described as main ligands in Volumes **D 1** to **D 4** occur only as subheadings in the formula index of this volume. Water and alcohols of solvates are not included as additional ligands.

The last column lists the pertinent pages. The boldfaced **D 1**, **D 3**, and **D 4** before the page numbers refer to those complex compounds described in the volumes "Manganese" **D 1** to **D 4** in which ligands of the present volume occur as additional ligands.

Empirical formulas and names of ligands coordinated as deprotonated acids are given in their nondeprotonated form. Ligands occurring in tautomeric equilibria are only presented in the form commonly known.

List of abbreviations used in the volume:

en	ethylenediamine	DMSO	dimethyl sulfoxide
py	pyridine	THF	tetrahydrofuran
bpy	2,2'-bipyridine = 2,2'-bipyridyl	Hacac	acetylacetone
phen	1,10-phenanthroline	H_4edta	ethylenediamine-N,N,N',N'-tetraacetic acid
ur	urea		
dmf = DMF	dimethylformamide	H_4cdta	*trans*-1,2-cyclohexanediamine-tetraacetic acid

$C_6H_8N_4O_4$	$H_2NC(O)NHNHCH_2$–R, R = 6-nitro-2-furyl		208
$C_6H_8N_{12}O_2$	R–N=N–CH($COOC_2H_5$)–N=N–R, R = 1*H*-tetrazol-5-yl		310
$C_6H_9NO_2$	(*E, E*)–CH_3CH=CH–CH=CHC(O)NHOH		212
$C_6H_9NO_6$	$N(CH_2COOH)_3$		15/7, 20/1
	and $C_2H_5NO_2$	H_2NCH_2COOH	18
	and $C_3H_7NO_3$	$HOCH_2CH(NH_2)COOH$	18
	and $C_4H_7NO_2$	CH_2=CHCH(NH_2)COOH	18
	and $C_4H_7NO_4$	$HOOCCH(NH_2)CH_2COOH$	18
	and $C_4H_8N_2$	$H_2NCH_2C(O)NHCH_2COOH$	19
	and $C_5H_9NO_2$	proline	18
	and $C_5H_9NO_4$	$HOOCCH(NH_2)CH_2CH_2COOH$	18
	and $C_6H_9N_3O_2$	histidine	19
	and $C_6H_{14}N_4O_2$	HN=C(NH_2)NH$(CH_2)_3$CH(NH_2)COOH	19
	and $C_7H_{15}NO_2$	$(CH_3)_2CHCH(NH_2)COOC_2H_5$	18
	and C_9H_7NO	8-hydroxyquinoline = 8-quinolinol	18/9
	and $C_{10}H_8N_2$	2,2′-bipyridine (= bpy)	18/9
	and $C_{10}H_{16}N_2O_8$	$(HOOCCH_2)_2NCH_2CH_2N(CH_2COOH)_2$ (= H_4edta)	54
	and $C_{10}H_{16}N_5O_{13}P_3$	adenosine triphosphate	18/9
$C_6H_9N_3O_2$	3-Methylene-2,6-piperidinedione dioxime		271
$C_6H_{10}N_2O_2$	1,2-Cyclohexanedione dioxime		266/7
$C_6H_{10}N_2O_5$	$H_2NC(O)CH_2N(CH_2COOH)_2$		21
$C_6H_{11}NO_4S$	$HSCH_2CH_2N(CH_2COOH)_2$		9
$C_6H_{11}NO_5$	$HOCH_2CH_2N(CH_2COOH)_2$		7/9, 14
$C_6H_{12}N_2O_2$	$(CH_3)_2NC(O)C(O)N(CH_3)_2$		125/6
	cyclo-C_6H_{11}–N(–N=O)OH		276
$C_6H_{12}N_2O_4$	$H_2NCH_2CH_2N(CH_2COOH)_2$		10
	$[CH_2NHCH_2COOH]_2$		24
	and $C_2H_8N_2$	$H_2NCH_2CH_2NH_2$ (= en)	24
	and $C_{10}H_8N_2$	2,2′-bipyridyl (= bpy)	24
$C_6H_{13}NO$	$CH_3C(O)N(C_2H_5)_2$		100/1
$C_6H_{14}N_4$	C_2H_5NHC(=NH)C(=NH)NHC_2H_5		160
$C_6H_{14}N_4O_2$	$H_2NNHC(O)(CH_2)_4C(O)NHNH_2$		205/7

$C_7H_8N_2O$	$(6\text{-}CH_3)2\text{-}C_5H_3N\text{–}CH{=}NOH$					261
	and C_5H_5N	pyridine (= py)				261/2
	and C_6H_7N	methylpyridine				261/2
	and C_7H_9N	ethylpyridine				261/2
	$2\text{-}C_5H_4N\text{–}C(CH_3){=}NOH$					263
	$C_6H_5C(NH_2){=}NOH$					264
$C_7H_8N_2O_2$	$3\text{-}C_5H_4N\text{–}C(O)NHCH_2OH$					116/7
	$(2\text{-}HO)C_6H_4C(O)NHNH_2$					179/80
	and $C_6H_3N_3O_7$	$(O_2N)_3C_6H_2(OH)$				180
	and $C_6H_9N_3O_2$	histidine				181/2
	and $C_6H_9N_3O_2$	histidine	and H_3N	NH_3		182
	and $C_6H_9N_3O_2$	histidine	and C_3H_6O	$CH_3C(O)CH_3$		182
	and $C_7H_8N_2O$	$C_6H_5C(O)NHNH_2$				181
	and $C_7H_8N_2O$	$C_6H_5C(O)NHNH_2$	and H_3N	NH_3		181
	and $C_7H_8N_2O$	$C_6H_5C(O)NHNH_2$	and C_3H_6O	$CH_3C(O)CH_3$		181
	$(3\text{-}HO)C_6H_4C(O)NHNH_2$					183
	$(H_2N)C_6H_4C(O)NHOH$					216/7
$C_7H_9N_5O_2$	$H_2NNHC(O)\text{–}R\text{–}C(O)NHNH_2$, R = 2,6-pyridinediyl					207/8
$C_7H_{11}NO_6$	$HOOCCH_2CH_2CH(COOH)NHCH_2COOH$					22
	$HOOCCH_2CH_2N(CH_2COOH)_2$					22/3
$C_7H_{13}NO_4S$	$CH_3SCH_2CH_2N(CH_2COOH)_2$					9
$C_7H_{13}NO_5$	$CH_3OCH_2CH_2N(CH_2COOH)_2$					9
$C_7H_{13}N_2O_3^{\bullet}$	3-Hydroxy-2,2,5,5-tetramethyl-4-oxo-1-imidazolidinyloxy radical					230
$C_7H_{14}N_2O_2$	$(CH_3)_2NC(O)CH_2C(O)N(CH_3)_2$					125/6
$C_7H_{16}N_2O$	$C_6H_{13}C(O)NHNH_2$					173
C_8						
$C_8H_5NO_2$	Phthalimide					129
	and H_3N	NH_3				129
	and CH_5N	H_2NCH_3				129
	and C_2H_7N	$H_2NC_2H_5$ or $HN(CH_3)_2$				129
	and C_3H_9N	$H_2NC_3H_7$ or $H_2NCH(CH_3)_2$				129

$C_8H_5NO_2$	Phthalimide		
	and $C_4H_{11}N$	$H_2NC_4H_9$ or $H_2NCH_2CH(CH_3)_2$ or $HN(C_2H_5)_2$	129
	and $C_5H_{13}N$	$H_2NC_5H_{11}$ or $H_2NCH_2CH_2CH(CH_3)_2$	129
$C_8H_7Br_2NO_3$	$(HO)_2(Br)_2C_6HC(CH_3){=}NOH$		240/1
$C_8H_7ClN_2O_2$	$HON{=}CHC(O)NHC_6H_4(Cl)$		265
$C_8H_7Cl_2NO_2$	$(HO)(Cl)_2C_6H_2C(CH_3){=}NOH$		240/1
$C_8H_7IN_2O_2$	$HON{=}CHC(O)NHC_6H_4(I)$		265
$C_8H_7NO_4$	$HOOCC_6H_4C(O)NHOH$		218/9
$C_8H_7N_3O_2$	1,3-Isoindolinedione dioxime		271
$C_8H_8BrNO_2$	$(HO)(Br)C_6H_3C(CH_3){=}NOH$		240/1
$C_8H_8ClNO_2$	$CH_3C(O)N(OH)C_6H_4(Cl)$		219
	$(HO)(Cl)C_6H_3C(CH_3){=}NOH$		240/1
$C_8H_8INO_2$	$(HO)(I)C_6H_3C(CH_3){=}NOH$		240/1
$C_8H_8N_2O_2$	$HON{=}CHC(O)NHC_6H_5$		265
	$HON{=}C(C_6H_5)CH{=}NOH$		266/7
$C_8H_8N_2O_3$	$CH_3C(O)NHC_6H_3(OH)(N{=}O)$		253
$C_8H_8N_2O_4$	$HOHNC(O)C_6H_4C(O)NHOH$		218/9
	$(HO)(O_2N)C_6H_3C(CH_3){=}NOH$		240/1
$C_8H_8N_6O$	R–N=N–$C_6H_3(OH)(CH_3)$, R = 1 *H*-tetrazol-5-yl		310
$C_8H_9NO_2$	$C_6H_5CH_2C(O)NHOH$		212/3
	and $C_2H_5NO_2$	H_2NCH_2COOH	212/3
	and $C_3H_7NO_2$	$H_2NCH(CH_3)COOH$ or $H_2NCH_2CH_2COOH$	212/3
	and $C_6H_9N_3O_2$	histidine	212/3
	and $C_9H_{11}NO_2$	$C_6H_5CH_2CH(NH_2)COOH$	212/3
	and $C_{10}H_8N_2$	2,2′-bipyridine (= bpy)	213
	$CH_3C(O)N(OH)C_6H_5$		219
	$(HO)(CH_3)C_6H_3CH{=}NOH$		235, 237
	$(HO)C_6H_4C(CH_3){=}NOH$		240, 242
$C_8H_9NO_3$	$(HO)(CH_3)C_6H_3C(O)NHOH$		215
	$(HO)(CH_3O)C_6H_3CH{=}NOH$		236/7
	$(HO)_2C_6H_3C(CH_3){=}NOH$		240/1

$C_{10}H_8N_2O$	R–CH=NOH, R = 2-quinolinyl or 3-isoquinolinyl	262/3
$C_{10}H_8N_2O_2$	R–C(O)NHOH, R = 2-quinolyl	218
	HON=C(–R)CH=NOH, R = 2-benzofuryl	266/7
	R–N(–N=O)OH, R = 1-naphthyl	276
$C_{10}H_8N_2O_4$	HON=C(R)C(R)=NOH, R = 2-furyl	266/7
$C_{10}H_8N_2O_4S$	2-Hydroxy-1-nitroso-3-naphthalenesulfonamide	254/5
$C_{10}H_8N_4O_4$	3-Methyl-1-(4-nitrophenyl)-1*H*-pyrazole-4,5-dione 4-oxime	250/1
$C_{10}H_9NO_4$	(*Z*)-2,2′-Furoin oxime	247
$C_{10}H_9N_3OS$	R–N=N–$C_6H_3(OH)(CH_3)$, R = 2-thiazolyl	312
$C_{10}H_9N_3O_2$	3-Methyl-1-phenyl-1*H*-pyrazole-4,5-dione 4-oxime	248/51
$C_{10}H_{10}BrNO_2$	$CH_3C(OH)=C(Br)C(O)NHC_6H_5$	106/7
	$H_2C=C(CH_3)C(O)N(OH)C_6H_4(Br)$	221
$C_{10}H_{10}BrNO_3$	$CH_3C(O)CH_2C(O)N(OH)C_6H_4(Br)$	219/20
$C_{10}H_{10}ClNO_2$	$H_2C=C(CH_3)C(O)N(OH)C_6H_4(Cl)$	221
$C_{10}H_{10}ClNO_3$	$CH_3C(O)CH_2C(O)N(OH)C_6H_4(Cl)$	219/20
$C_{10}H_{10}N_4O_2$	R–N=N–$C_6H_4(OH)$, R = 2,4-dihydro-5-methyl-3-oxo-3*H*-pyrazol-4-yl	307/8
$[C_{10}H_{10}N_2O_4]_n$	$-[C_6H_2(OH)(COOH)CH_2NHC(O)NHCH_2]-_n$	155
$C_{10}H_{11}NO_2$	$CH_3C(O)CH_2C(O)NHC_6H_5$	106
	(*E*)-$CH_3CH=CHC(O)N(OH)C_6H_5$	219/20
	$H_2C=C(CH_3)C(O)N(OH)C_6H_5$	221
	$H_2C=CHC(O)N(OH)C_6H_4(CH_3)$	221
$C_{10}H_{11}NO_3$	$CH_3C(O)CH_2C(O)N(OH)C_6H_5$	219/20
$C_{10}H_{11}NO_4$	$C_6H_5N(CH_2COOH)_2$	8/9
$C_{10}H_{11}NO_5$	$(HO)C_6H_4N(CH_2COOH)_2$	10, 14/5
$C_{10}H_{11}N_5O$	R–N=N–$C_6H_4CH_3$, R = 5-amino-2,4-dihydro-3-oxo-3*H*-pyrazol-4-yl	306
$C_{10}H_{11}N_5O_2$	R–N=N–$C_6H_4OCH_3$, R = 5-amino-2,4-dihydro-3-oxo-3*H*-pyrazol-4-yl	306
$C_{10}H_{12}ClNO_2$	$(HO)(Cl)C_6H_3C(C_3H_7)=NOH$	240/1
$C_{10}H_{12}N_2O_2$	$HON=CHC(O)N(C_2H_5)C_6H_5$	265

$C_{10}H_{12}N_2O_3$	$HON{=}CHC(O)NHC_6H_4(OC_2H_5)$	265
$C_{10}H_{12}N_2O_4$	$2\text{-}C_5H_4N\text{–}CH_2N(CH_2COOH)_2$	8, 11
$C_{10}H_{13}NO_2$	$C_3H_7C(O)N(OH)C_6H_5$	219/20
	$(HO)(CH_3)_2C_6H_2C(CH_3){=}NOH$	240/3
$C_{10}H_{14}N_2O$	$3\text{-}C_5H_4N\text{–}C(O)N(C_2H_5)_2$	117/9
$C_{10}H_{14}N_2O_2$	$(CH_3)C_6H_4NHNHCH(CH_3)COOH$	89
$C_{10}H_{15}N_3O$	$(CH_3)C_6H_4N{=}N\text{–}N(OH)C_3H_7$	318/9
$C_{10}H_{16}N_2O_8$	$(HOOCCH_2)_2NCH_2CH_2N(CH_2COOH)_2$ (= H_4edta)	28/52
	and H_4N_2 ... H_2NNH_2	53
	and $C_2H_5NO_2$ ... H_2NCH_2COOH	54
	and $C_2H_8N_2$ ... $H_2NCH_2CH_2NH_2$ (= en)	53, 54
	and $C_4H_7NO_4$ ... $HN(CH_2COOH)_2$	54
	and $C_6H_9NO_6$ ... $N(CH_2COOH)_3$	54
	and $C_6H_9N_3O_2$ ... histidine	54
	$[CH_2NHCH(COOH)CH_2COOH]_2$	58/9
$C_{10}H_{17}NO_5$	$R\text{–}CH_2N(CH_2COOH)_2$, R = 4-oxiranyl	9
$C_{10}H_{18}N_2O_7$	$HOOCCH_2N(C_2H_4OH)CH_2CH_2N(CH_2COOH)_2$	26/8
	and $C_4H_6O_3$ ... $(CH_3CO)_2O$	28
$C_{10}H_{19}NO_4$	$C_6H_{13}N(CH_2COOH)_2$	9
	$(CH_3)_3CCH_2CH_2N(CH_2COOH)_2$	9
$C_{10}H_{20}N_6O_4$	$[CH_2N(CH_2C(O)NH_2)_2]_2$	131
$C_{10}H_{21}N_5O_5$	R–NH–C(=NH)–NH–C(=NH)–NR′R″, R = glucosyl, R′ = R″ = CH_3 or R′ = H, R″ = C_2H_5	165/6
$C_{10}H_{22}N_4$	$C_4H_9NH\text{–}C({=}NH)C({=}NH)\text{–}NHC_4H_9$	160
$C_{10}H_{22}N_4O_2$	$H_2NNHC(O)(CH_2)_8C(O)NHNH_2$	205/7
$C_{10}H_{24}N_{10}$	$R\text{–}(CH_2)_6\text{–}R$, R = 1-biguanidino	165
C_{11}		
$C_{11}H_8N_6O$	$R\text{–}N{=}N\text{–}C_{10}H_6(OH)$, R = 1*H*-tetrazol-5-yl	310
$C_{11}H_9NO_2$	$C_{10}H_7\text{–}C(O)NHOH$, $C_{10}H_7$ = 1-naphthyl	215/6
	R–CH=NOH, R = 2-hydroxy-1-naphthyl	239
	and C_5H_5N ... pyridine (= py)	239

$C_{11}H_9NO_3$	R–C(O)N(OH)C_6H_5, R = 2-furyl	228
	and $C_{10}H_8N_2$ .... 2,2′-bipyridine (= bpy)	228
	and $C_{12}H_8N_2$ .... 9,10-phenanthroline (= phen)	228
	2-Hydroxy-3-methyl-1,4-naphthoquinone 1-oxime	245
$C_{11}H_9NO_4$	7-Hydroxy-4-methyl-2*H*-1-benzopyrane-8-carbaldehyde oxime	240
$C_{11}H_9N_3O$	2-C_5H_4N–N=N–C_6H_4(OH)	291
$C_{11}H_9N_3O_2$	2-C_5H_4N–N=N–$C_6H_3(OH)_2$	292/3
$C_{11}H_9N_3O_4$	1-Tolyl-2,4,5,6(1*H*,3*H*)-pyrimidinetetrone 5-oxime	248/9, 252
$C_{11}H_{10}ClN_3OS$	R–N=N–C_6H_3(OH)(Cl), R = dimethyl-2-thiazolyl	312/3
$C_{11}H_{10}N_2O$	R–C(O)$NHNH_2$, R = 1-naphthyl	185
$C_{11}H_{10}N_4OS$	R–N=N–C_6H_5, 4-hydroxy-2-mercapto-6-methyl-5-pyrimidyl	309
$C_{11}H_{11}NO_3$	H_2C=CHC(O)N(OH)C_6H_4C(O)CH_3	221
$C_{11}H_{11}NO_6$	(HOOC)C_6H_4N$(CH_2COOH)_2$	22/3
$C_{11}H_{11}N_3OS$	R–N=N–C_6H_3(OH)(CH_3), R = methyl-2-thiazolyl	312
	R–N=N–C_6H_2(OH)$(CH_3)_2$, R = 2-thiazolyl	312
$C_{11}H_{11}N_3OS$	R–N=N–C_6H_3(OH)(C_2H_5), R = 2-thiazolyl	312/3
$C_{11}H_{11}N_3O_2$	Derivative of 1*H*-pyrazole-4-carbaldehyde oxime	246
$C_{11}H_{12}N_2O_4$	HON=CHC(O)NHC_6H_4($COOC_2H_5$)	265
$C_{11}H_{12}N_4O_2$	H_2NNHCOOCH_2CH_2–R, R = 1-phthalazinyl	89
$C_{11}H_{13}NO_2$	H_2C=C(CH_3)C(O)N(OH)C_6H_4(CH_3)	221
$C_{11}H_{13}NO_3$	CH_3C(O)CH_2C(O)N(OH)C_6H_4(CH_3)	219/20
$C_{11}H_{13}NO_4$	$C_6H_5CH_2$N$(CH_2COOH)_2$	9
$C_{11}H_{13}NO_6$	$(HO)_2C_6H_3CH_2$N$(CH_2COOH)_2$	10
$C_{11}H_{14}N_2O_4$	(CH_3)C_5H_3N–CH_2N(CH_2COOH)	11
$C_{11}H_{15}NO_3$	$(HO)_2C_6H_3$C(C_4H_9)=NOH	240/3
$C_{11}H_{16}N_2O_{10}$	$(HOOCCH_2)_2$NCH_2CH(COOH)N$(CH_2COOH)_2$	62
$C_{11}H_{17}NO_2$	R–C(O)NHOH, R = adamantyl	214
$C_{11}H_{17}NO_8S$	HOOC$CH_2S^{\oplus}$(CH_3)CH_2CH_2CH($COO^{\ominus}$)N$(CH_2COOH)_2$	22

$C_{12}H_{11}NO_3$	R–C(O)N(OH)$C_6H_4(CH_3)$, R = 2-furyl	228
$C_{12}H_{11}NO_9$	$(HOOC)_2(HO)C_6H_2N(CH_2COOH)_2$	22
$C_{12}H_{11}N_3O$	2-C_5H_4N–N=N–$C_6H_3(OH)(CH_3)$	291
$C_{12}H_{11}N_3O_2$	$(HO)(H_2N)C_6H_3$–N=N–$C_6H_4(OH)$	279
$C_{12}H_{11}N_3O_4S$	$(HO_3S)C_6H_4$–N=N–N(OH)C_6H_5	318/9
$C_{12}H_{12}N_4OS$	R–N=N–C_6H_5, R = 4-hydroxy-6-methyl-2-(methylthio)-5-pyrimidyl	309
$C_{12}H_{12}N_4O_4S$	$(HO)(HO_3S)C_6H_3$–N=N–$C_6H_3(NH_2)_2$	279
$C_{12}H_{13}NO_3$	H_2C=C(CH_3)C(O)N(OH)$C_6H_4C(O)CH_3$	221
$C_{12}H_{13}N_3OS$	R–N=N–$C_6H_2(OH)(CH_3)_2$, R = methyl-2-thiazolyl	312
	R–N=N–$C_6H_3(OH)(CH_3)$, R = dimethyl-2-thiazolyl	312/3
$C_{12}H_{13}N_3O_2S$	R–N=N–$C_6H_3(OH)(OCH_3)$, R = dimethyl-2-thiazolyl	312/3
$C_{12}H_{15}NO_4$	$(CH_3)_2C_6H_3N(CH_2COOH)_2$	15
$C_{12}H_{15}NO_5$	$(HO)(CH_3)C_6H_3CH_2N(CH_2COOH)_2$	10
$C_{12}H_{16}N_2O_8$	$(HOOCCH_2)_2NCH_2C{\equiv}CCH_2N(CH_2COOH)_2$	64
$C_{12}H_{18}N_2O_8$	$(HOOCCH_2)NCH_2CH{=}CHCH_2N(CH_2COOH)_2$	64
$C_{12}H_{20}N_2O_8$	$[CH_2N(CH_2COOH)CH(CH_3)COOH]_2$	56
	$[CH_2N(CH_2COOH)CH_2CH_2COOH]_2$	57
	$[CH_2NHCH(COOH)CH_2CH_2COOH]_2$	58
	$(HOOCCH_2)_2NCH_2CH(C_2H_5)N(CH_2COOH)_2$	59
	$(HOOCCH_2)_2NCH_2C(CH_3)_2N(CH_2COOH)_2$	61
	$[CH(CH_3)N(CH_2COOH)_2]_2$	61/2
	$(HOOCCH_2)_2N(CH_2)_4N(CH_2COOH)_2$	64/5
$C_{12}H_{20}N_2O_8S$	$(HOOCCH_2)_2NCH_2CH_2SCH_2CH_2N(CH_2COOH)_2$	65
$C_{12}H_{20}N_2O_9$	$(HOOCCH_2)_2NCH_2CH_2OCH_2CH_2N(CH_2COOH)_2$	65
$C_{12}H_{21}N_3O_6$	1,4,7-Triazanonane-1,4,7-triacetic acid	84, 85/6
$C_{12}H_{22}N_4O_6$	$H_2NC(O)CH_2N(CH_2COOH)CH_2CH(C_2H_5)N(CH_2COOH)CH_2C(O)NH_2$	131
	$[CH(CH_3)N(CH_2COOH)CH_2C(O)NH_2]_2$	131
$C_{12}H_{23}N_3O_6$	$C_2H_5N(R)CH_2CH_2N(R)CH_2CH_2NH(R)$, R = CH_2COOH	81
$C_{12}H_{25}N_5O_5$	R–NH–C(=NH)–NH–C(=NH)–N$(C_2H_5)_2$, R = glucosyl	165/6

C_{13}

$C_{13}H_9ClN_2O_2$	$(Cl)C_6H_4{-}N{=}N{-}C_6H_3(OH)(CHO)$	281
$C_{13}H_9N_3OS$	$R{-}N{=}N{-}C_{10}H_6(OH)$, R = 2-thiazolyl	313/4
$C_{13}H_9N_3O_2S$	$R{-}N{=}N{-}C_6H_3(OH)_2$, R = 2-benzothiazolyl	316/7
$C_{13}H_9N_3O_4S_2$	$R{-}N{=}N{-}C_{10}H_5(OH)(SO_3H)$, R = 2-thiazolyl	314
$C_{13}H_{10}BrNO_2$	$(Br)C_6H_4C(O)N(OH)C_6H_5$	222/5
	$(HO)(Br)C_6H_3C(C_6H_5){=}NOH$	244/5
$C_{13}H_{10}ClNO_2$	$C_6H_5C(O)N(OH)C_6H_4(Cl)$	222/4
	$(Cl)C_6H_4C(O)N(OH)C_6H_5$	222/5
$C_{13}H_{10}FNO_2$	$(F)C_6H_4C(O)N(OH)C_6H_5$	222/4
$C_{13}H_{10}INO_2$	$(I)C_6H_4C(O)N(OH)C_6H_5$	222/5
$C_{13}H_{10}N_2O_4$	$(O_2N)C_6H_4C(O)N(OH)C_6H_5$	222/4
	$(HO)(O_2N)C_6H_3C(C_6H_5){=}NOH$	244/5
$C_{13}H_{10}N_2O_5$	$(HO)_2(O_2N)C_6H_2C(C_6H_5){=}NOH$	244/5
$C_{13}H_{10}N_2O_5S$	$(HO_3S)C_6H_4{-}N{=}N{-}C_6H_3(OH)(CHO)$	282
$C_{13}H_{10}N_2O_6S$	$C_6H_5{-}N{=}N{-}C_6H_2(OH)(COOH)(SO_3H)$	282
	$(HO_3S)C_6H_4{-}N{=}N{-}C_6H_3(OH)(COOH)$	283
$C_{13}H_{10}N_4OS$	$R{-}N{=}N{-}C_{10}H_6(OH)$, R = methyl-1,3,4-thiadiazol-2-yl	317
$C_{13}H_{11}NO_2$	$C_6H_5OC(O)NHC_6H_5$	133
	$C_6H_5C(O)N(OH)C_6H_5$	222/7
	$(HO)C_6H_4C(C_6H_5){=}NOH$	244/5
$C_{13}H_{11}NO_3$	$(HO)_2C_6H_3C(C_6H_5){=}NOH$	244/5
$C_{13}H_{11}NO_4$	2-Hydroxy-3-methyl-1,4-naphthoquinone 1-(*O*-acetyl)oxime	246
$C_{13}H_{11}N_3O_2$	$3{-}C_5H_4N{-}C(O)NHNHC(O)C_6H_5$	199
$C_{13}H_{11}N_3O_5S$	$(HO_3S)C_6H_4{-}N{=}N{-}C_6H_3(OH)(CH{=}NOH)$	238
$C_{13}H_{12}N_2O_2$	$(HO)(CH_3)C_6H_3{-}N{=}N{-}C_6H_4(OH)$	279
$C_{13}H_{13}NO_2$	$(HO)C_{10}H_6C(C_2H_5){=}NOH$	245

$C_{13}H_{13}N_3O$	$(CH_3)C_6H_4$–N=N–N(OH)C_6H_5	318/9
$C_{13}H_{13}N_3O_2$	$(CH_3O)C_6H_4$–N=N–N(OH)C_6H_5	318/9
$C_{13}H_{14}N_4$	2-C_5H_4N–N=N–$C_6H_4N(CH_3)_2$	294
$C_{13}H_{14}N_4O$	2-C_5H_4N–N=N–$C_6H_3(OH)N(CH_3)_2$	294/5
$C_{13}H_{14}N_4OS$	R–N=N–C_6H_5, R = 2-(ethylthio)-4-hydroxy-6-methyl-5-pyrimidyl	309
$C_{13}H_{15}NO_6$	$HOOCCH(CH_2C_6H_5)N(CH_2COOH)_2$	21
$C_{13}H_{15}N_3OS$	R–N=N–$C_6H_3(OH)(C_2H_5)$, R = dimethyl-2-thiazolyl	312/3
	R–N=N–$C_6H_2(OH)(CH_3)_2$, R = dimethyl-2-thiazolyl	312/3
$C_{13}H_{15}N_3O_2S$	R–N=N–$C_6H_2(OH)(OCH_3)(CH_3)$, R = dimethyl-2-thiazolyl	312/3
$C_{13}H_{16}N_2O_4$	HON=CHC(O)NH$C_6H_4(COOC_4H_9)$	265
$C_{13}H_{16}N_4OS$	R–N=N–$C_6H_3(OH)N(CH_3)_2$, R = dimethyl-2-thiazolyl	312/3
$C_{13}H_{17}NO_2$	*cyclo*-$C_6H_{11}C(O)N(OH)C_6H_5$	219/20
$C_{13}H_{22}N_2O_8$	$(HOOCCH_2)_2NCH_2CH[CH(CH_3)_2]N(CH_2COOH)_2$	59
	$(HOOCCH_2)_2NCH(CH_3)CH_2CH(CH_3)N(CH_2COOH)_2$	62, 63
	$(HOOCCH_2)_2N(CH_2)_5N(CH_2COOH)_2$	64
C_{14}		
$C_{14}H_9NO_2$	9,10-Phenanthrenedione monooxime	248
$C_{14}H_{10}BrNO_3$	$(HOOC)C_6H_4C(O)NHC_6H_4(Br)$	130
$C_{14}H_{10}ClNO_3$	$(HOOC)C_6H_4C(O)NHC_6H_4(Cl)$	130
$C_{14}H_{10}N_2O_5$	$(HOOC)C_6H_4C(O)NHC_6H_4(NO_2)$	130
$C_{14}H_{10}N_4O_4S$	R–N=N–$C_{10}H_5(OH)(SO_3H)$, R = 2-pyrimidyl	309
$C_{14}H_{11}NO_2$	(*E*)–$C_6H_5C(O)C(=NOH)C_6H_5$	249/50
$C_{14}H_{11}NO_3$	$(HOOC)C_6H_4C(O)NHC_6H_5$	130
$C_{14}H_{11}N_3O_2$	3-C_5H_4N–$CH_2OC(O)NHC_6H_4(CN)$	133
$C_{14}H_{12}BrNO_2$	$(Br)C_6H_4C(O)N(OH)C_6H_4(CH_3)$	222/4

$C_{14}H_{12}ClNO_2$	$(Cl)C_6H_4C(O)N(OH)C_6H_4(CH_3)$	222/4
$C_{14}H_{12}FNO_2$	$(F)C_6H_4C(O)N(OH)C_6H_4(CH_3)$	222/4
$C_{14}H_{12}N_2O_2$	$C_6H_5C(O)NHNHC(O)C_6H_5$	197
	$HON{=}C(C_6H_5)C(C_6H_5){=}NOH$	266/7
	$(CH_3)C_6H_4{-}N{=}N{-}C_6H_3(OH)(CHO)$	281
$C_{14}H_{12}N_2O_3$	$(HO)C_6H_4C(O)NHNHC(O)C_6H_5$	197/8
	$(CH_3)(HO)C_6H_3{-}N{=}N{-}C_6H_4(COOH)$	284
$C_{14}H_{12}N_2O_4$	$(HO)C_6H_4C(O)NHNHC(O)C_6H_4(OH)$	198
	$(O_2N)C_6H_4C(O)N(OH)C_6H_4(CH_3)$	222/4
$C_{14}H_{13}NO_2$	$(C_6H_5)_2CHC(O)NHOH$	212, 214
	and $C_{10}H_8N_2$ .... 2,2′-bipyridine (= bpy)	213
	$C_6H_5CH_2C(O)N(OH)C_6H_5$	219/21
	$C_6H_5C(O)N(OH)C_6H_4(CH_3)$	222/4, 227
	$(CH_3)C_6H_4C(O)N(OH)C_6H_5$	222/5, 227
	$(HO)(CH_3)C_6H_3C(C_6H_5){=}NOH$	244/5
$C_{14}H_{13}NO_3$	$(CH_3O)C_6H_4C(O)N(OH)C_6H_5$	222/5
	$(HO)(CH_3O)C_6H_3C(C_6H_5){=}NOH$	244/5
$C_{14}H_{13}N_3O_2$	$CH_3C(O)C_6H_4{-}N{=}N{-}N(OH)C_6H_5$	318/9
$C_{14}H_{14}N_4O_2$	$2{-}C_5H_4N{-}C(O)NHCH_2CH_2NHC(O){-}C_5H_4N{-}2$	112/4
	$CH_3C(O)NHC_6H_4{-}N{=}N{-}N(OH)C_6H_5$	318/9
$C_{14}H_{15}Cl_2N_3O$	$R{-}C(O)N(i{-}C_3H_7)CH_2C_6H_3(Cl)_2$, R = 1*H*-imidazol-1-yl	132
$C_{14}H_{15}N_3O$	$(CH_3)C_6H_4{-}N{=}N{-}N(OH)C_6H_4(CH_3)$	318/9
$C_{14}H_{16}N_2O_8$	$(HOOCCH_2)_2N{-}C_6H_4{-}N(CH_2COOH)_2$	76/7
$C_{14}H_{18}N_4O_3$	$CH_3OC(O)NH{-}R$, R = 1-[(butylamino)carbonyl]-1*H*-benzimidazol-2-yl	133
$C_{14}H_{21}N_5O_5$	$R{-}NH{-}C({=}NH){-}NH{-}C({=}NH){-}NHC_6H_5$, R = glucosyl	166
$C_{14}H_{22}ClN_3O_2$	$(H_2N)(Cl)(CH_3O)C_6H_2C(O)NHCH_2CH_2N(C_2H_5)_2$	110
$C_{14}H_{22}N_2O_8$	$(HOOCCH_2)_2N{-}R{-}N(CH_2COOH)_2$, R = *trans*-1,2-cyclohexanediyl	30, 66/75
	$(HOOCCH_2)_2N{-}R{-}N(CH_2COOH)_2$, R = 1,4-cyclohexanediyl	75
$C_{14}H_{23}N_3O_{10}$	$HOOCCH_2N[CH_2CH_2N(CH_2COOH)_2]_2$	82/3

$C_{14}H_{24}N_2O_8$	$(CH_3OOCCH_2)_2NCH_2CH_2N(CH_2COOCH_3)_2$	55
	$\text{+}CH_2N(CH_2COOH)CH(C_2H_5)COOH]_2$	56
	$\text{+}CH_2N(CH_2CH_2COOH)_2]_2$	58
	$\text{+}CH_2N[CH(CH_3)COOH]_2]_2$	58
	$(HOOCCH_2)_2NCH_2CH[CH_2CH(CH_3)_2]N(CH_2COOH)_2$	60
	$(HOOCCH_2)_2N(CH_2)_6N(CH_2COOH)_2$	64
$C_{14}H_{24}N_2O_{10}$	$\text{+}CH_2OCH_2CH_2N(CH_2COOH)_2]_2$	65/6
C_{15}		
$C_{15}H_{10}BrN_3O$	$(4\text{-}Br)2\text{-}C_5H_3N\text{–}N{=}N\text{–}C_{10}H_6(OH)$	295, 297
$C_{15}H_{10}ClN_3O$	$(4\text{-}Cl)2\text{-}C_5H_3N\text{–}N{=}N\text{–}C_{10}H_6(OH)$	295, 297
	$2\text{-}C_5H_4N\text{–}N{=}N\text{–}C_{10}H_5(OH)(Cl)$	296
$C_{15}H_{10}N_2O_3$	$R\text{–}N{=}N\text{–}C_6H_5$, R = 4-hydroxycoumarin-3-yl	290/1
$C_{15}H_{10}N_4O_3$	$(4\text{-}O_2N)2\text{-}C_5H_4N\text{–}N{=}N\text{–}C_{10}H_6(OH)$	295
$C_{15}H_{10}N_4O_4$	1-(4-Nitrophenyl)-3-phenyl-1*H*-pyrazole-4,5-dione 4-oxime	250/1
$C_{15}H_{10}N_4O_6S$	$R\text{–}N{=}N\text{–}C_6H_4(NO_2)$, R = 8-hydroxy-5-sulfo-7-quinolyl	303
$C_{15}H_{11}N_3O$	$2\text{-}C_5H_4N\text{–}N{=}N\text{–}C_{10}H_6(OH)$	295/9
	$R\text{–}N{=}N\text{–}C_6H_5$, R = 8-hydroxy-5-quinolyl	303
	$R\text{–}N{=}N\text{–}C_6H_5$, R = 8-hydroxy-7-quinolyl	303
$C_{15}H_{11}N_3O_2$	1,3-Diphenyl-1*H*-pyrazole-4,5-dione 4-oxime	250/1
	$[2\text{-}C_5H_4N(O)]\text{–}N{=}N\text{–}C_{10}H_6(OH)$	299/300
	$R\text{–}N{=}N\text{–}C_6H_4(OH)$, R = 8-hydroxy-5-quinolyl	303
	$R\text{–}N{=}N\text{–}C_6H_3(OH)_2$, R = 2-quinolyl	303
$C_{15}H_{11}N_3O_4S$	$2\text{-}C_5H_4N\text{–}N{=}N\text{–}C_{10}H_5(OH)(SO_3H)$	296
	$R\text{–}N{=}N\text{–}C_6H_5$, R = 8-hydroxy-5-sulfo-7-quinolyl	303
$C_{15}H_{12}N_2O_4$	$C_6H_5C(O)N\text{–}C_6H_4(NO_2)\text{–}C(O)CH_3$	110
$C_{15}H_{13}NO_2$	$C_6H_5CH{=}CHC(O)N(OH)C_6H_5$	221
	$CH_3C(O)N(OH)R$, R = 9*H*-fluoren-2-yl	221
$C_{15}H_{13}NO_3$	$(HOOC)C_6H_4C(O)NHC_6H_4(CH_3)$	130
$C_{15}H_{13}NO_4$	$(HOOC)C_6H_4C(O)NHC_6H_4(OCH_3)$	130
$C_{15}H_{14}N_2O_4$	$(HON{=}CH)(HO)C_6H_3\text{–}CH_2\text{–}C_6H_3(OH)(CH{=}NOH)$	268/9

C_{26} to C_{29}

$C_{26}H_{18}N_4O_5$ (OHC)(HO)C_6H_3–N=N–C_6H_4–C_6H_4–N=N–C_6H_3(OH)(COOH) 283

$C_{26}H_{18}N_4O_6S$ R–SO_2–R, R = C_6H_4–N=N–C_6H_3(OH)(CHO) 282

$C_{26}H_{20}N_4O_8S_2$ RCH=CHR, R = $C_6H_3(SO_3H)$–N=N–C_6H_4(OH) 284

$C_{26}H_{20}N_6O_4$ (HON=CH)(HO)C_6H_3–N=N–C_6H_4–C_6H_4–N=N–C_6H_3(OH)(CH=NOH) 268/9

$C_{26}H_{25}NO_9S$ Semixylenol orange 8, 11

$C_{26}H_{38}N_2O_4$ $(C_3H_7)_2$NC(O)CH_2O–R–OCH_2C(O)N$(C_3H_7)_2$, R = naphthylene 128

$C_{28}H_{24}N_6O_4$ [HON=C(CH_3)](HO)C_6H_3–N=N–C_6H_4–C_6H_4–N=N–C_6H_3(OH)[C(CH_3)=NOH] 268/9

$C_{28}H_{40}N_2O_4$ $(C_3H_7)_2$NC(O)CH_2O–R–OCH_2C(O)N$(C_3H_7)_2$, R = acenaphthenylene 128

$C_{29}H_{29}N_5O_3$ R–N=N–$C_{10}H_5$(OH)C(O)NHC_6H_4(OCH_3), R = 1-methylanabasyl 302

C_{30} to C_{38}

$C_{30}H_{20}N_6O_2$ R–N=N–C_6H_4–C_6H_4–N=N–R, R = 8-hydroxy-5-quinolyl 305

$C_{30}H_{20}N_6O_{22}S_6$ 5-[(1,8-Dihydroxy-3,6-disulfo-2-naphthalenyl)azo]-4-hydroxy-3-[(8-hydroxy-3,6-disulfo-1-naphthalenyl)azo]-2,7-naphthalenedisulfonic acid 289

$C_{30}H_{56}N_2O_8$ ‡CH_2N(CH_2COOH)CH($C_{10}H_{21}$)COOH$]_2$ 56

$C_{31}H_{32}N_2O_{13}S$ Xylenol orange 80/1

$C_{32}H_{32}N_2O_{12}$ Phthalein complexone 79/80

$C_{33}H_{35}N_2NaO_{13}S$ Methylxylenol blue 80

$C_{35}H_{32}N_6O$ [5,14-Dihydro-6,8,15,17-tetramethyldibenzo[b,i][1,4,8,11]-tetraazacyclotetradecin-7-yl)-[4-(phenylazo)phenyl]methanone 311

$C_{37}H_{40}N_2Na_4O_{13}S$ Methylthymol blue 80

$C_{38}H_{44}N_2O_{12}$ Thymolphthalein complexone 79/80

Table of Conversion Factors

Following the notation in Landolt-Börnstein [7], values which have been fixed by convention are indicated by a bold-face last digit. The conversion factor between calorie and Joule that is given here is based on the thermochemical calorie, $cal_{th\,ch}$, and is defined as 4.184**0** J/cal. However, for the conversion of the "Internationale Tafelkalorie", cal_{IT}, into Joule, the factor 4.1868 J/cal is to be used [1, p. 147]. For the conversion factor for the British thermal unit, the Steam Table Btu, BTU_{ST}, is used [1, p. 95].

Force	N	dyn	kp
1 N (Newton)	1	10^{5}	0.1019716
1 dyn	10^{-5}	1	1.019716×10^{-6}
1 kp	9.8066**5**	9.8066**5** $\times10^{5}$	1

Pressure	Pa	bar	kp/m²	at	atm	Torr	lb/in²
1 Pa (Pascal) = 1 N/m²	1	10^{-5}	1.019716×10^{-1}	1.019716×10^{-5}	0.986923×10^{-5}	0.750062×10^{-2}	145.0378×10^{-6}
1 bar = 10^{6} dyn/cm²	10^{5}	1	10.19716×10^{3}	1.019716	0.986923	750.062	14.50378
1 kp/m² = 1 mm H_2O	9.8066**5**	0.98066**5** $\times10^{-4}$	1	10^{-4}	0.967841×10^{-4}	0.735559×10^{-1}	1.422335×10^{-3}
1 at = 1 kp/cm²	0.98066**5** $\times10^{5}$	0.98066**5**	10^{4}	1	0.967841	735.559	14.22335
1 atm = 76**0** Torr	1.0132**5** $\times10^{5}$	1.0132**5**	1.033227×10^{4}	1.033227	1	760	14.69595
1 Torr = 1 mm Hg	133.3224	1.333224×10^{-3}	13.59510	1.359510×10^{-3}	1.315789×10^{-3}	1	19.33678×10^{-3}
1 lb/in² = 1 psi	6.89476×10^{3}	68.9476×10^{-3}	703.069	70.3069×10^{-3}	68.0460×10^{-3}	51.7149	1

Work, Energy, Heat	J	kWh	kcal	Btu	MeV
1 J (Joule) = 1 Ws = 1 Nm = 10^7 erg	1	2.778×10^{-7}	2.39006×10^{-4}	9.4781×10^{-4}	6.242×10^{12}
1 kWh	$3.\mathbf{6} \times 10^{6}$	1	860.4	3412.14	2.247×10^{19}
1 kcal	4184.**0**	1.1622×10^{-3}	1	3.96566	2.6117×10^{16}
1 Btu (British thermal unit)	1055.06	2.93071×10^{-4}	0.25164	1	6.5858×10^{15}
1 MeV	1.602×10^{-13}	4.450×10^{-20}	3.8289×10^{-17}	1.51840×10^{-16}	1

1 eV ≙ 23.0578 kcal/mol = 96.473 kJ/mol

Power	kW	PS	kp·m/s	kcal/s
1 kW = 10^{10} erg/s	1	1.35962	101.972	0.239006
1 PS	0.73550	1	**75**	0.17579
1 kp·m/s	$9.80\mathbf{665} \times 10^{-3}$	0.01333	1	2.34384×10^{-3}
1 kcal/s	4.184**0**	5.6886	426.650	1

References:

[1] A. Sacklowski, Die neuen SI-Einheiten, Goldmann, München 1979. (Conversion tables in an appendix.)
[2] International Union of Pure and Applied Chemistry, Manual of Symbols and Terminology for Physicochemical Quantities and Units, Pergamon, London 1979; Pure Appl. Chem. **51** [1979] 1/41.
[3] The International System of Units (SI), National Bureau of Standards Spec. Publ. 330 [1972].
[4] H. Ebert, Physikalisches Taschenbuch, 5th Ed., Vieweg, Wiesbaden 1976.
[5] Kraftwerk Union Information, Technical and Economic Data on Power Engineering, Mülheim/Ruhr 1978.
[6] E. Padelt, H. Laporte, Einheiten und Größenarten der Naturwissenschaften, 3rd Ed., VEB Fachbuchverlag, Leipzig 1976.
[7] Landolt-Börnstein, 6th Ed., Vol. II, Pt. 1, 1971, pp. 1/14.
[8] ISO Standards Handbook 2, Units of Measurement, 2nd Ed., Geneva 1982.

Key to the Gmelin System of Elements and Compounds

HCl $ZnCl_2$ $CrCl_2$ $ZnCrO_4$

System Number	Symbol	Element
1		Noble Gases
2	H	Hydrogen
3	O	Oxygen
4	N	Nitrogen
5	F	Fluorine
6	**Cl**	**Chlorine**
7	Br	Bromine
8	I	Iodine
	At	Astatine
9	S	Sulfur
10	Se	Selenium
11	Te	Tellurium
12	Po	Polonium
13	B	Boron
14	C	Carbon
15	Si	Silicon
16	P	Phosphorus
17	As	Arsenic
18	Sb	Antimony
19	Bi	Bismuth
20	Li	Lithium
21	Na	Sodium
22	K	Potassium
23	NH_4	Ammonium
24	Rb	Rubidium
25	Cs	Caesium
	Fr	Francium
26	Be	Beryllium
27	Mg	Magnesium
28	Ca	Calcium
29	Sr	Strontium
30	Ba	Barium
31	Ra	Radium
32	**Zn**	**Zinc**
33	Cd	Cadmium
34	Hg	Mercury
35	Al	Aluminium
36	Ga	Gallium

System Number	Symbol	Element
37	In	Indium
38	Tl	Thallium
39	Sc, Y La—Lu	Rare Earth Elements
40	Ac	Actinium
41	Ti	Titanium
42	Zr	Zirconium
43	Hf	Hafnium
44	Th	Thorium
45	Ge	Germanium
46	Sn	Tin
47	Pb	Lead
48	V	Vanadium
49	Nb	Niobium
50	Ta	Tantalum
51	Pa	Protactinium
52	**Cr**	**Chromium**
53	Mo	Molybdenum
54	W	Tungsten
55	U	Uranium
56	Mn	Manganese
57	Ni	Nickel
58	Co	Cobalt
59	Fe	Iron
60	Cu	Copper
61	Ag	Silver
62	Au	Gold
63	Ru	Ruthenium
64	Rh	Rhodium
65	Pd	Palladium
66	Os	Osmium
67	Ir	Iridium
68	Pt	Platinum
69	Tc	Technetium[1]
70	Re	Rhenium
71	Np,Pu...	Transuranium Elements

Material presented under each Gmelin System Number includes all information concerning the element(s) listed for that number plus the compounds with elements of lower System Number.

For example, zinc (System Number 32) as well as all zinc compounds with elements numbered from 1 to 31 are classified under number 32.

[1] A Gmelin volume titled "Masurium" was published with this System Number in 1941.

A Periodic Table of the Elements with the Gmelin System Numbers is given on the Inside Front Cover